Edited by
Wolfgang Weigand and
Philippe Schollhammer

Bioinspired Catalysis

Related Titles

Loos, K. (ed.)

Biocatalysis in Polymer Chemistry

2011
Print ISBN: 978-3-527-32618-1
also available in electronic formats

Bullock, R.M. (ed.)

Catalysis without Precious Metals

2010
Print ISBN: 978-3-527-32354-8
also available in electronic formats

Zhou, Q. (ed.)

Privileged Chiral Ligands and Catalysts

2011
Print ISBN: 978-3-527-32704-1
also available in electronic formats

Mizuno, N. (ed.)

Modern Heterogeneous Oxidation Catalysis

Design, Reactions and Characterization

2009
Print ISBN: 978-3-527-31859-9
also available in electronic formats

Edited by Wolfgang Weigand and Philippe Schollhammer

Bioinspired Catalysis

Metal-Sulfur Complexes

WILEY-VCH
Verlag GmbH & Co. KGaA

Editors

Wolfgang Weigand
Friedrich-Schiller-Universität
Inst.f.Anorg.+ Analyt.Chemie
August-Bebel-Str. 2
07743 Jena
Germany

Prof. Philippe Schollhammer
Université de Bretagne
Occidentale
Avenue Victor le Gorgeu 6
29285 Brest
France

Cover
The structures shown on the cover have the PDB ID SF4 and 3U7Q (DOI:10.1126/SCIENCE.1214025) and were visualized using VMD (Humphrey, W., Dalke, A. and Schulten, K., 'VMD - Visual Molecular Dynamics', J. Molecular Graphics, 1996, vol. 14, pp. 33–38)

Library of Congress Card No.: applied for

British Library Cataloguing-in-Publication Data
A catalogue record for this book is available from the British Library.

Bibliographic information published by the Deutsche Nationalbibliothek
The Deutsche Nationalbibliothek lists this publication in the Deutsche Nationalbibliografie; detailed bibliographic data are available on the Internet at <http://dnb.d-nb.de>.

Print ISBN: 978-3-527-33308-0
ePDF ISBN: 978-3-527-66419-1
ePub ISBN: 978-3-527-66418-4
Mobi ISBN: 978-3-527-66417-7
oBook ISBN: 978-3-527-66416-0

Cover-Design Grafik-Design Schulz, Fußgönheim, Germany
Typesetting Laserwords Private Limited, Chennai, India
Printing and Binding Markono Print Media Pte Ltd., Singapore

Printed on acid-free paper

Contents

List of Contributors

Ulf-Peter Apfel
Ruhr-University Bochum
Inorganic Chemistry
I/Bioinorganic Chemistry
Universitätsstrasse 150
44801 Bochum
Germany

Vincent Artero
Université Grenoble Alpes
CNRS, CEA Life Science
Division
Laboratoire de Chimie et
Biologie des Métaux
17 Av. des Martyrs
38054 Grenoble Cedex 9
France

Frédéric Barrière
Université de Rennes I
Institut des Sciences Chimiques
de Rennes
UMR CNRS 6226 – Campus de
Beaulieu – Bâtiment 10 C
35042 Rennes Cedex
France

Ian Dance
University of New South Wales
School of Chemistry
Sydney 2052
Australia

Joe Dawson
The University of Nottingham
School of Chemistry
University Park
Nottingham NG7 2RD
UK

Luca De Gioia
University of Milano-Bicocca
Department of Biotechnology
and Biosciences
Piazza della Scienza, 2
20126 Milan
Italy

Marc Fontecave
Université Grenoble Alpes
CNRS, CEA Life Science
Division
Laboratoire de Chimie et
Biologie des Métaux
17 Av. des Martyrs
38054 Grenoble Cedex 9
France

and

Collège de France
CNRS
Laboratoire de Chimie des
Processus Biologiques
11 place Marcelin Berthelot
75005 Paris
France

Juan C. Fontecilla-Camps
Institut de Biologie Structurale
'Jean-Pierre Ebel'
Metalloproteins Unit
71, Avenue des Martyrs
CS 10090, 38044
Grenoble Cedex 9
France

Ashta Chandra Ghosh
Institut für Biochemie
Bioorganische Chemie
Felix-Hausdorffstr. 4
17487 Greifswald
Germany

Craig A. Grapperhaus
University of Louisville
Department of Chemistry
2320 South Brook Street
Louisville
KY 40292
USA

Claudio Greco
University of Milano-Bicocca
Department of Earth and
Environmental Sciences
Piazza della Scienza, 1
20126 Milan
Italy

Richard A. Henderson
Newcastle University
School of Chemistry
Newcastle upon Tyne, NE1 7RU
UK

Yilin Hu
University of California
Irvine, Department of Molecular
Biology and Biochemistry
2236 McGaugh Hall
Irvine
CA 92697-3900
USA

Davinder Kumar
University of Louisville
Department of Chemistry
2320 South Brook Street
Louisville
KY 40292
USA

Chi Chung Lee
University of California
Irvine, Department of Molecular
Biology and Biochemistry
2236 McGaugh Hall
Irvine
CA 92697-3900
USA

Jonathan McMaster
The University of Nottingham
School of Chemistry
University Park
Nottingham NG7 2RD
UK

John A. Murphy
University of Strathclyde
Department of Pure and Applied Chemistry
Thomas Graham Building
295 Cathedral Street
Glasgow
G1 1XL Scotland
UK

Yvain Nicolet
Institut de Biologie Structurale 'Jean-Pierre Ebel'
Metalloproteins Unit
71, Avenue des Martyrs
CS 10090, 38044
Grenoble Cedex 9
France

Carlo Perotto
The University of Nottingham
School of Chemistry
University Park
Nottingham NG7 2RD
UK

François Y. Pétillon
Université de Bretagne Occidentale
UMR CNRS 6521 "Chimie, Electrochimie Moléculaires et Chimie Analytique"
UFR Sciences et Techniques
6 avenue Le Gorgeu
CS 93837
29238 Brest-Cedex 3
France

Christopher J. Pickett
University of East Anglia
Energy Materials Laboratory
School of Chemistry
Norwich Research Park
Norwich NR4 7TJ
UK

Markus W. Ribbe
University of California
Irvine, Department of Molecular Biology and Biochemistry
2236 McGaugh Hall
Irvine
CA 92697-3900
USA

Philippe Schollhammer
Université de Bretagne Occidentale
UMR CNRS 6521 "Chimie, Electrochimie Moléculaires et Chimie Analytique"
UFR Sciences et Techniques
6 avenue Le Gorgeu
CS 93837
29238 Brest-Cedex 3
France

Martin Schröder
The University of Nottingham
School of Chemistry
University Park
Nottingham NG7 2RD
UK

Carola Schulzke
Institut für Biochemie
Bioanorganische Chemie
Felix-Hausdorffstr. 4
17487 Greifswald
Germany

Callum Scullion
University of Strathclyde
Department of Pure and Applied Chemistry
Thomas Graham Building
295 Cathedral Street
Glasgow G1 1XL Scotland
UK

Jean Talarmin
Université de Bretagne Occidentale
UMR CNRS 6521 "Chimie, Electrochimie Moléculaires et Chimie Analytique"
UFR Sciences et Techniques
6 avenue Le Gorgeu
CS 93837
29238 Brest-Cedex 3
France

Phong D. Tran
Nanyang Technological University
Energy Research Institute
ERI@N
50 Nanyang Drive 637553
Singapore

Gönter Wächtershäuser
8 Weinstr. Munich
80333
Gemany

and

209 Mill Race Drive
Chapel Hill
NC 27512
USA

Wolfgang Weigand
Friedrich-Schiller Universität Jena
Institut für Anorganische und Analytische Chemie
Humboldtstraße 8
07743 Jena
Germany

Jared A. Wiig
University of California
Irvine, Department of Molecular Biology and Biochemistry
2236 McGaugh Hall
Irvine
CA 92697-3900
USA

Joseph A. Wright
University of East Anglia
Energy Materials Laboratory
School of Chemistry
Norwich Research Park
Norwich NR4 7TJ
UK

Preface

In nature, metal–sulfur clusters provide the active sites of several enzymes which catalyze important reactions such as, for example, H^+/H_2 conversion by hydrogenases and nitrogen fixation by the nitrogenases. The understanding of the functioning of these metalloenzymes and the design of active bioinspired molecules are great challenges for chemists, and they require a multidisciplinary approach at the interface of chemistry and biology.

The focus of this book is to highlight the more recent developments in metal–sulfur complexes involved in bioinspired catalysis.

As an introductory part, the first chapter shows the possible importance of metal-sulfides in the chemautotrophic origin of life. In this context, the iron–sulfur world theory is emphasized. This introductory part is completed by the second chapter concerning biogenesis and redox, catalytic, and regulatory properties of Fe–S clusters. The second part deals with impressive advances that have been reported during the past 15 years in the molecular chemistry of models of the active sites of the different classes of hydrogenases. Advances into the electrocatalytic H_2 production and uptake are reported. A third part is intended to provide recent insights concerning N_2-fixation by nitrogenases for which understanding their functioning is still one of the most exciting and intricate modern challenges in bioinorganic chemistry. Recent X-ray analyses of the structure of the FeMo cofactor have led to a kind of renaissance of this field. In the two previous parts, syntheses as well as analytical and theoretical aspects of structural and functional models of these enzymes are described. The fourth part deals with metal–sulfur-containing enzymes activating nitriles (nitrile hydratase) of molybdenum and tungsten enzymes, which contain the pyranopterin cofactor and perform a wide variety of biological functions. They catalyze a diversity of mostly two-electron oxidation–reduction reactions crucial in the metabolism of nitrogen, sulfur, and carbon (xanthine oxidase, dimethyl sulfoxide reductase, sulfite oxidase, and nitrate reductases). In the final chapter, some applicative perspectives of bioinspired complexes, as electrode materials, and artificial photosynthetic systems, are illustrated.

Finally, we thank the authors for their contributions to this book and their efforts and we hope that the readers will find in it a source of recent knowledge in this area of chemistry and of inspiration.

Wolfgang Weigand
Friedrich-Schiller-Universität,
Institut für Anorganische &
Analytische Chemie
Jena, Germany

and

Philippe Schollhammer
Université de Bretagne Occidentale
France

Part I
Primordial Metal–Sulfur-Mediated Reactions

Bioinspired Catalysis: Metal-Sulfur Complexes, First Edition.
Edited by Wolfgang Weigand and Philippe Schollhammer.

1 From Chemical Invariance to Genetic Variability

Günter Wächtershäuser

> *The archaeologist of nature is at liberty to go back to the traces that remain of nature's earliest revolutions, and, appealing to all he knows or can conjecture about its mechanism, to trace the origin of that great family of creatures … down even to mosses and lichens, and finally down to the lowest perceivable stage of nature, to crude matter. From this and from the forces within, by mechanical laws, like those that are at work in the formation of crystals, seems to be derived the whole technique of nature.*
>
> *Immanuel Kant [1]*

1.1 Heuristic of Biochemical Retrodiction

Darwin (1863) wrote in a letter to Hooker [2]: "It is mere rubbish, thinking at present of the origin of life; one might as well think of the origin of matter." Studies of nucleosynthesis are now quite advanced, but research into the origin of life is still an immature science. The problem of early evolution of life is unique and requires its own heuristic. A commonly used heuristic consists of one-to-one back-extrapolations of individual biochemical features (Figure 1.1a), for which Lipmann [3] coined the term *backward projection*. More and more backward projections add evermore ingredients to the recipe. Inevitably, this way of thinking leads to the notion of a "primordial broth." No one has ever spelled out all that what would or would not have been in the broth and how precisely the organization of life could have come about within such a chaotic situation.

This conceptual hodgepodge is overcome by a heuristic of convergent back-extrapolation, termed *biochemical retrodiction* (Figure 1.1b) [4]. Some extant features are still projected all the way back to the origin. Typically, however, several extant biochemical features are related to one simpler common functional precursor feature and several precursor features are related to a still deeper common precursor. This pattern is applied over and over again, drawing in more and more extant features, progressing to ever deeper, fewer, and simpler precursor features, and generating an overall pattern of backward convergence. Ultimately,

Bioinspired Catalysis: Metal-Sulfur Complexes, First Edition.
Edited by Wolfgang Weigand and Philippe Schollhammer.

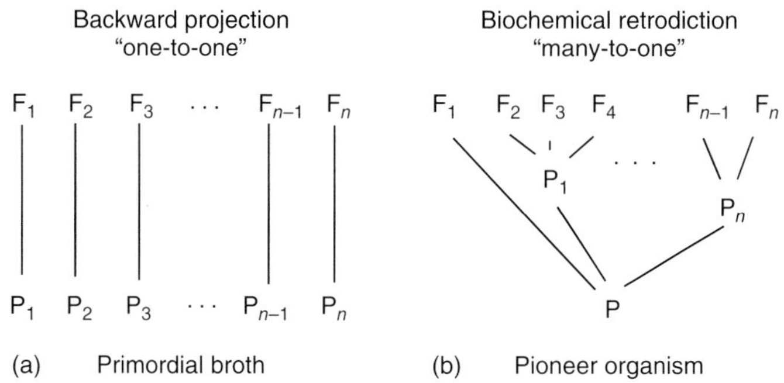

Figure 1.1 Heuristics of (a) parallel backward projection and (b) convergent biochemical retrodiction (F1, F2, etc., extant features; P, P1, P2,etc., precursor features). (From Ref [47] © (2010), Springer.)

the heuristic of biochemical retrodiction aims at a restricted set of chemical compounds and processes, which cooperate to form a distinct chemical entity with the ability to reproduce and evolve: the "pioneer organism." Its chemistry is intrinsically synthetic, thereby imposing from the start directionality from simple to complex on the overall process of evolution. In this sense, the pioneer organism paves the way for all future evolution, hence its name.

Specifically, we include in our platform for retrodiction extant biochemical features, which combine aspects of evolutionary change with aspects that have been largely invariant over time by the universal laws of chemistry. It is precisely the aspect of chemical invariance within evolved biochemical features that provides directionality and allows us to unravel evolutionary history backward to the very pioneer organism of life. For the biochemical retrodiction of multistep biosynthetic pathways we employ in addition the Florkin–Granick rule [5, 6] that earlier steps in a pathway have greater evolutionary antiquity than later steps. This rule is based on the assumption that biosynthetic pathways evolve by terminal extensions. We should apply caution, however, because pathway evolution comprises also lateral branchings, recruitments, reversals, and eliminations [7].

The results of biochemical retrodiction are evaluated empirically by chemical experiments and theoretically by quantum-chemical calculations [8], with the perceived geochemical scenario determining the parameters. Biochemical retrodiction suggests chemical experiments and experimental results inform revised retrodictions, and such iterative procedure promises a progressive exploration of the pioneer organism. Finally, when our procedure leads to competing hypotheses, we prefer the one with the greater explanatory power, that is, the ability to explain a greater number of extant biochemical facts with fewer evolutionary assumptions [9]. We shall now use this methodology for a step-by-step reconstruction of the pioneer organism, beginning at the simplest level: the elements of life.

1.2 Retrodicting the Elements of Life

The elements of central biochemistry [10] fall into two distinct subsets. (i) The main group nonmetal bioelements (H, C, N, O, P, S, Se) make up the bulk of the biomass with mostly structural roles. They originate deep in the mantle of the Earth and form volcanic gases (H_2, N_2, CO_2, CO, CH_4, NH_3, H_2O, SO_2, H_2S, H_2Se, COS, HCN, CH_3SH, P_4O_{10}). (ii) The transition metal bioelements (Fe, Co, Ni, V, Mo, W, Mn, Cu, Zn) occur in organisms only in trace amounts, with mostly catalytic functions. Together with the main group biometals Mg, Ca they form essentially stationary crustal minerals. These two classes of bioelements come into close encounter at volcanic-hydrothermal flow sites in the presence of liquid water [11], cf. [12].

Four main group nonmetal bioelements (H, C, N, O) form the structural basis for biochemistry. They are involved in all biochemical reactions, each one with its unique roles, indispensable for life, down to the pioneer organism. Sulfur and its companion selenium have diverse biocatalytic and bioenergetic functions. They are projected into the pioneer organism. Phosphorus is indispensible in genetics and bioenergetics, but limited to phosphate group chemistry. It must have been acquired after the origin of life. Therefore, the pioneer organism is defined *prima facie* by main group system H–C–O–N–S–Se.

Among the transition metal bioelements we find some of the most crucial, indispensible catalysts of central anaerobic biochemistry. Iron in the form of moderately soft ferrous ions has diverse biochemical functions. It is the most abundant transition metal in aqueous, anaerobic, volcanic-hydrothermal settings. In the same settings, hydrogen sulfide (H_2S), the source for soft thio ligands is a ubiquitous volcanic gas. These two locally coinciding components have a high bonding affinity for each other, as evident from the abundance of iron–sulfur clusters in extant metalloenzymes [13] and of iron–sulfur minerals in extant volcanic-hydrothermal flow sites. Therefore, the world of the pioneer organism has been dubbed "*iron–sulfur world*" [7]. The Fe–S bonding strength under anaerobic conditions serves as a gauge for the suitability of other transition metal bioelements, notably Co and Ni. These three iron group metals form the catalytic core of the pioneer organism.

Cu and Mo were unavailable for the anaerobic pioneer organism, because they form extremely insoluble sulfides (Cu_2S and MoS_2). They could have entered the biosphere only after oxygenation of the oceans. Zn is also discounted. Under sulfidic conditions, it forms highly insoluble ZnS and it is not redox-active. Mn (as Mn^{2+}) is too hard for the iron–sulfur world. Cr, the group companion of Mo, exists under volcanic-hydrothermal conditions as hard cation (Cr^{3+}) and had no chance to enter early biochemistry. Compared to its extremely low overall abundance in the Solar System, W is highly enriched in the walls of hydrothermal flow ducts and it does not form an extremely insoluble sulfide [14]. It would have been available for the pioneer organism. Vanadium has a remarkable chemical similarity to its diagonal neighbor Mo [15] and a high crustal abundance without being trapped as an insoluble sulfide. We conclude that the pioneer organism was catalytically defined *prima facie* by the transition metal system Fe–Co–Ni–W–(V).

1.3 Retrodicting Pioneer Catalysis

Extant biocatalysis is dominated by enzymes. These classify into metalloenzymes and nonmetalloenzymes. Nonmetalloenzymes require a large number of weak group interactions to fold and to stabilize transition states. This means high sequence accuracy, that is, late evolutionary arrival. The accuracy need of metalloenzymes is more relaxed. They typically exhibit a few strong coordination bonds to transition metals, reacting molecules, or transition states in addition to weak group interactions. Effectiveness of such coordination bonds is relatively insensitive to sequence variations. Therefore, a few coordination bonds may have been sufficient for protein folding and enzyme catalysis at the beginning of translation.

Next we note that evolutionary variability is not uniform throughout the structure of a metalloenzyme. Variability increases and invariance decreases from the (innermost) transition metal through the (inner) ligators to the (outer) ligand moieties. Hence, in the course of evolution central transition metals and ligators, once established remained invariant, with rare replacements (Fe → Mo, V in nitrogenases; W → Mo in tungsto/molybdopterins; Fe → Co, Ni in tetrapyrroles; S → Se or S → O in Fe–S clusters). The outer protein ligands evolved to modulate the catalytic properties of otherwise invariant transition metals and ligators.

We now apply the heuristic of biochemical retrodiction. We replace the protein ligands and any highly evolved organic ligands by simpler small-molecule ligands with the same ligator (e.g., CysS → HS). We do so formally without much concern for proof of chemical stabilities of retrodicted intermediates. We arrive ultimately at transition metal complexes/clusters with inorganic ligands that may derive directly from volcanic gases. These small structures are characterized by a low ratio of bridging to nonbridging ligands and by a total lack of periodicity so that the use of the term *nanocrystal* for such a small structure would be utterly misleading. How do these small structures relate to large transition metal minerals with their high connectivity and short-range to long-range periodicity (crystallinity)? In order to find an answer to this question, we should appreciate that volcanic-hydrothermal flow settings are characterized by mineral transformations: thermal dehydration, ligand exchange, hydrolysis, and mobilization by leaching, carbonylation, and cyanidation. These transformations are governed by the Ostwald–Volmer step rule, whereby the withdrawal of energy from a chemical system that can exist in several states of density will proceed stepwise from low to high density (typically = stability). Mineral transformations that obey the principle of minimal structural (topological) change may occur by solid-state reactions. All others occur via mobilization and recrystallization. Aqueous recrystallization begins with endergonic nucleation by transient, fluctuational aggregation of small, aperiodic clusters. When a critical size is reached, exergonic growth to (nano)crystals sets in. We now are ready to answer our question. We simply view ligand evolution as an adaptation to the intermediate structures *en route* to mineral nucleation. Nucleation intermediates are transient, yet invariant in type. They cannot be directly observed, but they become fixed by protein ligands and thereby become observable. Protein ligands are variable in sequence and they adapt to catalytically competent complexes/clusters. Mineral catalysis is heterogeneous surface catalysis. Metalloenzyme catalysis, from the point of view of the tiny catalytic centers, is homogeneous catalysis. And the evolutionary transition from mineral catalysis to metalloenzyme catalysis must have proceeded from versatile (multifunctional) heterogeneous catalysis on large mineral surfaces to specialized (oligofunctional) complex/cluster catalysis. This evolutionary transition backtraces mineral formation from endergonic nucleation to exergonic crystal growth.

In another vein, we see the evolution of the multistep cyclical process of extant metalloenzyme catalysis as the result of a progression from linear to cyclical transformations. Primordial metallocatalysis is seen as a linear transformation with few reaction steps and as evolving later into a catalytic cycle with an increased number of reaction steps (much like industrial development of cyclical catalysis that also begins typically with the discovery of a simpler, linear transformation). Let us briefly look at some examples.

- *[Fe–S] proteins* contain a family of simple, structurally related [Fe–S] complexes [13] that are biosynthetically connected as follows: [2Fe–2S]$CysS_3L \rightarrow$ [4Fe-4S]$CysS_3L$ (L = CysS or RO); [4Fe-4S]$CysS_3L' \rightarrow$ [Fe3-S4]$CysS_3$ ($L' = RO$) (R = SerO, AspO, or the like). The retrodiction of analogous inorganic complexes (e.g., CysS ⇐ SH; RO ⇐ OH) has experimental support [16, 17].
- *Nitrogenases* evolved in the direction [FeFe] ⇒ {[FeMo] or [FeV]} nitrogenase. Their catalytic Fe–S–C clusters derive biosynthetically from an ancestral [FeFe] cluster, which in turn derives from two [4Fe-4S] clusters and from AdoMet as C source [18]. They all have the same redox-active iron–sulfur P cluster. We project all nitrogenases back to an inorganic [Fe–S] system of volcanic-hydrothermal provenance.
- The *[Fe–Ni–S] enzymes* carbon monoxide dehydrogenase (anaerobic) (CODH) and acetyl-CoA synthase (ACS) [19] may be traced back to an inorganic [Fe–Ni–S] system, with ligands derived from H_2S, H_2O, NH_3 [20].
- *Molybdopterin* and *tungstopterin* enzymes of extant biochemistry trace back to an ancestral family of common precursor enzymes that were exclusively endowed with oxygen-sensitive tungstopterin cofactors [14]. With oxygenation of the surface of the Earth the fastidiously anaerobic tungstopterin enzymes receded into anaerobic niches and the aerobic world became molybdenized [14]. Among the tungstoenzymes that are holdovers to this day, we find extraordinary catalytic abilities, for example, direct reduction of carboxylic acids [21] or hydration of acetylene [22]. Tungstopterins are traced back to inorganic tetrathiotungstate dianions $[WS_4]^{2-}$ [4] with coexistent thiovanadate anions ($VO_xS_{4-x}{}^{3-}$, $x = 0-3$).

Such retrodictions are specific enough to benefit from, and find support in the study of (aqueous or nonaqueous) model compounds [23, 24]. In a broader context, biocatalyst retrodictions should be correlated with global geochemical developments: for example, cooling of the mantle and crust; depletion of nickel in the ocean; oxygenation of the ocean and atmosphere [25]; onset of aerobiosis; and onset of molybdenization of the biosphere [14]. In another broad sense, we see two causes for complications: (i) habitats became dislocated away from volcanic-hydrothermal sites and (ii) formerly nutritious volcanic-hydrothermal components became toxic (e.g., H_2S). Both causes led to the emergence of strategies for biosynthesis of nutrients from metabolites (e.g., from cysteine by desulfurases [26]) or from other nutrients; or strategies for the conversion of one ligand into another ligand – all under highly controlled scaffolding conditions. Let us look at the biochemistry of H_2.

- The three *hydrogenases* have deep dissimilarities, but also startling commonalities in terms of active site Fe–(CO)–(CN) coordination. It is frequently concluded that their commonalities are "convergent" and not due to a common history. This conclusion is obscure and, strictly speaking, not supported by facts. Surely, the protein structures do not reveal common ancestry at the level of the last universal common ancestor (LUCA). Their catalytic clusters differ in terms of structure and biosynthesis. A deeper, pretranslational common ancestry, however, may well be a realistic assumption (Figure 1.2). We shall attempt to trace the evolution of the three hydrogenases back into invariant inorganic chemistry.
- *[Fe] hydrogenase* (anaerobic) converts H_2 in *one* step into H^+ and a H^- ligand that is transferred directly and reversibly to the target substrate (methenyl-H_4MPT) [28]. The redox-inactive Fe center has two CO ligands, which are derived from CO_2 via an internal CO pool [29]. The bidentate organic guanylylpyridinol ligand may be a replacement for inorganic ligands: pyridine-*N* for cyano [30] and acyl-CO for CO [28]. This leads us to a retrodicted ancestral $Fe(CO)_3(CN)(SH)$ complex that may be formally related to the carbonyl $Fe(CO)_5$ (Figure 1.2).
- *[FeFe] hydrogenase* (strictly anaerobic) has a redox-active [Fe–Fe] center and converts H_2 into *two* e^- (via intermediate H^-). These are transferred efficiently to a [4Fe–4S] cluster that is covalently attached via its CysS ligand as the bridge [31]. The set of inorganic ligands comprises CN^-, CO, and -S-CH_2-NH-CH_2-S-. They all derive from dehydroglycine (formed from tyrosine by an AdoMet-dependent radical mechanism) [32, 33]. Retrodiction of an inorganic ancestral cluster (Figure 1.2) is based on chemical synthesis (Rauchfuss reaction) [34]. Incidentally, dehydroglycine is related to the formation of glycine from a CN/CO ligand system [27].
- *[NiFe] hydrogenases* have a redox-active [Ni–Fe] center. It converts H_2 into *two* e^- (via intermediate H^-) that are transferred via [4Fe-4S] clusters. Its catalytic $Fe(CN)_2(CO)$ moiety is preassembled on a scaffold [35], with CN^- (from carbamoyl phosphate [36]) added before CO (from a metabolite-derived internal

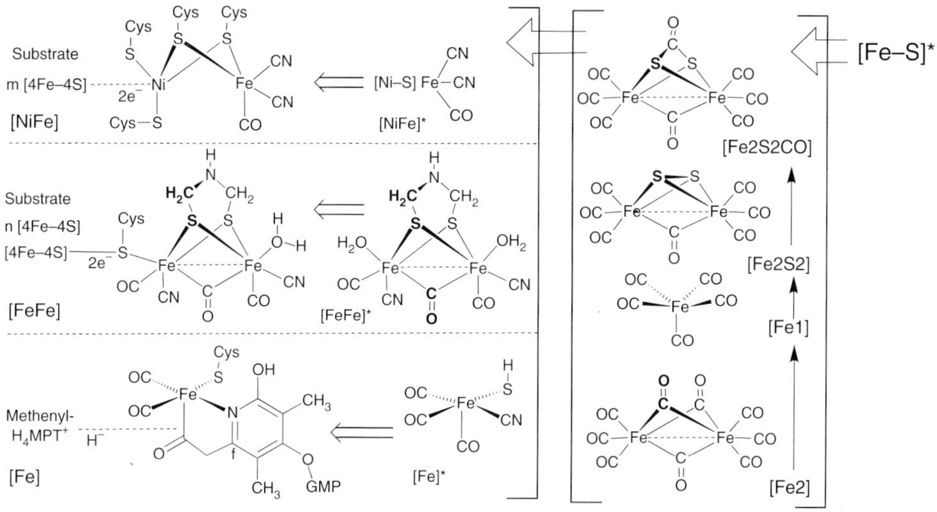

Figure 1.2 Common origin of hydrogenases in a set of carbonylated precursors ([Fe1] = $Fe(CO)_5$) in pseudorotation; [Fe2S2CO] forms from [Fe2] under primordial conditions [27].

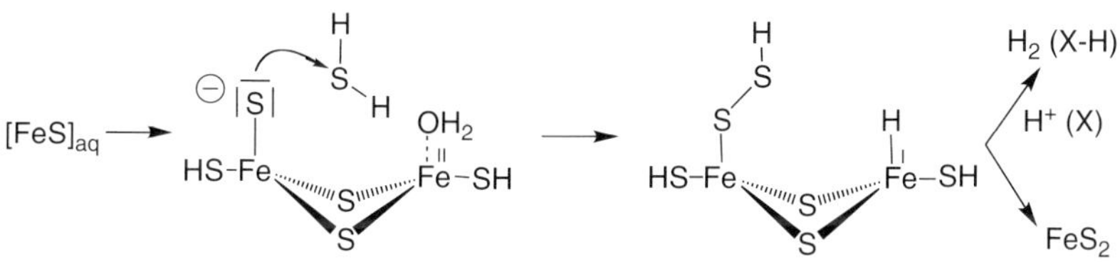

Figure 1.3 Notional mechanism of pyrite formation from FeS/H_2S (X = substrate).

CO pool [37]). This leads us to the ancestral structure $[Ni-S]Fe(CN)_2(CO)$ (Figure 1.2).

All three hydrogenases may ultimately go back to the conversion of FeS/H_2S to FeS_2/H_2 [38, 39], for which a mechanism is postulated (Figure 1.3) that starts from aqueous $[FeS]_{aq}$ [40] and assumes a nucleophilic attack of a sulfide ligand on the sulfur atom of H_2S to form a persulfide ligand (as source for FeS_2) [40] in combination with either formation of H_2 by reaction of a hydride ligand with H^+ or reduction of a substrate X by hydride or electron transfer. From this ur-hydrogenase/reductase chemistry, the evolution of hydrogenases began by abandonment of irreversible pyrite formation in favor of catalytic cycling, reversal of H_2-formation into H_2-uptake, and substrate reduction (e.g., persulfide cleavage).

It proceeded to carbonylation with volcanic CO (cf. carbonyl transformations shown in brackets [41, 42]). The next evolutionary step would have been cyanidation by an exogenous cyanide source or by an unknown endogenous ligand conversion, with the effect of stabilizing the CO/SR ligand sphere [43, 44].

1.4
Retrodicting Metabolic Reproduction and Evolution

Extant organisms reproduce by sequential phases of metabolic reproduction, genetic reproduction, and cellular reproduction. We retrodict an early stage of life when all reproduction was metabolic reproduction. For conceptualizing the evolution of metabolic reproduction, we subdivide the metabolism into pathways and catalysts (Figure 1.4). Pathway evolution is broken down into elementary transformations: terminal extension, lateral branching, recruitment of nutrients or intermediates, pathway cyclization, pathway reversal, and eliminations [6]. Further, we project all enzymes into metalloenzymes, these into metallocatalysts with inorganic and simple organic ligands, and these finally into an inorganic starter catalyst system, which converts volcanic gases into organic compounds.

Let us assume one of the produced organic compounds turns into a ligand of a transition metal center of the starter catalyst system to augment its catalytic activity, cf. [45]. Consider terminal pathway extension and a proximal ligand effect of an organic product B (Figure 1.5). There are two possibilities. (i) The organic product B feeds back as ligand to enhance catalyst K_m for producing the same organic product B (metabolic reproduction). (ii) The organic product B feeds forward as ligand to enhance a catalyst K'_m for converting the organic product B into another product C. It has the effect of boosting the concentration of C to a level that is sufficient for eliciting a new ligand function (metabolic evolution) [46, 47]. Similar feed-forward effects concern lateral branching, recruitment, or cyclization. In addition, ligand effects may recruit new catalytic transition metals or colonize new chemical habitats. Feed-forward effects may also work over a distance.

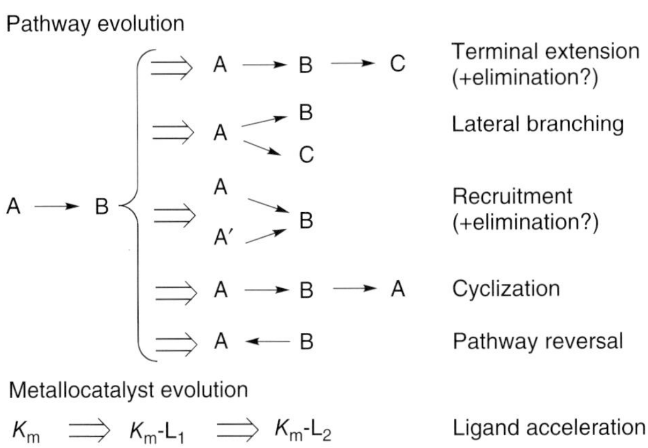

Figure 1.4 Metabolic evolution (A, A′: nutrients; B, C, organic products; ⇒: evolutionary transformation; K_m: catalytic transition metal center; and L_1, L_2: organic ligands), with permission from Springer.

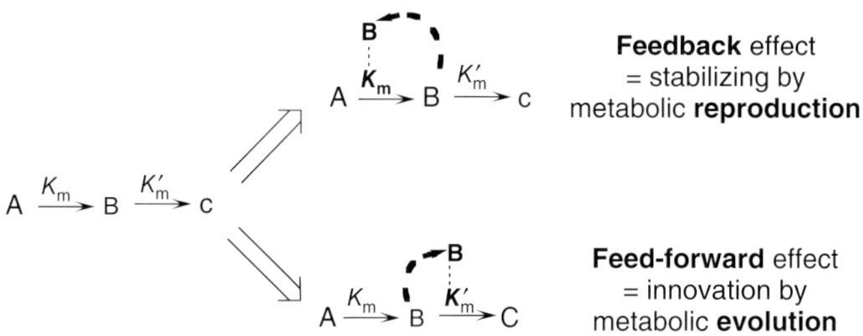

Figure 1.5 Product ligand effects (dashed arrow: ligand effect and ⇒: evolutionary change).

Metabolic expansions by feed-forward effects weaken the preexistent reaction network and require adequate stabilization by feedback effects. Metabolic stability requires multiple product effects, one product enhancing at least two transition metals, or one transition metal that catalyzes at least two reaction steps. With this realization we arrive at the notion of a metabolic avalanche breakthrough. Moreover, by bonding to transition metal centers, organic products may be stabilized against decomposition or hydrolysis (self-selection).

With every feedback effect the metabolism deepens its autonomy from the environment. Eventually, the sum total of all feedback effects will cause the metabolism to run even under conditions that would no longer permit its *de novo* initiation. This marks the beginning of the precariousness of life and the potential for death. It may be said that the emerging organisms are "alive" to the extent that they are "mortal." Chemistry, by acquiring historicity, turns into biology. Each catalytic product effect, as ligand or otherwise, constitutes an instance of memory, or dynamic inheritance. All subsequent evolution is a concatenation of memory effects, progressing from dynamic inheritance of "analog" feedback loop information to genetic inheritance of "digital" sequence information.

1.5 Retrodicting Pioneer-Metabolic Reactions

We now turn to the viability of retrodicted starter catalysts and reducing agents for reactions at or upstream from the location of the pioneer organism. With regard to reaction *temperature*, experiments have proved that the catalytic burden of biochemical reactions suffers a steep increase with decreasing temperature [48]. Therefore, primitive primordial catalysts could generate required reaction rates only at sufficiently high reaction temperatures (e.g., 130–160 °C), determined by the slowest step in a multistep pathway. Subsequent evolution proceeded irreversibly down the temperature scale to the extent that catalytic activities could increase sufficiently by ligand modification to keep up with the increasing catalytic burden (Wolfenden theorem). The *pH* of the pioneer organism is determined by the minimum value (lower at higher temperatures) for aqueous H_2S, CH_3SH, or NH_3 to exist as reactive free bases. Primordial rocks were ultramafic (<45% SiO_2), that is, orthosilicates, generating alkaline conditions (pH 9–12). Acidic volcanic gases caused a neutralization front to move in flow direction. Thus, alkalinity decreased from pH 12 to 9 and beyond by leaching first $Ca(OH)_2$ and then $Mg(OH)_2$. Later approach of neutrality permitted condensation reactions and RNA. Thus, the homestead of life was characterized as a "place with liquid water having a nearly neutral pH … where hot volcanic exhalations clash with a circulating hydrothermal water flow" [7, p. 480].

Reductants are needed for carbon/nitrogen fixation. Volcanic-hydrothermal H_2 is available, but quite inert. It requires a transition metal ur-hydrogenase. As to the reductant CO, water gas equilibration between CO/H_2O and CO_2/H_2 replaces CO by H_2, increasingly so with decreasing temperature. In hot liquid water, this occurs without a catalyst, but not below about 200 °C [49]. Minerals also serve as reductants, notably $Fe(OH)_2$ (from hydrolysis of orthosilicates) [50] and FeS/H_2S (Figure 1.3) for driving metabolic reactions.

Inorganic conversions are foundational for the pioneer organism. The system FeS/H_2S has been revealed as ur-nitrogenase ($^{15}N_2 \rightarrow {}^{15}NH_3$) (Weigand reaction) [51], as reductant for forming NH_3 from NO_3^- (due to hydrolysis of NO_x that forms in a $CO_2/N_2/H_2O$ atmosphere by lightning) [52], and as reductant for converting CO_2 to CH_3SH via COS [53]. The aqueous system $H_2S/CO/(NiFe)S$ (~neutral pH) generates COS, CH_3SH, and $CH_3CO(SCH_3)$ [20]. The water-free system $FeS/HCOOH/C_9H_{19}SH$ converts at 250 °C (2000 bar) to system $FeS/H_2O/CO/C_9H_{19}SH$ to generate $C_9H_{19}COOH$, and pyruvate [54]. Iterative carbon fixation (up to C5) was demonstrated for $CO/H_2S/NiS$: $C_n\text{-SH} \rightarrow C_n\text{-COSH} \rightarrow C_n\text{-CHO} \rightarrow C_{n+1}\text{-SH}$ [55]. Aqueous double carbonylation was demonstrated in the presence of $Ca(OH)_2$ (H^+ scavenger) for RSH ($R=CH_3$, $PhCH_2$) and $CO/Co_2(CO)_8/NiS$ to form RCO-COOH and RCHOH-COOH [27]. Double carbonylation of $PhCH_2$-SH was also shown for CO/(Co,Ni)(OH,S), but dramatically higher productivity was found for $CO/(Co,Ni)(CN)_2$ or CO/(Co,Ni)(OH)(CN) along two pathways: (i) double CO fixation and (ii) CO fixation for CHOH and cyanidation for COOH [27].

Pairs of *α-amino/α-hydroxy acids* and derivatives of formula R-CHA-COY (A = OH, NH_2, $NHCH_3$, NHCOOH; Y = OH, NH_2, $NHCH_3$; R = H, CH_3, CH_2OH, C_2H_5) were formed in good yield by direct, one-pot reaction in the aqueous system (CO or H_2)/Ni(OH)(CN) at 100–280 °C, optimally 130–180 °C (70 bar) with $(Ca, Mg)(OH)_2$ as H^+ scavenger (Figure 1.6). Some higher homologs and α-keto acids were also found. CN ligands were the main C source and CO was e^- source and minor C source. The reaction system comprises layered $Ni(OH)_2$ and $[Ni(CN)_4]^{2-}$. When glycine or alanine as products of this reaction were added to the chemical system (CO/Ni(OH)(CN)) for double carbonylation of $PhCH_2$-SH, the yield of $PhCH_2$-CHOH-COOH was significantly increased – a model for the postulated feed-forward effect [27].

α-Amino acids are activated as *N*-carboxy-aminoacyl anhydrides (L) with COS/NiS (derived from $CO/H_2S/NiS$) at pH ~ 9, which are drawn into a *peptide*

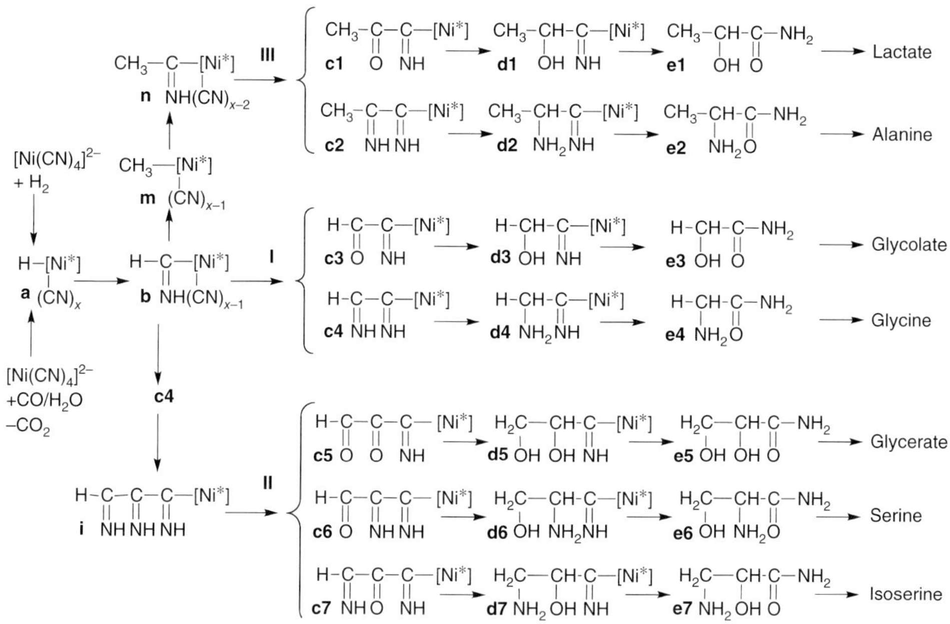

Figure 1.6 Notional mechanism of formation of α-amino/α-hydroxy acids by serial H–N=C insertion [56] ([Ni*], Ni center of unknown nuclearity, oxidation state, and ligand sphere; CO insertion not shown). (From Ref [27] © (2012), Wiley.)

cycle [57, 58] (Figure 1.7). They react with the free amino group of another α-amino acid to form a dipeptide, or with the free amino group of the dipeptide to form a tripeptide, and so forth. The peptides also react with the system COS/NiS to acquire an N-terminal hydantoin ring that hydrolyzed via an N-terminal urea moiety to a free amino acid and a peptide that is shortened N-terminally by one amino acid. By this peptide cycle, the sequence space is scanned for the most stable peptide structures (self-selecting peptide library).

1.6 Early Evolution in a Spatiotemporal Flow Context

An ocean may have existed 4.4 billion years ago, as evidenced by the composition of particles of $ZrSiO_4$ that were formed at that time and preserved in much younger sedimentary rocks [59]. Therefore, the pioneer organism may well date back deep into the Hadean eon, when the Earth had just emerged from its globally molten state. The crust was thin and frequently fragmented by large impactors, generating a great variety of chemical conditions and a bed of debris with a maze of flow ducts for hydrothermal venting and volcanic outgassing [60]. The styles of volcanic-hydrothermal typology differed drastically from those of the present day,

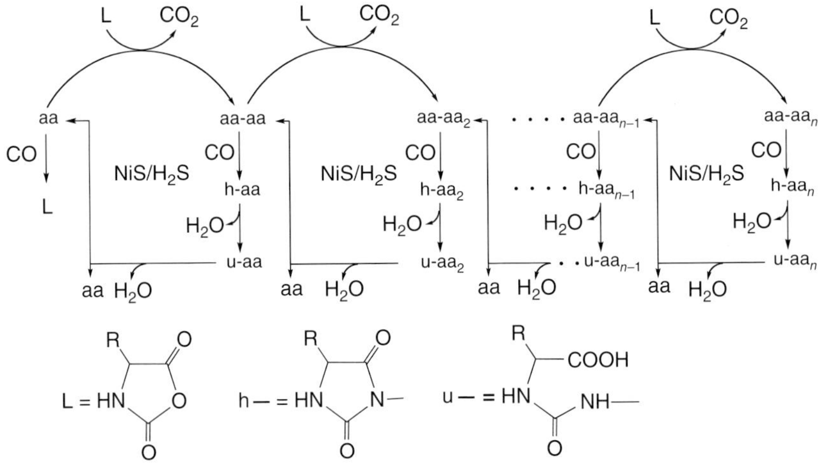

Figure 1.7 Peptide cycle driven by COS derived from CO/H_2S (aa, amino acid), with permission from Wiley.

and the ocean experienced continuous cleansing by intense sedimentation of volcanic ashes and hydrothermal percolation. In short, uniformitarianism does not apply.

The Hadean mantle must have had temperatures well above the melting point of komatiite (1700 °C). Under conditions of graphite saturation, the molar equilibrium ratio CO:CO_2 increases with increasing temperature and decreasing pressure. Hence, it must have been higher than the ratio 1 : 1, which exists at 1200 °C (2 kbar) [61]. By mixing with cooler hydrothermal water the hot volcanic fluids were quenched, preserving a high activity of CO.

A Hadean volcanic flow path is subdivided into three regions. (i) A deep flow region with a temperature above the critical temperature of water operates as a gas/solid reactor. (ii) A medium flow region with a temperature below the critical temperature of water operates as a gas/liquid/solid trickle-bed reactor. (iii) An uppermost flow region operates as a chromatographic reactor with a stationary packed bed of catalytic particles and a mobile phase of hydrothermal water and dissolved (volcanic) reactants. It exhibits gradients of temperature, pressure, and pH and chemical zoning, as well as chemical reactions in sync with chromatographic separation of organic products (by differential coordination bonding) in flow direction. The strongest surface bonders are the slowest travelers. They concentrate in early "ligand zones" to engage in a surface metabolism. In such a location, the pioneer organism starts by ligand-based feedback and feed-forward effects. It renders the pioneer organism a spatially and temporally coherent entity.

While reactive interactions occur on a molecular scale, larger dimensions are also of interest. Kuhn [62] considered rocks with an open pore structure for size exclusion separation of RNA molecules in an (RNA world) origin of life. Generally speaking, porosity and surface irregularities increase catalytic activity and cause molecular sieve effects. An ingenious experiment-based "chemical garden" proposal sees early evolution inside a mound of open FeS-bounded "cells." A FeS membrane precipitated at the top of the mound, where hot (e.g., 150 °C), alkaline, sulfidic (e.g., 0.5 M HS^-) vent fluid contacts warm (e.g., 90 °C), acidic, carbonated (~10 bar CO_2) ocean water with dissolved Fe^{2+} (e.g., 0.5 M). The mound grew by breaking and remaking of the FeS membrane. A pH gradient across the top FeS membrane fueled the synthetic reactions inside the mound by chemiosmotically driven ATP synthesis [63]. As the mound grew in height (up to 10 m), the metabolism inside the open-cell structure evolved from FeS-catalyzed carbon fixation at the bottom via the RNA world in the center to LUCA higher up – with full-fledged cells of bacteria and archaea breaking loose at the top [64]. Unfortunately, however, the experimentally detected cell structure is a freeze-drying artifact [46]; RNA cannot exist under hot alkaline conditions; and an experimental FeS membrane "behaved more like a permeable reactive barrier than a membrane" [40].

A volcanic-hydrothermal flow duct is a local affair, which means that the origin of life must have occurred in many locations. Owing to the invariant laws of chemistry, the essential origin chemistry must have been the same everywhere, slight parameter-dependent differences of composition notwithstanding. With the beginning of evolution as a historic process, locally different varieties with different catalytic endowments would have arisen. There was, however, a grand uniformization process at work: large impactors ejected chunks of crustal material high into the atmosphere or even into orbit. Inside these ejected (orbiting) chunks, cellular organisms are believed to have survived until later reentry [60]. By the same argument (sets of) catalysts composed of transition metal centers and evolved organic ligands would have survived *a fortiori* – a truly global lateral transfer of catalytic endowment. Global lateral transfer by this and other mechanisms continued unabated until the advent of LUCA. Lateral transfer of catalytic endowment by physical scattering of ejected crustal material has another important consequence. Fluid flow through volcanic-hydrothermal flow ducts from deep hot locations into the ocean is of too short a duration to accommodate a long period of evolution. Intense scattering of crustal materials, however, and their settlement in renewed flow beds would restart the process time and again with evermore advanced catalytic material. With the formation of lipids by carbon fixation and subsequent cellularization, the metabolism became an internal affair, segregated from the chemical environment – the beginning of an increasing autonomy from the mineral base of life.

The start of the genetic machinery is marked by the advent of organophosphate groups for hard ligation. The components of the genetic machinery (phospho)-sugars, bases, (oligo)nucleos(t)ides, and an ever-expanding set of oligopeptides are all ligand formers that earned their keep with ligand effects. Up to this point, evolution is a *direct* process. New metallocatalysts lead to new organic products that turn *directly* into ligands for more advanced metallocatalysts. With the beginning of ribosomal translation, evolution turns into an *indirect* affair. Mutated polynucleotide sequences are translated into new peptides and these then become ligands for more advanced metallocatalysts. The ribosome is to this day a metallocatalyst with a seemingly inorganic magnesium-phosphate core, controlled by invariant laws of coordination and surrounded by a gigantic ligand sphere that expanded layer by variable layer over time in radial direction [65]. With an increasingly dominant role of the genetic machinery, the inorganic origins of life in coordination/organometal chemistry became more and more overshadowed. But to this day, the genetic machinery remains what it was at the beginning: the great randomizer of life – invented by life for adaptation for a continued existence in regions evermore remote from the original mineral homestead in a volcanic-hydrothermal flow setting. The genetic machinery works by stochastic variations, which give us the impression of evolution as a directionless, random walk. But in its core and from the beginning the evolution of life is chemically unique, predetermined by universal laws and progressing along the arrow of time.

Acknowledgments

Support by Deutsche Forschungsgemeinschaft for experimental evaluation is gratefully acknowledged.

References

1. Kant, I. (1790) Kritik der Urteilskraft. §81, Translation by Meredith, J.C. (1952) *The Critique of Judgment*, Clarendon Press, Oxford.
2. Darwin, C. (1863) Letter to J.D. Hooker, 29 March, 1863.
3. Lipmann, F.A. (1965) in *The Origin of Prebiological Systems and of Their Molecular Matrices* (ed, S.W. Fox), Academic Press, New York, pp. 259–280.
4. Wächtershäuser, G. (1997) The origin of life and its methodological challenge. *J. Theor. Biol.*, **187**, 483–494.
5. Florkin, M. (1949) *Biochemical Evolution*, Academic Press, New York.
6. Granick, S. (1957) Speculations on the origin and evolution of photosynthesis. *Ann. N. Y. Acad. Sci.*, **69**, 292–308.
7. Wächtershäuser, G. (1992) Groundworks for an evolutionary biochemistry: the iron-sulphur world. *Prog. Biophys. Mol. Biol.*, **58**, 85–201.
8. Marx, D. and Hutter, J. (2009) *Ab Initio Molecular Dynamics: Basic Theory and Advanced Methods*, Cambridge University Press, Cambridge.
9. Popper, K.R. (1963) *Conjectures and Refutations: The Growth of Scientific Knowledge*, Routledge & Kegan Paul, London.
10. Fraústo da Silva, J.J.R. and Williams, R.J.P. (1991) *The Biological Chemistry of the Elements*, Clartendon Press, Oxford.
11. Wächtershäuser, G. (1988) Before enzymes and templates: theory of surface metabolism. *Microbiol. Rev.*, **52**, 452–484.
12. Corliss, J.B., Baross, J.A., and Hoffman, S.E. (1981) An hypothesis concerning the relationship between submarine hot springs and the origin of life on Earth. *Oceanol. Acta* (Proceedings of the 26th International Geological Conference), **4**, 59–69.
13. Johnson, D.C., Dean, D.R., Smith, A.D., and Johnson, M.K. (2005) Structure, function, and formation of biological iron-sulfur clusters. *Annu. Rev. Biochem.*, **74**, 247–281.
14. Kletzin, A. and Adams, M.W.W. (1996) Tungsten in biological systems. *FEMS Microbiol. Rev.*, **18**, 5–63.
15. Crans, D.C., Smee, J.J., Gaidamauskas, E., and Yang, L. (2004) The chemistry and biochemistry of vanadium and the

biological activities exerted by vanadium compounds. *Chem. Rev.*, **104**, 849–902.
16. Malkin, R. and Rabinowitz, J.C. (1966) The reconstitution of clostridial ferredoxin. *Biochem. Biophys. Res. Commun.*, **23**, 822–827.
17. Müller, A. and Schladerbeck, N.H. (1986) Einfache aerobe Bildung eines $[Fe_4S_4]^{2+}$-Clusterzentrums. *Naturwissenschaften*, **73**, 669–670.
18. Wiig, J.A., Hu, Y., Lee, C.C., and Ribbe, M.W. (2012) Radical SAM-dependent carbon insertion into the nitrogenase M-cluster. *Science*, **337**, 1672–1675.
19. Ragsdale, S.W. (2009) Nickel-based enzyme systems. *J. Biol. Chem.*, **284**, 18571–18575.
20. Huber, C. and Wächtershäuser, G. (1997) Activated acetic acid by carbon fixation on (Fe,Ni)S under primordial conditions. *Science*, **276**, 245–247.
21. White, H., Strobl, G., Feicht, R., and Simon, H. (1989) Carboxylic acid reductase: a new tungsten enzyme which catalyzes the reduction of non-activated carboxylic acids to aldehydes. *Eur. J. Biochem.*, **184**, 89–96.
22. Seiffert, G.B., Ullmann, G.M., Messerschmidt, A., Schink, B., Kroneck, P.M.H., and Einsle, O. (2007) Structure of the non-redox-active tungsten/[4Fe:4S] enzyme acetylene hydratase. *Proc. Natl. Acad. Sci. U.S.A.*, **104**, 3073–3077.
23. Groysman, S. and Holm, R.H. (2009) Biomimetic chemistry of iron, nickel, molybdenum and tungsten in sulfur-ligated protein sites. *Biochemistry*, **48**, 2310–2320.
24. Hsieh, C.-H., Erdem, Ö.F., Harman, S.D., Singleton, M.L., Reijerse, E., Lubitz, W., Popescu, C.V., Reibenspies, J.H., Brothers, S.M., Hall, M.B., and Darensbourg, M.Y. (2012) Structural and spectroscopic features of mixed valent $Fe^{II}Fe^{I}$ complexes and factors related to the rotated configuration of diiron hydrogenase. *J. Am. Chem. Soc.*, **134**, 13089–13102.
25. Kornhauser, K.O., Pecoits, E., Lalonde, S.V., Papineau, D., Nisbet, E.G., Barley, M.E., Arndt, N.T., Zahnle, K., and Kamber, B.S. (2009) Oceanic nickel depletion and a methanogen famine before the Great Oxidation Event. *Nature*, **458**, 750–753.
26. Py, B. and Barras, F. (2010) Building Fe—S proteins: bacterial strategies. *Nat. Rev. Microbiol.*, **8**, 436–446.
27. Huber, C., Kraus, F., Hanzlik, M., Eisenreich, W., and Wächtershäuser, G. (2012) Elements of metabolic evolution. *Chem. Eur. J.*, **18**, 2063–2080.
28. Dey, S., Das, P.K., and Dey, A. (2013) Mononuclear iron hydrogenase. *Coord. Chem. Rev.*, **257**, 42–63.
29. Schick, M., Xie, X., Ataka, K., Kahnt, J., Linne, U., and Shima, S. (2012) Biosynthesis of the iron-guanylylpyridinol cofactor of [Fe]-hydrogenase in methanogenic archaea as elucidated by stable-isotope labeling. *J. Am. Chem. Soc.*, **134**, 3271–3280.
30. Shima, S., Pilak, O., Vogt, S., Schick, M., Stagni, M.S., Meyer-Klaucke, W., Warkentin, E., Thauer, R.K., and Ermler, U. (2008) The crystal structure of [Fe]-hydrogenase reveals the geometry of the active site. *Science*, **321**, 572–575.
31. Rauchfuss, T.B. (2010) Unraveling the biosynthesis of nature's fastest hydrogenase. *Angew. Chem. Int. Ed.*, **49**, 4166–4168.
32. Pilet, E., Nicolet, Y., Mathevon, C., Douki, T., Fontecilla-Camps, J.C., and Fontecave, M. (2009) The role of the maturase HydG in [FeFe]-hydrogenase active site synthesis and assembly. *FEBS Lett.*, **583**, 506–511.
33. Kuchenreuther, J.M., George, S.J., Grady-Smith, C.S., Cramer, S.P., and Swartz, J.R. (2011) Cell-free H-cluster synthesis and [FeFe] hydrogenase activation: all five CO and CN^- ligands derive from tyrosine. *PLoS One*, **6**, e20346.
34. Li, H. and Rauchfuss, T.B. (2002) Iron carbonyl sulfides, formaldehyde, and amines condense to give the proposed azadithiolate cofactor of the Fe-only hydrogenases. *J. Am. Chem. Soc.*, **124**, 726–727.
35. Bürstel, I., Siebert, E., Winter, G., Hummel, P., Zebger, I., Friedrich, B., and Lenz, O. (2012) A universal scaffold for synthesis of the $Fe(CN)_2(CO)$ moiety of [NiFe] hydrogenase. *J. Biol. Chem.*, **287**, 38845–38853.

36. Böck, A., King, P.W., Blokesch, M., and Posewitz, M.C. (2006) Maturation of hydrogenases. *Adv. Microb. Physiol.*, **51**, 1–71.
37. Bürstel, I., Hummel, P., Siebert, E., Wisitruangsakul, N., Zebger, I., Friedrich, B., and Lenz, O. (2011) Probing the origin of the metabolic precursor of the CO ligand in the catalytic center of [NiFe] hydrogenase. *J. Biol. Chem.*, **286**, 44937–44944.
38. Taylor, P., Rummery, T.E., and Owen, D.G. (1979) Reactions of iron monosulfide solids with aqueous hydrogen sulfide up to 160 °C. *J. Inorg. Nucl. Chem.*, **41**, 1683–1687.
39. Drobner, E., Huber, H., Wächtershäuser, G., Rose, D., and Stetter, K.O. (1990) Pyrite formation linked with hydrogen evolution under anaerobic conditions. *Nature*, **346**, 742–744.
40. Rickard, D. and Luther, G.W. III, (2007) Chemistry of iron sulfides. *Chem. Rev.*, **107**, 514–562.
41. Hieber, W., Beck, W., and Braun, G. (1960) Anionische Kohlenoxidkomplexe. *Angew. Chem.*, **72**, 795–874.
42. Nametkin, N.S., Kolobkov, B.I., Tyurin, V.D., Muratov, A.N., Nekhaev, A.I., Mavlonov, M., Sideridu, A.Y., Aleksandrov, G.G., Lebedev, A.V., Tashev, M.T., and Dustov, H.B. (1984) Synthesis and structure of dithiodiiron heptacarbonyl. *J. Organomet. Chem.*, **276**, 393–397.
43. Rauchfuss, T.B., Contakes, S.M., Hsu, S.C.N., Reynolds, M.A., and Wilson, S.R. (2001) The influence of cyanide on the carbonylation of iron(II): synthesis of Fe—SR—CN—CO centers related to the hydrogenase active sites. *J. Am. Chem. Soc.*, **123**, 6933–6934.
44. Lyon, E.J., Georgakaki, I.P., Reibenspies, J.H., and Darensbourg, M.Y. (2001) Coordination sphere flexibility of active-site models for Fe-only hydrogenase: studies in intra- and intermolecular diatomic ligand exchange. *J. Am. Chem. Soc.*, **123**, 3268–3278.
45. Berrisford, D.J., Bolm, C., and Sharpless, K.B. (1995) Ligand accelerated catalysis. *Angew. Chem. Int. Ed. Engl.*, **34**, 1059–1070.
46. Wächtershäuser, G. (2006) From volcanic origins of chemoautotrophic origin of life to bacteria, archaea and eukarya. *Philos. Trans. R. Soc. London, Ser. B*, **361**, 1787–1808.
47. Wächtershäuser G. 2010. In: (Eds.) L. L. Barton, M. Mandl, A. Loy *Geomicrobiology: Molecular and Environmental Perspective*. Springer, Dordrecht, pp. 1-35.
48. Stockbridge, R.B., Lewis, C.A. Jr., and Wolfenden, R. (2010) Impact of temperature on the time required for the establishment of primordial biochemistry, and for the evolution of enzymes. *Proc. Natl. Acad. Sci. U.S.A.*, **107**, 22102–22105.
49. Seewald, J.S., Zolotov, M.Y., and McCollom, T. (2006) Investigation of single carbon compounds under hydrothermal conditions. *Geochim. Cosmochim. Acta*, **70**, 446–460.
50. Huber, C. and Wächtershäuser, G. (2003) Primordial reductive amination revisited. *Tetrahedron Lett.*, **44**, 1695–1697.
51. Dörr, M., Käßbohrer, J., Grunert, R., Kreisel, G., Brand, W.A., Werner, R.A., Geilmann, H., Apfel, C., Robl, C., and Weigand, W. (2003) A possible prebiotic formation of ammonia from dinitrogen on iron-sulfide surfaces. *Angew. Chem. Int. Ed.*, **42**, 1540–1543.
52. Blöchl, E., Keller, M., Wächtershäuser, G., and Stetter, K.O. (1992) Reactions depending on iron sulfide and linking geochemistry with biochemistry. *Proc. Natl. Acad. Sci. U.S.A.*, **89**, 8117–8120.
53. Heinen, W. and Lauwers, A.M. (1996) Organic sulfur compounds resulting from the interaction of iron sulfide, hydrogen sulfide and carbon dioxide in an anaerobic aqueous environment. *Orig. Life Evol. Biosph.*, **26**, 131–150.
54. Cody, G.D., Boctor, N.Z., Filley, T.R., Hazen, R.M., Scott, J.H., Sharma, A., and Yoder, H.S. Jr., (2000) Primordial carbonylated iron-sulfur compounds and the synthesis of pyruvate. *Science*, **289**, 1337–1340.
55. Loison, A., Dubant, S., Adam, P., and Albrecht, P. (2010) Elucidation of an iterative process of carbon-carbon bond

formation of prebiotic significance. *Astrobiology*, **10**, 973–988.
56. Fehlhammer, W.P. and Fritz, M. (1993) Emergence of CNH and cyano complex based organometallic chemistry. *Chem. Rev.*, **93**, 1243–1280.
57. Huber, C. and Wächtershäuser, G. (1998) Peptides by activation of amino acids on (Fe,Ni)S surfaces: implications for the origin of life. *Science*, **281**, 670–672.
58. Huber, C., Eisenreich, W., Hecht, S., and Wächtershäuser, G. (2003) A possible primordial peptide cycle. *Science*, **301**, 938–940.
59. Harrison, T.M. (2009) The Hadean crust: evidence from >4 Ga zircons. *Annu. Rev. Earth Planet. Sci.*, **37**, 479–505.
60. Cockell, C.S. (2006) The origin and emergence of life under impact bombardment. *Philos. Trans. R. Soc. London, Ser. B*, **361**, 1845–1875.
61. Holloway, J.R. and Blank, J.G. (1994) Application of experimental results to C-O-H species in natural melts. *Rev. Mineral.*, **30**, 187–230.
62. Kuhn, H. and Waser, J. (1981) Molecular self-organization and the origin of life. *Angew. Chem., Int. Ed. Engl.*, **20**, 500–520.
63. Russell, M.J. and Hall, A.J. (1997) The emergence of life from iron monosulphide bubbles at a submarine hydrothermal redox and pH front. *J. Geol. Soc.*, **154**, 377–402.
64. Martin, W. and Russell, M.J. (2003) On the origin of cells: an hypothesis for the evolutionary transitions from abiotic geochemistry to chemoautotrophic prokaryotes, and from prokaryotes to nucleated cells. *Philos. Trans. R. Soc. London, Ser. B*, **358**, 27–85.
65. Hsiao, C., Mohan, S., Kalahar, B.K., and Williams, L.D. (2009) Peeling the onion: ribosomes are ancient molecular fossils. *Mol. Biol. Evol.*, **26**, 2415–2425.

2
Fe–S Clusters: Biogenesis and Redox, Catalytic, and Regulatory Properties

Yvain Nicolet and Juan C. Fontecilla-Camps

2.1
Introduction

It is now generally accepted that conditions on the primordial Earth were very harsh. During eons meteorite showers hit the planet, making its surface inhospitable and the bottom of the oceans the most likely environment for prebiotic chemistry and the emergence of life [1]. Both NiS and FeS could have been expelled from hydrothermal hot acidic springs as has been observed in contemporary black smokers. Because of the almost total absence of molecular oxygen and the acidity of the Hadean ocean, these and other sulfides would have remained in solution [1]. In the "iron sulfur world" theory of the origin of life proposed by Wächtershäuser [2], FeS is supposed to have played a central role by reacting with H_2S to yield pyrite and molecular hydrogen ("pyrite-pulled metabolism"). In a series of experiments, Huber and Wächtershäuser [3] used a combination of NiS and FeS to demonstrate the catalytic formation of acetate from CO and CH_3SH at high temperature. In addition, pyruvate has been synthesized from alkyl thiolates and CO in the presence of FeS at high temperatures and pressures [4].

Along the same lines, Russell [5] proposed that catalytic metal sulfides could have been involved in abiotic carbon fixation. In the absence of oxygen, the precipitation of H_2S with Fe^{2+} would have resulted in the formation of semipermeable membranes consisting mostly of FeS, which could have been filled with hydrothermal fluids. Prebiotic reactions would have been possible between simple compounds such as CO, CN^-, NH_3, CH_3SH, and HCHO within bubbles confined by the mineral membranes [5]. With time, reactions taking place within these structures could have become more complex [6]. Eventually, peptide bonds would have been synthesized, as it was experimentally observed using CO and a FeS/NiS mixture [7]. The early involvement of such a mixture of metal sulfides in protobiological reactions is supported, at least partially, by the conspicuous presence of metal sulfur clusters in metalloenzymes that catalyze reactions with gases, which may be considered primordial. They include nitrogenase, carbon monoxide dehydrogenase (CODH), acetyl coenzyme A synthase (ACS), and the NiFe- and FeFe-hydrogenases [8]. Notably, CODH, ACS, and one of the hydrogenases contain

Bioinspired Catalysis: Metal-Sulfur Complexes, First Edition.
Edited by Wolfgang Weigand and Philippe Schollhammer.

combinations of nickel, iron, and sulfur in their active sites [9]. In general, one puzzling observation is that although, relatively speaking, the metal cluster can be simple, the protein matrices have a tendency to be very complex. Consequently, it is not easy to understand how complex proteins became associated with these metal active sites. There are two possibilities: either the metal cluster was originally captured from the inorganic environment by simple peptides being later synthesized by increasingly complex enzymes during an evolutive process, or we are dealing with a case of a chemical constraint where metal-sulfide species are the best suited to catalyze gas-based reactions both inorganically and biologically.

By far, the most abundant and plastic metal-containing clusters are of the Fe–S type. The most common ones are [4Fe–4S] and [2Fe–2S] clusters, but they can either be more complex as in the case of nitrogenase [10, 11] or be modified by other metal ions such as Ni in CODH [12]. Such a variety of structures is not generated spontaneously but requires a mechanism for their assembly. Examples of associations between iron and sulfur in proteins are given in Figure 2.1.

Because it is often possible to carry out the assembly of [4Fe–4S] and [2Fe–2S] clusters *in vitro* by simply adding Na_2S, iron, and a reducing agent to a solution containing a suitable cysteine-containing apo-protein, it was assumed that the same simple process applied to biological systems. However, it is now known that a sophisticated arrangement of gene products in three equivalent machineries take care of the assembly of Fe–S clusters in relevant proteins [17–21].

Because of their intrinsic lability when exposed to reactive oxygen species (ROS) or reactive nitrogen species (RNS), [4Fe–4S] clusters are used in a series of transcriptional regulators to sense their environment such as, for instance, the transition between aerobic and anaerobic metabolisms, which depends on the fumarate nitrate reduction regulator (FNR) [22, 23]. Regulation can also take place at the mRNA level as exemplified by cytoplasmic aconitase/iron regulatory protein 1 (IRP1) (cytosolic aconitase (c-acon)), which controls iron homeostasis [24]. Another example is IscR, a regulator of one of the Fe–S assembly machinery that can bind a [2Fe–2S] cluster, directly senses the level of Fe–S clusters in the cell [22, 23]. The various redox, catalytic, and regulatory properties of Fe–S clusters, as well as their biogenesis, are discussed subsequently.

2.2
Fe–S Cluster Biogenesis and Trafficking

Fe–S clusters contain iron ions both in the ferrous and ferric states and sulfur atoms as sulfide ions. As mentioned, it is possible to correctly assemble a [2Fe–2S] or a [4Fe–4S] cluster *in vitro* in a suitable purified apo-protein simply by adding ferrous or ferric ions, sulfide ions, and a reductant. However, *in vivo* Fe–S cluster assembly requires dedicated machineries in order to notably control the fluxes of the required elements, the number of Fe–S clusters, and to prevent the potentially deleterious effects of free iron and sulfide [13, 17, 25]. The first evidence for the existence of dedicated machineries for Fe–S cluster biosynthesis

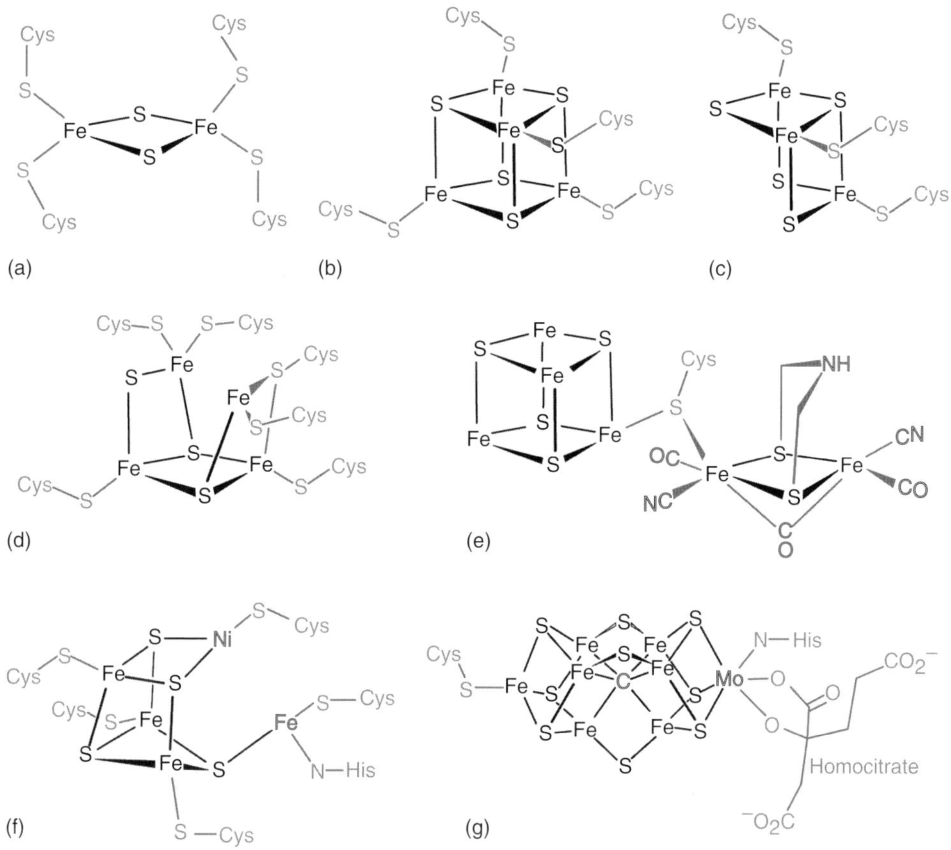

Figure 2.1 Fe–S clusters found in metalloproteins. (a–c) Standard Fe–S clusters [13]. (d) The proximal cluster of O_2-tolerant NiFe hydrogenases [14–16]. (e–g) The active sites of FeFe hydrogenase, CODH, and nitrogenase [8].

was obtained from studies on the synthesis of FeMo-co (Figure 2.1g), the active site of FeMo-nitrogenase in *Azotobacter vinelandii* and the discovery of the *nif* (nitrogen fixation) operon [19]. Subsequently, the *isc* (iron–sulfur cluster) and *suf* (sulfur mobilization) operons were discovered [20, 21], first in prokaryotes and later in yeast and mammals, underscoring their ubiquity [17, 18]. These two systems seem to be specific: in *Escherichia coli*, the *isc* operon is responsible for the assembly machinery of Fe–S clusters under normal growth conditions, whereas the *suf* operon appears to function when bacteria are exposed to stress [13]. Homologs of the *suf* operon are also found in plastids and cyanobacteria [26, 27].

The three systems, *nif*, *isc*, and *suf*, use homologous enzymes for the production of the sulfide ions of the Fe–S cluster. These enzymes are collectively called *cysteine desulfurase* and specifically *NifS*, *IscS*, and *SufS*. In *E. coli*, IscS is a key enzyme that also provides sulfur atoms to several pathways unrelated to Fe–S cluster biosynthesis, such as thiamin or molybdopterin biosyntheses and tRNA modification [28]. Conversely, SufS activity is likely to be restricted to Fe–S biogenesis [28]. Several X-ray structures of NifS, IscS, or SufS have been determined in different states: free, in complex with substrate or analogs, and wild type as well as mutants [29–32]. They all have a very similar fold and form dimers, essential for activity. Indeed, each of the two active site cavities is defined by residues that belong to both monomers. Cysteine desulfurases use pyridoxal phosphate as a cofactor to abstract the sulfur atom of a free L-cysteine. The liberated S binds the active cysteine at the cys-loop in the active site of the enzyme, with concomitant release of L-alanine (Figure 2.2) [28, 33]. Subsequently, the cys-loop swings out of the active site and transfers the sulfur atom from the active persulfided cysteine to a target protein. The main difference between IscS and SufS is the length of their respective cys-loops [28]. In the former, the loop is longer, which presumably increases its flexibility and range for S transfer [28–32]. Indeed, the extra cys-loop length may be required because IscS transfers sulfur atoms to several partners that bind to different sites on the desulfurase [28, 34]. Conversely, in SufS, the cys-loop does not need such flexibility because this enzyme has fewer partners and only two preferential orientations are required for S transfer. The length difference between the two desulfurases may explain why the cys-loop is disordered in all IscS structures reported so far (except when in complex with holo-IscU [35]), whereas it is well ordered in SufS [28–32, 34].

The second feature common to the three Fe–S cluster assembly machineries is the use of a scaffold protein that forms a complex with the desulfurase. NifU, IscU, and most probably the SufBCD complex, play this role serving as a transient site for Fe–S cluster assembly [36–39]. The scaffold needs to interact with both sulfur and iron donors to have a chemical and structural environment well suited to correctly assemble the Fe–S cluster and to be able to transfer it to client proteins [25, 40]. Scaffold proteins are also found in biosynthetic pathways of complex metalloenzyme active sites such as those found in nitrogenase [41–43] and FeFe hydrogenase [44–46] (Figure 2.1e,g). IscU and related scaffolds have an assembly site that contains residues of three cysteines, one histidine, and one aspartate [13, 25]. Mutation of the latter in alanine impairs both IscS–IscU complex dissociation and Fe–S cluster transfer [36, 40]. We used this property of the Asp-to-Ala IscU mutant to solve the crystal structure of the stabilized [2Fe–2S]-containing IscS–IscU complex from *Archaeoglobus fulgidus* [35]. A comparison of its assembly site with the one in the apo Isc–IscU complex from *E. coli* shows that the three cysteines and the histidine of the assembly site are very similarly oriented in the two cases [34, 35]. In addition, there is clear room for the missing –COOH group of the mutated aspartate. This observation indicates that the scaffold site will only undergo minimal conformational changes upon Fe–S cluster assembly and suggests that the aspartate residue does not participate directly in this process [35].

Although the sulfur donor and the function of the cysteine desulfurase have been extensively characterized, the direct source of iron remains elusive. Initially, the protein IscA, which can bind iron with high affinity, was proposed to fulfill this role [47, 48]. However, IscA does not interact with the cysteine desulfurase or with the scaffold protein [49]. This observation drastically reduces its chances

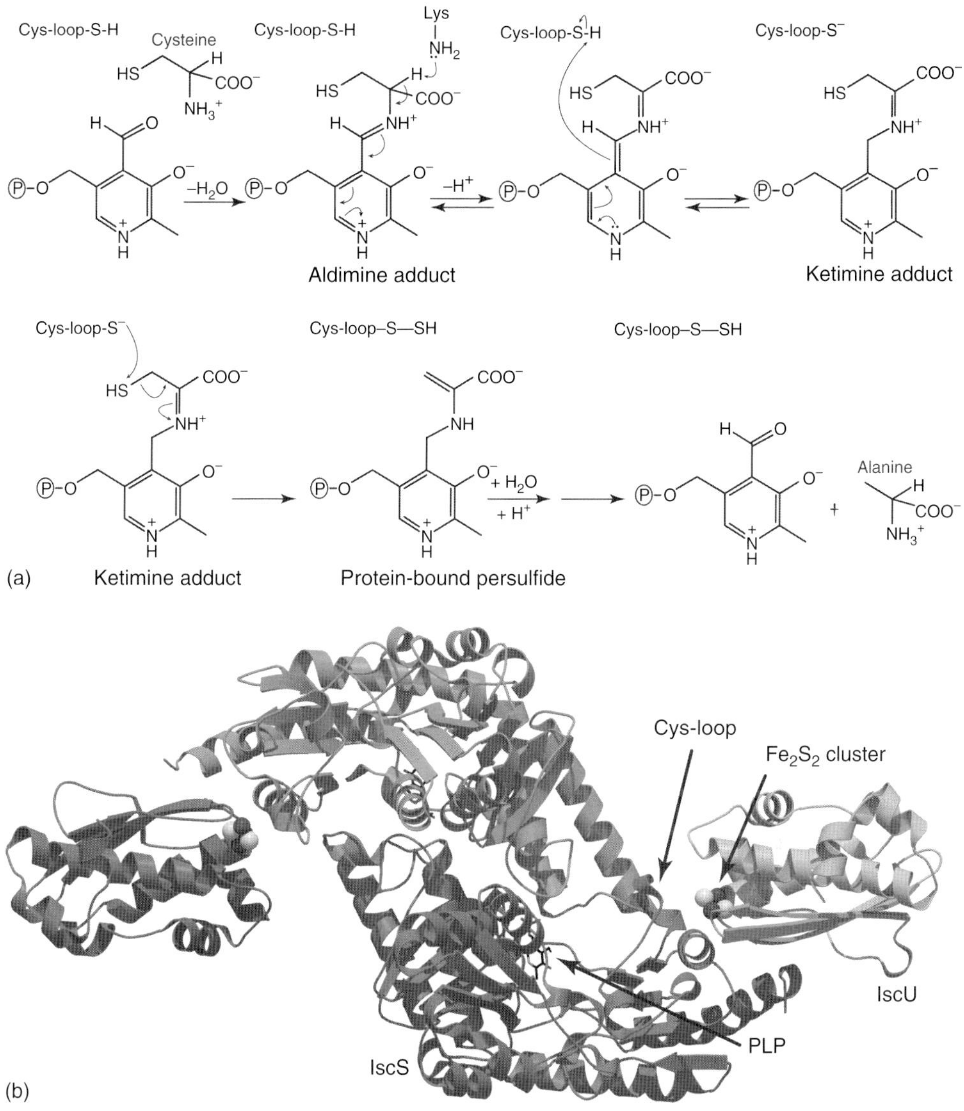

Figure 2.2 (a) The catalytic cycle of L-cysteine desulfurase [13]. The final product of this cycle is a persulfide form of the active site cysteine residue. (b) Crystal structure of the holo form of the (IscS–IscU)$_2$ complex from *Archaeoglobus fulgidus* [35].

of being the iron donor. Conversely, the prokaryotic protein CyaY and its eukaryotic homolog frataxin are more likely candidates to accomplish this function [18]. Indeed, CyaY interacts with both IscS and IscU [34, 50, 51]. Frataxin even forms a stable complex with both partners in mitochondria [52]. CyaY can bind iron with low affinity [53–56], an advantage when delivering it, but a clear disadvantage when carrying the metal ion to the scaffold assembly site. Importantly, mutation of the *cyay* gene does not affect the activity of Fe–S-cluster-containing proteins [57]. Conversely, mutations in frataxin may cause severe diseases, such as Friedreich's ataxia, or perturbations in mitochondrial iron homeostasis [58, 59]. Unexpectedly, although frataxin activates the eukaryotic cysteine desulfurase, CyaY inhibits its bacterial counterpart [60]. This reported difference in the roles of these two homologous proteins appears controversial and one may wonder whether CyaY and frataxin are directly involved in either iron donation or iron sensing [60]. Further characterization of the function of the two proteins will be required before the process of iron donation to the Fe–S cluster assembly site is elucidated.

As briefly mentioned earlier, the structural core for Fe–S cluster assembly is the complex between the cysteine desulfurase and the scaffold protein. Structural characterization of complexes from *E. coli* in its apo form and *A. fulgidus* in its holo form has shown that two IscU proteins bind at opposite sides of the IscS dimer (Figure 2.2) [34, 35]. Because sulfur is provided as S^0 by the persulfided active cysteine and iron is delivered as ferrous ions, the overall reaction to generate a [2Fe–2S] cluster is

$$2Fe^{2+} + 2S^0 + 2e^- \rightarrow Fe_2S_2]^{2+} \tag{2.1}$$

Two of the four required electrons are likely to originate from the oxidation of the two ferrous irons to ferric and the remaining two may be provided by ferredoxin, another component of the ISC machinery. The temporal order of events taking place in the scaffold site during cluster assembly has been a subject of debate [13]. It is known that cysteine desulfurase can directly transfer one or more S^0 atoms to the cysteine thiolates of the scaffold protein [61]. However, the persulfided scaffold cannot assemble a Fe–S cluster *in vitro* when ferrous ions are added and has been described as a dead-end product [62]. Conversely, native IscU does not bind ferrous ions with high affinity in the absence of S and a reductant [62, 63]. In conclusion, ferrous ions seem incapable of reducing cysteine persulfides and, in the absence of free L-cysteine, IscU cannot bind iron at the scaffold site in the $(IscS–IscU)_2$ complex. As mentioned earlier, in the apo IscS–IscU complex from *E. coli* the scaffold site is already preformed to accommodate a Fe_2S_2 cluster. These observations would favor a concerted mechanism with concomitant arrival of iron and sulfur. However, further investigation will be required to fully understand at the electronic level the reactions involved in the early steps of Fe–S cluster assembly.

Once [2Fe–2S] clusters are correctly assembled in the IscU scaffold site, the $(IscS–IscU)_2$ complex rapidly dissociates with concomitant production of IscU dimers with different Fe and S contents. As mentioned, an aspartic residue from IscU (Asp35 in *E. coli*) seems involved in the [2Fe–2S]-containing $(IscS–IscU)_2$ complex dissociation process [35, 36]. At first, IscU dimers contain mostly one [2Fe–2S] cluster per dimer, but with time, this species decreases in favor of a two-[2Fe–2S]-cluster-containing IscU dimer [64]. The latter can be rapidly converted into a one-[4Fe–4S]-cluster-containing IscU dimer when exposed to reducing conditions [64, 65].

Although IscU can directly transfer a Fe–S cluster to a target protein, such as aconitase *in vitro* [40], *in vivo*, accessory proteins serve as relays between the Fe–S cluster assembly core (i.e., the IscS–IscU complex), and the client proteins [17, 25]. One of the best-studied examples corresponds to the A-type carrier IscA. Even though IscA can bind Fe–S clusters [66], it cannot interact with the cysteine desulfurase IscS [49]. This rules out a scaffold function for this protein. Conversely, although IscA cannot transfer a Fe–S cluster to the scaffold protein IscU, it can receive it from IscU and transfer it to client proteins [67]. This observation suggests that IscA acts as an intermediate carrier between IscU and the client apoproteins. Any given organism has several A-type carriers and although they can partially complement each other, they do not carry out the exact same function [25]. These differences may reflect some degree of specialization because it is difficult to conceive that one protein will be able to recognize a very large number of different targets.

2.3 Redox Properties of Fe–S Clusters

One of the main functions of Fe–S clusters in proteins is electron transfer. The most common types are the [4Fe–4S] clusters and [2Fe–2S] clusters, but other combinations of Fe and S are also found. [4Fe–4S] clusters can display either +2/+1 or +2/+3 redox pairs such as, for instance, in hydrogenases, and HiPiPs, respectively [68]. Depending on the cluster redox state, the Fe ions can be formally ferrous or ferric: +2 state: 2 Fe^{3+} and 2 Fe^{2+}; +1 state: 1 Fe^{3+} and 3 Fe^{2+}; and +3 state: 3 Fe^{3+} and 1 Fe^{2+}. In addition, electronic delocalization may occur between ions. For example, in the case where the [4Fe–4S] is coordinated by four cysteine thiolates and in the +2 redox state there are two mixed-valence pairs: $Fe^{2.5+}$ $Fe^{2.5+}$ antiferromagnetically coupled [69]. [4Fe–4S] clusters often form "wires" connecting the active site of an enzyme to its surface. The intercluster distances are typically in the 10–14 Å range. Less common are [3Fe–4S] clusters, which have 0/+1 redox states. An example of a [4Fe–4S]–[3Fe–4S]–[4Fe–4S] "wire" is found in NiFe hydrogenases [70]. The reason for having the medial [3Fe–4S] cluster is not well understood; a mutant where this cluster was transformed into a [4Fe–4S] cluster, which changed its redox potential by over +200 mV, had a catalytic rate very similar to the one of the wild-type enzyme. However, the mutant hydrogenase was more sensitive to oxygen-induced damage [71].

As discussed subsequently (see Section 2.5), the emergence of increasing amounts of O_2 in the atmosphere forced the existing anaerobic organisms to either restrict their niches to anoxic settings or to evolve solutions to cope with this gas. NiFe hydrogenases are good examples of the latter solution. Although standard hydrogenases are inactivated to an "unready," difficult-to-reactivate state upon oxygen exposure [72], both O_2-resistant and O_2-tolerants enzymes have evolved.

NiFeSe hydrogenases, which belong to the former class, have one of the cysteine residues that coordinate Ni at the active site substituted by a seleno-cysteine [73]. Although these enzymes are inactivated by O_2 concentrations >2–3% they can be very rapidly reactivated under reducing conditions [74]. The O_2-tolerant class of NiFe hydrogenases is unique among these enzymes because (i) its active site and Fe–S clusters have more positive redox potentials than those of standard hydrogenases and (ii) the proximal cluster (PC) (relative to the active site) has unusual structure, composition, and redox properties (Figure 2.1d). Instead of having the classical four labile sulfide and four cysteine ligands, the PC has three labile sulfides and six cysteine ligands and is highly asymmetrical [14–16]. Functionally, it can undergo two successive oxidations: +3/+4 and +4/+5 (equivalent to +1/+2 in standard [4Fe–4S] clusters and +2/+3, in HiPiPs). The ability to send two successive electrons to the active site prevents the formation of partially reduced ROS at that site and protects it from their damaging effects. Thus, O_2-tolerant NiFe hydrogenases can oxidize H_2 under air [75]. Structurally, the +4/+5 superoxidation involves the migration of one cluster Fe toward a deprotonated main chain amido N to form a Fe–N bond [15, 16]. We have shown that this superoxidation is possible because the PC asymmetry allows for full valence localization in the three redox states [76]. The plasticity of PC in O_2-tolerant NiFe hydrogenases contrasts with the rigidity and pair-wise electronic delocalization of the same cluster in their O_2-sensitive counterparts, which forbids two successive redox events.

The other common Fe–S cluster type is the [2Fe–2S] species (Figure 2.1a). This cluster is found, for example, in plant ferredoxins and FeFe hydrogenases from *Clostridia*. Its simpler structure makes it more stable than [4Fe–4S] clusters. In fact, both [4Fe–4S] cluster assembly and its degradation have [2Fe–2S] intermediates. The [2Fe–2S] clusters can have Fe^{3+}–Fe^{3+} or Fe^{3+}–Fe^{2+} valence states, equivalent to +2 and +1 redox values for the whole cluster.

2.4 Fe–S Clusters and Catalysis

2.4.1 Redox Catalysis

Owing to their redox properties, Fe–S clusters are also involved in redox catalysis in numerous proteins. However, as was discussed earlier, standard Fe–S clusters, such as [2Fe–2S] and [4Fe–4S] centers transfer one electron at a time and, consequently, are well adapted to participate in radical chemical reactions similar to the ones catalyzed by *S*-adenosyl-L-methionine (SAM)-dependent enzymes. These proteins use a $[4Fe–4S]^{1+}$ cluster to perform the reductive cleavage of SAM, leading to the production of methionine and the highly reactive 5′-deoxyadenosyl radical species (5′-dA•) (Figure 2.3). Radical SAM enzymes use this species to perform various chemically difficult reactions, such as sulfur atom insertions,

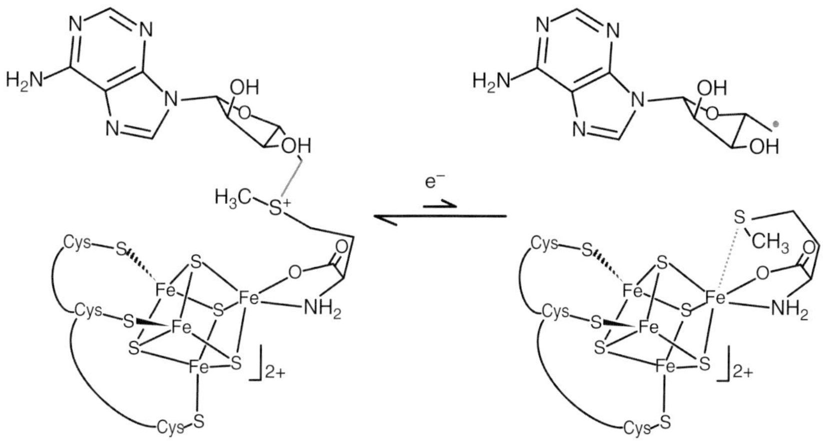

Figure 2.3 The cleavage of *S*-adenosyl-L-methionine (SAM) by the one-electron transfer from the conserved [4Fe–4S] cluster and generation of the 5′-dA• radical in radical SAM enzymes [77, 89].

oxidations, cyclizations, methylations, and radical transfer to proteins [77, 78]. Their substrates are also very diverse, ranging from small organic molecules to proteins, RNA, or DNA. The interest in studying these proteins has significantly grown in the past 20 years, mainly because they are present in the three kingdoms of life and usually participate in key steps in the biosynthetic pathways of major cofactors. Furthermore, it has been shown recently that radical SAM enzymes are involved in the synthesis of many antibiotics [78] and in RNA (thio)methylation, either to increase the specificity of the translational process [79, 80] or to induce resistance to antibiotics [81–84]. Many groups have contributed to the unveiling of the mechanism by which these enzymes use a [4Fe–4S] cluster to generate a radical species [77, 78]. All radical SAM enzymes bind the [4Fe–4S] cluster using three cysteine residues, usually found in a consensus sequence motif [85]. The fourth [4Fe–4S] cluster iron atom is bound to the methionine moiety of SAM via an amino carboxylate bidentate linkage [86–88]. Some of these enzymes use SAM as a substrate and others as a cofactor, depending on whether a new SAM molecule needs to bind to the enzyme to restore its activity, or the cofactor is regenerated at the end of the radical reaction. Consequently, two possible mechanisms have been postulated: a sulfur-based electron transfer between the Fe–S cluster and SAM in the former case and an iron-based electron transfer in the latter [77]. However, all available X-ray structures of radical SAM enzymes indicated, in agreement with previous electron-nuclear double resonance (ENDOR) data [86, 88], that the interaction between SAM and the [4Fe–4S] cluster is the same whether SAM is used as a cofactor or a substrate [89]. Theoretical calculations indicated that the electron transfer pathway depends only on the energy levels of the molecular orbitals of the [4Fe–4S] cluster and the SAM sulfonium ion, and, consequently, it should always proceed through the same mechanism, which we found to be iron-based [89]. Whether SAM is used as a substrate or a cofactor depends only on the characteristics of the radical reaction and the protein matrix, which, depending on its affinity for SAM cleavage products, either releases them (SAM is a substrate) or keeps them in place and regenerates SAM (SAM is a cofactor) [90]. Radical SAM enzymes were initially identified as a superfamily of proteins sharing a three-dimensional fold [85, 91, 92]. Numerous new members with novel folds have been recently added to this superfamily, enlarging our concept of what is a radical SAM enzyme [93]. For instance, ThiC, which is involved in thiamine biosynthesis, has a fold dissimilar from that of standard radical SAM enzymes and binds the [4Fe–4S] cluster differently; nevertheless, it still produces the 5′-dA• radical [94]. Dph2 departs even further from the archetypical enzymes because it has unique [4Fe–4S] cluster and SAM-binding sites. Dph2 generates a 3-amino-3-carboxypropyl radical, instead of the canonical 5′-dA• [95]. Theoretical calculations have indicated that, depending on the orientation of SAM relative to the unique iron atom of the [4Fe–4S] cluster, a different S^+–C bond is cleaved [96]. Thus, different relative bindings of the [4Fe–4S] cluster and SAM allow for a wide selection of reactions, underscoring the plasticity of SAM-based radical chemistry.

The protein IspH ((*E*)-4-hydroxy-3-methyl-but-2-enyl diphosphate reductase), involved in isoprenoid biosynthesis, uses a reduced [4Fe–4S] cluster to perform a two-electron reduction, generating an oxidized HiPiP-like intermediate [97]. This is one of the few examples of such redox chemistry catalyzed by [4Fe–4S]-cluster-containing proteins. In order to perform other redox-based catalyses, proteins can also use modified clusters such as the ones observed in FeFe hydrogenase [98], nitrogenase [11], or CODH-acetylCoA synthase [99], with the addition of unusual ligands and/or the use of modified Fe–S cluster arrangements with additional metal ions (Figure 2.1).

2.4.2
Nonredox Fe–S Cluster-Based Catalysis

A large number of proteins use Fe in Fe–S clusters as a Lewis acid to perform nonredox catalysis. Among them, one of the most extensively studied is mitochondrial aconitase, which uses a [4Fe–4S] cluster to catalyze the conversion of citrate into isocitrate (Figure 2.4), [100, 101]. As in the case of radical SAM enzymes, three cysteine residues bind the [4Fe–4S] cluster, leaving the fourth iron atom available for substrate binding and the Lewis-based activation of a water molecule. These conserved features are found in numerous dehydratases such as IspG [102], Fumarase A [103, 104], or quinolinate synthase [105, 106]. These proteins adopt different folds, illustrating the importance of such Fe–S-cluster-based chemistry that most probably has emerged from multiple convergent evolution events. Another example of a non-redox use of a [4Fe–4S] cluster has been reported for the protein MoaA, involved in the biosynthesis of molybdopterin [107].

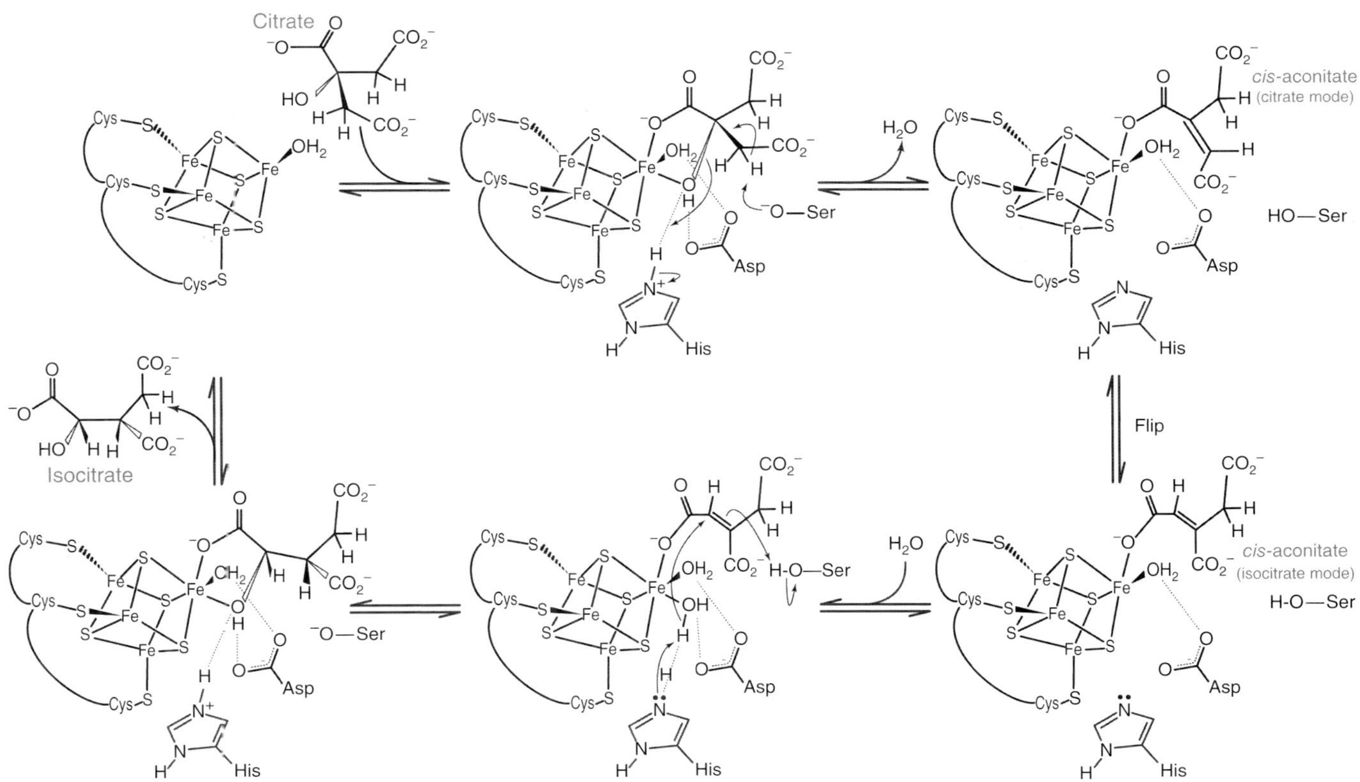

Figure 2.4 The catalytic cycle of aconitase. Nonredox, Lewis-type catalysis by a [4Fe–4S] cluster Fe ion [101, 108].

2.5
Fe–S Clusters and Oxidative Stress

Oxidation of a [4Fe–4S] cluster leads to the formation of a metastable form in the +3 oxidation state that rapidly converts into a $Fe_3S_4]^+$with concomitant release of one ferrous ion according to Reactions 2.2a (or 2.2b) and 2.3 [104, 109].

$$Fe_4S_4]^{2+} + O_2^{\cdot -} + 2H^+ \rightarrow Fe_4S_4]^{3+} + H_2O_2 \quad (2.2a)$$

$$Fe_4S_4]^{2+} + H_2O_2 + H^+ \rightarrow Fe_4S_4]^{3+} + H_2O + HO^{\cdot} \quad (2.2b)$$

$$Fe_4S_4]^{3+} \rightarrow Fe_3S_4]^+ + Fe^{2+} \quad (2.3)$$

Because the lost iron is essential for catalysis, the oxidized enzymes are inactive. Oxygen itself can directly oxidize a [4Fe–4S] cluster according to Reaction 2.4, as observed in the FNR (see subsequent text) [110–112]. This reaction produces a highly reactive superoxide radical species, which, in turn, will cause additional oxidative damage:

$$Fe_4S_4]^{2+} + O_2 \rightarrow Fe_4S_4]^{3+} + O_2^{\cdot -} \quad (2.4)$$

Furthermore, the release of ferrous ions is deleterious for the cell, because they can generate new ROS, according to the so-called Fenton's reaction [109]:

$$Fe^{2+} + H_2O_2 \rightarrow Fe^{3+} + HO^- + HO^{\cdot} \quad (2.5)$$

The hydroxyl radical is the second most powerful oxidant in nature, after fluorine, and can damage DNA. Superoxide ions, hydrogen peroxide, and peroxynitrite are some of the other physiological oxidants that can react with, and inactivate, Fe–S-cluster-containing proteins [113].

Although many Fe–S-cluster-containing proteins are oxygen-sensitive, they are essential and must be fully functional under aerobic conditions. These proteins evolved when the atmosphere of our planet was anoxic. With the appearance of atmospheric oxygen resulting from photosynthesis about 2 billion years ago, many living organisms had to develop adaptation strategies. Whenever possible, Fe–S-cluster-containing proteins were replaced by oxygen-insensitive ones. One example is the *de novo* production of deoxyribonucleoside triphosphates. The required ribonucleotide reductase (RNR) activity is found in three evolutionary unrelated proteins [114, 115]. Class III RNRs require a radical SAM enzyme, which contains an oxygen-sensitive [4Fe–4S] cluster essential for its activity [115, 116]. Conversely, Class II RNRs use 5′-deoxyadenosylcobalamin, a much less oxygen-sensitive cofactor [117] and Class I RNRs possess a binuclear iron center and require oxygen for their function [115]. Another example of the adaptation to oxygen is the synthesis of dehydroglycine (DHG), an intermediate in the synthesis of the thiazole ring of thiamine. DHG is produced by two different enzymes. Similar to Class III RNRs, ThiH is oxygen-sensitive and belongs to the radical SAM superfamily, which requires a [4Fe–4S] cluster for activity

[118, 119]. Conversely, ThiO is a FAD-dependent glycine oxidase that can function in the presence of molecular oxygen [120, 121].

In many other cases, however, living organisms have conserved their Fe–S-cluster-containing enzymes and have concomitantly developed protective mechanisms to prevent the deleterious effects of ROS and RNS. Examples are (i) superoxide dismutase, which catalyzes the dismutation of superoxide ions into oxygen and hydrogen peroxide and is consequently an important part of the cell's defense against free radicals [122], and (ii) catalase, which converts hydrogen peroxide into water and oxygen and thus plays an important role in preventing oxidative damage [123, 124]. These two protective systems are not directly involved in the recovery of damaged Fe–S clusters. Intriguingly, it has been observed that superoxide-stressed cells recover their Fe–S-cluster-dependent activities within minutes when placed in an anaerobic environment, without requiring *de novo* protein synthesis [125]. Because neither ISC nor SUF, the main cellular Fe–S cluster assembly machineries, are responsible for this repair, a specific *ad hoc* system must exist [125]. In fact, this would be the most parsimonious and cost-effective solution to oxidative damage because cells recovering from oxidative stress conditions will have a hard time affording a complete reassembly of their Fe–S clusters. A transcriptomic analysis of *E. coli* upon nitric oxide stress showed that *ytfE* was one of the highly induced genes [126–128]. Furthermore, deleting *ytfE* leads to a decrease in the activity of Fe–S-cluster-containing enzymes [129], a higher sensitivity to oxidative damage and an inability to recover when the source of stress is removed [129, 130]. The addition of purified YftE protein to a crude extract of an *E. coli*$^{\Delta yftE}$ strain, previously exposed to nitric oxide stress, results in complete recovery of aconitase B and fumarase A activities at rates similar to those of the wild-type strain [128]. YtfE appears to be a major component of a recently discovered machinery called *repair of iron centers* (*RIC*) [130, 131]. Paralogs of *ytfE* are found in eukaryotes, but the function of their gene products remains uncharacterized to this date [130, 132]. RIC proteins are about 25 kDa and can form homodimers through formation of intermolecular disulfide bridges [129, 130]. However, dimerization does not seem to be necessary for Fe–S cluster repair activity. Biochemical and spectroscopic characterization showed that RIC proteins contain two iron atoms per monomer, arranged in a non-heme binuclear center with histidine and carboxylate ligands [129, 133]. A secondary structure prediction study suggests that YtfE forms a four-helix bundle [130, 132]. Further characterization will be required to understand how RIC proteins can mobilize iron for Fe–S cluster repair and to determine the molecular basis of their activity.

2.6
Regulation of Protein Expression by Fe–S Clusters

As was discussed earlier, Fe–S clusters are sensitive to oxidative damage and when possible, organisms have replaced their Fe–S-cluster-containing enzymes by oxygen-resistant counterparts. However, as Fe–S clusters are found in several key enzymes of central metabolism, such as aconitase or fumarate reductase, organisms must have adapted to survive. For instance, as illustrated, they took advantage of the oxygen sensitivity of Fe–S clusters and the bioavailability of its components to use them as sensors of their environment.

2.6.1
Eukaryotic Iron Regulatory Protein 1 (IRP1)

Iron is an essential component of proteins participating in key cellular processes, such as respiration, DNA synthesis, and oxygen transport. An excess of cellular iron can generate free radicals that damage biological molecules in general (Reaction 2.5). Conversely, low cellular iron levels block cell proliferation. Iron is transported in its oxidized Fe^{3+} state that can bind to plasma transferrin (Tf) with very high affinity. Most cells acquire iron in its diferric Tf (Tf-Fe^{3+} form), which binds to cell surface transferrin receptor 1 (TfR1). The Tf-Fe^{3+}–TfR1 complex is then internalized by endocytosis [134].

Iron uptake, storage, and utilization are regulated by IRP1 and IRP2 [135, 136]. IRPs control the translation of proteins involved in these processes by binding to the corresponding mRNAs at specific noncoding sequences called *iron-responsive elements* (*IREs*) [135]. The tight interaction between protein and RNA preserves iron homeostasis in animal cells containing insufficient iron. When body iron levels are appropriate, its export is reduced through ferroportin degradation and iron is stored in ferritin [134]. When cellular levels of iron are low, IRPs bind tightly to (i) 5′ IREs in ferritin and ferroportin mRNAs repressing translation and (ii) to 3′ IREs in TfR1 mRNA, thereby blocking its degradation [136]. Conversely, excess iron prevents the binding of IRPs to IREs, which increases ferritin and ferroportin synthesis and promotes TfR1 mRNA degradation. The IRP's concerted regulation of iron uptake, storage, and export results in adequate cellular iron levels, preventing the generation of toxic levels of this metal ion. The presence of IREs in a variety of mRNAs unrelated to iron metabolism indicates that processes other than iron homeostasis are also regulated by IRPs [134].

Under iron-replete conditions, IRP1, but not IRP2, binds a [4Fe–4S] cluster, and functions as a c-acon, converting citrate to isocitrate [24]. The balance between the IRE-binding and the enzymatic activity of IRP1 is regulated in a complex way. Normally, IRP1 is mostly present as c-acon [137] possibly protecting the cell against oxidative stress by generating isocitrate, which is oxidized by NADP-dependent isocitrate dehydrogenase. Glutathione reductase uses the resulting NADPH to produce reduced glutathione. This species is, in turn, oxidized by glutathione peroxidase, removing oxidative-stress-induced H_2O_2. The disassembly of the [4Fe–4S] cluster of IRP1, caused by either iron starvation or oxidative stress, triggers its conversion to the IRE-binding form [138]. In this conformation, IRP1 augments iron accessibility and uptake by both impairing the translation of ferritin mRNA and protecting the Tf receptor mRNA from degradation [24]. Nevertheless, too much cellular iron may be dangerous to the cell because of Fenton-type redox chemistry that leads to the generation of highly reactive oxygen species (ROS) (Reaction 2.5) [139]. Because low iron levels are also detrimental to the cell, the IRE-binding activity of IRP1 is strictly regulated.

2.6.1.1 IRP1 and Fe–S Cluster Biogenesis

IRP1 and cellular Fe–S cluster biogenesis are linked through the formation and loss of the [4Fe–4S] cluster of the former. Biogenesis of Fe–S clusters and their insertion into cytosolic proteins, such as IRP1, initially requires the import of iron to the mitochondrion [134, 140]. There, a Fe–S cluster is first formed on the scaffold protein ISCU within a complex that also includes the putative iron-binding protein frataxin, the sulfide-donor cysteine desulfurase NFS1 and ISD11, a protein that stimulates NFS1 activity [141–146]. The Fe–S cluster assembled on ISCU, or other scaffold proteins [147], is transferred to apo-protein clients with the help of protein chaperones and electron donors such as ferredoxin [17]. The mitochondrial Fe–S cluster biogenesis machinery and the export machinery components are needed to activate the cytosolic Fe–S cluster assembly (CIA) pathway. This pathway was first discovered using IRP1 [148].

In mammalian cells, the activity of both mitochondrial and cytosolic Fe–S proteins is impaired if components of the mitochondrial Fe–S biogenesis pathway are disrupted [58, 141, 143, 149–154]. The resulting reduction in c-acon activity correlates with IRP1 activation, which, in turn, is associated with an increase in TfR1 and diminished ferritin expression. These experiments confirm the dependence of cytosolic Fe–S cluster biogenesis on active mitochondrial cluster formation in mammalian cells. They also show that the consequences of disrupted mitochondrial Fe–S biogenesis may in part involve indirect effects on IRPs or IRP targets.

2.6.1.2 Reactive Oxygen Species and IRP1 Fe–S Cluster Stability

The finding that the RNA-binding activity of IRP1 depends on the disassembly of a [4Fe–4S] cluster from c-acon posed the question of how this process is elicited. Low levels of cellular iron were not considered to be sufficient to induce cluster loss. Conversely, aconitases are known to be reactive toward ROS and RNS because their [4Fe–4S] cluster is solvent exposed and ROS and RNS can trigger cluster loss [101]. Although the fact that NO can activate IRP1 may play an important role in macrophage activation and/or action, there is no clear physiological role for ROS. It is known that superoxide can start Fe–S cluster disruption in aconitases by generating their [3Fe–4S] non-RNA-binding form. Also, H_2O_2 causes the transformation of c-acon into IRP1, although the cluster loss mechanism is not well understood [155, 156]. It is known, nevertheless, that a signaling pathway possibly involving protein kinase C-dependent phosphorylation of serine 138 is required for IRP1 activation by H_2O_2 [157, 158]. Indeed, c-acon phosphomimetic (i.e., aspartate or glutamate) mutants of Ser138 are much more prone to O_2-dependent cluster disassembly than the wild-type protein [159, 160]. They can even disassemble their Fe–S cluster spontaneously without the action of an effector [160]. These experiments indicate that IRE-binding activity can be rapidly generated either by cluster-perturbing effectors or by phosphorylation-induced modifications of Fe–S cluster stability. In conclusion, ROS may be physiologically important because they constantly convert a small fraction of c-acon into IRP1. The c-acon/IRP1 interconversion, which would depend on cellular iron levels, could thus regulate IRE-containing mRNAs.

Another regulatory role for IRPs is indicated by the presence of a functional IRE in the 5′ untranslated region of the mRNA encoding hypoxia-inducible transcription factor-2α (HIF-2α). Thus, IRPs are likely to be involved in the adaptive response to hypoxia and low cellular iron levels [161]. Indeed, it has been reported that the HIF-2α IRE binds IRP1 [162]. This regulation of HIF-2α mRNA by IRP1 indicates that, under hypoxic conditions and sufficient iron, IRP1 would be found in its non-RNA-binding c-acon form, making sure that HIF-2α is synthesized and the production of red blood cells is increased.

2.6.1.3 X-Ray Structural Studies of IRP1–IRE Complexes

Walden *et al.* [163, 164] have reported the crystal structures of IRP1 bound to a ferritin IRE (Ftn H) and to transferring receptor 1B (TfR B). Together with the previous structure of c-acon [165], the IRP1–Ftn H co-crystal structure defined how apo-IRP1 binds with high affinity to IREs. C-acon is composed of four domains with residues from each of them contributing to the active site, including the three cysteines that coordinate the [4Fe–4S] cluster. In the IRP1–Ftn H structure, domain 4 adopts a position very different from the one it occupies in c-acon (Figure 2.5). Domain 3 has also moved, whereas domains 1 and 2 form an invariable core. The movements of domains 3 and 4 relative to the core expose a hydrophilic cavity where the ferritin IRE binds (Figure 2.5).

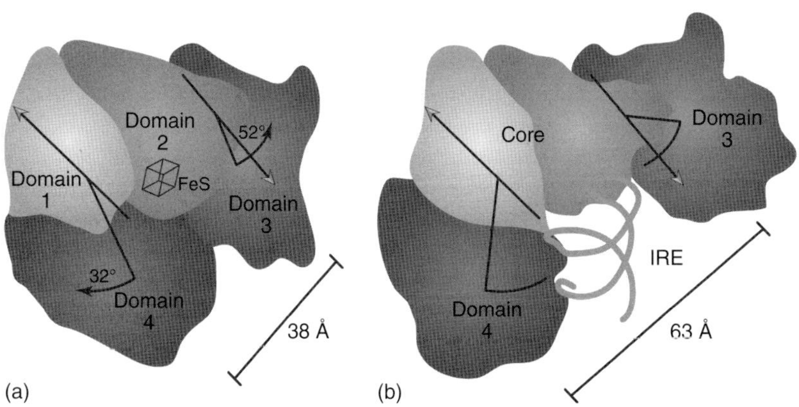

Figure 2.5 Domain positions in c-acon (a) [165] and the IRP1:IRE (Ftn H)-RNA complex [164]. The core in the IRP1:IRE complex is formed by domains 1 and 2 of c-acon (b). The two structures are displayed in the same orientation. The internal [4Fe–4S] cluster in c-acon is shown as a cube and the IRE as a light gray ribbon. The axes of rotation around which domains 3 and 4 pivot and the arcs showing the motion of the centers of mass of these domains are also depicted. (Adapted from Ref. [166].)

IRP1 binds to structural features shared by the TfR B and Ftn H IREs with essentially the same protein:RNA bonding patterns (not shown). This probably reflects the induced fit character and plasticity of the IRP:IRE interaction. The conformational change of IRP1 in going from the c-acon-cluster-containing form to the apo-protein exposed a molecular surface that was occasionally bound by mRNAs during evolution. Those mRNAs that contained a domain structure complementary to the newly exposed surface of apo aconitase became proto-IREs and were subsequently selectively retained, because they endowed the organisms with the advantage of being able to respond to iron deprivation. Thus, the IRP regulatory system did not involve the invention of new genes. Instead, it used aconitase, likely to be one of the oldest enzymes, to generate a new regulatory function [166].

2.6.2
Bacterial Fumarate Nitrate Reduction Regulator (FNR)

FNR is used by facultative anaerobic microorganisms as a molecular switch between aerobic and anaerobic metabolisms and has been extensively characterized over the years [110–112, 167–172]. Under anaerobic conditions, FNR is a dimer containing one [4Fe–4S] cluster per monomer. This dimer can bind to specific palindromic DNA sequences to either repress or activate a vast number of genes [23, 173, 174]. Upon exposure to low oxygen levels, the [4Fe–4S] cluster is rapidly converted into a [2Fe–2S] cluster via a transient [3Fe–4S] center [110, 170]. The [2Fe–2S]-cluster-containing form of FNR of gram-negative bacteria cannot dimerize and, consequently, it does not bind DNA [170, 175]. Thus, this oxygen-induced cluster conversion triggers the switch between anaerobic and aerobic metabolism in *E. coli*. Very recently, Raman resonance (RR) spectroscopy data indicated that, instead of a [3Fe–4S]- to [2Fe–2S] cluster conversion with concomitant release of one ferrous and two sulfide ions [167, 169], its most probable mechanism involves sulfur-centered oxidations with production of cysteine persulfide ligands [169]. It has been demonstrated that, upon addition of ferrous ions and dithiothreitol (DTT) as reductant, it is possible to convert back the cysteine persulfide-bound [2Fe–2S] cluster to a [4Fe–4S] one, supporting the existence of repair mechanisms, such as RIC, that do not require additional sulfide ions [169]. As [4Fe–4S] clusters are cubes and [2Fe–2S] clusters are rhombs (Figure 2.1a,b), such Fe–S cluster conversion necessarily induces structural rearrangements. These rearrangements are very likely responsible for the dimer dissociation, which impairs DNA binding and triggers the switch between anaerobic and aerobic metabolisms. However, no structural data are available to date and further work will be required to fully understand the connection between oxygen-induced degradation of the Fe–S cluster and structural rearrangements.

2.6.3
The ISC Assembly Machinery Regulator IscR

IscR, the ISC assembly machinery regulator, directly senses Fe–S cluster levels in the cell. IscR is a dimeric protein that can bind [2Fe–2S] clusters via three conserved cysteines and one not yet identified residue [23, 176]. In *E. coli*, IscR controls the expression of over 40 genes, which are either involved in the synthesis of enzymes required in anaerobic metabolism, such as periplasmic nitrate reductase and hydrogenases 1 and 2, or in Fe–S cluster assembly, that is, the ISC and SUF machineries [22, 177, 178]. Apo and holo forms of IscR bind to two distinct DNA sequences called *Type 1* and *Type 2 sites* [25, 177]. Under anaerobic conditions and sufficient iron, IscR exists mostly as a [2Fe–2S]-cluster-containing protein capable of downregulating the expression of the ISC operon. Conversely, when iron levels are low or under oxidative stress, IscR exists mainly in its apo form. Presumably owing to structural changes associated with this conversion, IscR cannot repress the expression of the ISC operon under these conditions. Instead, it only binds Type 2 sites, inducing the activation of the SUF machinery [177, 179], which can assemble Fe–S clusters under oxidative conditions. IscR can thus directly sense the level of Fe–S clusters in the cell and control their production. However, it has been shown that [2Fe–2S]-cluster-containing IscR can also bind Type 2 sites [179]. This challenges a simple Manichean view of its regulation process, which is certainly subtler than initially envisioned. Further studies will be required to fully understand these regulation pathways.

2.7
Conclusion

In this chapter, we have reviewed the multiple properties of Fe–S clusters with special emphasis on the ubiquitous [2Fe–2S] and [4Fe–4S] species (more complex Fe–S-containing active sites are discussed in other chapters of this book). These rather simple Fe–S clusters may be the evolutionary descendants of mineral Fe^{2+} sulfides, which are likely to have played a central role during the emergence of life as proposed by Wächterhäuser *et al.* and Russel *et al.* This notion is reinforced by the presence of Fe–S-cluster-containing proteins in many fundamental life processes, such as respiration and several primordial anaerobic-gas-based reactions catalyzed by microbes, such as nitrogen reduction to ammonia, CO_2 reduction to acetate, and hydrogen uptake. Biological [4Fe–4S] and [2Fe–2S] clusters are both structurally and functionally plastic, being involved in both redox and nonredox catalysis, aerobic/anaerobic metabolic switches, homeostasis, and electron transfer. A remarkable accomplishment of evolution is the use of the oxygen sensitivity of the [4Fe–4S] clusters to regulate microbial metabolism. Maybe unexpectedly, the biogenesis of [2Fe–2S] and [4Fe–4S] clusters, which is being actively studied, involves complicated processes that are just being unveiled.

Understanding the fundamental structural and functional basis of the many processes that involve Fe–S clusters still remains a challenging goal that will continuously require the combined efforts of spectroscopists, biochemists, structural biologists, and theoretical chemists.

References

1. Russell, M.J., Hall, A.J., Boyce, A.J., and Fallick, A.E. (2005) 100th anniversary special paper: on hydrothermal convection systems and the emergence of life. *Econ. Geol.*, **100** (3), 419–438.
2. Wächtershäuser, G. (1988) Pyrite formation, the 1st energy source for life – a hypothesis. *Syst. Appl. Microbiol.*, **10** (3), 207–210.
3. Huber, C. and Wächtershäuser, G. (1997) Activated acetic acid by carbon fixation on (Fe,Ni)S under primordial conditions. *Science*, **276** (5310), 245–247.
4. Cody, G.D., Boctor, N.Z., Filley, T.R., Hazen, R.M., Scott, J.H., Sharma, A., and Yoder, H.S. (2000) Primordial carbonylated iron-sulfur compounds and the synthesis of pyruvate. *Science*, **289** (5483), 1337–1340.
5. Russell, M.J. (2003) The importance of being alkaline. *Science*, **302** (5645), 580–581.
6. Russell, M.J. and Hall, A.J. (1997) The emergence of life from iron monosulphide bubbles at a submarine hydrothermal redox and pH front. *J. Geol. Soc.*, **154**, 377–402.
7. Huber, C., Eisenreich, W., Hecht, S., and Wachtershauser, G. (2003) A possible primordial peptide cycle. *Science*, **301** (5635), 938–940.
8. Fontecilla-Camps, J.C., Amara, P., Cavazza, C., Nicolet, Y., and Volbeda, A. (2009) Structure-function relationships of anaerobic gas-processing metalloenzymes. *Nature*, **460**, 814–822.
9. Volbeda, A. and Fontecilla-Camps, J.C. (2006) Catalytic nickel-iron-sulfur clusters: from minerals to enzymes. *Top. Organomet. Chem.*, **17**, 57–82.
10. Einsle, O., Tezcan, F.A., Andrade, S.L., Schmid, B., Yoshida, M., Howard, J.B., and Rees, D.C. (2002) Nitrogenase MoFe-protein at 1.16 A resolution: a central ligand in the FeMo-cofactor. *Science*, **297** (5587), 1696–1700.
11. Hoffman, B.M., Lukoyanov, D., Dean, D.R., and Seefeldt, L.C. (2013) Nitrogenase: a draft mechanism. *Acc. Chem. Res.* doi: 10.1021/ar300267m
12. Volbeda, A. and Fontecilla-Camps, J.C. (2005) Structural bases for the catalytic mechanism of Ni-containing carbon monoxide dehydrogenases. *Dalton Trans.*, **7** (21), 3443–3450.
13. Johnson, D.C., Dean, D.R., Smith, A.D., and Johnson, M.K. (2005) Structure, function, and formation of biological iron-sulfur clusters. *Annu. Rev. Plant. Physiol. Plant. Mol. Biol.*, **74**, 247–281.
14. Fritsch, J., Scheerer, P., Frielingsdorf, S., Kroschinsky, S., Friedrich, B., Lenz, O., and Spahn, C.M. (2011) The crystal structure of an oxygen-tolerant hydrogenase uncovers a novel iron-sulphur centre. *Nature*, **479** (7372), 249–252.
15. Shomura, Y., Yoon, K.S., Nishihara, H., and Higuchi, Y. (2011) Structural basis for a [4Fe-3S] cluster in the oxygen-tolerant membrane-bound [NiFe]-hydrogenase. *Nature*, **479** (7372), 253–256.
16. Volbeda, A., Amara, P., Darnault, C., Mouesca, J.M., Parkin, A., Roessler, M.M., Armstrong, F.A., and Fontecilla-Camps, J.C. (2012) X-ray crystallographic and computational studies of the O_2-tolerant [NiFe]-hydrogenase 1 from *Escherichia coli*. *Proc. Natl. Acad. Sci. U.S.A.*, **109** (14), 5305–5310.
17. Lill, R. (2009) Function and biogenesis of iron-sulphur proteins. *Nature*, **460** (7257), 831–838.
18. Lill, R., Hoffmann, B., Molik, S., Pierik, A.J., Rietzschel, N., Stehling, O., Uzarska, M.A., Webert, H., Wilbrecht, C., and Muhlenhoff, U. (2012) The role of mitochondria in cellular iron-sulfur protein biogenesis and iron

metabolism. *Biochim. Biophys. Acta, Mol. Cell. Res.*, **1823** (9), 1491–1508.

19. Jacobson, M.R., Cash, V.L., Weiss, M.C., Laird, N.F., Newton, W.E., and Dean, D.R. (1989) Biochemical and genetic analysis of the *nifUSVWZM* cluster from *Azotobacter vinelandii. Mol. Gen. Genet.*, **219** (1-2), 49–57.
20. Zheng, L.M., Cash, V.L., Flint, D.H., and Dean, D.R. (1998) Assembly of iron-sulfur clusters – identification of an *iscSUA-hscBA-fdx* gene cluster from *Azotobacter vinelandii. J. Biol. Chem.*, **273** (21), 13264–13272.
21. Takahashi, Y. and Tokumoto, U. (2002) A third bacterial system for the assembly of iron-sulfur clusters with homologs in archaea and plastids. *J. Biol. Chem.*, **277** (32), 28380–28383.
22. Crack, J.C., Green, J., Hutchings, M.I., Thomson, A.J., and Le Brun, N.E. (2012) Bacterial iron-sulfur regulatory proteins as biological sensor-switches. *Antioxid. Redox Signal.*, **17** (9), 1215–1231.
23. Crack, J.C., Green, J., Thomson, A.J., and Le Brun, N.E. (2012) Iron-sulfur cluster sensor-regulators. *Curr. Opin. Chem. Biol.*, **16** (1-2), 35–44.
24. Haile, D.J., Rouault, T.A., Tang, C.K., Chin, J., Harford, J.B., and Klausner, R.D. (1992) Reciprocal control of RNA-binding and aconitase activity in the regulation of the iron-responsive element binding protein: role of the iron-sulfur cluster. *Proc. Natl. Acad. Sci. U.S.A.*, **89** (16), 7536–7540.
25. Py, B. and Barras, F. (2010) Building Fe–S proteins: bacterial strategies. *Nat. Rev. Microbiol.*, **8** (6), 436–446.
26. Kessler, D. and Papenbrock, J. (2005) Iron-sulfur cluster biosynthesis in photosynthetic organisms. *Photosynth. Res.*, **86** (3), 391–407.
27. Balk, J. and Pilon, M. (2011) Ancient and essential: the assembly of iron-sulfur clusters in plants. *Trends Plant Sci.*, **16** (4), 218–226.
28. Hidese, R., Mihara, H., and Esaki, N. (2011) Bacterial cysteine desulfurases: versatile key players in biosynthetic pathways of sulfur-containing biofactors. *Appl. Microbiol. Biotechnol.*, **91** (1), 47–61.
29. Kaiser, J.T., Clausen, T., Bourenkow, G.P., Bartunik, H.D., Steinbacher, S., and Huber, R. (2000) Crystal structure of a NifS-like protein from *Thermotoga maritima*: implications for iron sulphur cluster assembly. *J. Mol. Biol.*, **297** (2), 451–464.
30. Mihara, H., Fujii, T., Kato, S., Kurihara, T., Hata, Y., and Esaki, N. (2002) Structure of external aldimine of *Escherichia coli* CsdB, an IscS/NifS homolog: implications for its specificity toward selenocysteine. *J. Biochem.*, **131** (5), 679–685.
31. Cupp-Vickery, J.R., Urbina, H., and Vickery, L.E. (2003) Crystal structure of IscS, a cysteine desulfurase from *Escherichia coli. J. Mol. Biol.*, **330** (5), 1049–1059.
32. Tirupati, B., Vey, J.L., Drennan, C.L., and Bollinger, J.M. (2004) Kinetic and structural characterization of Slr0077/SufS, the essential cysteine desulfurase from *Synechocystis sp* PCC 6803. *Biochemistry*, **43** (38), 12210–12219.
33. Zheng, L.M., White, R.H., Cash, V.L., Jack, R.F., and Dean, D.R. (1993) Cysteine desulfurase activity indicates a role for NifS in metallocluster biosynthesis. *Proc. Natl. Acad. Sci. U.S.A.*, **90** (7), 2754–2758.
34. Shi, R., Proteau, A., Villarroya, M., Moukadiri, I., Zhang, L.H., Trempe, J.F., Matte, A., Armengod, M.E., and Cygler, M. (2010) Structural basis for Fe–S cluster assembly and tRNA thiolation mediated by IscS protein–protein interactions. *PLoS Biol.*, **8** (4), e1000354.
35. Marinoni, E.N., de Oliveira, J.S., Nicolet, Y., Raulfs, E.C., Amara, P., Dean, D.R., and Fontecilla-Camps, J.C. (2012) (IscS-IscU)2 complex structures provide insights into Fe_2S_2 biogenesis and transfer. *Angew. Chem. Int. Ed.*, **51** (22), 5439–5442.
36. Raulfs, E.C., O'Carroll, I.P., Dos Santos, P.C., Unciuleac, M.C., and Dean, D.R. (2008) *In vivo* iron-sulfur cluster formation. *Proc. Natl. Acad. Sci. U.S.A.*, **105** (25), 8591–8596.
37. Layer, G., Gaddam, S.A., Ayala-Castro, C.N., Choudens, S.O., Lascoux, D., Fontecave, M., and Outten, F.W. (2007) SufE transfers sulfur from SufS to SufB

for iron-sulfur cluster assembly. *J. Biol. Chem.*, **282** (18), 13342–13350.

38. Chahal, H.K., Dai, Y.Y., Saini, A., Ayala-Castro, C., and Outten, F.W. (2009) The SufBCD Fe–S scaffold complex interacts with SufA for Fe–S cluster transfer. *Biochemistry*, **48** (44), 10644–10653.
39. Outten, F.W., Wood, M.J., Munoz, F.M., and Storz, G. (2003) The SufE protein and the SufBCD complex enhance SufS cysteine desulfurase activity as part of a sulfur transfer pathway for Fe–S cluster assembly in *Escherichia coli*. *J. Biol. Chem.*, **278** (46), 45713–45719.
40. Unciuleac, M.C., Chandramouli, K., Naik, S., Mayer, S., Huynh, B.H., Johnson, M.K., and Dean, D.R. (2007) *In vitro* activation of apo-aconitase using a 4Fe-4S cluster-loaded form of the IscU Fe–S cluster scaffolding protein. *Biochemistry*, **46** (23), 6812–6821.
41. Dean, D.R. and Brigle, K.E. (1985) *Azotobacter vinelandii nifD*-encoded and *nifE*-encoded polypeptides share structural homology. *Proc. Natl. Acad. Sci. U.S.A.*, **82** (17), 5720–5723.
42. Rubio, L.M. and Ludden, P.W. (2005) Maturation of nitrogenase: a biochemical puzzle. *J. Bacteriol.*, **187** (2), 405–414.
43. Rubio, L.M. and Ludden, P.W. (2008) Biosynthesis of the iron-molybdenum cofactor of nitrogenase. *Annu. Rev. Microbiol.*, **62**, 93–111.
44. McGlynn, S.E., Shepard, E.M., Winslow, M.A., Naumov, A.V., Duschene, K.S., Posewitz, M.C., Broderick, W.E., Broderick, J.B., and Peters, J.W. (2008) HydF as a scaffold protein in [FeFe] hydrogenase H-cluster biosynthesis. *FEBS Lett.*, **582** (15), 2183–2187.
45. Nicolet, Y. and Fontecilla-Camps, J.C. (2012) Structure-function relationships in FeFe -hydrogenase active site maturation. *J. Biol. Chem.*, **287** (17), 13532–13540.
46. Mulder, D.W., Shepard, E.M., Meuser, J.E., Joshi, N., King, P.W., Posewitz, M.C., Broderick, J.B., and Peters, J.W. (2011) Insights into [FeFe]-hydrogenase structure, mechanism, and maturation. *Structure*, **19** (8), 1038–1052.
47. Ding, H. and Clark, R.J. (2004) Characterization of iron binding in IscA, an ancient iron-sulphur cluster assembly protein. *Biochem. J.*, **379** (Pt. 2), 433–440.
48. Ding, H., Clark, R.J., and Ding, B. (2004) IscA mediates iron delivery for assembly of iron-sulfur clusters in IscU under the limited accessible free iron conditions. *J. Biol. Chem.*, **279** (36), 37499–37504.
49. Tokumoto, U., Nomura, S., Minami, Y., Mihara, H., Kato, S., Kurihara, T., Esaki, N., Kanazawa, H., Matsubara, H., and Takahashi, Y. (2002) Network of protein–protein interactions among iron-sulfur cluster assembly proteins in *Escherichia coli*. *J. Biochem.*, **131** (5), 713–719.
50. Layer, G., Ollagnier-de Choudens, S., Sanakis, Y., and Fontecave, M. (2006) Iron-sulfur cluster biosynthesis: characterization of *Escherichia coli* CYaY as an iron donor for the assembly of [2Fe-2S] clusters in the scaffold IscU. *J. Biol. Chem.*, **281** (24), 16256–16263.
51. Adinolfi, S., Iannuzzi, C., Prischi, F., Pastore, C., Iametti, S., Martin, S.R., Bonomi, F., and Pastore, A. (2009) Bacterial frataxin CyaY is the gatekeeper of iron-sulfur cluster formation catalyzed by IscS. *Nat. Struct. Mol. Biol.*, **16** (4), 390–396.
52. Colin, F., Martelli, A., Clemancey, M., Latour, J.M., Gambarelli, S., Zeppieri, L., Birck, C., Page, A., Puccio, H., and Ollagnier de Choudens, S. (2013) Mammalian frataxin controls sulfur production and iron entry during *de Novo* Fe4S4 cluster assembly. *J. Am. Chem. Soc.*, **135** (2), 733–740.
53. Ding, H., Yang, J., Coleman, L.C., and Yeung, S. (2007) Distinct iron binding property of two putative iron donors for the iron-sulfur cluster assembly: IscA and the bacterial frataxin ortholog CyaY under physiological and oxidative stress conditions. *J. Biol. Chem.*, **282** (11), 7997–8004.
54. Nair, M., Adinolfi, S., Pastore, C., Kelly, G., Temussi, P., and Pastore, A. (2004) Solution structure of the bacterial frataxin ortholog, CyaY: mapping the

iron binding sites. *Structure*, **12** (11), 2037–2048.

55. Bou-Abdallah, F., Adinolfi, S., Pastore, A., Laue, T.M., and Dennis Chasteen, N. (2004) Iron binding and oxidation kinetics in frataxin CyaY of *Escherichia coli*. *J. Mol. Biol.*, **341** (2), 605–615.
56. Pastore, C., Franzese, M., Sica, F., Temussi, P., and Pastore, A. (2007) Understanding the binding properties of an unusual metal-binding protein--a study of bacterial frataxin. *FEBS J.*, **274** (16), 4199–4210.
57. Li, D.S., Ohshima, K., Jiralerspong, S., Bojanowski, M.W., and Pandolfo, M. (1999) Knock-out of the *cyaY* gene in *Escherichia coli* does not affect cellular iron content and sensitivity to oxidants. *FEBS Lett.*, **456** (1), 13–16.
58. Sheftel, A., Stehling, O., and Lill, R. (2010) Iron-sulfur proteins in health and disease. *Trends Endocrinol. Metab.*, **21** (5), 302–314.
59. Pandolfo, M. and Pastore, A. (2009) The pathogenesis of Friedreich ataxia and the structure and function of frataxin. *J. Neurol.*, **256** (Suppl. 1), 9–17.
60. Bridwell-Rabb, J., Iannuzzi, C., Pastore, A., and Barondeau, D.P. (2012) Effector role reversal during evolution: the case of frataxin in Fe–S cluster biosynthesis. *Biochemistry*, **51** (12), 2506–2514.
61. Urbina, H.D., Silberg, J.J., Hoff, K.G., and Vickery, L.E. (2001) Transfer of sulfur from IscS to IscU during Fe/S cluster assembly. *J. Biol. Chem.*, **276** (48), 44521–44526.
62. Nuth, M., Yoon, T., and Cowan, J.A. (2002) Iron-sulfur cluster biosynthesis: characterization of iron nucleation sites for assembly of the $[2Fe\text{-}2S]^{+}$ cluster core in IscU proteins. *J. Am. Chem. Soc.*, **124** (30), 8774–8775.
63. Agar, J.N., Yuvaniyama, P., Jack, R.F., Cash, V.L., Smith, A.D., Dean, D.R., and Johnson, M.K. (2000) Modular organization and identification of a mononuclear iron-binding site within the NifU protein. *J. Biol. Inorg. Chem.*, **5** (2), 167–177.
64. Agar, J.N., Krebs, C., Frazzon, J., Huynh, B.H., Dean, D.R., and Johnson, M.K. (2000) IscU as a scaffold for iron-sulfur cluster biosynthesis: sequential assembly of [2Fe-2S] and [4Fe-4S] clusters in IscU. *Biochemistry*, **39** (27), 7856–7862.
65. Chandramouli, K., Unciuleac, M.C., Naik, S., Dean, D.R., Huynh, B.H., and Johnson, M.K. (2007) Formation and properties of 4Fe-4S clusters on the IscU scaffold protein. *Biochemistry*, **46** (23), 6804–6811.
66. Krebs, C., Agar, J.N., Smith, A.D., Frazzon, J., Dean, D.R., Huynh, B.H., and Johnson, M.K. (2001) IscA, an alternate scaffold for Fe–S cluster biosynthesis. *Biochemistry*, **40** (46), 14069–14080.
67. Ollagnier-de-Choudens, S., Sanakis, Y., and Fontecave, M. (2004) SufA/IscA: reactivity studies of a class of scaffold proteins involved in Fe–S cluster assembly. *J. Biol. Inorg. Chem.*, **9** (7), 828–838.
68. Carter, C.W. Jr., Kraut, J., Freer, S.T., Alden, R.A., Sieker, L.C., Adman, E., and Jensen, L.H. (1972) A comparison of Fe_4S_4 clusters in high-potential iron protein and in ferredoxin. *Proc. Natl. Acad. Sci. U.S.A.*, **69** (12), 3526–3529.
69. Mouesca, J.M. and Lamotte, B. (1998) Iron-sulfur clusters and their electronic and magnetic properties. *Coord. Chem. Rev.*, **178**, 1573–1614.
70. Volbeda, A., Charon, M.H., Piras, C., Hatchikian, E.C., Frey, M., and Fontecilla-Camps, J.C. (1995) Crystal structure of the nickel-iron hydrogenase from *Desulfovibrio gigas*. *Nature*, **373** (6515), 580–587.
71. Rousset, M., Montet, Y., Guigliarelli, B., Forget, N., Asso, M., Bertrand, P., Fontecilla-Camps, J.C., and Hatchikian, E.C. (1998) [3Fe-4S] to [4Fe-4S] cluster conversion in *Desulfovibrio fructosovorans* [NiFe] hydrogenase by site-directed mutagenesis. *Proc. Natl. Acad. Sci. U.S.A.*, **95** (20), 11625–11630.
72. Chapman, A., Cammack, R., Hatchikian, C.E., McCracken, J., and Peisach, J. (1988) A pulsed EPR study of redox-dependent hyperfine interactions for the nickel centre of *Desulfovibrio gigas* hydrogenase. *FEBS Lett.*, **242** (1), 134–138.

73. Patil, D.S. (1994) Nickel-iron-selenium hydrogenase. *Methods Enzymol.*, **243**, 68–94.
74. Parkin, A., Goldet, G., Cavazza, C., Fontecilla-Camps, J.C., and Armstrong, F.A. (2008) The difference a Se makes? Oxygen-tolerant hydrogen production by the [NiFeSe]-hydrogenase from *Desulfomicrobium baculatum. J. Am. Chem. Soc.*, **130** (40), 13410–13416.
75. Pandelia, M.-E., Fourmond, V., Tron-Infossi, P., Lojou, E., Bertrand, P., Léger, C., Giudici-Orticoni, M.-T., and Lubitz, W. (2010) Membrane-bound hydrogenase I from the hyperthermophilic bacterium aquifex aeolicus: enzyme activation, redox intermediates and oxygen tolerance. *J. Am. Chem. Soc.*, **132** (20), 6991–7004.
76. Mouesca, J.M., Fontecilla-Camps, J.C., and Amara, P. (2013) The structural plasticity of the proximal [4Fe3S] cluster is responsible for the O_2 tolerance of membrane-bound [NiFe] hydrogenases. *Angew. Chem. Int. Ed.*, **52** (7), 2002–2006.
77. Frey, P.A., Hegeman, A.D., and Ruzicka, F.J. (2008) The radical SAM superfamily. *Crit. Rev. Biochem. Mol. Biol.*, **43** (1), 63–88.
78. Challand, M.R., Driesener, R.C., and Roach, P.L. (2011) Radical *S*-adenosylmethionine enzymes: mechanism, control and function. *Nat. Prod. Rep.*, **28** (10), 1696–1721.
79. Hernandez, H.L., Pierrel, F., Elleingand, E., Garcia-Serres, R., Huynh, B.H., Johnson, M.K., Fontecave, M., and Atta, M. (2007) MiaB, a bifunctional radical-S-adenosylmethionine enzyme involved in the thiolation and methylation of tRNA, contains two essential [4Fe-4S] clusters. *Biochemistry*, **46** (17), 5140–5147.
80. Perche-Letuvee, P., Kathirvelu, V., Berggren, G., Clemancey, M., Latour, J.M., Maurel, V., Douki, T., Armengaud, J., Mulliez, E., Fontecave, M., Garcia-Serres, R., Gambarelli, S., and Atta, M. (2012) 4-Demethylwyosine synthase from *Pyrococcus abyssi* is a radical-S-adenosyl-L-methionine enzyme with an additional $[4Fe\text{-}4S]^{+2}$ cluster that interacts with the pyruvate co-substrate. *J. Biol. Chem.*, **287** (49), 41174–41185.
81. Boal, A.K., Grove, T.L., McLaughlin, M.I., Yennawar, N.H., Booker, S.J., and Rosenzweig, A.C. (2011) Structural basis for methyl transfer by a radical SAM enzyme. *Science*, **332** (6033), 1089–1092.
82. Grove, T.L., Benner, J.S., Radle, M.I., Ahlum, J.H., Landgraf, B.J., Krebs, C., and Booker, S.J. (2011) A radically different mechanism for S-adenosylmethionine-dependent methyltransferases. *Science*, **332** (6029), 604–607.
83. Booth, M.P., Challand, M.R., Emery, D.C., Roach, P.L., and Spencer, J. (2010) High-level expression and reconstitution of active Cfr, a radical-SAM rRNA methyltransferase that confers resistance to ribosome-acting antibiotics. *Protein Expr. Purif.*, **74** (2), 204–210.
84. Yan, F., LaMarre, J.M., Rohrich, R., Wiesner, J., Jomaa, H., Mankin, A.S., and Fujimori, D.G. (2010) RlmN and Cfr are radical SAM enzymes involved in methylation of ribosomal RNA. *J. Am. Chem. Soc.*, **132** (11), 3953–3964.
85. Sofia, H.J., Chen, G., Hetzler, B.G., Reyes-Spindola, J.F., and Miller, N.E. (2001) Radical SAM, a novel protein superfamily linking unresolved steps in familiar biosynthetic pathways with radical mechanisms: functional characterization using new analysis and information visualization methods. *Nucleic Acids Res.*, **29** (5), 1097–1106.
86. Walsby, C.J., Hong, W., Broderick, W.E., Cheek, J., Ortillo, D., Broderick, J.B., and Hoffman, B.M. (2002) Electron-nuclear double resonance spectroscopic evidence that S-adenosylmethionine binds in contact with the catalytically active $[4Fe\text{-}4S]^{+}$ cluster of pyruvate formate-lyase activating enzyme. *J. Am. Chem. Soc.*, **124** (12), 3143–3151.
87. Walsby, C.J., Ortillo, D., Broderick, W.E., Broderick, J.B., and Hoffman, B.M. (2002) An anchoring role for FeS clusters: chelation of the amino acid moiety of S-adenosylmethionine to the unique iron site of the [4Fe-4S] cluster of pyruvate formate-lyase activating

enzyme. *J. Am. Chem. Soc.*, **124** (38), 11270–11271.

88. Walsby, C.J., Ortillo, D., Yang, J., Nnyepi, M.R., Broderick, W.E., Hoffman, B.M., and Broderick, J.B. (2005) Spectroscopic approaches to elucidating novel iron-sulfur chemistry in the "Radical-SAM" protein superfamily. *Inorg. Chem.*, **44** (4), 727–741.
89. Nicolet, Y., Amara, P., Mouesca, J.M., and Fontecilla-Camps, J.C. (2009) Unexpected electron transfer mechanism upon AdoMet cleavage in radical SAM proteins. *Proc. Natl. Acad. Sci. U.S.A.*, **106** (35), 14867–14871.
90. Challand, M.R., Ziegert, T., Douglas, P., Wood, R.J., Kriek, M., Shaw, N.M., and Roach, P.L. (2009) Product inhibition in the radical S-adenosylmethionine family. *FEBS Lett.*, **583** (8), 1358–1362.
91. Nicolet, Y. and Drennan, C.L. (2004) AdoMet radical proteins – from structure to evolution – alignment of divergent protein sequences reveals strong secondary structure element conservation. *Nucleic Acids Res.*, **32** (13), 4015–4025.
92. Vey, J.L. and Drennan, C.L. (2011) Structural insights into radical generation by the radical SAM superfamily. *Chem. Rev.*, **111** (4), 2487–2506.
93. Dowling, D.P., Vey, J.L., Croft, A.K., and Drennan, C.L. (2012) Structural diversity in the AdoMet radical enzyme superfamily. *Biochim. Biophys. Acta*, **1824** (11), 1178–1195.
94. Chatterjee, A., Li, Y., Zhang, Y., Grove, T.L., Lee, M., Krebs, C., Booker, S.J., Begley, T.P., and Ealick, S.E. (2008) Reconstitution of ThiC in thiamine pyrimidine biosynthesis expands the radical SAM superfamily. *Nat. Chem. Biol.*, **4** (12), 758–765.
95. Zhang, Y., Zhu, X., Torelli, A.T., Lee, M., Dzikovski, B., Koralewski, R.M., Wang, E., Freed, J., Krebs, C., Ealick, S.E., and Lin, H. (2010) Diphthamide biosynthesis requires an organic radical generated by an iron-sulphur enzyme. *Nature*, **465** (7300), 891–896.
96. Kampmeier, J.A. (2010) Regioselectivity in the homolytic cleavage of *S*-adenosylmethionine. *Biochemistry*, **49** (51), 10770–10772.
97. Wang, W., Wang, K., Span, I., Jauch, J., Bacher, A., Groll, M., and Oldfield, E. (2012) Are free radicals involved in IspH catalysis? An EPR and crystallographic investigation. *J. Am. Chem. Soc.*, **134** (27), 11225–11234.
98. Fontecilla-Camps, J.C., Volbeda, A., Cavazza, C., and Nicolet, Y. (2007) Structure/function relationships of [NiFe]- and [FeFe]-hydrogenases. *Chem. Rev.*, **107** (10), 4273–4303.
99. Kung, Y. and Drennan, C.L. (2011) A role for nickel-iron cofactors in biological carbon monoxide and carbon dioxide utilization. *Curr. Opin. Chem. Biol.*, **15** (2), 276–283.
100. Flint, D.H. and Allen, R.M. (1996) Iron-sulfur proteins with nonredox functions. *Chem. Rev.*, **96** (7), 2315–2334.
101. Beinert, H., Kennedy, M.C., and Stout, C.D. (1996) Aconitase as iron-sulfur protein, enzyme, and iron-regulatory protein. *Chem. Rev.*, **96** (7), 2335–2374.
102. Lee, M., Grawert, T., Quitterer, F., Rohdich, F., Eppinger, J., Eisenreich, W., Bacher, A., and Groll, M. (2010) Biosynthesis of isoprenoids: crystal structure of the [4Fe-4S] cluster protein IspG. *J. Mol. Biol.*, **404** (4), 600–610.
103. Flint, D.H., Emptage, M.H., and Guest, J.R. (1992) Fumarase a from *Escherichia coli*: purification and characterization as an iron-sulfur cluster containing enzyme. *Biochemistry*, **31** (42), 10331–10337.
104. Flint, D.H., Tuminello, J.F., and Emptage, M.H. (1993) The inactivation of Fe–S cluster containing hydro-lyases by superoxide. *J. Biol. Chem.*, **268** (30), 22369–22376.
105. Ollagnier-de Choudens, S., Loiseau, L., Sanakis, Y., Barras, F., and Fontecave, M. (2005) Quinolinate synthetase, an iron-sulfur enzyme in NAD biosynthesis. *FEBS Lett.*, **579** (17), 3737–3743.
106. Rousset, C., Fontecave, M., and Ollagnier de Choudens, S. (2008) The [4Fe-4S] cluster of quinolinate synthase from *Escherichia coli*: investigation of cluster ligands. *FEBS Lett.*, **582** (19), 2937–2944.
107. Lees, N.S., Hänzelmann, P., Hernandez, H.L., Subramanian, S., Schindelin, H.,

Johnson, M.K., and Hoffman, B.M. (2009) ENDOR spectroscopy shows that guanine N1 binds to [4Fe–4S] cluster II of the S-adenosylmethionine-dependent enzyme MoaA: mechanistic implications. *J. Am. Chem. Soc.*, **131** (26), 9184–9185.
108. Lloyd, S.J., Lauble, H., Prasad, G.S., and Stout, C.D. (1999) The mechanism of aconitase: 1.8 Å resolution crystal structure of the S642A: citrate complex. *Protein Sci.*, **8** (12), 2655–2662.
109. Imlay, J.A. (2003) Pathways of oxidative damage. *Annu. Rev. Microbiol.*, **57**, 395–418.
110. Crack, J., Green, J., and Thomson, A.J. (2004) Mechanism of oxygen sensing by the bacterial transcription factor fumarate-nitrate reduction (FNR). *J. Biol. Chem.*, **279** (10), 9278–9286.
111. Crack, J.C., Gaskell, A.A., Green, J., Cheesmant, M.R., Le Brun, N.E., and Thomson, A.J. (2008) Influence of the environment on the $[4Fe\text{-}4S]^{2+}$ to $[2Fe\text{-}2S]^{2+}$ cluster switch in the transcriptional regulator FNR. *J. Am. Chem. Soc.*, **130** (5), 1749–1758.
112. Crack, J.C., Green, J., Cheesman, M.R., Le Brun, N.E., and Thomson, A.J. (2007) Superoxide-mediated amplification of the oxygen-induced switch from 4Fe-4S to 2Fe-2S clusters in the transcriptional regulator FNR. *Proc. Natl. Acad. Sci. U.S.A.*, **104** (7), 2092–2097.
113. Imlay, J.A. (2006) Iron-sulphur clusters and the problem with oxygen. *Mol. Microbiol.*, **59** (4), 1073–1082.
114. Torrents, E., Aloy, P., Gibert, I., and Rodriguez-Trelles, F. (2002) Ribonucleotide reductases: divergent evolution of an ancient enzyme. *J. Mol. Evol.*, **55** (2), 138–152.
115. Nordlund, N. and Reichard, P. (2006) Ribonucleotide reductases. *Annu. Rev. Plant. Physiol. Plant. Mol. Biol.*, **75**, 681–706.
116. Logan, D.T., Andersson, J., Sjoberg, B.M., and Nordlund, P. (1999) A glycyl radical site in the crystal structure of a class III ribonucleotide reductase. *Science*, **283** (5407), 1499–1504.
117. Sintchak, M.D., Arjara, G., Kellogg, B.A., Stubbe, J., and Drennan, C.L. (2002) The crystal structure of class II ribonucleotide reductase reveals how an allosterically regulated monomer mimics a dimer. *Nat. Struct. Biol.*, **9** (4), 293–300.
118. Challand, M.R., Martins, F.T., and Roach, P.L. (2010) Catalytic activity of the anaerobic tyrosine lyase required for thiamine biosynthesis in *Escherichia coli*. *J. Biol. Chem.*, **285** (8), 5240–5248.
119. Kriek, M., Martins, F., Challand, M.R., Croft, A., and Roach, P.L. (2007) Thiamine biosynthesis in *Escherichia coli*: identification of the intermediate and by-product derived from tyrosine. *Angew. Chem. Int. Ed.*, **46** (48), 9223–9226.
120. Park, J.H., Dorrestein, P.C., Zhai, H., Kinsland, C., McLafferty, F.W., and Begley, T.P. (2003) Biosynthesis of the thiazole moiety of thiamin pyrophosphate (Vitamin B1). *Biochemistry*, **42** (42), 12430–12438.
121. Settembre, E.C., Dorrestein, P.C., Park, J.H., Augustine, A.M., Begley, T.P., and Ealick, S.E. (2003) Structural and mechanistic studies on ThiO, a glycine oxidase essential for thiamin biosynthesis in *Bacillus subtilis*. *Biochemistry*, **42** (10), 2971–2981.
122. Miller, A.F. (2012) Superoxide dismutases: ancient enzymes and new insights. *FEBS Lett.*, **586** (5), 585–595.
123. Nicholls, P. (2012) Classical catalase: ancient and modern. *Arch. Biochem. Biophys.*, **525** (2), 95–101.
124. Mishra, S. and Imlay, J. (2012) Why do bacteria use so many enzymes to scavenge hydrogen peroxide? *Arch. Biochem. Biophys.*, **525** (2), 145–160.
125. Djaman, O., Outten, F.W., and Imlay, J.A. (2004) Repair of oxidized iron-sulfur clusters in *Escherichia coli*. *J. Biol. Chem.*, **279** (43), 44590–44599.
126. Mukhopadhyay, P., Zheng, M., Bedzyk, L.A., LaRossa, R.A., and Storz, G. (2004) Prominent roles of the NorR and Fur regulators in the *Escherichia coli* transcriptional response to reactive nitrogen species. *Proc. Natl. Acad. Sci. U.S.A.*, **101** (3), 745–750.
127. Justino, M.C., Vicente, J.B., Teixeira, M., and Saraiva, L.M. (2005) New genes implicated in the protection of anaerobically grown *Escherichia coli* against

nitric oxide. *J. Biol. Chem.*, **280** (4), 2636–2643.

128. Justino, M.C., Almeida, C.C., Teixeira, M., and Saraiva, L.M. (2007) *Escherichia coli* di-iron YtfE protein is necessary for the repair of stress-damaged iron-sulfur clusters. *J. Biol. Chem.*, **282** (14), 10352–10359.

129. Justino, M.C., Almeida, C.C., Goncalves, V.L., Teixeira, M., and Saraiva, L.M. (2006) *Escherichia coli* YtfE is a di-iron protein with an important function in assembly of iron-sulphur clusters. *FEMS Microbiol. Lett.*, **257** (2), 278–284.

130. Justino, M.C., Baptista, J.M., and Saraiva, L.M. (2009) Di-iron proteins of the RIC family are involved in iron-sulfur cluster repair. *Biometals*, **22**, 99–108.

131. Vine, C.E., Justino, M.C., Saraiva, L.M., and Cole, J. (2010) Detection by whole genome microarrays of a spontaneous 126-gene deletion during construction of a *ytfE* mutant: confirmation that a *ytfE* mutation results in loss of repair of iron-sulfur centres in proteins damaged by oxidative or nitrosative stress. *J. Microbiol. Methods*, **81** (1), 77–79.

132. Overton, T.W., Justino, M.C., Li, Y., Baptista, J.M., Melo, A.M.P., Cole, J.A., and Saraiva, L.M. (2008) Widespread distribution in pathogenic bacteria of di-iron proteins that repair oxidative and nitrosative damage to iron-sulfur centers. *J. Bacteriol.*, **190** (6), 2004–2013.

133. Todorovic, S., Justino, M.C., Wellenreuther, G., Hildebrandt, P., Murgida, D.H., Meyer-Klaucke, W., and Saraiva, L.M. (2008) Iron-sulfur repair YtfE protein from *Escherichia coli*: structural characterization of the di-iron center. *J. Biol. Inorg. Chem.*, **13** (5), 765–770.

134. Anderson, C.P., Shen, M., Eisenstein, R.S., and Leibold, E.A. (2012) Mammalian iron metabolism and its control by iron regulatory proteins. *Biochim. Biophys. Acta*, **1823** (9), 1468–1483.

135. Rouault, T.A. (2006) The role of iron regulatory proteins in mammalian iron homeostasis and disease. *Nat. Chem. Biol.*, **2** (8), 406–414.

136. Wallander, M.L., Leibold, E.A., and Eisenstein, R.S. (2006) Molecular control of vertebrate iron homeostasis by iron regulatory proteins. *Biochim. Biophys. Acta*, **7**, 668–689.

137. Meyron-Holtz, E.G., Ghosh, M.C., and Rouault, T.A. (2004) Mammalian tissue oxygen levels modulate iron-regulatory protein activities in vivo. *Science*, **306** (5704), 2087–2090.

138. Pantopoulos, K., Mueller, S., Atzberger, A., Ansorge, W., Stremmel, W., and Hentze, M.W. (1997) Differences in the regulation of iron regulatory protein-1 (IRP-1) by extra- and intracellular oxidative stress. *J. Biol. Chem.*, **272** (15), 9802–9808.

139. Hentze, M.W., Muckenthaler, M.U., and Andrews, N.C. (2004) Balancing acts: molecular control of mammalian iron metabolism. *Cell*, **117** (3), 285–297.

140. Schultz, I.J., Chen, C.Y., Paw, B.H., and Hamza, I. (2010) Iron and porphyrin trafficking in heme biogenesis. *J. Biol. Chem.*, **285** (35), 26753–26759.

141. Fosset, C., Chauveau, M.J., Guillon, B., Canal, F., Drapier, J.C., and Bouton, C. (2006) RNA silencing of mitochondrial m-Nfs1 reduces Fe–S enzyme activity both in mitochondria and cytosol of mammalian cells. *J. Biol. Chem.*, **281** (35), 25398–25406.

142. Gerber, J., Muhlenhoff, U., and Lill, R. (2003) An interaction between frataxin and Isu1/Nfs1 that is crucial for Fe/S cluster synthesis on Isu1. *Embo Rep.*, **4** (9), 906–911.

143. Li, K.Y., Besse, E.K., Ha, D., Kovtunovych, G., and Rouault, T.A. (2008) Iron-dependent regulation of frataxin expression: implications for treatment of Friedreich ataxia. *Hum. Mol. Genet.*, **17** (15), 2265–2273.

144. Schmucker, S., Martelli, A., Colin, F., Page, A., Wattenhofer-Donze, M., Reutenauer, L., and Puccio, H. (2011) Mammalian frataxin: an essential function for cellular viability through an interaction with a preformed ISCU/NFS1/ISD11 iron-sulfur assembly complex. *PLoS One*, **6** (1), e16199.

145. Tsai, C.L. and Barondeau, D.P. (2010) Human frataxin is an allosteric switch

that activates the Fe–S cluster biosynthetic complex. *Biochemistry*, **49** (43), 9132–9139.

146. Zhang, Y., Lyver, E.R., Nakamaru-Ogiso, E., Yoon, H., Amutha, B., Lee, D.W., Bi, E., Ohnishi, T., Daldal, F., Pain, D., and Dancis, A. (2008) Dre2, a conserved eukaryotic Fe/S cluster protein, functions in cytosolic Fe/S protein biogenesis. *Mol. Cell. Biol.*, **28** (18), 5569–5582.
147. Muhlenhoff, U., Richter, N., Pines, O., Pierik, A.J., and Lill, R. (2011) Specialized function of yeast Isa1 and Isa2 proteins in the maturation of mitochondrial 4Fe-4S proteins. *J. Biol. Chem.*, **286** (48), 41205–41216.
148. Roy, A., Solodovnikova, N., Nicholson, T., Antholine, W., and Walden, W.E. (2003) A novel eukaryotic factor for cytosolic Fe–S cluster assembly. *Embo J.*, **22** (18), 4826–4835.
149. Condo, I., Malisan, F., Guccini, I., Serio, D., Rufini, A., and Testi, R. (2010) Molecular control of the cytosolic aconitase/IRP1 switch by extramitochondrial frataxin. *Hum. Mol. Genet.*, **19** (7), 1221–1229.
150. Shi, Y.B., Ghosh, M., Kovtunovych, G., Crooks, D.R., and Rouault, T.A. (2012) Both human ferredoxins 1 and 2 and ferredoxin reductase are important for iron-sulfur cluster biogenesis. *Biochim. Biophys. Acta*, **1823** (2), 484–492.
151. Song, D.S., Tu, Z., and Lee, F.S. (2009) Human ISCA1 interacts with IOP1/NARFL and functions in both cytosolic and mitochondrial iron-sulfur protein biogenesis. *J. Biol. Chem.*, **284** (51), 35297–35307.
152. Tong, W.H. and Rouault, T.A. (2006) Functions of mitochondrial ISCU and cytosolic ISCU in mammalian iron-sulfur cluster biogenesis and iron homeostasis. *Cell Metab.*, **3** (3), 199–210.
153. Uhrigshardt, H., Singh, A., Kovtunovych, G., Ghosh, M., and Rouault, T.A. (2010) Characterization of the human HSC20, an unusual DnaJ type III protein, involved in iron-sulfur cluster biogenesis. *Hum. Mol. Genet.*, **19** (19), 3816–3834.
154. Ye, H., Jeong, S.Y., Ghosh, M.C., Kovtunovych, G., Silvestri, L., Ortillo, D., Uchida, N., Tisdale, J., Camaschella, C., and Rouault, T.A. (2010) Glutaredoxin 5 deficiency causes sideroblastic anemia by specifically impairing heme biosynthesis and depleting cytosolic iron in human erythroblasts. *J. Clin. Invest.*, **120** (5), 1749–1761.
155. Mueller, S. (2005) Iron regulatory protein 1 as a sensor of reactive oxygen species. *Biofactors*, **24** (1-4), 171–181.
156. Wang, J., Chen, G.H., Filebeen, C., and Pantopoulos, K. (2008) Insights on regulation and function of the iron regulatory protein 1 (IRP1). *Hemoglobin*, **32** (1-2), 109–115.
157. Eisenstein, R.S., Tuazon, P.T., Schalinske, K.L., Anderson, S.A., and Traugh, J.A. (1993) Iron responsive element binding protein: phosphorylation by protein kinase C. *J. Biol. Chem.*, **268** (36), 27363–27370.
158. Pantopoulos, K. and Hentze, M.W. (1998) Activation of iron regulatory protein-1 by oxidative stress in vitro. *Proc. Natl. Acad. Sci. U.S.A.*, **95** (18), 10559–10563.
159. Brown, N.M., Anderson, S.A., Steffen, D.W., Carpenter, T.B., Kennedy, M.C., Walden, W.E., and Eisenstein, R.S. (1998) Novel role of phosphorylation in Fe–S cluster stability revealed by phosphomimetic mutations at Ser-138 of iron regulatory protein 1. *Proc. Natl. Acad. Sci. U.S.A.*, **95** (26), 15235–15240.
160. Deck, K.M., Vasanthakumar, A., Anderson, S.A., Goforth, J.B., Kennedy, M.C., Antholine, W.E., and Eisenstein, R.S. (2009) Evidence that phosphorylation of iron regulatory protein 1 at serine 138 destabilizes the 4Fe-4S cluster in cytosolic aconitase by enhancing 4Fe–3Fe cycling. *J. Biol. Chem.*, **284** (19), 12701–12709.
161. Sanchez, M., Galy, B., Muckenthaler, M.U., and Hentze, M.W. (2007) Iron-regulatory proteins limit hypoxia-inducible factor-2 alpha expression in iron deficiency. *Nat. Struct. Mol. Biol.*, **14** (5), 420–426.
162. Zimmer, M., Ebert, B.L., Neil, C., Brenner, K., Papaioannou, L., Melas, A.,

Tolliday, N., Lamb, J., Pantopoulos, K., Golub, T., and Iliopoulos, O. (2008) Small-molecule inhibitors of HIF-2a translation link its 5′ UTR iron-responsive element to oxygen sensing. *Mol. Cell*, **32** (6), 838–848.

163. Walden, W.E., Selezneva, A., and Volz, K. (2012) Accommodating variety in iron-responsive elements: crystal structure of transferrin receptor 1 B IRE bound to iron regulatory protein 1. *FEBS Lett.*, **586** (1), 32–35.

164. Walden, W.E., Selezneva, A.I., Dupuy, J., Volbeda, A., Fontecilla-Camps, J.C., Theil, E.C., and Volz, K. (2006) Structure of dual function iron regulatory protein 1 complexed with ferritin IRE-RNA. *Science*, **314** (5807), 1903–1908.

165. Dupuy, J., Volbeda, A., Carpentier, P., Darnault, C., Moulis, J.M., and Fontecilla-Camps, J.C. (2006) Crystal structure of human iron regulatory protein 1 as cytosolic aconitase. *Structure*, **14** (1), 129–139.

166. Rouault, T.A. (2006) If the RNA fits, use it. *Science*, **314** (5807), 1886–1887.

167. Crack, J.C., Green, J., Le Brun, N.E., and Thomson, A.J. (2006) Detection of sulfide release from the oxygen-sensing 4Fe-4S cluster of FNR. *J. Biol. Chem.*, **281** (28), 18909–18913.

168. Jervis, A.J., Crack, J.C., White, G., Artymiuk, P.J., Cheesman, M.R., Thomson, A.J., Le Brun, N.E., and Green, J. (2009) The O_2 sensitivity of the transcription factor FNR is controlled by Ser24 modulating the kinetics of 4Fe-4S to 2Fe-2S conversion. *Proc. Natl. Acad. Sci. U.S.A.*, **106** (12), 4659–4664.

169. Zhang, B., Crack, J.C., Subramanian, S., Green, J., Thomson, A.J., Le Brun, N.E., and Johnson, M.K. (2012) Reversible cycling between cysteine persulfide-ligated [2Fe-2S] and cysteine-ligated [4Fe-4S] clusters in the FNR regulatory protein. *Proc. Natl. Acad. Sci. U.S.A.*, **109** (39), 15734–15739.

170. Khoroshilova, N., Popescu, C., Munck, E., Beinert, H., and Kiley, P.J. (1997) Iron-sulfur cluster disassembly in the FNR protein of *Escherichia coli* by O_2: 4Fe-4S to 2Fe-2S conversion with loss of biological activity. *Proc. Natl. Acad. Sci. U.S.A.*, **94** (12), 6087–6092.

171. Lazazzera, B.A., Beinert, H., Khoroshilova, N., Kennedy, M.C., and Kiley, P.J. (1996) DNA binding and dimerization of the Fe–S-containing FNR protein from *Escherichia coli* are regulated by oxygen. *J. Biol. Chem.*, **271** (5), 2762–2768.

172. Sutton, V.R., Mettert, E.L., Beinert, H., and Kiley, P.J. (2004) Kinetic analysis of the oxidative conversion of the $[4Fe\text{-}4S]^{2+}$ cluster of FNR to a $[2Fe\text{-}2S]^{2+}$ cluster. *J. Bacteriol.*, **186** (23), 8018–8025.

173. Fleischhacker, A.S. and Kiley, P.J. (2011) Iron-containing transcription factors and their roles as sensors. *Curr. Opin. Chem. Biol.*, **15** (2), 335–341.

174. Green, J., Crack, J.C., Thomson, A.J., and LeBrun, N.E. (2009) Bacterial sensors of oxygen. *Curr. Opin. Microbiol.*, **12** (2), 145–151.

175. Kiley, P.J. and Beinert, H. (1998) Oxygen sensing by the global regulator, FNR: the role of the iron-sulfur cluster. *FEMS Microbiol. Rev.*, **22** (5), 341–352.

176. Schwartz, C.J., Giel, J.L., Patschkowski, T., Luther, C., Ruzicka, F.J., Beinert, H., and Kiley, P.J. (2001) IscR, an Fe–S cluster-containing transcription factor, represses expression of *Escherichia coli* genes encoding Fe–S cluster assembly proteins. *Proc. Natl. Acad. Sci. U.S.A.*, **98** (26), 14895–14900.

177. Giel, J.L., Rodionov, D., Liu, M., Blattner, F.R., and Kiley, P.J. (2006) IscR-dependent gene expression links iron-sulphur cluster assembly to the control of O_2-regulated genes in *Escherichia coli*. *Mol. Microbiol.*, **60** (4), 1058–1075.

178. Wu, Y. and Outten, F.W. (2009) IscR controls iron-dependent biofilm formation in *Escherichia coli* by regulating type I fimbria expression. *J. Bacteriol.*, **191** (4), 1248–1257.

179. Nesbit, A.D., Giel, J.L., Rose, J.C., and Kiley, P.J. (2009) Sequence-specific binding to a subset of IscR-regulated promoters does not require IscR Fe–S cluster ligation. *J. Mol. Biol.*, **387** (1), 28–41.

Part II
Model Complexes of the Active Site of Hydrogenases – Proton and Dihydrogen Activation

Bioinspired Catalysis: Metal-Sulfur Complexes, First Edition.
Edited by Wolfgang Weigand and Philippe Schollhammer.

3
[NiFe] Hydrogenases

Joe Dawson, Carlo Perotto, Jonathan McMaster, and Martin Schröder

3.1
Introduction

Research into analogs of the active site of [NiFe] hydrogenase is underpinned by a desire to develop industrially viable catalysts for the cleavage of dihydrogen (H_2) to protons or the evolution of H_2 from reduction of protons (Scheme 3.1) [1–3].

$$H_2 \rightleftharpoons 2H^+ + 2e^-$$

Scheme 3.1 The reversible two-electron oxidation of H_2.

Metal complex catalysts often operate at elevated temperatures and/or pressure regimes, and in organic solvents [4]. Thus, an industrial catalyst must be stable under relatively harsh reaction conditions, but should preferably be relatively inexpensive to synthesize via a readily accessible route. These factors often prevent the replacement of these chemical systems with biological catalysts. However, biological enzymes provide key advantages as they usually function at ambient temperature and pressure, and operate in water as a solvent [5, 6]. Furthermore, biological catalysts often exhibit extremely high turnovers [5] with high regio-, stereo-, and enantioselectivity [7], and their directed specificity generally removes the need for protecting groups [6].

The advantages of both *in vivo* and *in vitro* catalysts may be incorporated into a single system via the design, synthesis, and study of low-molecular weight metal complexes that mimic the functioning components of active biosites. This approach can provide an optimum balance of properties and affords insights to the structural, mechanistic, and spectroscopic features, and the changes that take place at the active site during catalysis [6, 8]. It also affords routes to the design and synthesis of metal complexes and clusters via new ligands and variations of metal center(s) that could serve as catalysts for the production or consumption of H_2.

Bioinspired Catalysis: Metal-Sulfur Complexes, First Edition.
Edited by Wolfgang Weigand and Philippe Schollhammer.

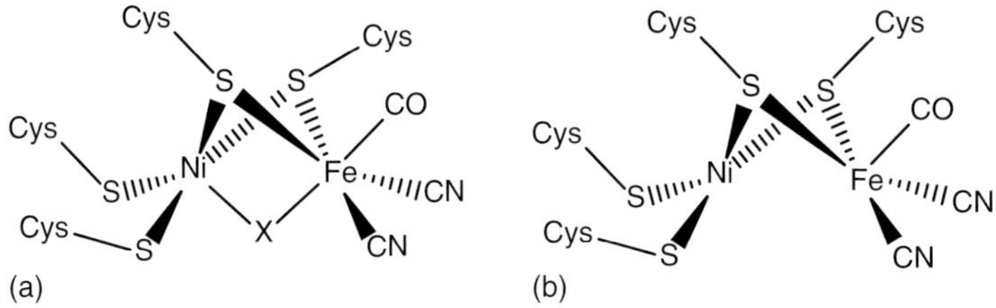

Figure 3.1 Schematic view of the active sites of [NiFe] hydrogenase from *D. vulgaris* Myazaki F in the (a) oxidized and (b) reduced forms (b) [9, 12, 13]. The oxidized form of the active site shows the X bridging species, which is currently postulated as peroxide (OOH^-) or hydroxide (OH^-) [21, 22].

3.2 Introduction to [NiFe] Hydrogenases

In 1995, Volbeda *et al.* [3] reported the first X-ray crystal structure of a [NiFe] hydrogenase obtained from the oxidized form of *Desulfovibrio gigas.* [NiFe] hydrogenases incorporate Ni and one Fe atom at the active site [3, 9–16], with the Ni center bound by four S donors from cysteine residues, two of which bridge the Ni and Fe centers (Figure 3.1). Infrared (IR) spectroscopy has confirmed that the coordination sphere of the Fe center is completed by three nonprotein ligands [1], either two cyanide groups and one carbonyl group [10, 17, 18], three carbonyl groups [13], or a S-monoxide, a carbonyl, and a cyanide group [9, 19]. In each example, the Fe center possesses a distorted square-pyramidal geometry [20].

3.3 Nickel Thiolate Complexes as Analogs of [NiFe] Hydrogenase

Low-molecular weight models of the active sites of the [NiFe] hydrogenases seek to mimic the structural, electronic, redox, and functional chemistry of these metal centers. The aim of this review is to place this work in context and to highlight recent advances in developing new *in vitro* catalysts for reversible H_2 oxidation and production. Early work on the modeling of the active sites of the [NiFe] hydrogenases predates the solution of the first crystal structure of the enzyme [23]. Initially, it was suggested that the active site of the enzyme consisted of a mono- or binuclear Ni-thiolate system, with extended X-ray absorption fine structure (EXAFS) studies indicating that the ligand environment about the Ni center was a tetragonal $[Ni(SR)_4]^{2-}$ unit and Fe was believed to be present only in [4Fe–4S] electron-transfer relays [24–28]. Therefore, much research focused on the preparation of complexes of type $[Ni(SR)_4]^{2-}$ [29–32], which were compared to the known features of the [NiFe] hydrogenases [24].

Rosenfield *et al.* [33] reported a series of mononuclear Ni(II) complexes that incorporated four monodentate thiolate ligands to form a distorted tetrahedral

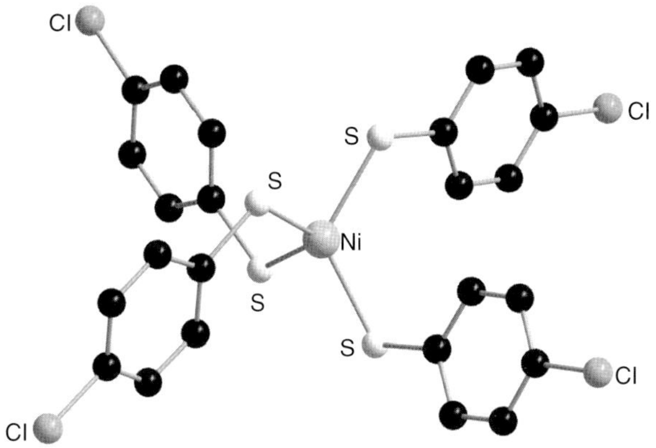

Figure 3.2 View of the single-crystal X-ray structure of $[Ni(S\text{-}p\text{-}C_6H_4Cl)_4]^{2-}$ from the series of complexes $(R_4N)_2[Ni(SAr)_4]$ (Ar = C_6H_5, p-C_6H_4Cl, p-$C_6H_4CH_3$, and m-C_6H_4Cl) [33].

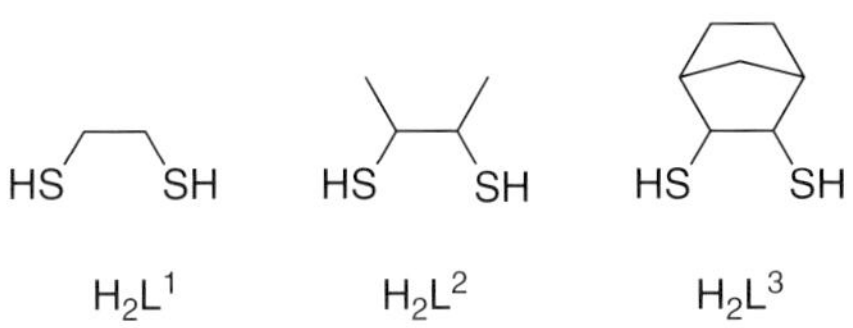

Figure 3.3 The chelating bidentate thiol ligands ethane-1,2-dithiol (H_2L^1) [26, 35–37], butane-2,3-dithiol (H_2L^2) [36], and norbornane dithiol (H_2L^3) [28].

species, $(R_4N)_2[Ni(SAr)_4]$ (Ar = C_6H_5, p-C_6H_4Cl, p-$C_6H_4CH_3$, and m-C_6H_4Cl) (Figure 3.2). In the solid state, $[Ni(SAr)_4]^{2-}$ possesses two S–Ni–S angles that are close to 90° and an average Ni–S distance of 2.281(1) Å, which closely resemble the coordination at the Ni site in the [NiFe] hydrogenase enzyme [34]. This work was extended via the complexation of chelating bidentate thiolate ligands derived from ethane-1,2-dithiol (H_2L^1) [26, 35–37], butane-2,3-dithiol (H_2L^2) [36], and norbornane dithiol (H_2L^3) [28] (Figure 3.3) to Ni(II).

$(Ph_4P)_2[Ni(L^1)_2]\cdot 4H_2O$ and $K_2[Ni(L^2)_2]\cdot C_2H_5OH$ can be synthesized by the reaction of $NiCl_2\cdot 6H_2O$ with an excess of the respective thiolate ligands. The X-ray crystal structure of $[Ni(L^1)_2]^{2-}$ confirms a square–planar complex (Figure 3.4) [36], which contrasts with the distorted tetrahedral $[Ni(SAr)_4]^{2-}$ centers. This may reflect the lower basicity and weaker ligand field of thiophenolates compared to alkyl thiolates [33]. The redox properties of the complexes $[Ni(L^1)_2]^{2-}$ and $[Ni(L^2)_2]^{2-}$ are significantly different [36]. $[Ni(L^1)_2]^{2-}$ degrades at the electrode surface upon oxidation from Ni(II) to the formal Ni(III) state in the range −0.60 to −0.46 V versus standard hydrogen electrode (SHE), while $[Ni(L^2)_2]^{2-}$ undergoes reversible oxidation from Ni(II) to Ni(III) at −0.51 V versus SHE.

Thiolate groups have a strong tendency to form bridges between two or three metal centers, and therefore simple alkyl thiolates readily form oligonuclear or

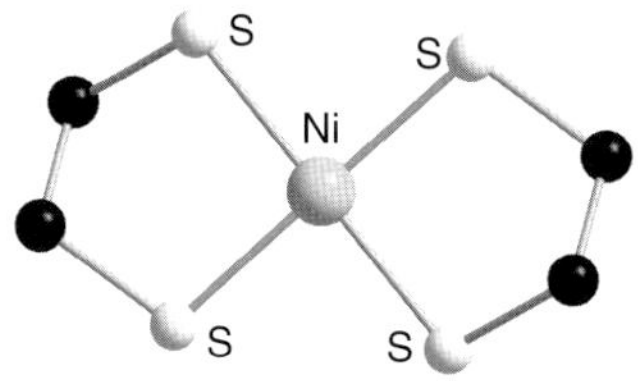

Figure 3.4 View of the single-crystal X-ray structure of $[Ni(L^1)_2]^{2-}$ [36].

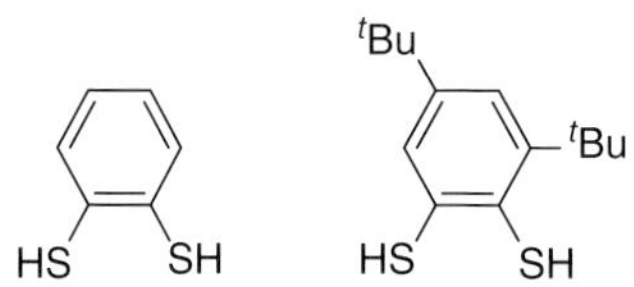

H_2L^4 ^{t}Bu-H_2L^4

Figure 3.5 H_2L^4 and ^{t}Bu-H_2L^4 reported by Cha *et al.* [39, 40].

polynuclear complexes with Ni(II) salts [26, 35]. Thus, $[Ni(L^n)_2]^{2-}$ rapidly converts to binuclear $[Ni_2(L^n)_3]^{2-}$ ($n = 1$, 2) complexes in protic or wet aprotic solvents [36]. To inhibit this reactivity, Fox *et al.* [28] developed the chelating bidentate norbornane dithiol ligand, $[H_2L^3]$ (Figure 3.3). Oligomerization of the complex $[Ni(L^3)_2]^{2-}$ is reduced because the norbornane unit enforces a rigid conformation on the S–C–C–S chelate.

In general, aromatic thiolates are less basic than their alkyl counterparts and are thus more likely to form mononuclear Ni(II) complexes, although the formation of polynuclear complexes with phenyl thiolates can still occur [38]. The mononuclear complexes $[Ni(L^4)_2]^{2-}$ and $[Ni(^tBu\text{-}L^4)_2]^{2-}$ can be generated by the reaction of $[NiCl_2(PPh)_3]$ with H_2L^4 or ^{t}Bu-H_2L^4, respectively (Figure 3.5) [39]. The formal Ni(III) complex $[Ni(L^4)_2]^-$ is formed by oxidation of the Ni(II) precursor at $E_{1/2} = -0.23$ V versus SHE [40]. $[Ni(^tBu\text{-}L^4)_2]^{2-}$ can be oxidized to $[Ni(^tBu\text{-}L^4)_2]^-$ and $[Ni(^tBu\text{-}L^4)_2]^0$, and crystallographic analysis suggests that these oxidation processes are "$[NiS_4]$" based [41]. However, the aromatic ligand H_2L^4 reacts with Ni(II) salts to give the neutral trinuclear cluster $\{Ni_3[\mu_2\text{-}L^4][\mu,\mu'\text{-}L^4]_2(PPh_3)_2\}$ (Figure 3.6), which in coordinating solvents readily fragments to charged species and other polynuclear Ni-thiolate complexes [39].

To avoid undesired oligomerization, Sellmann *et al.* [27, 41–45] adopted an alternative approach involving polychelating ligands containing both thiophenolate and thioether donors to form a series of Ni(II) complexes. The coordination geometries at the metal centers in these species are controlled by the flexible polydentate ligands (Figure 3.7).

The reaction of the extended ligand $[L^5]^{2-}$ ($n = 5$) (Figure 3.7) with Ni(II) forms the binuclear complex $[Ni_2(L^5)_2]$ (Figure 3.8a), with the ligands bridging M(II) ions rather than chelating a single metal center. A similar binuclear complex $[Ni_2(L^6)_2]$ was isolated from the reaction of $[L^6]^{2-}$ with Ni(II) (Figure 3.8b) [27].

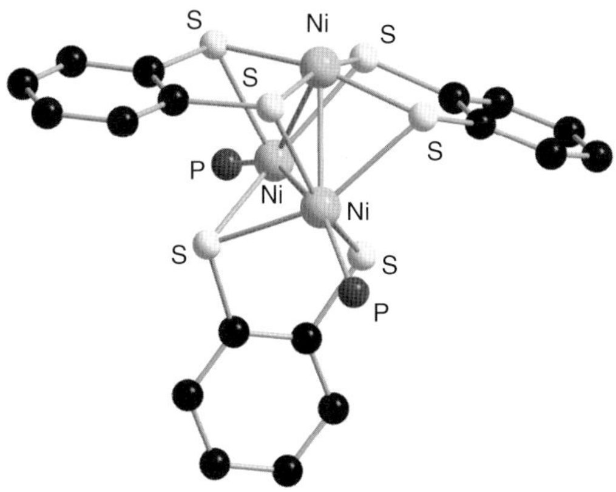

Figure 3.6 View of the single-crystal X-ray structure of $\{Ni_3[(\mu)_2\text{-}L^4][(\mu, \mu')\text{-}L^4]_2(PPh_3)_2\}$; the phenyl rings of each PPh_3 ligand have been omitted for clarity [39].

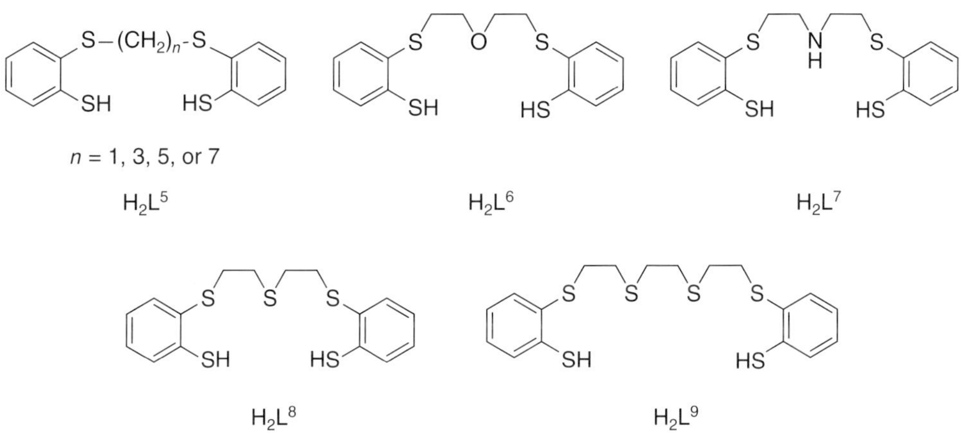

Figure 3.7 Selection of polychelate ligands used by Sellmann *et al.* [27, 41–45].

In contrast, the Ni(II) complexes $[Ni(L^7)]$ and $[Ni(L^8)]$ are mononuclear five-coordinate species with the central NH and S centers of the polychelate involved in binding to the metal ion [27]. $[Ni(L^8)]$ was the first example of a Ni(II) complex with five S-donor atoms, and its single-crystal X-ray structure shows an average Ni–S thioether bond length of 2.21 Å, which is shorter than typical Ni–S(thioether) bond lengths ($|Ni–S| = 2.36$ Å; $\sigma = 0.11$ Å) [46]. The central S-thioether donor is only weakly bound to Ni(II) with a bond length of 2.74 Å (Figure 3.9) [27]. In contrast, $[Ni(L^9)]$ is a mononuclear square–planar diamagnetic complex where the two central thioether S atoms do not coordinate to the metal center [27].

Macrocyclic ligands (Figure 3.10) have also been employed to study the interaction of S-donors with Ni(II) centers [47–50]. These ligands stabilize Ni(I) and Ni(III) centers and the resultant complexes can be assessed structurally and

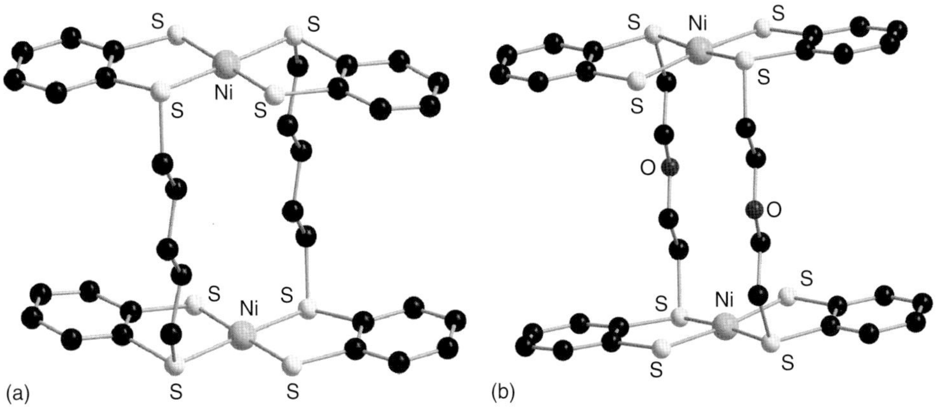

Figure 3.8 Views of the single-crystal X-ray structures of the binuclear complexes (a) $[Ni_2(L^5)_2]$ and (b) $[Ni_2(L^6)_2]$ [27].

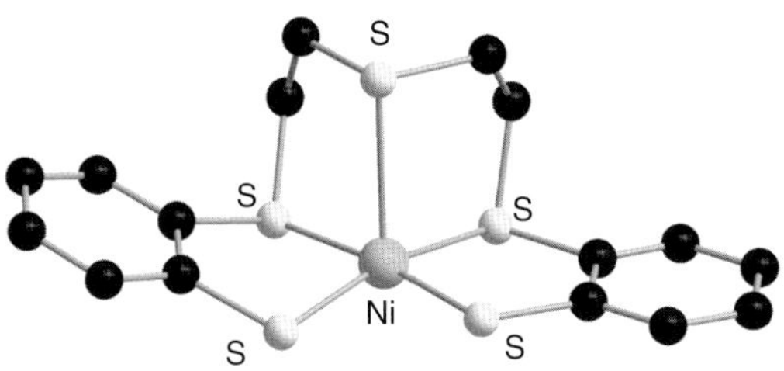

Figure 3.9 View of the single-crystal X-ray structure of the five-coordinate complex $[Ni(L^8)]$ [27].

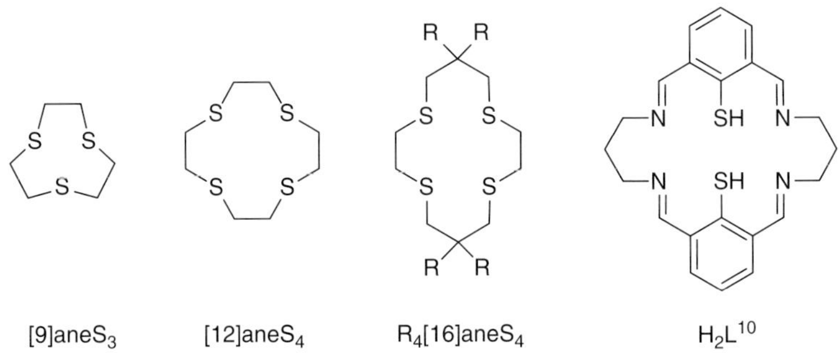

Figure 3.10 A selection of sulfur-containing macrocyclic ligands (R = alkyl group) [47–50].

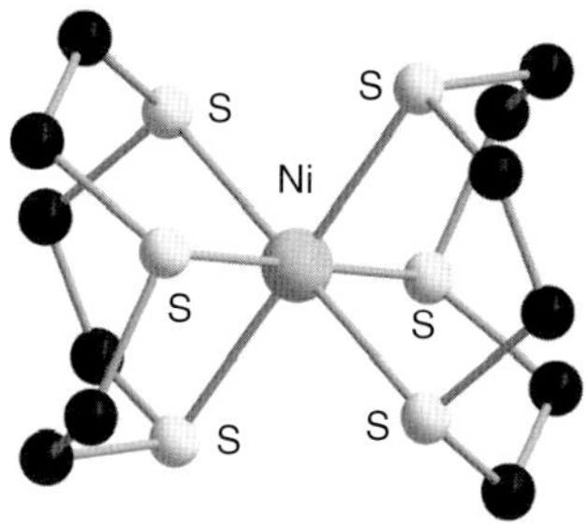

Figure 3.11 View of the single-crystal X-ray structure of $[Ni([9]aneS_3)_2]^{3+}$ [49].

spectroscopically. Although these macrocyclic complexes incorporate thioether rather than thiolate S-donors, the electronic structures of the Ni(I) and Ni(III) redox products can be related to the paramagnetic Ni(I) and Ni(III) intermediate states of the active forms of [NiFe] hydrogenase [51]. Thus, the complex $[Ni([9]aneS_3)_2]^{3+}$ (Figure 3.11) was the first homoleptic octahedral Ni(III) species to be reported [49]. It shows an average Ni–S bond length of 2.3129(25) Å with the electron paramagnetic resonance (EPR) spectrum showing $g_1 = 2.093$, $g_2 = 2.075$, $g_3 = 2.027$.

The mixed valence Ni(II)/Ni(III) complex $[Ni_2(L^{10})(MeCN)_4]^{3+}$ has been prepared by both electrochemical and chemical oxidation of the Ni(II)Ni(II) precursor [52, 53] and analyzed by multifrequency and pulsed EPR spectroscopy. This confirms that 91% of the spin density is found at the Ni(III) center, of which 72% is in the $3d_{z^2}$ orbital and 16% in the $3d_{xy}$ orbital with the Ni(II) containing −3.5% spin density, and the axial MeCN ligands 7.5% spin density [50].

To probe the Ni(III) state that has been proposed for the active site of the [NiFe] hydrogenases, Holm and Kruger [54] investigated the reaction of pyridine-2,6-bis(monothiocarboxylic acid) (H_2pdtc) with $[Ni(OAc)_2 \cdot 4H_2O]$ in a mixture of EtOH and Et_4NOH to give $(Et_4N)_2[Ni(pdtc)_2]$ (Figure 3.12). $(Et_4N)_2[Ni(pdtc)_2]$ contains two tridentate $pdtc^{2-}$ ligands that adopt a meridional configuration incorporating a planar NiS_4 center with two mutually trans axial pyridyl N-donors to give an overall tetragonally distorted octahedral stereochemistry at Ni(II). $(Et_4N)_2[Ni(pdtc)_2]$ is readily oxidized by iodine to form $(Et_4N)[Ni(pdtc)_2]$ and EPR studies confirm this complex incorporates Ni(III) in a d_{z^2} ground-state electronic configuration, similar to the oxidized form of

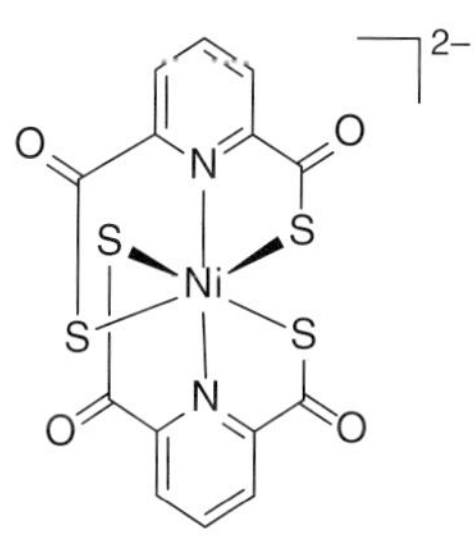

Figure 3.12 View of the complex $[Ni(pdtc)_2]^{2-}$ [54].

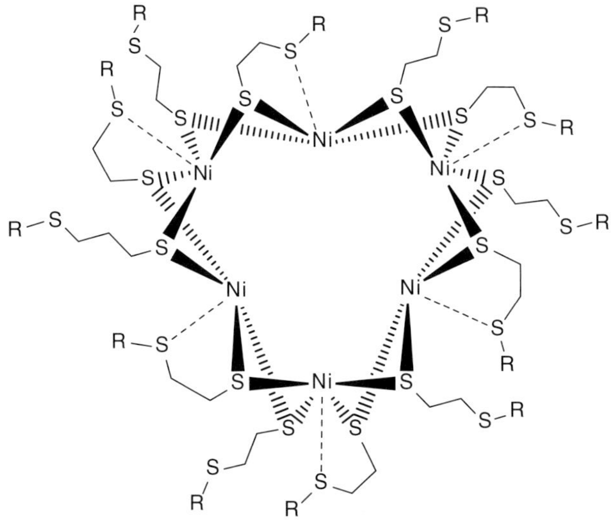

Figure 3.13 View of the metallacrown complex $[Ni_6L_{12}]$ ($L^- = {}^-S\text{-}CH_2\text{-}CH_2\text{-}S\text{-}C_6H_4\text{-}Cl$) synthesized by Bouwman *et al.* [55, 56].

the [NiFe] hydrogenase enzyme [54]. The metal-centered redox processes of $(Et_4N)[Ni(pdtc)_2]$ are strongly influenced by the ligand properties and resulted in the relatively low potential of −0.085 V versus Fc^+/Fc for the Ni(III/II) couple in dimethylformamide (DMF). Moreover, $(Et_4N)_2[Ni(pdtc)_2]$ and $(Et_4N)[Ni(pdtc)_2]$ were the first pair of structurally defined Ni(II/III) complexes incorporating anionic S-donors.

Bouwman *et al.* have reported [55, 56] a metallacrown complex $[Ni_6L_{12}]$ ($L^- = {}^-S\text{-}CH_2\text{-}CH_2\text{-}S\text{-}C_6H_4\text{-}Cl$) (Figure 3.13) that demonstrates a functional resemblance to the [NiFe] hydrogenase. $[Ni_6L_{12}]$ is readily protonated at the thioether sulfurs and this robust protonated species has been studied by ^1H NMR and EPR spectroscopies. The protonated form of $[Ni_6L_{12}]$ was shown to catalyze the reduction of H^+ to H_2 in the presence of protic acids at a potential of −1.5 V versus Ag/AgCl following a mechanism similar to that proposed for [NiFe] hydrogenases [55].

More recently, Sarkar *et al.* [57] reported a Ni(II) radical–ligand complex $[PPh_4][Ni^{II}(L^2)(L^{-\bullet})]$ (Figure 3.14) that catalyzes the reduction of protons from *p*-toluene sulfonic acid (TsOH). Spectral and electrochemical investigations on $[PPh_4][Ni^{II}(L^{2-})(L^{-\bullet})]$ confirm that TsOH initially protonates a sulfur donor site in the complex. Importantly, this system electrocatalyzes H_2 evolution at −0.69 V versus Ag/AgCl in CH_3CN, the lowest reduction potential reported thus far in the literature. It is argued that $[Ni^{II}(L^{2-})(L^{-\bullet})][PPh_4]$ also exhibits EPR parameters and reduction potentials similar to the Ni–C state of the enzyme [57]. However, the assignment of redox states in this Ni(II) radical–ligand system is the subject of some debate [58, 59].

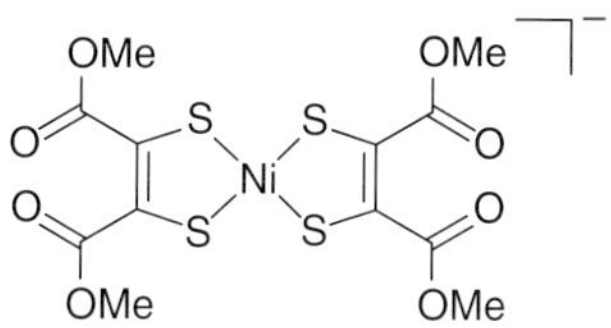

Figure 3.14 View of the complex $[Ni^{II}(L^{2-})(L^{-\bullet})]^-$ reported by Sarkar *et al.* [57].

3.4
[NiFe] Hydrogenase Model Complexes

The principal strategy to develop complexes that mimic the active site in [NiFe] hydrogenase enzymes has been to use a Ni(II) subunit ligated by four thiolate or related S-donors, two of which bridge to an organometallic CO/CN^- ligated Fe subunit [3, 9–14, 16, 60]. Thus, the simplest strategy to generate these systems is to synthesize first a Ni(II) complex with chelating S ligands and use this subunit as a metallothiolate ligand to an organometallic CO/CN^- ligated Fe center [61]. However, this approach can be troublesome as metal-thiolate complexes tend to form homometallic oligomers and clusters and, therefore, mixed donor environments were used in early synthetic protocols [62–65]. Nevertheless, a few examples of structures possessing [NiFe] motifs with pertinent ligation around the metal centers are known, in stark contrast to the extensive and detailed number of structural models for [FeFe] hydrogenases [66]. In the following sections, we discuss a selection of the various mixed donor [NiFe] systems that have been reported, and we also refer the reader to a number of excellent comprehensive reviews of [NiFe] model complexes that have been published previously [29, 34, 61, 62, 66–69].

3.4.1
Amine $[N_2Ni(\mu\text{-}S_2)Fe]$ Complexes

Darensbourg *et al.* [64] synthesized the first thiolate-bridged [NiFe] compound $[Ni(L)Fe(CO)_4]$ (L = *N,N′*-bis(2-mercaptoethyl)-1,5-diazacyclooctane) in 1996 (Figure 3.15a). The isolation of this complex demonstrates that a heterobimetallic complex can be generated using N_2S_2-co-ordination at Ni(II) with a terminal thiolate donor binding in a monodentate manner to a $Fe(CO)_4$ fragment. The single thiolate bridge results in a relatively long Ni–Fe separation of 3.76 Å, too long for any direct interaction between the metal centers, and longer than that observed at the active sites of the [NiFe] hydrogenases, which show a Ni–Fe separation of 2.5–2.6 Å in the reduced form [11] and about 2.9 Å in the oxidized form of the enzyme [3, 10]. Pohl *et al.* [70] reported the first example of a [NiFe] complex with two thiolate bridges in 1997 (Figure 3.15b). This unique double thiolate bridge results in a Ni–Fe bond length of 2.8 Å, similar to that observed in the oxidized form of the enzyme [3, 10].

Schröder *et al.* have reported two structural models based on mixed amine-thiolate polychelate scaffolds. The complex $[Ni(tsalen)Fe(CO)_3]$ [71] (Figure 3.15c)

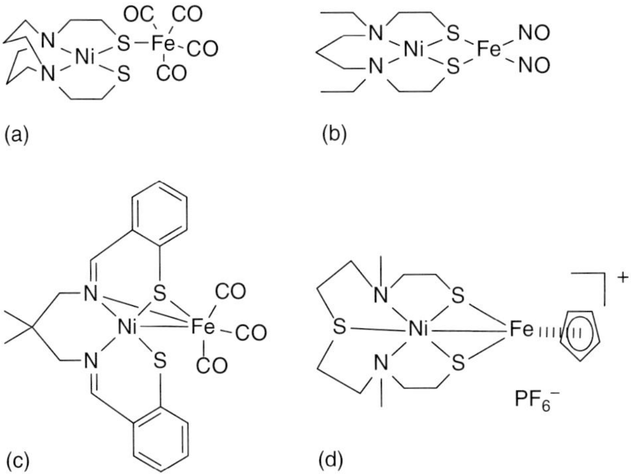

Figure 3.15 View of structures incorporating the $[N_2Ni(\mu\text{-}S_2)Fe]$ ligand motifs reported by (a) Darensbourg *et al.* [64], (b) Pohl *et al.* [70], and (c,d) Schröder *et al.* [65, 71].

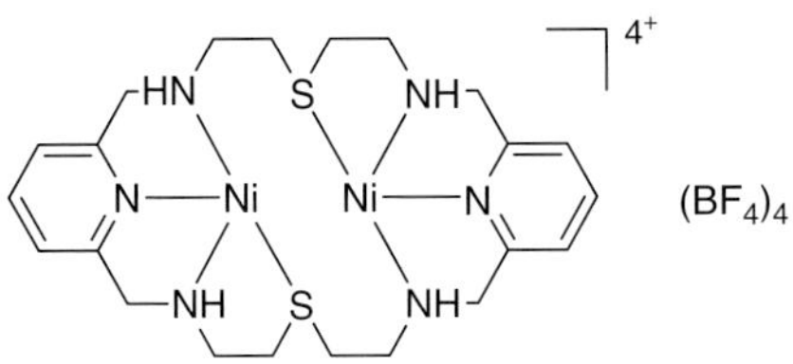

Figure 3.16 View of $[Ni_2(L)](BF_4)_4$ ($L = Py_2N_4S_2$) reported by Lodeiro *et al.* [72].

uses a Schiff-base ligand to introduce imine π-bonds and bridging thiolates, while an alternative approach protects the Fe center within a $Fe(Cp)^+$ (Cp, cyclopentadienyl) moiety that can bind to a Ni(II) thiolate complex to give a distorted square-pyramidal geometry at Ni(II) (Figure 3.15d) [65]. Both of these complexes show relatively short Ni–Fe bond lengths of 2.539(4) and 2.8924(6) Å, respectively, suggesting potential metal–metal interactions in these systems.

Lodeiro *et al.* [72] have described the use of the $Py_2N_4S_2$ macrocycle [73] (Figure 3.16) to successfully bind two metallic centers, either as $[Ni_2(L)]^{4+}$ (Figure 3.16), $[Fe_2(L)]^{4+}$ or $[NiFe(L)]^{4+}$. The ligand binds to the metal ions via a mixed N/S donor set and each metal center is coordinated further by solvent or anionic ligands to generate a distorted octahedral geometry about each metal center.

3.4.2 Phosphine $[P_2Ni(\mu\text{-}S_2)Fe]$ Complexes

Replacement of N_2S_2 ligation about the Ni(II) center by mixed soft P- and S donors seeks to better replicate the cysteinate environment of the Ni center at

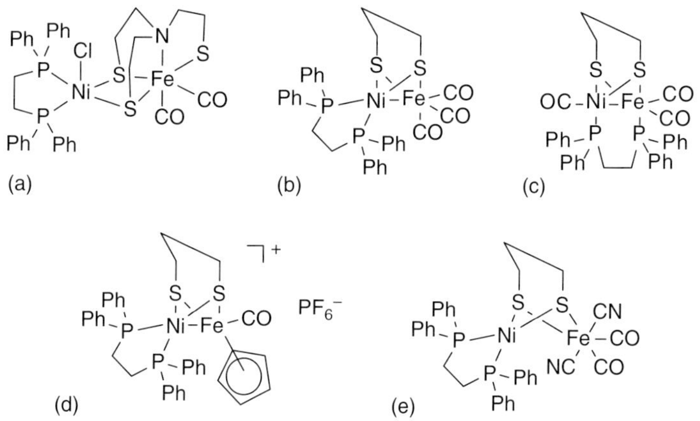

Figure 3.17 Views of the complexes (a) {[Fe(*NS*$_3$)(CO)$_2$-*S,S'*]NiCl(dppe)} (*NS*$_3$ = [N(CH$_2$CH$_2$S)$_3$]$^{3-}$; dppe, 1,2-bis(diphenylphosphino)ethane) reported by Evans *et al.* [74, 75], (b) [(dppe)Ni(μ-pdt)Fe(CO)$_3$], (c) [(OC)Ni(μ-dppe)(μ-pdt)Fe(CO)$_2$], (d) [(dppe)Ni(pdt)Fe(CO)(Cp)]PF$_6$ reported by Schröder *et al.* [65], and (e) [(dppe)Ni(pdt)Fe(CO)$_2$(CN)$_2$] reported by Jiang *et al.* [76].

the active site of the [NiFe] hydrogenases. Evans *et al.* [74, 75] have reported the synthesis and characterization of {[Fe(*NS*$_3$)(CO)$_2$-*S,S'*]NiCl(dppe)} (*NS*$_3$ = [N(CH$_2$CH$_2$S)$_3$]$^{3-}$; dppe, 1,2-bis(diphenylphosphino)ethane) (Figure 3.17a), which has structural features similar to the active site of [NiFe] hydrogenase. The IR spectrum of the complex shows a ν_{CO} stretching vibration at 1944 cm^{-1}, similar to that observed for the inactive form of [NiFe] hydrogenase from *D. gigas* (ν_{CO} = 1947 cm^{-1}) [10]. Reaction of [Ni(dppe)(pdt)] (H$_2$pdt, 1,3-propanedithiol) with [Fe(CO)$_3$(bda)] (bda, benzylidene acetone) or [Fe$_3$(CO)$_{12}$] affords [(dppe)Ni(μ-pdt)Fe(CO)$_3$] (Figure 3.17b) [65] in which the Ni(II) center adopts a distorted tetrahedral P$_2$S$_2$ geometry bridging to a square-based pyramidal Fe(0) center (Figure 3.17b). The complex [(dppe)Ni(μ-pdt)Fe(CO)$_3$] converts to [(OC)Ni(μ-dppe)(μ-pdt)Fe(CO)$_2$] (Figure 3.17c) in solution. Both [(dppe)Ni(μ-pdt)Fe(CO)$_3$] and [(OC)Ni(μ-dppe)(μ-pdt)Fe(CO)$_2$] show butterfly arrangements at the binuclear core [65], with the Ni–Fe separation of 2.5 Å consistent with the presence of a formal Ni–Fe bond. This is consistent with the density functional theory (DFT) calculations, which suggest that a similar Ni–Fe bond may be important in the reduced active form of [NiFe] hydrogenases. Furthermore, the acute dihedral angle <Ni–(μ-S)$_2$–Fe of 80.3° found in [(dppe)Ni(μ-pdt)Fe(CO)$_3$] is comparable to that observed in the enzyme in its reduced active form (88.95°) [11]. Reaction of [Fe(Cp)(CO)$_2$I] with [Ni(dppe)(pdt)] affords [(dppe)Ni(pdt)Fe(Cp)(CO)]PF$_6$ (Figure 3.17d), which displays a square-planar Ni(II) core bound to a [Fe(Cp)(CO)]$^+$ unit via two bridging μ$_2$-thiolate centers [65]. The Ni–Fe distance of 2.7795(9) Å is longer than that in [(dppe)Ni(μ-pdt)Fe(CO)$_3$] [65]. The complex [(dppe)Ni(pdt)Fe(CO)$_2$(CN)$_2$] (Figure 3.17e) has been synthesized through the sequential reaction of *fac*-[Fe(CN)$_2$(CO)$_3$I]$^-$ with K$_2$pdt and [Ni(dppe)Cl$_2$] [76]. [(dppe)Ni(pdt)Fe(CO)$_2$(CN)$_2$] features a Ni center

in a square-planar P_2S_2 environment, a Ni–Fe distance of 2.809(1) Å, and an unusual semibridging CN^- ligand.

Although more than 50 [NiFe] hydrogenase models have now been reported [66], very few have been shown to display functional activity. Often, it is not clear whether these model compounds have been assayed for H_2 oxidation or proton reduction. One reason for this may be that it is generally believed that H_2 activation occurs at the Ni site and very few stable Ni hydrides are known [23, 77–79]. However, Dubois *et al.* have shown that Ni can be an excellent hydride acceptor in the presence of a pendant base (see Section 3.5).

Rauchfuss *et al.* [80] have shown that $[(dppe)Ni(\mu\text{-}pdt)Fe(CO)_3]$ [65] in CH_2Cl_2 readily protonates in the presence of $HBF_4 \cdot Et_2O$ to form $[(dppe)Ni(\mu\text{-}pdt)(\mu\text{-}H)Fe(CO)_3]BF_4$, in which the Ni and Fe sites are bridged by a hydride, with quasi-octahedral coordination at Fe and square-pyramidal geometry at Ni. The complex $[(dppe)Ni(\mu\text{-}pdt)(\mu\text{-}H)Fe(CO)_3]BF_4$ represents the first example of a [NiFe] thiolate hydride complex [81]. $[(dppe)Ni(\mu\text{-}pdt)(\mu\text{-}H)Fe(CO)_3]BF_4$ is also an active catalyst for the reduction of protons to H_2 at potentials near −1.37 V versus Fc^+/Fc [80]. $[(dppe)Ni(\mu\text{-}pdt)(\mu\text{-}H)Fe(CO)_3]BF_4$ reacts with phosphines by thermal and photochemical substitution to form monosubstituted complexes of the type $[(dppe)Ni(\mu\text{-}pdt)(\mu\text{-}H)Fe(PR_3)(CO)_2]BF_4$ ($PR_3 = P(OPh)_3$, PPh_3, PPh_2Py) [82], which show quasi-reversible redox behavior and catalyze the reduction of H^+ to H_2. The neutral complexes $[(dppe)Ni(\mu\text{-}pdt)Fe(PR_3)(CO)_2]$ ($PR_3 = P(OPh)_3$, PPh, PPh_2Py) can be isolated by direct deprotonation of $[(dppe)Ni(\mu\text{-}pdt)(\mu\text{-}H)Fe(PR_3)(CO)_2]BF_4$ with NaOMe. More recently, a new synthetic route, based on the reaction of the [Ni(diphosphine)(dithiolate)] precursor with $[Fe(CO)_4I_2]$ followed by a two-electron reduction with $[Co(Cp)_2]$, has allowed the preparation of four novel complexes of the type $[(dxpe)Ni(\mu\text{-}xdt)Fe(CO)_3]$ (dxpe = dppe, dcpe; $xdt^{2-} = pdt^{2-}$, edt^{2-}; dcpe, 1,2-bis(dicyclohexylphosphino)ethane, H_2pdt, 1,3-propanedithiol, H_2edt, 1,2-ethanedithiol [83]. Protonation of the neutral series $[(dxpe)Ni(\mu\text{-}xdt)Fe(CO)_3]$ affords the four respective hydride-containing complexes that have the general formula $[(dxpe)Ni(xdt)(\mu\text{-}H)Fe(CO)_3]BF_4$, while treatment of $[(dppe)Ni(edt)(\mu\text{-}H)Fe(CO)_3]BF_4$ with PPh_3 results in the formation of $[(dppe)Ni(edt)(\mu\text{-}H)Fe(CO)_2(PPh_3)]BF_4$. Acid–base measurements indicate that on going from [Ni(dppe)] to the more basic [Ni(dcpe)] subunit, the acidity of the cationic hydride complexes decreases by 2.5 pK_a^{PhCN} units, while the reduction potential of the hydride species shifts by ~−250 mV. Chemical oxidation of $[(dxpe)Ni^I(\mu\text{-}xdt)Fe^I(CO)_3]$ by reaction with ferrocenium salts allows the isolation of the paramagnetic species, $[(dxpe)Ni(\mu\text{-}xdt)Fe(CO)_3]BF_4$ [84]. Crystallographic characterization of $[(dcpe)Ni(\mu\text{-}pdt)Fe(CO)_3]BF_4$ shows the Ni center in a square-planar geometry, whereas the Fe center is situated in an approximate square-pyramidal environment. EPR spectroscopic measurements and DFT calculations have assigned the $S = 1/2$ system to a low-spin Fe(I)-centered d_{z^2} ground state and this confirmed the generation of a new Ni(II)Fe(I) state. The complexes $[(dcpe)Ni(\mu\text{-}pdt)Fe(CO)_3]BF_4$ and $[(dppe)Ni(\mu\text{-}pdt)Fe(CO)_3]BF_4$ readily combine with a variety of monodentate phosphines to afford the monosubstituted complexes $[(dxpe)Ni(\mu\text{-}pdt)Fe(CO)_2(PR_3)]BF_4$. Use

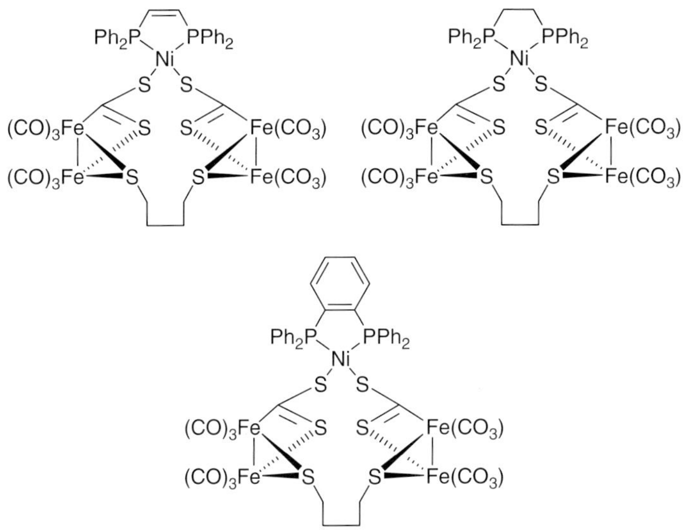

Figure 3.18 View of the series of [NiFe$_4$] macrocyclic compounds reported by Song *et al.* [85].

of bulky and strongly basic monophosphine ligands gave products incorporating distorted coordination geometries about the Fe centers. The structures of [(dppe)Ni(μ-pdt)Fe(CO)$_2$(PCy$_3$)]BF$_4$ and [(dcpe)Ni(μ-pdt)Fe(CO)$_2$(PCy$_3$)]BF$_4$ show a "rotated" Fe(I) center in which the PCy$_3$ ligand occupies a basal site and one CO ligand partially bridges the Ni and Fe centers [84]. Furthermore, cyclic voltammetric experiments indicate that highly basic phosphine ligands enable a second oxidation that corresponds to the [(dxpe)Ni(μ-pdt)Fe(CO)$_2$(PR$_3$)]$^{+/2+}$ redox couple. In this way, this compound closely resembles the Ni-SI$_a$ state of the natural enzyme and currently this is the best example of modeling this state of the hydrogenase enzyme in the literature.

Song *et al.* [85] have reported the synthesis of several novel macrocyclic [NiFe$_4$] complexes (Figure 3.18) that show Ni-bridged in quadruple-butterfly structures that are similar to the [NiFe$_2$S] butterfly motif found in the active site of [NiFe] hydrogenase. However, complexes in these studies have not been assayed for catalytic activity.

3.1.3
Thiolate [S$_x$Ni(μ-S$_y$)Fe] Complexes

Sellmann *et al.* reported [86] the first low-spin Ni(II) center bound in a S-only environment in which the [Ni(L^8)(L)]$^{2-}$ (L^{2-} = benzene-1,2-dithiolate) subunit is bound by two thiolate bridges to a [Fe^{II}CO(PMe$_3$)$_2$]$^{2+}$ fragment to give [(L)Ni(L^8)Fe(CO)(PMe$_3$)$_2$] (Figure 3.19a). It was suggested that the two PMe$_3$ ligands donate electron density in a way analogous to two CN$^-$ ligands. Perra

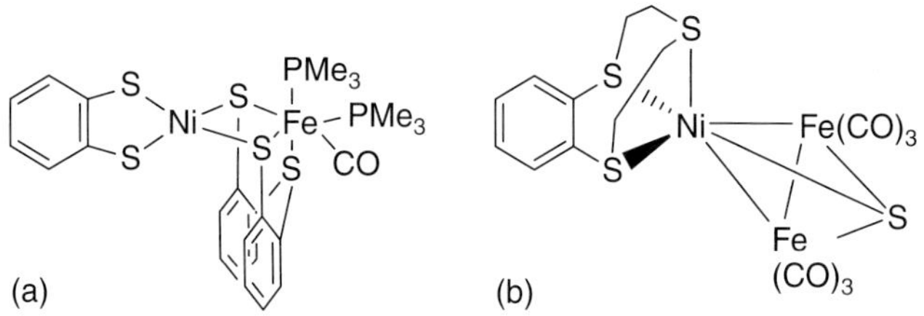

Figure 3.19 View of [NiFe] models reported by (a) Sellmann *et al.* [86] and (b) Perra *et al.* [87].

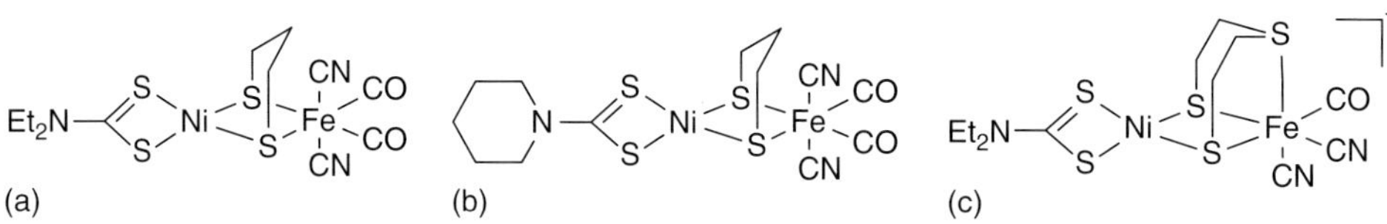

Figure 3.20 (a–c) View of [NiFe] models reported by Tatsumi *et al.* with CO and CN^- ligation [88, 90].

et al. [87] have reported the cluster $[Ni(L)SFe_2(CO)_6]$ (Figure 3.19b), which incorporates a bridging sulfide ligand and has parameters similar to those observed in the active site of the [NiFe] hydrogenase from *Desulfovibrio vulgaris* [9, 12, 13] and *Desulfovibrio desulfuricans* [14].

Two structural models of the [NiFe] hydrogenase biosites have been reported by Tatsumi *et al.* [88], in which square-planar dithiocarbamate Ni(II) derivatives are bridged by pdt^{2-} to a $Fe(CN)_2(CO)_2$ fragment (Figure 3.20a,b). Importantly, this is a rare system in which the Fe site is bound by both CO and CN^- ligation, with Ni–Fe and M–S bond distances closely matching those of the oxidized form of the [NiFe] hydrogenase enzymes of *D. gigas* and *Desulfovibrio fructosovorans* [3, 10, 89]. More recently, the same group reported the synthesis of a dithiolate-bridged binuclear complex, $[(Et_2NCS_2)Ni(\mu\text{-tpdt})Fe(CO)(CN)_2]^-$ ($tpdt^{2-}$, 3-thiapentanedithiolate) (Figure 3.20c) [90], which shows the desired *fac*-$Fe(CN)_2(CO)$ geometry at the Fe center. Furthermore, this complex reproduces the CN/CO stretching vibrations in the IR spectrum associated with the Ni-A, Ni-B, and Ni-SU states of the enzyme. However, the square-planar Ni(II) center in $[(Et_2NCS_2)Ni(\mu\text{-tpdt})Fe(CO)(CN)_2]^-$ is not an accurate structural match for the Ni-A, Ni-B, and Ni-SU states of [NiFe] hydrogenase, which are thought to contain penta-coordinated Ni(III) and Ni(II) centers [3, 91, 92].

Tatsumi *et al.* [93] also reported the thiolate-bridged binuclear [NiFe] complex $[(DmpS)Ni(\mu\text{-S}^tBu)_3Fe(CO)_3]$ (Dmp = C_6H_3-2,6-(mesityl)$_2$) (Scheme 3.2) comprising a $Fe(CO)_3$ unit coupled to a trigonal-pyramidal Ni center. The reversible binding of CO occurs at the Ni site to form $[(DmpS)(CO)Ni(\mu\text{-S}^tBu)_3Fe(CO)_3]$ (Scheme 3.2), which is structurally related to the CO-inhibited form of [NiFe] hydrogenase from *Chromatium vinosum* and *D. fructosovorans* [93]. However, this system incorporates three bridging metal-thiolates between the Ni and Fe centers, has no CN^- binding at Fe, and involves an elongated Ni–Fe separation of

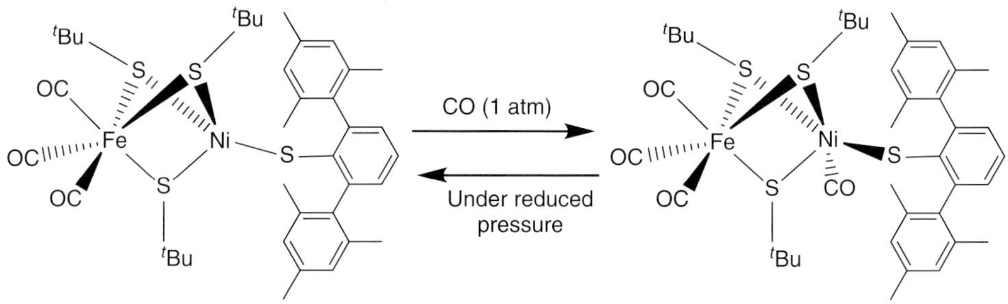

Scheme 3.2 The synthesis and reactivity of [(DmpS)Ni(μ-S^{t}Bu)$_3$Fe(CO)$_3$] (Dmp = C_6H_3-2,6-(mesityl)$_2$) reported by Tatsumi *et al.* [93].

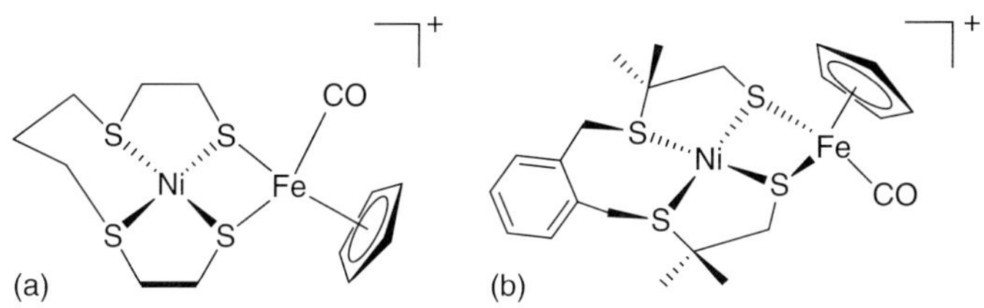

Figure 3.21 View of heterobinuclear [NiFe] complexes reported by (a) Schröder *et al.* [65] and (b) Artero *et al.* [94].

3.17 Å, substantially longer than that in the protein [13]. These disparities mean it is likely that the oxidation states of Fe and/or Ni are different from those found at the active sites of the [NiFe] hydrogenases [93].

Schröder *et al.* [65] reported a heterobinuclear [NiFe] system where the Ni center is bound by sulfur and Cp^-, with CO ligation at the Fe center (Figure 3.21a). The presence of Cp^- binding in this type of system may confer lower overpotentials to the catalyst during H_2 evolution and allow improved stability upon cycling through the catalytic process [94]. Artero *et al.* [94] reported a similar [NiFe] compound (Figure 3.21b), which is an electrocatalyst for H_2 evolution from trifluoroacetic acid (TFA) in DMF. It is believed that this analog passes through reactive intermediates similar to that of the enzyme during the catalytic cycle. Furthermore, the chelating nature of the ligands used in this complex enables the compound to undergo sustained H_2 production during bulk electrolysis experiments and this is the first reported example of such behavior.

3.4.4 Polymetallic [Ni(μ-S)$_z$Fe$_y$] Complexes

Two polymetallic [Ni(μ-S)$_z$Fe$_y$] systems have been reported to show electrocatalytic activity and, in both instances, this activity involves the reduction of protons to H_2 rather than H_2 oxidation [66]. Sellmann *et al.* [95] reported the complex {("S$_2$")[Ni(PMe$_3$)]$_2$Fe-(CO)("S$_2$")$_2$} (Figure 3.22), a trinuclear [Ni$_2$Fe] cluster that

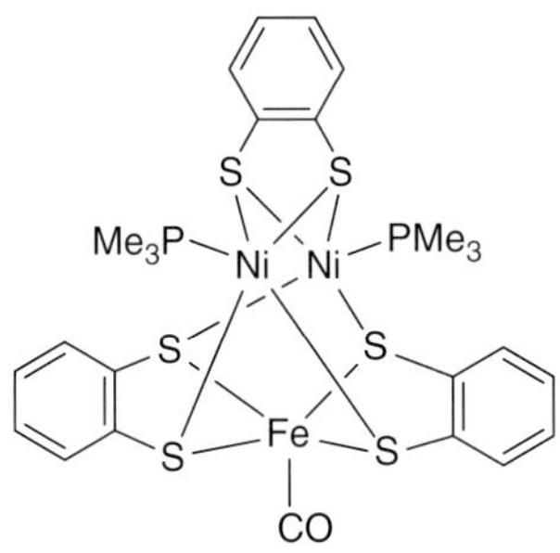

Figure 3.22 View of the complex {("S_2")[Ni(PMe_3)]$_2$Fe-(CO)("S_2")$_2$} reported by Sellmann *et al.* [95, 96].

shows structural similarities to the [NiFe] hydrogenase biosites with Fe and Ni centers in a coordination sphere dominated by thiolate S donors and with CO bound to Fe. The complex {("S_2")[Ni(PMe_3)]$_2$Fe-(CO)("S_2")$_2$} exhibits a reversible electrochemical process at −0.095 V versus SHE in CH_2Cl_2 and a low activity of 0.1 h^{-1} turnover frequency (ToF) for H_2 production in bulk electrolysis experiments [96]. It was proposed that this relatively anodic reduction potential was associated with the protonation of the system at the S center bound to Ni, an argument supported by DFT calculations.

Schröder *et al.* [97] reported two novel [NiFe] hydrogenase analogs [Ni(L^1)$Fe_2(CO)_6$] and [Ni($L^{1'}$)$Fe_2(CO)_6$] [$L^1 = (CH_3C_6H_3S_2)_2(CH_2)_3$, $L^{1'} = (C_6H_4S_2)_2(CH_2)_3$] (Figure 3.23). These heterotrinuclear clusters are synthesized from [Ni(L^1)] and [Ni($L^{1'}$)], respectively, by reaction with [$Fe_3(CO)_{12}$]. In both complexes, the Ni(II) ion is coordinated to two thioether and two thiolate donors, which bridge the Ni and Fe atoms. Each Fe center has a six-coordinate distorted octahedral coordination geometry involving interactions with a Ni atom, one Fe atom, one thiolate donor, and three terminal CO ligands. The Ni–Fe bond lengths are in the range 2.4789(9)–2.4936(8) Å [97] and are similar to those found in the active site of the [NiFe] hydrogenase from *Desulfomicrobium baculatum* (2.5 Å in the reduced form) and *D. vulgaris* Myazaki F (2.55 and 2.59 Å for the oxidized and reduced forms, respectively) [3, 9, 12, 97].

[Ni(L^1)$Fe_2(CO)_6$] undergoes an electrochemically reversible one-electron reduction at $E_{1/2} = -1.31$ V versus Fc^+/Fc to form the paramagnetic species [Ni(L^1)$Fe_2(CO)_6$]$^-$. The addition of TFA to a solution of [Ni(L^1)$Fe_2(CO)_6$]

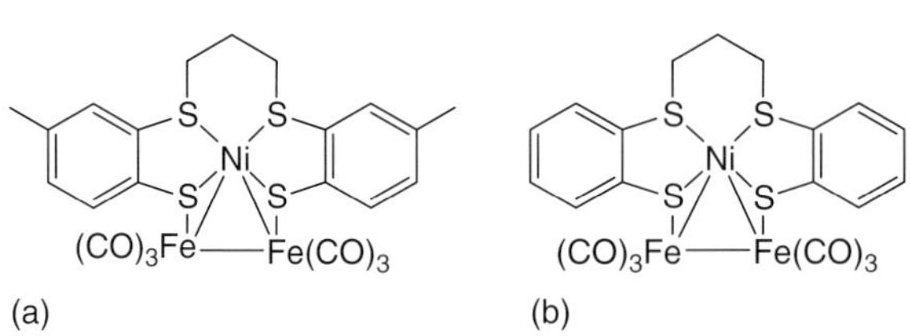

Figure 3.23 (a) [Ni(L^1)$Fe_2(CO)_6$] and (b) [Ni($L^{1'}$)$Fe_2(CO)_6$] reported by Schröder *et al.* [97].

in CH_2Cl_2 generates a catalytic cathodic current response in the cyclic voltammogram that is directly proportional to the concentration of TFA. Gas chromatography confirmed the production of H_2 in this experiment [98]. $[Ni(L^1)Fe_2(CO)_6]$ shows a turnover number (ToN) of $6\,h^{-1}$ in the presence of TFA (50 mM) at an applied potential of −1.64 V versus Fc^+/Fc [98], and a steady catalytic current could be maintained for periods of about 1 h, after which continued electrolysis results in a slow decolorization of the reaction mixture due to decomposition of $[Ni(L^1)Fe_2(CO)_6]$. Detailed EPR, IR spectroscopic, and DFT studies of $[Ni(L^1)Fe_2(CO)_6]^-$ confirm an in-plane delocalized singly occupied molecular orbital (SOMO) that is antibonding with respect to the metal–metal bonds within the $[NiFe_2]$ framework [97]. The SOMO in $[Ni(L^1)Fe_2(CO)_6]^-$ possesses a 24% Ni and 16–17% Fe character, with smaller contributions from the thioether (2.8%) and thiolate (1.4–1.7%), suggesting that the potential protonation site is likely to be metal- and not S-based. $[Ni(L^1)Fe_2(CO)_6]^{0/-}$ demonstrated for the first time that a heterobinuclear Ni–Fe cluster, containing biologically relevant ligands, could support the catalytic production of H_2 at comparable rates of turnover to the synthetic analogs of the [FeFe] hydrogenases at potentials that were considerably more anodic (about +900 mV) [98].

It has been suggested that the presence of several metal–metal interactions in $\{(\text{“}S_2\text{”})[Ni(PMe_3)]_2Fe(CO)(\text{“}S_2\text{”})_2\}$ and $[Ni(L^1)Fe_2(CO)_6]$ may stabilize their respective redox products, $\{(\text{“}S_2\text{”})[Ni(PMe_3)]_2Fe\text{-}(CO)(\text{“}S_2\text{”})_2\}^+$ and $[Ni(L^1)Fe_2(CO)_6]^-$ [61]. This inherent stability may account for the capability of these complexes to reduce protons to H_2. In addition to these compounds, the synthesis and characterization of the complex $[Ni(dppe)(\mu_3\text{-}S)Fe_2(CO)_6]$ has been reported by both Song *et al.* [99] and Evans *et al.* [100] but the reactivity of this species with acid has not been not reported.

3.5
Analogs of [NiFe] Hydrogenase Incorporating Proton Relays

A proton relay is a pendant base within a catalyst that may facilitate the transfer of protons between the solution and the catalytic site [101]. It is known that, in comparison to an organic basic site, a metal center can be slow to protonate [102]. Thus, a basic site in close proximity to the catalytic center may be ideally positioned to relay protons and promote the formation of the metal hydride intermediate required for the reversible oxidation of H_2 [103]. Inspiration for the incorporation of a pendant base as a proton relay in hydrogenase models can be attributed to the presence of a di(thiomethyl)amine moiety at the active site of the [FeFe] hydrogenases [104–106]. The activation of H_2 by the [FeFe] hydrogenases is proposed to occur via the binding of H_2 to the vacant site of the distal Fe center (Fe_D) (Scheme 3.3), followed by heterolytic cleavage to produce an Fe hydride and a protonated amine.

An understanding of how a basic site can influence the free energy of reaction for the heterolytic cleavage of H_2 can be obtained if we consider the individual

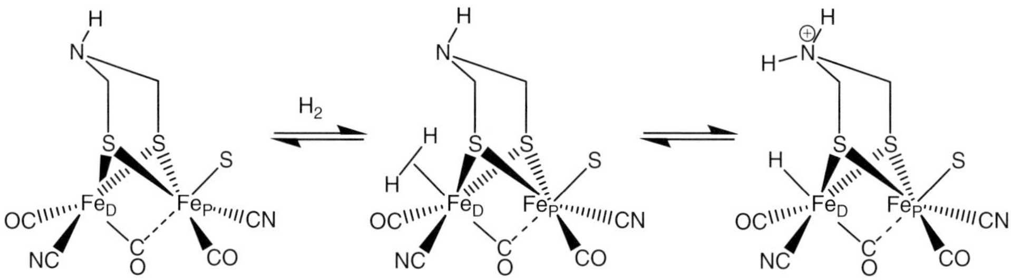

Scheme 3.3 A proposed mechanism of [FeFe] hydrogenase in reaction with H_2 [104–107].

Table 3.1 The thermodynamics of heterolytic cleavage of H_2 in the presence of a proton relay [101].

	$\Delta G°$ (kcal mol^{-1})
$H_2 \rightarrow H^+ + H^-$	76
$ML_n{}^{2+} + H^- \rightarrow HML_n{}^+$	$-G°_{H^-}$
$H^+ + Base \rightarrow HBase^+$	$-1.37\ pK_a$
$H_2 + ML_n{}^{2+} \rightarrow HML_n{}^+ + HBase^+$	$\Delta G° = 76 - \Delta G°_{H^-} - 1.37\ pK_a$

Heterolytic cleavage of H_2 requires 76 kcal mol^{-1} in CH_3CN [110].

reactions as the sum of their free energies equates to the free energy of the overall reaction (Table 3.1) [101]. If the protonated base shows a pK_a that complements the hydride acceptor ability of the metal center ($-G°_{H^-}$), the overall free energy of the reaction ($\Delta G°$) can then be close to zero (Table 3.1) [107–109]. In the following sections, we discuss hydrogenase analogs that incorporate functional proton relays.

3.5.1
Nickel Complexes Incorporating Protonation Sites

Curtis *et al.* [111] reported the complex $[Ni(PNP)_2]^{2+}$ (PNP = $Et_2PCH_2NMeCH_2PEt_2$) (Figure 3.24) that incorporates two basic amine groups within the ligand. These amine groups are noncoordinating because of their incorporation within a strained four-membered ring, which shows some structural similarities to the bridging di(thiomethyl)amine ligand in [FeFe] hydrogenase.

$[Ni(PNP)_2]^{2+}$ cleaves H_2 heterolytically and a cathodic shift of 0.6 V in the oxidation of $[HNi(PNP)_2]^+$ was observed compared to $[HNi(depp)_2]^+$ depp,1,3-bis(diethylphosphino)propane (Figure 3.24) [111]. This large shift indicates a fast (or coupled) intramolecular proton transfer from Ni to the amine of the PNP ligand upon oxidation of the metal from Ni(II)–H to Ni(III)–H [111]. The coupling of the proton- and electron-transfer events overcomes the large energy barrier associated with the formation of a Ni(III)–H center (Figure 3.25) [101]. These results demonstrate that the pendant base not only facilitates intramolecular

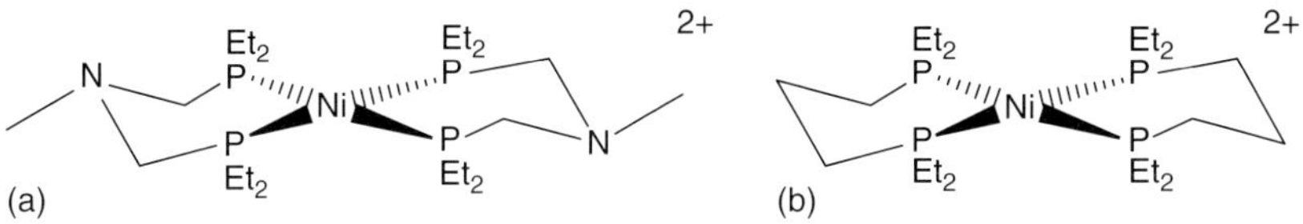

Figure 3.24 View of (a) $[Ni(PNP)_2]^{2+}$ containing two NMe basic groups and (b) the analogous complex $[Ni(depp)_2]^{2+}$ incorporating no amine sites.

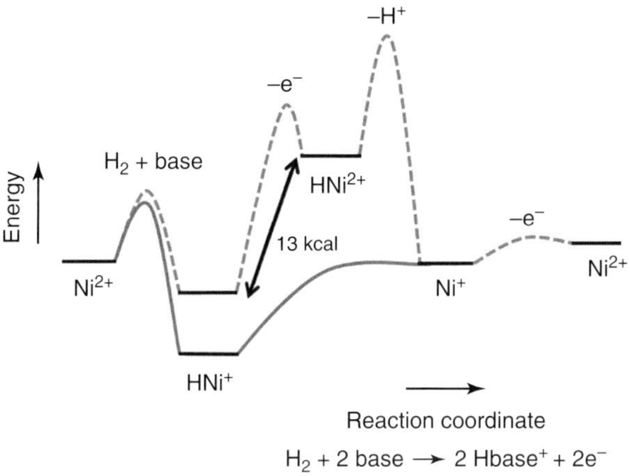

Figure 3.25 Comparison of the free energy reaction profiles for the electrocatalytic oxidation of H_2 using $[Ni(depp)_2]^{2+}$ (dashed line) and $[Ni(PNP)_2]^{2+}$ (solid line) under the same conditions [101].

heterolytic cleavage of H_2 but also promotes intermolecular exchange between the hydride ligand and protons in solution [101, 112, 113].

DuBois *et al.* have targeted [109, 114, 115] Ni(II) phosphine complexes incorporating N-donor basic sites, closely related to the di(thiomethyl)amine bridge at the active sites of the [FeFe] hydrogenases. $[Ni(P^{R}{}_2N^{R'}{}_2)_2]^{2+}$ incorporates basic amine groups orientated in a boat conformation, bringing them above the Ni center such that binding interactions to an incoming H_2 may be templated (Figure 3.26) [109, 114, 115].

These catalysts show relatively high rates of electrocatalysis that can be tuned for either H_2 oxidation or production by changing the substituents of the diphosphine ligands [101]. More bulky substituents on the P donors result in larger tetrahedral distortions and a greater affinity for hydride at the Ni center. Moreover, it has been proposed that H_2 binding for the derivatives of $[Ni(P^{R}{}_2N^{R'}{}_2)_2]^{2+}$ is stabilized by interaction of H_2 with both pendant bases (Figure 3.26), which increases the observed rates of catalysis [109, 112, 114, 115].

Pool and DuBois [116] have reported complexes $[Ni(P^{R}{}_2N^{R'}{}_2)_2]^{2+}$ (Figure 3.26) functionalized with thiophene derivatives at the R′ positions with the view of generating modified electrode surfaces. This complex system is very active toward H_2 production at a low overpotential of 280 mV. Introduction of *tert*-butyl

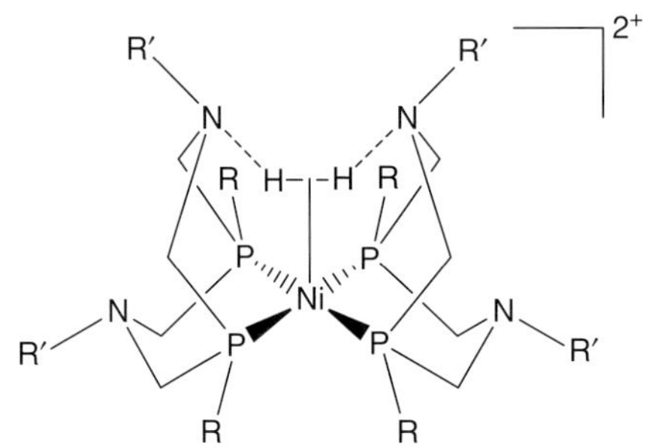

Figure 3.26 View of the complex $[Ni(P^R{}_2N^{R'}{}_2)_2]^{2+}$ (R = cyclohexyl; R′ = benzyl) showing a potential interaction with H_2 [109, 114, 115].

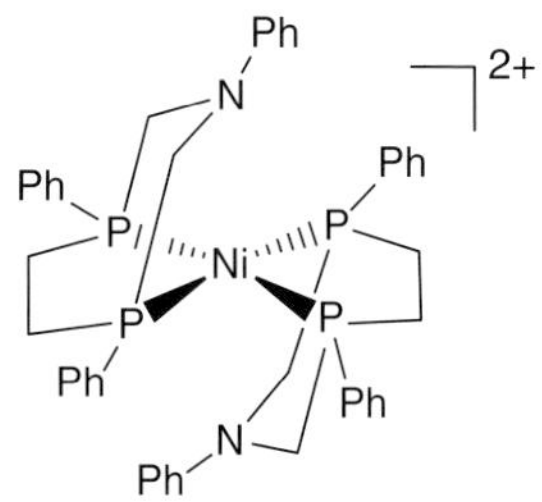

Figure 3.27 View of the complex $[Ni(P^{Ph}{}_2N^{Ph})_2]^{2+}$, the first catalyst to show rates for H_2 production greater than those of the [NiFe] hydrogenase enzyme [119].

groups into the R′ positions increases the basicity of the pendant amines in $[Ni(P^R{}_2N^{R'}{}_2)_2]^{2+}$ (Figure 3.26), and this result lowers the free energy of H_2 addition or cleavage, the rate-determining step for catalysis. Yang *et al.* [117] have identified $[Ni(P^R{}_2N^{R'}{}_2)_2]^{2+}$ (R′ = benzyl) as an active catalyst for H_2 oxidation with a ToF of 50 s^{-1}. This result shows that the driving force for H_2 addition can be tuned by variation of the ligand functionality [117, 118]. In 2011, a significant breakthrough was reported in which the complex $[Ni(P^{Ph}{}_2N^{Ph})_2]^{2+}$ ($P^{Ph}{}_2N^{Ph}$, 1,3,6-triphenyl-1-aza-3,6-diphosphacycloheptane) (Figure 3.27) was shown to act as an electrocatalyst for H_2 production with a ToF above 100 000 s^{-1} [119].

Fontecave *et al.* [120] have reported $[Ni(P^R{}_2N^{R'}{}_2)_2]^{2+}$ (R = Ph) (Figure 3.26) incorporated within an efficient and robust electrode material. This Ni-containing catalyst is both air- and moisture-stable and has been anchored to an amino-functionalized multiwalled carbon nanotube (MWCNT) electrode material [121] via the formation of stable covalent amide linkages (Figure 3.28) [120]. MWCNTs were chosen as the electrode material as they provide a large surface area for high catalyst loading, show high stability and high electrical conductivity [122, 123], and are versatile in terms of grafting molecular complexes to their surfaces [123, 124]. The electrocatalytic properties of the parent Ni complex are retained in the grafted material and an estimated ToF of 20 000 h^{-1} has been calculated for the material in the presence of [DMFH]OTf/DMF (1 : 1 mol/mol, 0.1 mol l^{-1}) (DMFH, dimethyl formamidium) in CH_3CN [120]. Practical applications of such molecular electrode materials require the use of aqueous electrolytes and, importantly, Fontecave *et al.* [120] have demonstrated that a similar Ni-functionalized

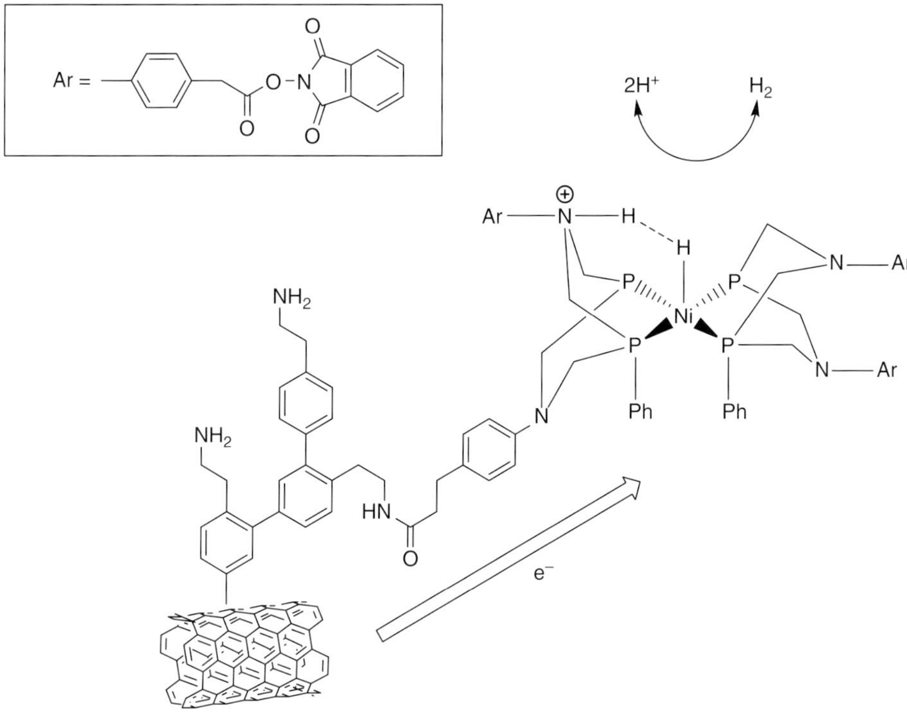

Figure 3.28 Representation of the $[Ni(P^R{}_2N^{R'}{}_2)_2]^{2+}$ derivative anchored to an MWCNTs functionalized with 4-(2-aminoethyl)phenyl groups [120].

MWCNT electrode can produce H_2 catalytically in dilute (pH 0–1) aqueous sulfuric acid solutions.

$[Ni(dppp)(NP^{Ph}{}_2N^{Bz}{}_2)]^{2+}$ ($P^{Ph}{}_2N^{Bz}{}_2$, 1,5-dibenzyl-3,7-diphenyl-1,5-diaza-3,7-diphosphacyclooctane; dppp, 1,3-bis(diphenylphosphino)propane) is an electrocatalyst for H_2 oxidation and for the first time, Ni–H and N–H intermediates could be identified spectroscopically [125]. However, $[Ni(dppp)(NP^{Ph}{}_2N^{Bz}{}_2)]^{2+}$ shows catalytic activity ~20 times lower than the analog of $[Ni(P^R{}_2N^{R'}{}_2)_2]^{2+}$ [125], this difference being attributed to the cooperative participation and positions of the two pendant bases in the latter system [125]. Thus, both the number and position of pendant bases play an important role in the design of this kind of molecular catalyst [125]. This observation has also proved relevant to O_2 reduction catalysts such as $[Ni(P^{Ph}{}_2N^{Me}{}_2)_2]^{2+}$ (Figure 3.29). $[Ni(P^{Ph}{}_2N^{Me}{}_2)_2]^{2+}$ catalyzes the reduction of O_2 with H_2 to form water and reduces O_2 in the presence of a weak acid [126]. In contrast, a Ni diphosphine compound closely related to this system, where no bases are incorporated into the coordination sphere, is inactive toward O_2 reduction. Catalysts of the type $[Ni(P^R{}_2N^{R'}{}_2)_2]^{2+}$ have been analyzed by DFT to elucidate the mechanism of proton reduction because $[Ni(P^R{}_2N^{R'}{}_2)_2]^{2+}$ systems have a variety of molecular conformations available and a range of potential protonation sites [127].

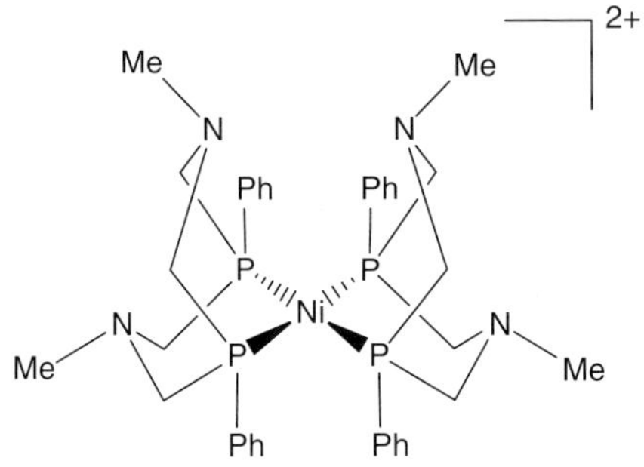

Figure 3.29 View of the complex $[Ni(P^{Ph}{}_2N^{Me}{}_2)_2]^{2+}$ reported by Yang *et al.* [126].

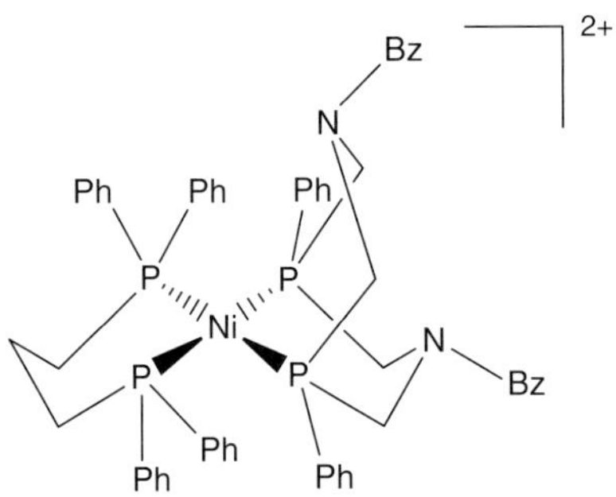

Figure 3.30 View of the complex $[Ni(dppp)(NPPh_2NBz_2)]^{2+}$ reported by Yang *et al.* [125].

It should be noted that Co analogs of $[Ni(P^R{}_2N^{R'}{}_2)_2]^{2+}$ and $[Ni(dppp)(NP^{Ph}{}_2N^{Bz}{}_2)]^{2+}$ (Figures 3.26 and 3.30) have also been reported by DuBois *et al.* [118, 128]. $[Co(PPh_2NPh_2)_2(CH_3CN)]^{2+}$ is not catalytically active toward the formation of H_2 [128]. However, a related complex $[Co(PPh_2NPh_2)(CH_3CN)_3]^{2+}$ incorporating a single cyclic diphosphine ligand is an active electrocatalyst for H_2 production with a ToF of 90 s^{-1} at an estimated overpotential of 285 mV [101, 128]. These values are generally comparable to those observed for the complexes $[Ni(P^R{}_2N^{R'}{}_2)_2]^{2+}$ [69, 129, 130]. For $[Co(PPh_2NPh_2)(CH_3CN)_3]^{2+}$, high catalytic activity is obtained with only one pendant amine base compared with the two used in corresponding Ni complexes [101]. Therefore, Co complexes containing a single diphosphine ligand with suitably positioned pendant base(s) represent a promising class of catalysts for H_2 production that are under active investigation [118].

3.5.2
[NiFe] Complexes Incorporating Protonation Sites

There has been much less focus on the functionalization of [NiFe] complexes containing protonation sites. Sun *et al.* [131] have reported the trinuclear $[NiFe_2]$ complexes $[Fe_2(CO)_6(\mu_3\text{-}S)_2Ni(Ph_2PCH_2)_2NR]$ ($R = {}^nBu$,

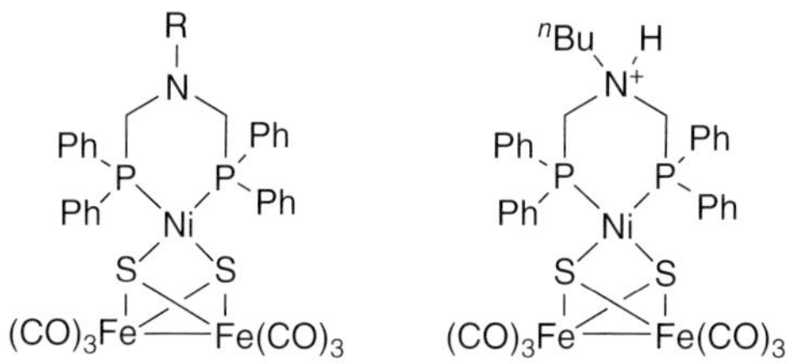

(a) (b)

Figure 3.31 The structures of (a) $[Fe_2(CO)_6(\mu_3\text{-}S)_2Ni(Ph_2PCH_2)_2NR]$ (R = ^{n}Bu, Ph) and (b) $[Fe_2(CO)_6(\mu_3\text{-}S)_2Ni(Ph_2PCH_2)_2NH(^n Bu)]$ [OTf] [131].

Ph) (Figure 3.31). Treatment of $[Fe_2(CO)_6(\mu_3\text{-}S)_2Ni(Ph_2PCH_2)_2N^n Bu]$ with trifluoromethanesulfonic acid (HOTf) produces a stable protonated form of the complex, $[Fe_2(CO)_6(\mu_3\text{-}S)_2Ni(Ph_2PCH_2)_2NH(^n Bu)]$[OTf], in which the [2Fe–2S] units adopt butterfly configurations with the Fe centers in square-pyramidal geometries, and the six-membered rings involving Ni(II) centers and the diphosphine ligands adopting chair conformations [131]. A comparison of $[Fe_2(CO)_6(\mu_3\text{-}S)_2Ni(Ph_2PCH_2)_2N(^n Bu)]$ and its protonated form confirms that protonation of the basic site has almost no influence on the metal framework of the complex, except for the spatial orientation of the ^{n}Bu on the N atom of the NCPNiPC chair conformation, which changes from equatorial to axial [131]. In this axial conformation, the bridgehead N-center is orientated such that it may play a role as a proton relay.

$[Fe_2(CO)_6(\mu_3\text{-}S)_2Ni(Ph_2PCH_2)_2N(^n Bu)]$ and $[Fe_2(CO)_6(\mu_3\text{-}S)_2Ni(dppe)]$ show similar electrochemical behavior (Table 3.2) [131]. However, upon addition of 2 equivalents of HOTf to a solution of $[Fe_2(CO)_6(\mu_3\text{-}S)_2Ni(Ph_2PCH_2)_2N(^n Bu)]$ and subsequent *in situ* formation of the protonated species, a new, more anodic reduction process is observed at $E_p^{\,c} = -1.40$ V versus Ag/Ag^+ (Table 3.2) [131]. This new process quickly disappears on the addition of aniline and the peaks associated with $[Fe_2(CO)_6(\mu_3\text{-}S)_2Ni(Ph_2PCH_2)_2N(^n Bu)]$ are recovered quantitatively. Thus, the basic site in this heterotrinuclear complex can be readily protonated,

Table 3.2 Redox potentials for $[Fe_2(CO)_6(\mu_3\text{-}S)_2Ni(Ph_2PCH_2)_2N(^n Bu)]$, $[Fe_2(CO)_6(\mu_3\text{-}S)_2Ni(Ph_2PCH_2)_2NH(^n Bu)]$[OTf], and $[Fe_2(CO)_6(\mu_3\text{-}S)_2Ni(dppe)]$ versus Ag/Ag^+ [131].

Complex	$E_p^{\,a}$ (V)	$E_p^{\,a}$ (V)	$E_p^{\,c}$ (V)
	Fe(II)Fe(I)/ Fe(II)Fe(II)	Fe(I)Fe(I)/ Fe(II)Fe(I)	Fe(I)Fe(I)/ Fe(0)Fe(I)
$[Fe_2(CO)_6(\mu_3\text{-}S)_2Ni(Ph_2PCH_2)_2N(^n Bu)]$	+0.52	+0.18	−1.73
$[Fe_2(CO)_6(\mu_3\text{-}S)_2Ni(Ph_2PCH_2)_2NH(^n Bu)]$[OTf]	—	+0.36	−1.40
$[Fe_2(CO)_6(\mu_3\text{-}S)_2Ni(dppe)]$	—	+0.19	−1.68

which effects the redox potentials of the Fe(I)Fe(I)/Fe(0)Fe(I) couple. However, it is not reported how the protonation sites influence the catalytic activity of these complexes [131].

3.6 Perspectives and Future Challenges

The design and synthesis of new catalysts and photocatalysts for the clean and sustainable generation of H_2 is a major target worldwide, and new catalysts for H_2 oxidation are required urgently for applications in fuel cells, especially to replace rare and expensive Pt-based catalysts. The development of the hydrogen economy hinges on the efficient and economic generation and utilization of H_2 as an energy vector. Increasing the catalytic turnover numbers, frequency, and rates of reaction via catalyst design, enhancing catalyst stability, and tuning the photo and redox properties of component systems all remain major challenges. Chemical design, synthesis, and functionalization lie at the heart of the future development of *in vitro* catalysts. Multicomponent systems that, for example, combine CdSe semiconductor nanocrystals with Ni-thiolate complexes can afford [132] robust catalysts for photogeneration of H_2 in water. The design and implementation of such multicomponent, polyfunctional aggregates has great potential to generate systems that approach or even surpass the activity of hydrogenase biosites. How nature uses H_2 as an energy source, and our current ability to mimic such complex *in vivo* reactivity, control, and efficiency using *in vitro* model complexes is developing apace, and remains a highly important challenge and target in the development of new sustainable energy sources.

Acknowledgments

We thank EPSRC and the University of Nottingham for support. MS gratefully acknowledges receipt of an ERC advanced grant.

References

1. Vignais, P.M. and Billoud, B. (2007) *Chem. Rev.*, **107**, 4206–4272.
2. De Lacey, A.L., Fernandez, V.M., Rousset, M., and Cammack, R. (2007) *Chem. Rev.*, **107**, 4304–4330.
3. Volbeda, A., Charon, M.H., Piras, C., Hatchikian, E.C., Frey, M., and Fontecilla-Camps, J.C. (1995) *Nature*, **373**, 580–587.
4. Kent, J.A. (2003) *Riegel's Handbook of Industrial Chemistry*, Plenum Publishers.
5. Bugg, T.D.H. (2004) *Introduction to Enzyme and Coenzyme Chemistry*, Blackwell Publishing Ltd.
6. Yi, L. (2006) *Angew. Chem. Int. Ed.*, **45**, 5588–5601.
7. Kaim, W. and Schwederski, B. (1994) *Bioinorganic Chemistry: Inorganic Elements in the Chemistry of Life*, John Wiley & Sons, Ltd, Chichester.
8. Holm, R.H. and Solomon, E.I. (1996) *Chem. Rev.*, **96**, 2237–2238.

9. Higuchi, Y., Yagi, T., and Yasuoka, N. (1997) *Structure*, **5**, 1671–1680.
10. Volbeda, A., Garcin, E., Piras, C., De Lacey, A.L., Fernandez, V.M., Hatchikian, E.C., Frey, M., and Fontecilla-Camps, J.C. (1996) *J. Am. Chem. Soc.*, **118**, 12989–12996.
11. Garcin, E., Vernede, X., Hatchikian, E.C., Volbeda, A., Frey, M., and Fontecilla-Camps, J.C. (1999) *Structure*, **7**, 557–566.
12. Higuchi, Y., Ogata, H., Miki, K., Yasuoka, N., and Yagi, T. (1999) *Structure*, **7**, 549–556.
13. Ogata, H., Mizoguchi, Y., Mizuno, N., Miki, K., Adachi, S., Yasuoka, N., Yagi, T., Yamauchi, O., Hirota, S., and Higuchi, Y. (2002) *J. Am. Chem. Soc.*, **124**, 11628–11635.
14. Matias, P.M., Soares, C.M., Saraiva, L.M., Coelho, R., Morais, J., Le Gall, J., and Carrondo, M.A. (2001) *J. Biol. Inorg. Chem.*, **6**, 63–81.
15. Montet, Y., Amara, P., Volbeda, A., Vernede, X., Hatchikian, E.C., Field, M.J., Frey, M., and Fontecilla-Camps, J.C. (1997) *Nat. Struct. Biol.*, **4**, 523–526.
16. Volbeda, A., Montet, Y., Vernede, X., Hatchikian, E.C., and Fontecilla-Camps, J.C. (2002) *Int. J. Hydrogen Energy*, **27**, 1449–1461.
17. Happe, R.P., Roseboom, W., Pierik, A.J., Albracht, S.P.J., and Bagley, K.A. (1997) *Nature*, **385**, 126.
18. Pierik, A.J., Roseboom, W., Happe, R.P., Bagley, K.A., and Albracht, S.P.J. (1999) *J. Biol. Chem.*, **274**, 3331–3337.
19. Higuchi, Y., Toujou, F., Tsukamoto, K., and Yagi, T. (2000) *J. Inorg. Biochem.*, **80**, 205–211.
20. Volbeda, A. and Fontecilla-Camps, J.C. (2005) *Coord. Chem. Rev.*, **249**, 1609–1619.
21. Ogata, H., Hirota, S., Nakahara, A., Komori, H., Shibata, N., Kato, T., Kano, K., and Higuchi, Y. (2005) *Structure*, **13**, 1635–1642.
22. Volbeda, A., Martin, L., Cavazza, C., Matho, M., Faber, B.W., Roseboom, W., Albracht, S.P.J., Garcin, E., Rousset, M., and Fontecilla-Camps, J.C. (2005) *J. Biol. Inorg. Chem.*, **10**, 239–249.
23. Heinekey, D.M. (2009) *J. Organomet. Chem.*, **694**, 2671–2680.
24. Yamamura, T., Miyamae, H., Katayama, Y., and Sasaki, Y. (1985) *Chem. Lett.*, **14**, 269–272.
25. Yamamura, T. (1986) *Chem. Lett.*, **15**, 801–804.
26. Snyder, B.S., Rao, C.P., and Holm, R.H. (1986) *Aust. J. Chem.*, **39**, 963–974.
27. Sellmann, D., Funfgelder, S., Pohlmann, G., Knoch, F., and Moll, M. (1990) *Inorg. Chem.*, **29**, 4772–4778.
28. Fox, S., Wang, Y., Silver, A., and Millar, M. (1990) *J. Am. Chem. Soc.*, **112**, 3218–3220.
29. Halcrow, M.A. and Christou, G. (1994) *Chem. Rev.*, **94**, 2421–2481.
30. Albracht, S.P.J. (1994) *Biochim. Biophys. Acta*, **1188**, 167–204.
31. Halcrow, M.A. (1995) *Angew. Chem. Int. Ed. Engl.*, **34**, 1193–1195.
32. Cammack, R., Frey, M., and Robson, R. (2001) *Hydrogen as a Fuel: Learning from Nature*, Taylor & Francis Group, London, New York.
33. Rosenfield, S.G., Armstrong, W.H., and Mascharak, P.K. (1986) *Inorg. Chem.*, **25**, 3014–3018.
34. Bouwman, E. and Reedijk, J. (2005) *Coord. Chem. Rev.*, **249**, 1555–1581.
35. Nicholson, J.R., Christou, G., Huffman, J.C., and Folting, K. (1987) *Polyhedron*, **6**, 863–870.
36. Baidya, N., Mascharak, P.K., Stephan, D.W., and Campagna, C.F. (1990) *Inorg. Chim. Acta*, **177**, 233–238.
37. Yamamura, T., Kurihara, H., Nakamura, N., Kuroda, R., and Asakura, K. (1990) *Chem. Lett.*, **19**, 101–104.
38. Silver, A. and Millar, M. (1992) *J. Chem. Soc., Chem. Commun.*, 948–949.
39. Cha, M., Sletten, J., Critchlow, S., and Kovacs, J.A. (1997) *Inorg. Chim. Acta*, **263**, 153–159.
40. Cha, M.Y., Catlin, C.L., Critchlow, S.C., and Kovacs, J.A. (1993) *Inorg. Chem.*, **32**, 5868–5877.
41. Sellmann, D., Binder, H., Haussinger, D., Heinemann, F.W., and Sutter, J. (2000) *Inorg. Chim. Acta*, **300**, 829–836.
42. Sellmann, D., Geipel, F., and Heinemann, F.W. (2000) *Eur. J. Inorg. Chem.*, **2000**, 271–279.

43. Sellmann, D., Geipel, F., and Heinemann, F. (2000) *Eur. J. Inorg. Chem.*, **2000**, 59–63.
44. Sellmann, D., Geipel, F., and Moll, M. (2000) *Angew. Chem. Int. Ed.*, **39**, 561–563.
45. Sellmann, D., Hofmann, T., and Knoch, F. (1994) *Z. Naturforsch., B: Chem. Sci.*, **49**, 821–826.
46. Allen, F. (2002) *Acta Crystallogr. B*, **58**, 380–388.
47. Blake, A.J., Halcrow, M.A., and Schröder, M. (1994) *J. Chem. Soc., Dalton Trans.*, 1463–1470.
48. Blake, A.J., Gould, R.O., Halcrow, M.A., and Schröder, M. (1993) *J. Chem. Soc., Dalton Trans.*, 2909–2920.
49. Blake, A.J., Gould, R.O., Halcrow, M.A., Holder, A.J., Hyde, T.I., and Schröder, M. (1992) *J. Chem. Soc., Dalton Trans.*, 3427–3431.
50. van Gastel, M., Shaw, J.L., Blake, A.J., Flores, M., Schröder, M., McMaster, J., and Lubitz, W. (2008) *Inorg. Chem.*, **47**, 11688–11697.
51. Lubitz, W., Reijerse, E., and van Gastel, M. (2007) *Chem. Rev.*, **107**, 4331–4365.
52. Atkins, A.J., Blake, A.J., and Schröder, M. (1993) *J. Chem. Soc., Chem. Commun.*, 1662–1665.
53. Atkins, A.J., Black, D., Blake, A.J., Marin-Becerra, A., Parsons, S., Ruiz-Ramirez, L., and Schröder, M. (1996) *Chem. Commun.*, 457–464.
54. Kruger, H.J. and Holm, R.H. (1990) *J. Am. Chem. Soc.*, **112**, 2955–2963.
55. Angamuthu, R. and Bouwman, E. (2009) *Phys. Chem. Chem. Phys.*, **11**, 5578–5583.
56. Angamuthu, R., Kooijman, H., Lutz, M., Spek, A.L., and Bouwman, E. (2007) *Dalton Trans.*, 4641–4643.
57. Begum, A., Moula, G., and Sarkar, S. (2010) *Chem. Eur. J.*, **16**, 12324–12327.
58. Stein, M. (2011) *Chem. Eur. J.*, **17**, 15046–15048.
59. Sarkar, S. (2011) *Chem. Eur. J.*, **17**, 15049–15050.
60. Margulis, L. (1996) *Proc. Natl. Acad. Sci. U.S.A.*, **93**, 1071–1076.
61. Canaguier, S., Artero, V., and Fontecave, M. (2008) *Dalton Trans.*, 315–325.
62. Marr, A.C., Spencer, D.J.E., and Schröder, M. (2001) *Coord. Chem. Rev.*, **219**, 1055–1074.
63. Osterloh, F., Saak, W., Haase, D., and Pohl, S. (1997) *Chem. Commun.*, 979–980.
64. Lai, C.H., Reibenspies, J.H., and Darensbourg, M.Y. (1996) *Angew. Chem., Int. Ed. Engl.*, **35**, 2390–2393.
65. Zhu, W.F., Marr, A.C., Wang, Q., Neese, F., Spencer, D.J.E., Blake, A.J., Cooke, P.A., Wilson, C., and Schröder, M. (2005) *Proc. Natl. Acad. Sci. U.S.A.*, **102**, 18280–18285.
66. Tard, C. and Pickett, C.J. (2009) *Chem. Rev.*, **109**, 2245–2274.
67. Georgakaki, I.P., Thomson, L.M., Lyon, E.J., Hall, M.B., and Darensbourg, M.Y. (2003) *Coord. Chem. Rev.*, **238**, 255–266.
68. Darensbourg, M.Y., Lyon, E.J., and Smee, J.J. (2000) *Coord. Chem. Rev.*, **206**, 533–561.
69. Evans, D.J. and Pickett, C.J. (2003) *Chem. Soc. Rev.*, **32**, 268–275.
70. Osterloh, F., Saak, W., and Pohl, S. (1997) *J. Am. Chem. Soc.*, **119**, 5648–5656.
71. Stenson, P.A., Marin-Becerra, A., Wilson, C., Blake, A.J., McMaster, J., and Schröder, M. (2006) *Chem. Commun.*, 317–319.
72. Nunez, C., Bastida, R., Macias, A., Valencia, L., Ribas, J., Capelo, J.L., and Lodeiro, C. (2010) *Dalton Trans.*, **39**, 7673–7683.
73. Steenland, M.W.A., Lippens, W., Herman, G.G., and Goeminne, A.M. (1993) *Bull. Soc. Chim. Belg.*, **102**, 239–242.
74. Davies, S.C., Evans, D.J., Hughes, D.L., Longhurst, S., and Sanders, J.R. (1999) *Chem. Commun.*, 1935–1936.
75. Smith, M.C., Barclay, J.E., Cramer, S.P., Davies, S.C., Gu, W.W., Hughes, D.L., Longhurst, S., and Evans, D.J. (2002) *J. Chem. Soc., Dalton Trans.*, 2641–2647.
76. Jiang, J., Maruani, M., Solaimanzadeh, J., Lo, W., Koch, S., and Millar, M. (2009) *Inorg. Chem.*, **48**, 6359–6361.
77. Matsumoto, T., Nagahama, T., Cho, J., Hizume, T., Suzuki, M., and Ogo, S. (2011) *Angew. Chem. Int. Ed.*, **50**, ISSN: 1433–7851. 10578–10580.

78. Tsay, C. and Peters, J.C. (2012) *Chem. Sci.*, **3**, 1313–1318.
79. Breitenfeld, J., Scopelliti, R., and Hu, X.L. (2012) *Organometallics*, **31**, 2128–2136.
80. Barton, B.E., Whaley, C.M., Rauchfuss, T.B., and Gray, D.L. (2009) *J. Am. Chem. Soc.*, **131**, 6942–6943.
81. Della Pergola, R., Fumagalli, A., Garlaschelli, L., Manassero, C., Manassero, M., Sansoni, M., and Sironi, A. (2008) *Inorg. Chim. Acta*, **361**, 1763–1769.
82. Barton, B.E. and Rauchfuss, T.B. (2010) *J. Am. Chem. Soc.*, **132**, 14877–14885.
83. Carroll, M.E., Barton, B.E., Gray, D.L., Mack, A.E., and Rauchfuss, T.B. (2011) *Inorg. Chem.*, **50**, 9554–9563.
84. Schilter, D., Nilges, M.J., Chakrabarti, M., Lindahl, P.A., Rauchfuss, T.B., and Stein, M. (2012) *Inorg. Chem.*, **51**, 490–495.
85. Song, L.C., Li, Y.L., Li, L., Gu, Z.C., and Hu, Q.M. (2010) *Inorg. Chem.*, **49**, 10174–10182.
86. Sellmann, D., Geipel, F., Lauderbach, F., and Heinemann, F.W. (2002) *Angew. Chem. Int. Ed.*, **41**, 632–634.
87. Perra, A., Wang, Q., Blake, A.J., Davies, E.S., McMaster, J., Wilson, C., and Schröder, M. (2009) *Dalton Trans.*, 925–931.
88. Li, Z.L., Ohki, Y., and Tatsumi, K. (2005) *J. Am. Chem. Soc.*, **127**, 8950–8951.
89. Romao, C.V., Pereira, I.A.C., Xavier, A.V., Le Gall, J., and Teixeira, M. (1997) *Biochem. Biophys. Res. Commun.*, **240**, 75–79.
90. Tanino, S., Li, Z.L., Ohki, Y., and Tatsumi, K. (2009) *Inorg. Chem.*, **48**, 2358–2360.
91. Fichtner, C., Laurich, C., Bothe, E., and Lubitz, W. (2006) *Biochemistry*, **45**, 9706–9716.
92. Roseboom, W., De Lacey, A.L., Fernandez, V.M., Hatchikian, E.C., and Albracht, S.P.J. (2006) *J. Biol. Inorg. Chem.*, **11**, 102–118.
93. Ohki, Y., Yasumura, K., Ando, M., Shimokata, S., and Tatsumi, K. (2010) *Proc. Natl. Acad. Sci. U.S.A.*, **107**, 3994–3997.
94. Canaguier, S., Field, M., Oudart, Y., Pecaut, J., Fontecave, M., and Artero, V. (2010) *Chem. Commun.*, **46**, 5876–5878.
95. Sellmann, D., Lauderbach, F., Geipel, F., Heinemann, F.W., and Moll, M. (2004) *Angew. Chem. Int. Ed.*, **43**, 3141–3144.
96. Lauderbach, F., Prakash, R., Gotz, A.W., Munoz, M., Heinemann, F.W., Nickel, U., Hess, B.A., and Sellmann, D. (2007) *Eur. J. Inorg. Chem.*, **2007**, 3385–3393.
97. Wang, Q., Barclay, J.E., Blake, A.J., Davies, E.S., Evans, D.J., Marr, A.C., McInnes, E.J.L., McMaster, J., Wilson, C., and Schröder, M. (2004) *Chem. Eur. J.*, **10**, 3384–3396.
98. Perra, A., Davies, E.S., Hyde, J.R., Wang, Q., McMaster, J., and Schröder, M. (2006) *Chem. Commun.*, 1103–1105.
99. Seyferth, D., Henderson, R.S., and Song, L.C. (1982) *Organometallics*, **1**, 125–133.
100. Lozano, A.A., Santana, M.D., Garcia, G., Barclay, J.E., Davies, S.C., and Evans, D.J. (2005) *Z. Anorg. Allg. Chem.*, **631**, 2062–2066.
101. Rakowski DuBois, M. and DuBois, D.L. (2009) *Chem. Soc. Rev.*, **38**, 62–72.
102. Kramarz, K.W. and Norton, J.R. (1994) *Progress in Inorganic Chemistry*, vol. 42, John Wiley & Sons, Inc., New York, pp. 1–65.
103. van der Vlugt, J.I., Rauchfuss, T.B., Whaley, C.M., and Wilson, S.R. (2005) *J. Am. Chem. Soc.*, **127**, 16012–16013.
104. Nicolet, Y., De Lacey, A.L., Vernede, X., Fernandez, V.M., Hatchikian, E.C., and Fontecilla-Camps, J.C. (2001) *J. Am. Chem. Soc.*, **123**, 1596–1601.
105. Fan, H.J. and Hall, M.B. (2001) *J. Am. Chem. Soc.*, **123**, 3828–3829.
106. Greco, C., Bruschi, M., De Gioia, L., and Ryde, U. (2007) *Inorg. Chem.*, **46**, 5911–5921.
107. Berning, D.E., Miedaner, A., Curtis, C.J., Noll, B.C., Rakowski DuBois, M., and Dubois, D.L. (2001) *Organometallics*, **20**, 1832–1839.
108. Curtis, C.J., Miedaner, A., Ellis, W.W., and DuBois, D.L. (2002) *J. Am. Chem. Soc.*, **124**, 1918–1925.

109. Fraze, K., Wilson, A.D., Appel, A.M., Rakowski DuBois, M., and DuBois, D.L. (2007) *Organometallics*, **26**, 3918–3924.
110. Ellis, W.W., Raebiger, J.W., Curtis, C.J., Bruno, J.W., and DuBois, D.L. (2004) *J. Am. Chem. Soc.*, **126**, 2738–2743.
111. Curtis, C.J., Miedaner, A., Ciancanelli, R., Ellis, W.W., Noll, B.C., Rakowski DuBois, M., and DuBois, D.L. (2003) *Inorg. Chem.*, **42**, 216–227.
112. DuBois, D.L. and Bullock, R.M. (2011) *Eur. J. Inorg. Chem.*, **2011**, 1017–1027.
113. O'Hagan, M., Shaw, W.J., Raugei, S., Chen, S.T., Yang, J.Y., Kilgore, U.J., DuBois, D.L., and Bullock, R.M. (2011) *J. Am. Chem. Soc.*, **133**, 14301–14312.
114. Wilson, A.D., Shoemaker, R.K., Miedaner, A., Muckerman, J.T., DuBois, D.L., and Rakowski DuBois, M. (2007) *Proc. Natl. Acad. Sci. U.S.A.*, **104**, 6951–6956.
115. Wilson, A.D., Newell, R.H., McNevin, M.J., Muckerman, J.T., Rakowski DuBois, M., and DuBois, D.L. (2006) *J. Am. Chem. Soc.*, **128**, 358–366.
116. Pool, D.H. and DuBois, D.L. (2009) *J. Organomet. Chem.*, **694**, 2858–2865.
117. Yang, J.Y., Chen, S.T., Dougherty, W.G., Kassel, W.S., Bullock, R.M., DuBois, D.L., Raugei, S., Rousseau, R., Dupuis, M., and Rakowski DuBois, M. (2010) *Chem. Commun.*, **46**, 8618–8620.
118. Wiedner, E.S., Yang, J.Y., Dougherty, W.G., Kassel, W.S., Bullock, R.M., Rakowski DuBois, M., and DuBois, D.L. (2010) *Organometallics*, **29**, 5390–5401.
119. Helm, M.L., Stewart, M.P., Bullock, R.M., Rakowski DuBois, M., and DuBois, D.L. (2011) *Science*, **333**, 863–866.
120. Le Goff, A., Artero, V., Jousselme, B., Tran, P.D., Guillet, N., Metaye, R., Fihri, A., Palacin, S., and Fontecave, M. (2009) *Science*, **326**, 1384–1387.
121. Griveau, S., Mercier, D., Vautrin-Ul, C., and Chausse, A. (2007) *Electrochem. Commun.*, **9**, 2768–2773.
122. Sgobba, V. and Guldi, D.M. (2009) *Chem. Soc. Rev.*, **38**, 165–184.
123. Jousselme, B., Bidan, G., Billon, M., Goyer, C., Kervella, Y., Guillerez, S., Abou Hamad, E., Goze-Bac, C., Mevellec, J.Y., and Lefrant, S. (2008) *J. Electroanal. Chem.*, **621**, 277–285.
124. Pinson, J. and Podvorica, F. (2005) *Chem. Soc. Rev.*, **34**, 429–439.
125. Yang, J.Y., Bullock, R.M., Shaw, W.J., Twamley, B., Fraze, K., Rakowski DuBois, M., and DuBois, D.L. (2009) *J. Am. Chem. Soc.*, **131**, 5935–5945.
126. Yang, J.Y., Bullock, R.M., Dougherty, W.G., Kassel, W.S., Twamley, B., DuBois, D.L., and DuBois, M.R. (2010) *Dalton Trans.*, **39**, 3001–3010.
127. Kachmar, A., Vetere, V., Maldivi, P., and Franco, A.A. (2010) *J. Phys. Chem. A*, **114**, 11861–11867.
128. Jacobsen, G.M., Yang, J.Y., Twamley, B., Wilson, A.D., Bullock, R.M., Rakowski DuBois, M., and DuBois, D.L. (2008) *Energy Environ. Sci.*, **1**, 167–174.
129. Gloaguen, F., Lawrence, J.D., Rauchfuss, T.B., Benard, M., and Rohmer, M.M. (2002) *Inorg. Chem.*, **41**, 6573–6582.
130. Cheah, M.H., Borg, S.J., Bondin, M.I., and Best, S.P. (2004) *Inorg. Chem.*, **43**, 5635–5644.
131. Duan, L.L., Wang, M., Li, P., Wang, N., Wang, F.J., and Sun, L.C. (2009) *Inorg. Chim. Acta*, **362**, 372–376.
132. Han, Z., Qiu, F., Eisenberg, R., Holland, P.L., and Krauss, T.D. (2012) *Science*, **338**, 1321–1324.

4
[FeFe] Hydrogenase Models: an Overview

Ulf-Peter Apfel, François Y. Pétillon, Philippe Schollhammer, Jean Talarmin, and Wolfgang Weigand

4.1
Introduction

The depletion of fossil fuel sources and the detrimental impact of these fuels on the environment promote society to develop alternative energy forms. Even though renewable energy sources are already accessible from wind, sun, geothermy, and hydroelectricity, an effective storage and transportation system is required for the successful application of these alternative forms in our modern industrial society. Hydrogen is a possible candidate for such a source, storage, and transportation system, and allows for environmentally friendly combustion with a high heating value of 142 MJ kg^{-1} s [1] (oil: 40–50 MJ kg^{-1} [2]). An inexpensive and robust catalyst that allows for transformation of cheap proton sources (e.g., water) into hydrogen with negligible energy deficiency is therefore required. In contrast to current industrial processes [3–7], powerful enzymatic systems reversibly transform protons into hydrogen in nature. The enzymes capable of doing this transformation are called *hydrogenases*, which either avoid a surplus of reducing power by H_2 depletion or provide H_2 as an energy source [8]. Among the different forms of hydrogenases [9, 10], the [FeFe] hydrogenases efficiently generate hydrogen under mild conditions (6000–9000 molecules H_2 s^{-1} per site [10–12], −0.4 V vs saturated calomel electrode (SCE) at pH 7 [13, 14]). In the late twentieth century, the groups of Peters [15] and Fontecilla-Camps [16] independently unveiled the topology of the active site of the [FeFe] hydrogenase. It consists of a di-iron subsite attached to a [4Fe4S] cluster via a bridging cysteine. The di-iron site, named $[2Fe]_H$, is of an organometallic nature because it contains carbonyl and cyanide ligands, besides a bridging dithiolate. The entire ensemble, containing the [2Fe2S] active site and [4Fe4S] cluster is generally referred to as the *H-cluster*. The H-cluster, isolated from *Desulfovibrio desulfuricans*, is presented in Figure 4.1 [16].

The formation of this unique [2Fe2S] cluster and its attached ligands was the subject of multiple investigations [17–26]. Mössbauer, electron paramagnetic resonance (EPR), Fourier transform infrared (FTIR) spectroscopic experiments, and theoretical calculations were performed to elucidate the proton reduction

Bioinspired Catalysis: Metal-Sulfur Complexes, First Edition.
Edited by Wolfgang Weigand and Philippe Schollhammer.

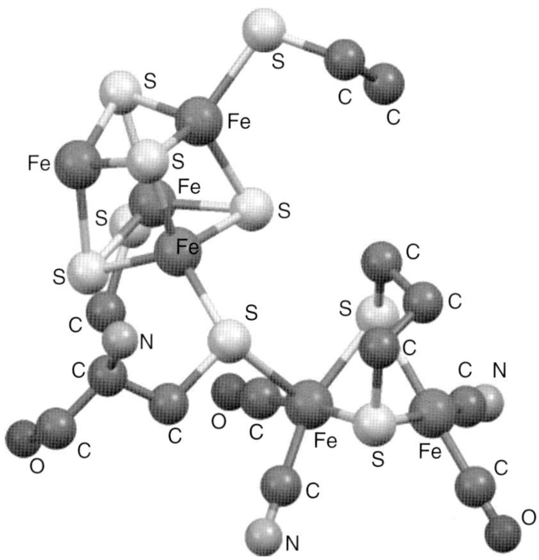

Figure 4.1 H-cluster of the [FeFe] hydrogenase as isolated from *Desulfovibrio desulfuricans* [16].

mechanism and the involved intermediates [27–36]. Recent reports strongly suggest that the dithiolate bridge is functionalized with an amine group, which is chemically relevant [37]. A simple, straightforward mechanism for the reversible enzymatic two-electron reduction process is presented in Scheme 4.1.

Notably, already in the early twentieth century, Reihlen [38], Hieber [39–42], and later, Huttner [43], King [44], Dahl and Wei [45], and Seyferth [46–50] described the synthesis of [2Fe2S] compounds showing structural analogy with the active site of the enzyme. Without the knowledge, however, of their biological relevance as models for the [FeFe] hydrogenases and their ability to form hydrogen, the proton reduction was not investigated. The discovery of the structure of the enzyme's active site and the subsequent findings of mechanistic details of the H_2 formation led to a renaissance of the [2Fe2S] cluster chemistry with a strong emphasis on their ability to generate hydrogen. Since the structural determination of the active site was reported, important progress has been made in this research field [9, 51]. The simplicity of the [2Fe2S] systems makes them an ideal target for the development of a novel proton reduction catalyst.

Because of the high number of publications in this field within the past 15 years, this chapter focuses especially on models for the mixed-valent H_{ox} state as well as protonation reactions of H_{red} models and on some significant insights into proton reduction chemistry. Additional insights on theoretical aspects and on mechanisms induced in biological hydrogen evolution or uptake are the subject of Chapters 6 and 7, respectively. The reader is also directed toward the references [51f,g] for insights concerning photochemical production of H_2 with di-iron models.

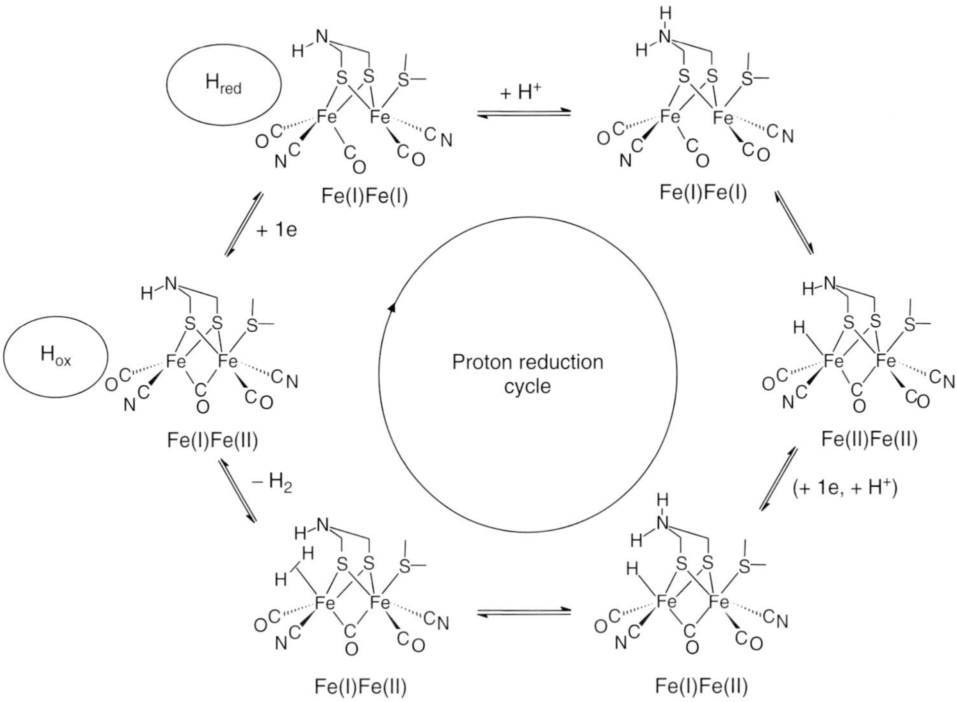

Scheme 4.1 Proposed catalytic mechanism for the consumption and formation of hydrogen (without global charges) [36] (Adapted from Ref [51h] © 2009, Elsevier.)

4.2 Synthetic Strategies toward [FeFe] Hydrogenase Model Complexes

Although numerous [FeFe] hydrogenase model complexes exist, synthetic access to iron-chalcogenolato compounds is limited to substitution reactions and oxidative additions (Scheme 4.2). The main challenge is therefore to synthesize the pertinent precursor molecules and to "pick the right pathway" to assemble the desired features. Typically, dithiols or disulfides and [$Fe_3(CO)_{12}$] react under reflux conditions to afford the corresponding hydrogenase mimics [52–56]. Likewise, chalcogenolato complexes with E = Se, Te are accessible via the same route with the respective chalcogen compounds [55–60]. If the precursors are unstable at elevated temperatures, [$Fe_3(CO)_{12}$] can be replaced by [$Fe_2(CO)_9$], enabling oxidative addition at room temperature [61, 62]. In a comparable pathway, dithiols/disulfides can be replaced by RECN (E = S, Se, Te) [59, 63]. A different synthetic pathway utilizes [$Fe_2(CO)_6(\mu\text{-}S_2)$] as starting material. The reaction with $LiEt_3BH$ generates a reactive dianionic intermediate, [$Fe_2(CO)_6(\mu\text{-}S)_2]^{2-}$, that reacts with halogenides [64–70] to give the desired [2Fe2S] cluster. Mannich reaction of [$Fe_2(CO)_6(\mu\text{-}SH)_2$] with formaldehyde and a secondary amine affords [$Fe_2(CO)_6$(μ-adt)] (adt (azadithiolate) = -$SCH_2N(R)CH_2S$-) [71]. In addition, replacement of one or two CO ligands by phosphines [72–81],

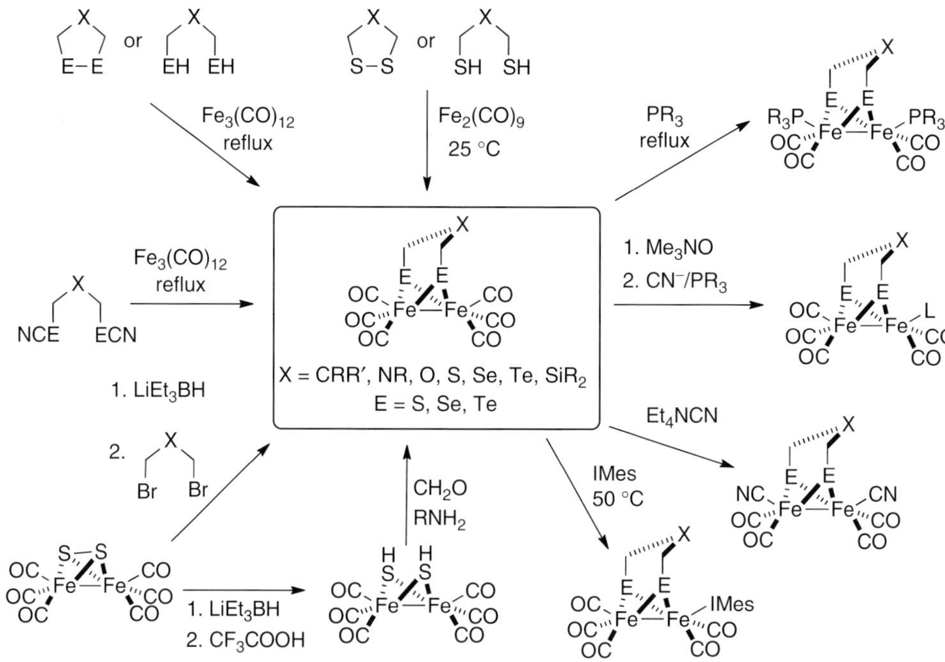

Scheme 4.2 Synthetic pathways toward [FeFe] hydrogenase mimics.

carbenes [82–92], and/or cyanides [66, 70, 93–96] allows tuning of the steric and electronic features of related clusters. The formation of mono- or symmetrically disubstituted derivatives depends both on the reaction conditions (temperature, presence of Me_3NO, etc.) and the sterical behaviors of the monodentate ligands [69, 70, 73, 83, 88, 96–99].

It is worth noting that only one model of the entire H-cluster was reported (Chart 4.1) [100].

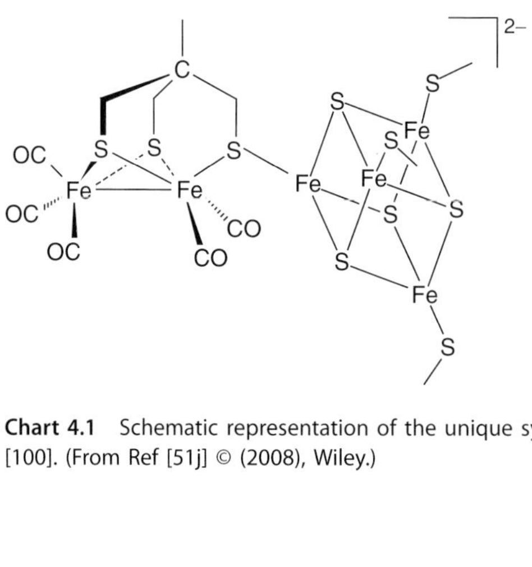

Chart 4.1 Schematic representation of the unique synthetic model of the entire H-cluster [100]. (From Ref [51j] © (2008), Wiley.)

4.3 Properties of Model Complexes

4.3.1 Biomimetic Models of the "Rotated State"

There is now a large consensus on the fact that H_{ox} is an important key intermediate in the catalytic cycle of the [FeFe] hydrogenase (Scheme 4.1). Notably, the H_{ox} state possesses a bridging carbonyl group and a mixed-valent $Fe^{I}Fe^{II}$ core. This state presents a specific structural feature named "rotated state" due to a formal rotation of one iron atom along the Fe–Fe bond to afford the bridging or semi-bridging carbonyl and causing a vacant coordination site on the same iron atom.

The first assembly containing a bridging carbonyl ligand revealing a structural conformity with the active site of the enzyme was reported by Pickett and coworkers [101]. The stability of this H_{ox} mimics was low, but increased when nitriles or mixtures of cyanides and phosphines replaced CO [102–104]. Of note, only one model complex exist and it comprises a moderately stable [2Fe3S] subsite structure with a bridging CO and two cyanide ligands (Chart 4.2a) [105, 106]. Contrary to the enzymatic system, this early mimics did not reveal a vacant coordination site, and a square pyramidal coordination environment of the distal iron atom was not observed.

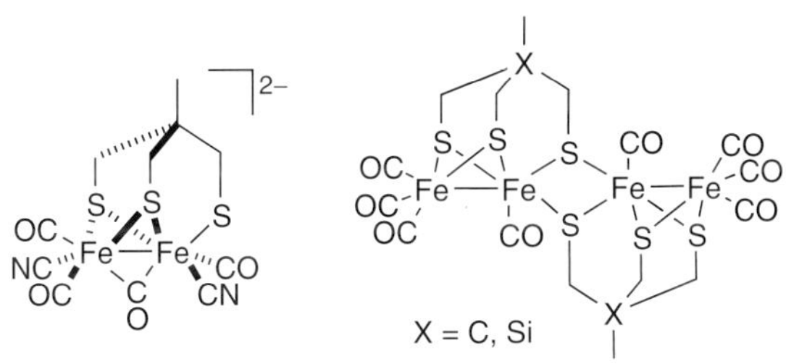

(a) (b)

Chart 4.2 Mixed-valent clusters (a) [105, 106] and (b) [55, 107].

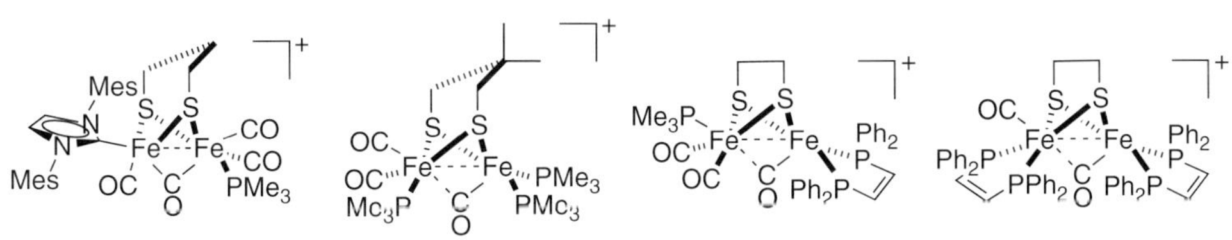

Chart 4.3 $Fe^{I}Fe^{II}$ systems featuring a rotated state [77, 88, 108, 109].

Original mixed-valent $Fe^{I}Fe^{II}\cdots Fe^{II}Fe^{I}$ tetra-iron complexes were also reported (Chart 4.2b) [55, 107] and were shown to be efficient catalysts for the generation of hydrogen. Density functional theory (DFT) calculations and spectroelectrochemical investigations showed that during the catalytic cycle a terminal CO group switches into a bridging mode and allows for the *in situ* formation of a rotated state [107]. The rate constant for the rate-determining step of the electrochemical H_2 formation was estimated to be $2000\,s^{-1}$. In comparison to this complex, $[Fe_2(CO)_6(\mu\text{-pdt})]$ (pdt, propyldithiolate) solely has a rate constant of $5\,s^{-1}$ for the electrochemical H_2 generation [135].

Recently, significant contributions of Darensbourg [88, 108] and Rauchfuss [77, 109] on mimicking the H_{ox} state were reported. Both groups observed stable H_{ox} model complexes when CO ligands were replaced with diphosphines or both monophosphines and carbenes (Chart 4.3). Especially, the need for a sterically bulky S-to-S linker in order to shield a vacant coordination site and to stabilize a "rotated state" structure was emphasized. In addition, the increased electron density introduced into the [FeFe] mimics by carbene or phosphine ligands leads to further stabilization of the mixed-valent "rotated state" and confirmed the previous computational predictions on the stabilization of H_{ox} models [110]. Although mixed-valent Fe^IFe^{II} species are accessible and play a crucial role in the enzymatic H_2 generation, current mimics predominantly possess a Fe^IFe^I center [9].

Sterical crowding of the bridge or use of nitrosyl ligands also induces a distorted geometry at one iron atom in Fe^I-Fe^I complexes [111], but to date there is only one example of a fully rotated Fe^I-Fe^I system that has been structurally characterized for the unsymmetrically disubstituted complex $[Fe_2(CO)_4(\kappa^2\text{-dmpe})\{\mu\text{-}(SCH_2)_2NBn\}]$ (dmpe = bis-dimethylphosphinoethane, Bn = benzyl) [112]. Theoretical investigations indicate that the presence of a remote agostic interaction together with desymmetrization of the di-iron system and the bulkiness of the dithiolate bridge allow stabilization of such a completely rotated geometry in the solid state (Chart 4.4). This very recent result opens up new perspectives for the design of synthetic di-iron dithiolates related to the active site of [FeFe] hydrogenases [112].

4.3.2 Electron Transfer in [FeFe] Hydrogenase Models

Depending on the experimental conditions and the nature of the catalyst, catalysis of proton reduction may be initiated by either a protonation or an electron-transfer step. When the latter situation occurs, it is essential to investigate the electrochemistry of the catalyst itself, as the reduction of metal–sulfur complexes is liable to produce M–S bond(s) cleavage that generates new sites for proton

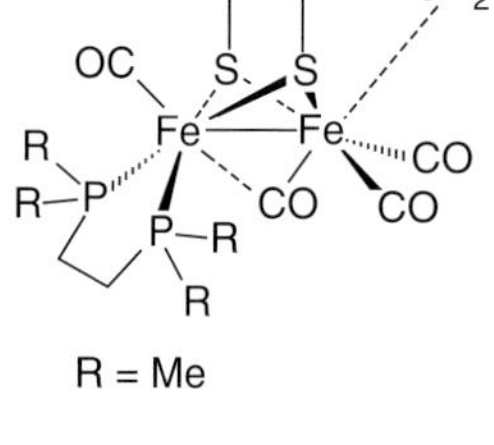

R = Me

Chart 4.4 Fully rotated $Fe^I–Fe^I$ system in solid state [112].

binding. Cyclic voltammetric studies of various $[Fe_2(CO)_6(\mu\text{-dithiolate})]$ complexes showed that the potential, the reversibility, and the electron stoichiometry of the reduction processes are highly dependent on the nature of the dithiolate ligand. Complexes with an aromatic ring or fused aromatic rings between the bridging sulfur atoms [113–118], with a SCH_2XCH_2S bridge where X is an amine NR ($R = CH_2CH_2OMe$, *i*Pr [117]), a SiR_2 group [55, 119], a sulfur atom [61], or complexes with a metalladithiolate bridge [120] were all shown to undergo a reversible or quasi-reversible single step, two-electron reduction. Complexes with Se, Te atoms [64], or phosphido groups [121, 122] at the bridging positions are also reduced in a quasi-reversible two-electron step. This arises from the fact that an electron-transfer-induced structural change thermodynamically favors the second electron transfer ($E°_2$) with respect to the first one ($E°_1$), a so-called potential inversion with $E°_2 - E°_1 > 0$ [123–125]. Theoretical investigations of the reduction processes by DFT allowed the structure change to be identified as the cleavage of a Fe–S bond and the shift of a CO ligand to a bridging position [55, 114, 118, 126, 127] (Scheme 4.3). An analog of the reduced form of the bdt-bridged complex (bdt = benzene dithiolate) was recently obtained and characterized [128].

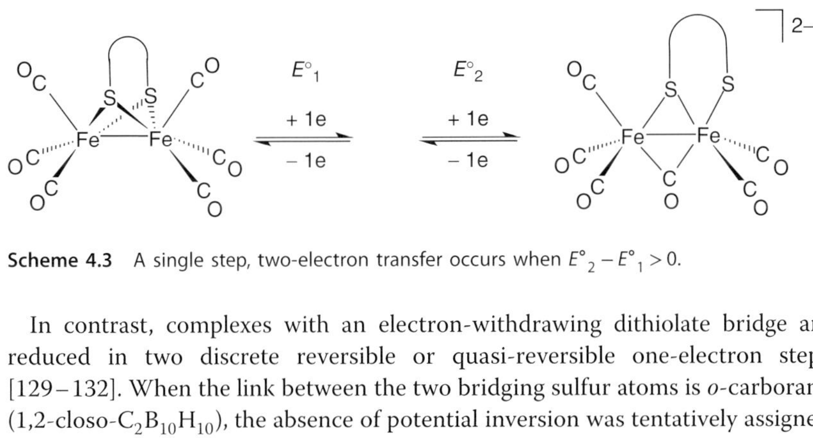

Scheme 4.3 A single step, two-electron transfer occurs when $E°_2 - E°_1 > 0$.

In contrast, complexes with an electron-withdrawing dithiolate bridge are reduced in two discrete reversible or quasi-reversible one-electron steps [129–132]. When the link between the two bridging sulfur atoms is *o*-carborane (1,2-closo-$C_2B_{10}H_{10}$), the absence of potential inversion was tentatively assigned to a partial delocalization of the negative charge resulting from the reduction on the carborane entity [129]. DFT calculations suggested that a Fe–S bond is cleaved in the one-electron reduced form of the complex with a biphenyl-2,2′-dithiolate bridge, but no indication was obtained for the formation of a CO bridge, and in this case, the structure change does not provoke a potential inversion [130]. The formation of a dimeric dianion was proposed to occur on a timescale longer than the cyclic voltammetric one [130].

The reduction of $[Fe_2(CO)_6(\mu\text{-pdt})]$ was reported to be a two-electron [133] or a one-electron process [134, 135] by cyclic voltammetry at moderate scan rates ($0.1\ V\ s^{-1}$), while two electrons are exchanged on longer timescales owing to the formation of daughter products [135–137]. Similar observations were also made for the edt analog ($edt = SCH_2CH_2S$) [135, 138], but for both $[Fe_2(CO)_6(\mu\text{-}S(CH_2)_nS)]$ complexes, fast-scan cyclic voltammetry simplified the reduction processes to a single one-electron step ($n = 2$, [138, 139]) or two successive one-electron transfers ($n = 3$, [126]).

Despite several theoretical investigations of the redox chemistry of simple model complexes [55, 114, 118, 121, 126, 127, 130, 138–140], the reasons for the different electron stoichiometries of their reduction processes are not fully understood.

4.3.3
Protonation Chemistry of [FeFe] Hydrogenase Models

The functioning of [FeFe] hydrogenases, at the molecular level, is closely related to the chemistry of metal-hydrides [35].Terminal hydrides are especially suggested to play a pivotal role during the enzymatic hydrogen generation [141, 142]. The protonation of the adt-bridge is also an important key feature of the enzyme's active site, allowing internal proton transfer from the adt-N to the [2Fe2S] core. Besides protonation at the di-iron site, the addition of protons to model complexes can lead to ligand protonated systems. The formation of a metal-hydride is thermodynamically preferred, whereas protonation reactions at the adt-linker are kinetically favored [143, 144]. Thus, the protonation of the adt-ligand [66, 68, 72, 143, 145–147], as well as other ligands [53, 55, 73, 148–151] is frequently observed for Fe^IFe^I models.

4.3.3.1 Hydride Formation

The direct protonation at the di-iron site requests a high electron density on the [2Fe2S] core. Protonated hexacarbonyl complexes $[Fe_2(CO)_6(\mu\text{-dithiolate})]$ (dithiolate ≠ adt) are commonly not accessible because of kinetic barriers for the direct protonation [51i]. Reduction to the Fe^0Fe^I or the Fe^0Fe^0 state is therefore mandatory to increase the basicity of the [2Fe2S] core and to form terminal Fe–H or bridging Fe–H–Fe complexes [51i]. Notably, direct protonation of $[Fe_2(CO)_6(\mu\text{-pdt})]$ in the presence of the strong acid $[SiEt_3][B(C_6F_5)_4]$/HCl was reported [152]. This system is, so far, the only reported example where direct protonation of a hexacarbonyl dithiolate complex in the Fe^IFe^I state was observed (dithiolate ≠ adt). High electron density at the di-iron core is commonly generated by replacement of CO ligands with electron-donating phosphines, carbenes, or cyanides [72, 103, 143, 145, 148, 149, 153–161]. The hydride species resulting from the protonation of these substituted complexes have higher stability compared to their unsubstituted counterparts, depending on the basicity of the di-iron site [162]. Typically, bridging-hydride resonances in ^1H NMR spectra of such derivatives are noted at high field, in about the range −8 to −20 ppm [103, 163]. Metal-hydride vibrations are generally weak and broad, and therefore are detected with difficulty in their spectra by IR spectroscopy; however, this method can be a powerful tool to characterize the related complexes via the carbonyl vibrations. In particular, the formation of a bridging-hydride ligand by protonation in these carbonyl di-iron derivatives clearly modifies the $\nu(CO)$ patterns, upon shifting the related peaks to higher wavenumbers (≈50 cm^{-1}) [53, 153, 164]. In other respects, numerous bridging-hydride compounds have been crystallographically characterized [79, 87, 145, 159]. Although the formation of bridging hydrides is common with di-iron molecules, biophysical and computational investigations on the enzyme suggest that such species are not pertinent in the biological processes [141, 142]. Indeed, these studies suggest that protons bind in a terminal and not a bridging mode. Notably, investigations on bridging hydrides revealed lower reactivity with protons and only a slow release of hydrogen [79, 103]. It is worth noting that bridging hydrides are reduced at lower potentials compared to their terminal analogs [79]. Terminal hydrides were only observed in a few model complexes comprising either an asymmetric substitution pattern or two sterically demanding diphosphines [79, 103, 160, 163, 165–171]. In contrast to their bridging counterparts, proton resonances of terminal hydrides were observed at about −3 to −5 ppm in such compounds. The formation of terminal hydrides was observed at low temperature during protonation processes of disymmetrically disubstituted diphosphine (Scheme 4.4) [165, 166] and bis N-heterocyclic carbene complexes [87]. When the temperature was raised, a rearrangement to the thermodynamically favored bridging mode was observed. The terminal/bridging-hydride isomerization is slowed down in the bis-chelated propanedithiolate compound $[Fe_2H(CO)_2(\kappa^2\text{-dppv})_2(\mu\text{-pdt})]^+$ (dppv = $Ph_2P(CH)_2PPh_2$) due to sterical clash between the propanedithiolate bridge and the diphosphine groups, and this terminal hydride is stable up to 20 °C [79].

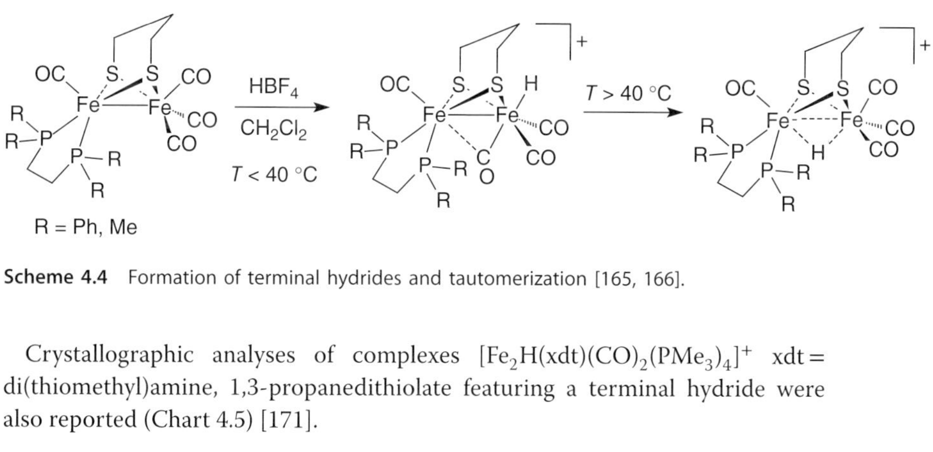

Scheme 4.4 Formation of terminal hydrides and tautomerization [165, 166].

Crystallographic analyses of complexes $[Fe_2H(xdt)(CO)_2(PMe_3)_4]^+$ xdt = di(thiomethyl)amine, 1,3-propanedithiolate featuring a terminal hydride were also reported (Chart 4.5) [171].

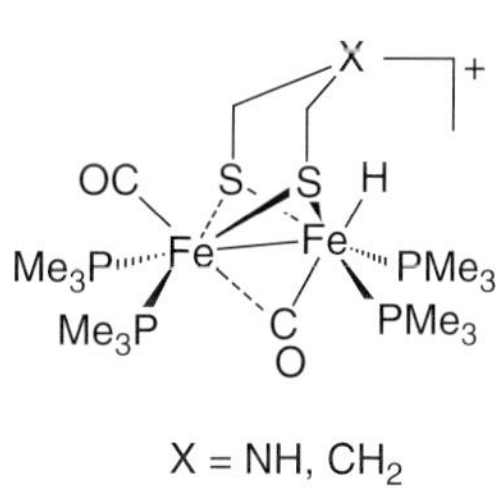

Chart 4.5 Structure of a terminal hydride diiron species.

4.3.3.2 Ligand Protonation and Proton Relays

Because the protonation at a metal center may be slow [172], particularly when isomerization processes are involved [149], numerous studies have been devoted to model complexes where a basic site was introduced in the second coordination sphere of the metals in order to speed up this reaction. Beyond efficiently capturing protons, the pendant base is expected to rapidly transfer them to the metal center(s). Different strategies have been used so far in di-iron dithiolate chemistry. The pendant base may consist of an NR function at the bridgehead position [66, 68, 72, 126, 143–145, 157, 173–184], as in the enzyme $[2Fe]_H$ subsite [180, 185, 186], or it may be introduced by means of a terminal or chelating ligand [53, 55, 73, 148–151, 187–190]. In the latter case, various amine-containing diphosphines ($R_2PCH_2NR'CH_2PR_2$ = PNP) were used following the pioneering studies of Rakowski DuBois and DuBois [191–194], showing the effectiveness of such tools for the production (or oxidation) of H_2 by mononuclear Ni or Fe complexes.

The use of chelating PNP ligands in di-iron dithiolate derivatives allows for the first objective to be achieved.[1)] Indeed, the protonation of these compounds is much faster than protonation of analogs containing ordinary chelating diphosphines [150, 151, 187–190]. However, the efficiency of the proton delivery to the metal center(s) is strongly influenced by the properties of the NR group of the PNP ligand and those of the bridge and other ligands, which impact the relative basicities of the pending base and the metal centers (Scheme 4.5) [150, 151, 190]. For a given bridging dithiolate, the ΔpK_a can be further modulated by adjusting the substituent R of the pendant amine [151, 190].

Similar observations were also made for adt-bridged complexes. N-protonation of the adt-ligand with a characteristic ^1H NMR resonance at about 5.9 ppm assigned to an ammonium species was reported [145]. The protonation of the adt-ligand hardly influences the υ(CO) vibrations and only a minor shift ($\approx 15\,cm^{-1}$) to higher wavenumbers was observed [146]. Even though the basicity of the adt-N is considerably lower, as expected for tertiary amines, protonation was already observed with moderately strong acids [143, 145, 146]. Theoretical computations revealed that the low basicity is a result of an orbital interaction between the adt-N lone-pair and the antibonding σ*(C–S) orbitals (Figure 4.2) [66] This interaction leads to a decrease in the electron density at the adt-N and to an increase in the C–S bond lengths, which was confirmed by crystallographic observations [145, 147].

While N-protonation is fast (even in the case of the hexacarbonyl species [66, 68, 71, 173–177]) if the ΔpK_a between the acid and the protonated amine [143] is sufficient, the occurrence of the N → Fe proton migration and the formation rate of the hydride complex are dependent on the relative basicities of both protonation sites. This tautomerization is fast in bis-diphosphine complexes with a NH or NCH_2Ph bridgehead group [79, 147, 170, 184]. In contrast, this tautomerization

1) However, it must be noted that the protonation of complexes where PNP is $R_2PN(R')PR_2$ leads to the bridging-hydride product, with no evidence of the formation of an ammonium intermediate, see Ref. 167 and references therein.

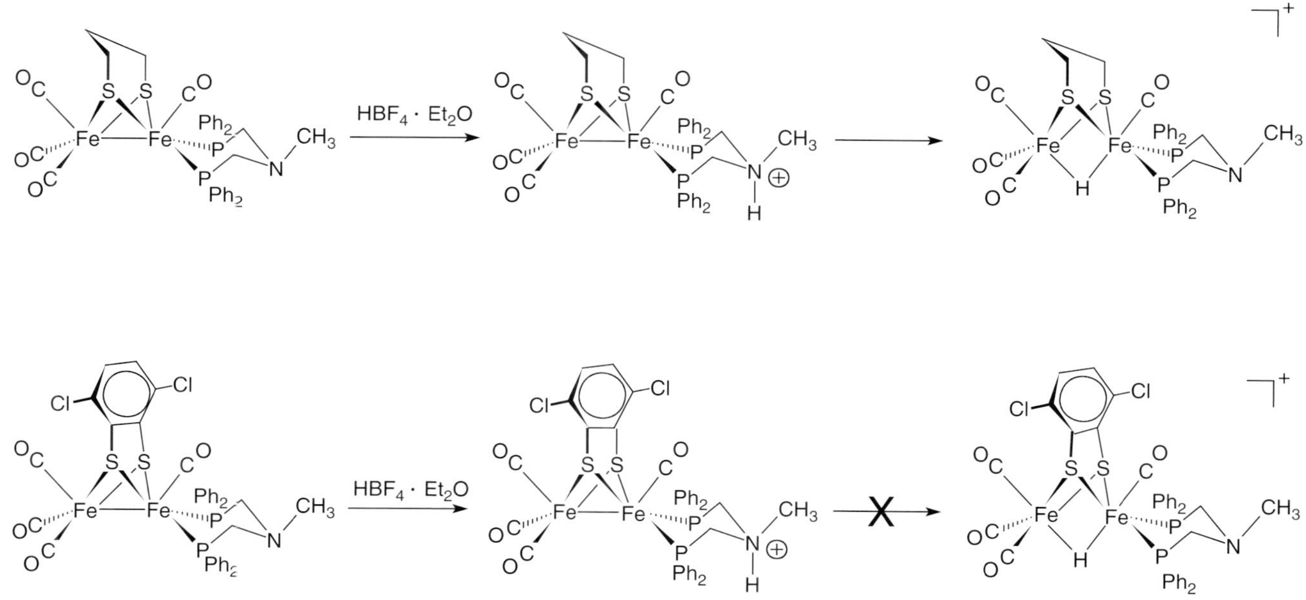

Scheme 4.5 Protonation of di-iron complexes with a pending base. (Adapted from Refs [150, 151, 190] © 2009, 2010, 2012 with permission from ACS, RSC and Wiley.)

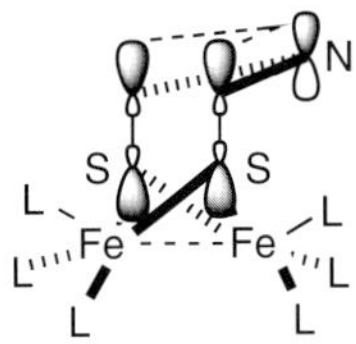

Figure 4.2 Hyperconjugation in adt-containing [FeFe] model complexes [66].

is not detected in an analog compound possessing only one chelating dppe and a *i*-Pr substituent at the nitrogen atom of the bridge (Scheme 4.6) [183].

In addition, migration of the N–H proton toward the metal center is affected by factors such as the nature of the anion of the acid [144, 147] or the solvent [150], which highlights the diversity of the parameters governing this key reaction. In the presence of a sufficiently strong acid, the N–H → Fe–H (or N–H → Fe(μ-H)Fe) migration may be followed by a new protonation of the pendant base [79, 147, 150, 188–190].[2)] Although this second protonation contributes to shift the reduction to less negative potentials (by about 0.2 V [144, 147, 187, 190]) with respect to the singly protonated precursor, it can make the reduction processes more complicated and the formation of a dicationic derivative is not necessarily an advantage from a kinetic point of view. Indeed, the reduction of the dication in the presence of acid may give rise to a low or inexistent catalytic current owing to the slow rate of the second protonation step [144, 190].

Importantly, recent structural characterization of a doubly protonated intermediate [147] (Chart 4.6) considerably supports the idea that the occurrence of a nitrogen atom in the dithiolate bridge together with the stabilization of the rotated geometry of one of the metal centers of the $[2Fe]_H$ subsite is vital to the efficiency of the biological catalyst.

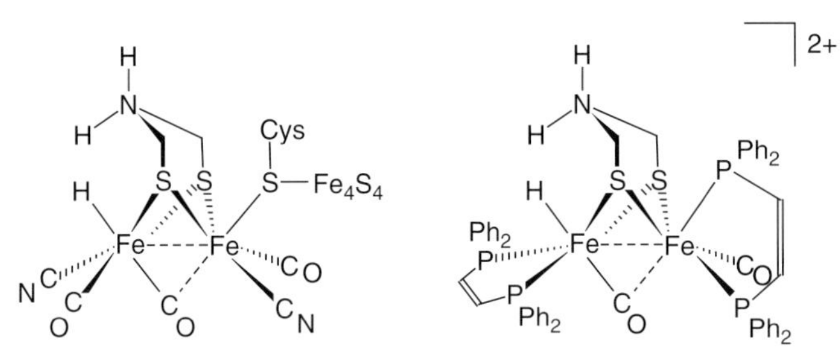

Chart 4.6 Representations of a postulated key-intermediate of the biological H^+/H_2 conversion (a, [185]) and an isolated diprotonated synthetic model (b, [147]).

2) Note that some complexes possessing two basic sites (at a terminal or chelating ligand or an aza-dithiolate bridge, and Fe–Fe bond) can be successively protonated at both sites [144, 145, 148]; the protonation order can be alternated by changing the acid [144, 145].

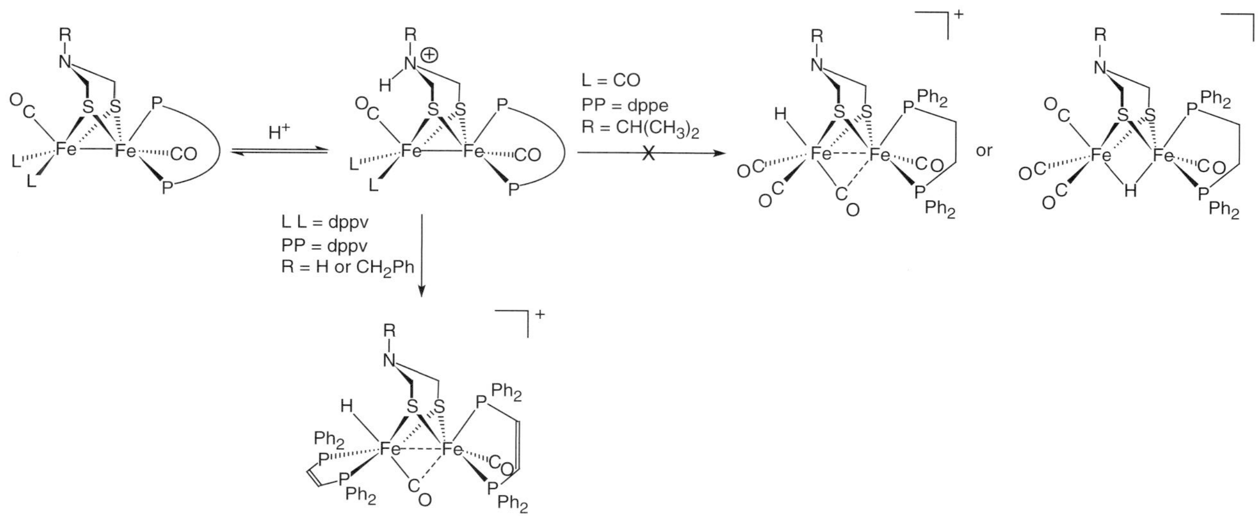

Scheme 4.6 Protonation processes in azadithiolate diiron compounds featuring chelated diphosphine.

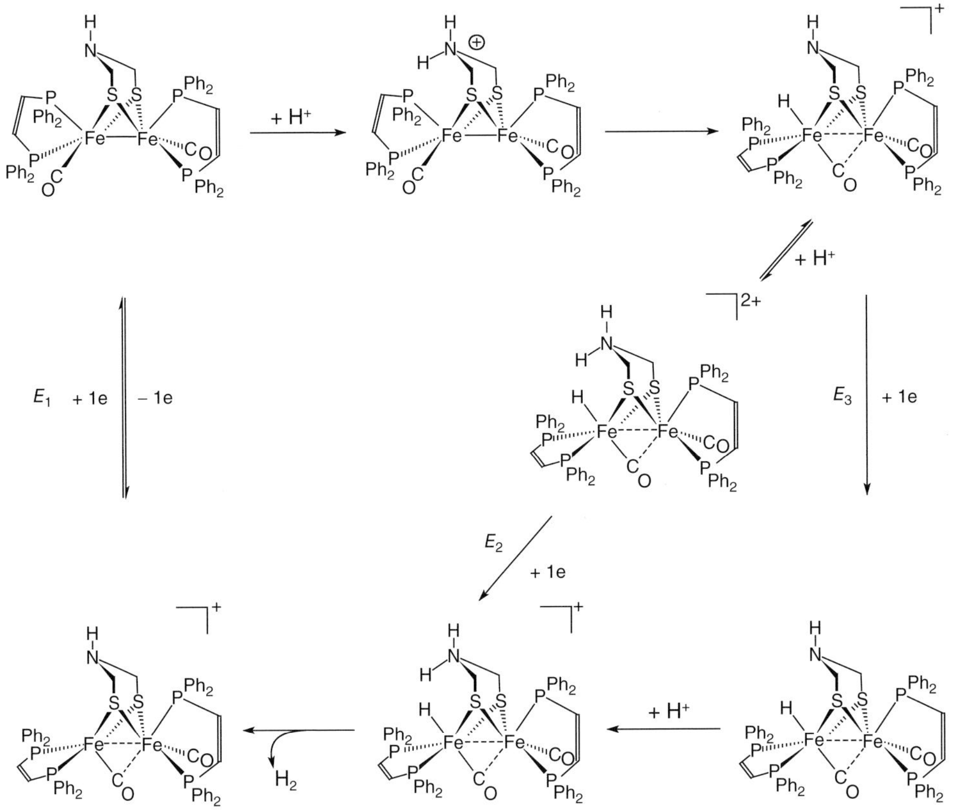

Scheme 4.7 Possible mechanism for the proton reduction electrocatalyzed by $[Fe_2(CO)_2(dppv)_2(\mu\text{-}SCH_2NHCH_2S)]$. (Adapted from Ref. [147] © 2012 with permission from American Chemical Society.)

Indeed, the rate constant of the catalytic reduction of CF_3CO_2H by this derivative ($k = 58\,000\ s^{-1}$ [147]) is among the highest reported thus far for various bioinspired electrocatalysts [195–199]. A possible mechanism for the electrocatalysis is illustrated in Scheme 4.7.

In case the acid is strong enough to protonate the N atom of the cationic hydride species and the potential negative enough to reduce the latter (i.e., applied potential $= E_2$), both the processes leading to the diprotonated cation might contribute to the electrocatalysis (E_2 and E_3 are −1.4 and −1.64 V vs Fc^+/Fc [147]). The release of H_2 from the diprotonated cation would generate the rotated, oxidized form of the starting material that should be instantly reduced at E_3, that is the reduction potential of the terminal hydride species.[3)]

3) $[Fe_2(CO)_2(dppv)_2(\mu\text{-pdt})]$ oxidizes at −0.86 V versus Fc^+/Fc [78]; it is likely that the adt complex oxidizes at a similar potential, so that the cation would be reduced at potential E_3 or E_2, the latter corresponding to the reduction of the doubly protonated dication.

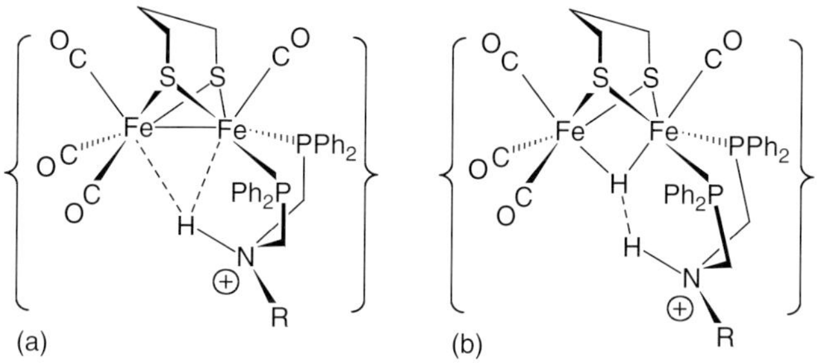

Chart 4.7 (a,b) Intermediates possibly involved in the reduction of protons by di-iron PNP complexes [150, 187–190].

In contrast, the di-iron derivatives possessing a PNP-type internal base are poorer electrocatalysts for proton reduction than the bis-dppv adt [190], even though the PNP chelate functions as an efficient proton relay and the reduction occurs at less negative potentials [150, 151, 187–190] than that of the bis-dppv species. The low catalytic efficacy of di-iron complexes with a PNP ligand may reside in the fact that, by construction, the NH proton is delivered to a bridging site (Chart 4.7a). Bridging hydrides are known to be less reactive toward protonation and harder to reduce than terminal hydrides [79, 103, 147]. The DFT analysis of the reduction processes of $[Fe_2(CO)_4(\mu\text{-pdt})(\kappa^2\text{-}Ph_2PCH_2NRCH_2PPh_2)]$ (R = Me, Ph) identified several possible pathways for the formation and release of H_2, but even in the case where intermediate b (Chart 4.7) was formed, a substantial barrier to the release of H_2 was found [190].

Numerous di-iron dithiolate complexes with a terminal hydride are known to rapidly isomerize to their thermodynamically more stable bridging counterparts [79, 87, 90, 147, 165–167, 170]. However, the mechanism proposed in Scheme 4.7 might still be valid if the reaction of the terminal hydride complex is faster than its isomerization, as suggested for the biological proton reduction process catalyzed by the [FeFe] hydrogenases [200].

Other parts of the molecule such as sulfur, CN^-, or pendant basic sites, however, might also play a crucial role within the catalytic mechanism of the hydrogen generation. In the absence of the adt-ligand, different protonation pathways were suggested. The protonation of CN^- in the anionic derivative $[Fe_2(CN)(CO)_4(PMe_3)(\mu\text{-pdt})]^-$ was reported but did not significantly alter catalytic H_2 generation and shifts of the CO bands of up to 15 cm^{-1} in the IR spectrum were observed [53, 149]. Recently, the bridging sulfur atoms were proposed to play a pivotal role as a possible proton-binding site in the enzymatic catalysis [201, 202]. Surprisingly, until now only a few S-protonated systems have been described (Chart 4.8) [55, 61, 73, 151, 171].

The formation of S–H protonated species, together with a bridging-hydride derivative, was proposed, on the basis of NMR and IR spectroscopy, to arise upon addition of HOTf to $[Fe_2(CO)_4(PMe_3)_2(\mu\text{-}\{CH_2\}_2NC_6H_4NO_2)]$ [73]. Electrochemical investigations on $[Fe_2(CO)_6(\mu\text{-sdt})]$ (sdt = -SCH_2SCH_2S-) suggested the likeliness of sulfur protonation but did not allow for a distinct

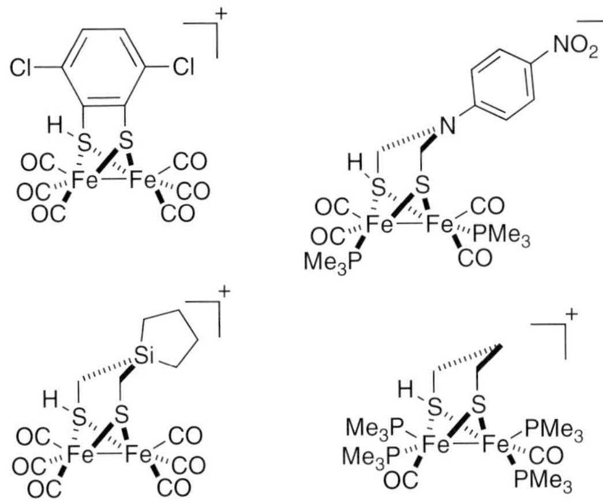

Chart 4.8 [FeFe] hydrogenase mimics comprising S–H protonation [55, 61, 73, 151, 171].

discrimination between the different sulfur atoms [61]. Subsequent studies [55, 151] further supported the hypothesis and provided spectroscopic evidence for S-protonation. On the basis of electrochemical investigations on $[Fe_2(CO)_6\{\mu\text{-}(SCH_2)_2Si(CH_2CH_2)_2\}]$, a participation of the formed S–H species in the catalytic hydrogen generation mechanism was proposed [55]. Computational studies, however, revealed that S–H formation is thermodynamically unfavored as opposed to the protonation of the metal centers [155]. The protonation chemistry of $[Fe_2(CO)_2(PMe_3)_4(\mu\text{-xdt})]$ (xdt = pdt or edt $\text{-}SCH_2CH_2S\text{-}$) was recently rationalized by involving a proton relay mechanism including S–H protonation [171]. The isolation and computational modeling of reduced intermediates, featuring a S-protonated thiolate, and bridging hydride, in catalytic proton reductions with $[Fe_2(CO)_6(\mu\text{-bdt})]$ (bdt = $C_6H_5S_2$) were also reported [127, 128]. Notably, S–H protonation is restricted to systems without other feasible proton-binding sites.

4.3.4
Water-Soluble Hydrogenase Mimics

The use of water as a proton source and as a solvent for the catalytic generation of hydrogen for industrial application is a highly desirable endeavor. The lack of solubility, however, of the current generation of [FeFe] hydrogenase mimics is a major drawback for these complexes to function as effective catalysts in water. As a consequence, only a few investigations report on the catalytic proton reduction processes in aqueous media. However, it has to be mentioned that the active site of the [FeFe] hydrogenase is deeply embedded within the protein in a hydrophobic environment and that the transport of water to the H-cluster is not biologically relevant.

It has been reported that the complex $[Fe_2(CO_5)(DAPTA)(\mu\text{-pdt})]$ (DAPTA, 3,7-diacetyl-1,3,7-triaza-5-phosphabicyclo[3.3.1]nonane) generates hydrogen in an acetonitrile/water mixture [203]. Subsequently, hydrogen formation was

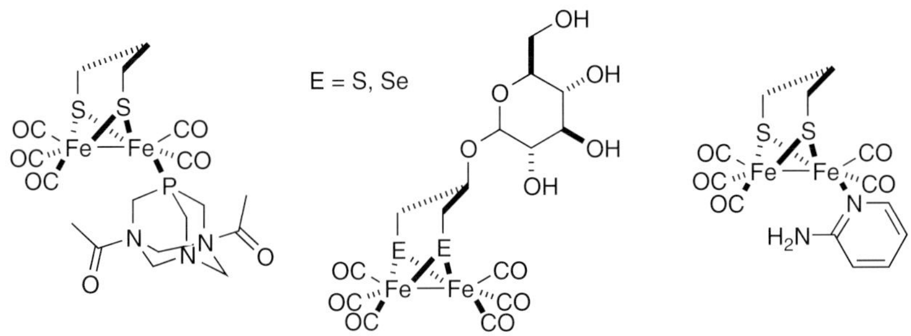

Chart 4.9 Water-reducing [FeFe] hydrogenase mimics [58, 203, 204].

also observed when $[Fe_2(CO_6)(\mu\text{-}E_2C_3H_5R)]$ (E = S, Se; R = glycoside) [58] or $[Fe_2(CO_5)(NH_2Py)(\mu\text{-pdt})]$ [204], respectively, were studied under similar electrocatalytic conditions (Chart 4.9). Although these systems allowed for the reduction of water, an organic solvent was required for their solubilization. Notably, higher stability of the compound $[Fe_2(CO_6)(\mu\text{-}E_2C_3H_5R)]$ was observed in water with the diselenolato bridge [58].

Solubilizing problems can be overcome by the use of modified electrodes [205–207]. With this goal, functionalized polyene-based polymers with a di-iron model complex were mounted on a multiwall carbon nanotube [205]. In addition, electrospun cellulose-acetate fibers with a di-iron species and polyvinyl pyrrolidine, have been prepared [206]. While the former system is instable, the second was fully operational in water. Compared to a Pt-disk electrode, electrodes containing the [FeFe] model complex revealed less catalytic efficiency. Previously, a dithiolate-bridged di-iron complex was grafted to an aminophenyl-modified electrode; unhappily, no electrochemical events that would be ascribed to the reduction of the di-iron complex were observed before the reduction of free protons from water [207]. In a host–guest approach, a sulfonate-group-containing complex was incorporated into β-cyclodextrin (Chart 4.10a) [208]. This assembly showed hydrogen generation in a pure aqueous environment but revealed problems associated with adsorption effects on the electrode. In a similar approach, a [FeFe] hydrogenase mimics and a Re(I)-complex were embedded into

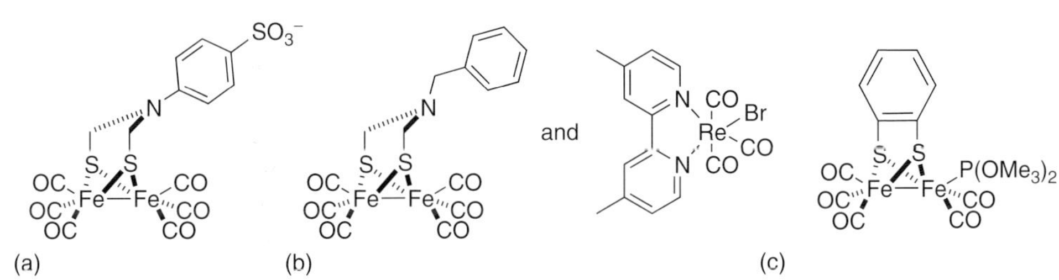

Chart 4.10 [FeFe] hydrogenase mimics utilized with host guest assemblies with β-cyclodextrins (a), in SDS micelles as photoactive system (b), and electrocatalytic system (c) [197, 198, 208, 209].

sodium dodecylsulfate (SDS) micelles in water (Chart 4.10b) [209]. The photo excitation of this construct with light ($\lambda > 400$ nm) for 1 h revealed hydrogen formation with a turnover number (TON) of 0.13. A similar system, in which $[Fe_2(CO)_5(P(OMe)_3)(\mu\text{-bdt})]$ was incorporated into SDS micelles, was also reported (Chart 4.10c) [197, 198]. Even though this system was deactivated in the presence of acetic acid, it was stable in a water/HCl mixture and showed catalytic turn over at <-0.7 V versus standard hydrogen electrode (SHE) at pH 3 with a catalytic efficiency of 4400 molecules $H_2\,s^{-1}$ [197]. Recent investigations on the catalytic activity of the adt model $[Fe_2(CO)_6(\mu\text{-}\{CH_2S\}_2NC_6H_4Br)]$ suggest that this species has one of the highest efficiencies (TON and TOF (turn over frequency)) in aqueous medium reported to date [196].

4.4 Conclusion

During the past two decades, significant development in modeling the active site of [FeFe] hydrogenases allowed to reveal the fundamental modes of action at the molecular level. The key structural features of the H-cluster have been identified and the challenge, which has to be faced, is to design efficient systems combining redox and base relay functionalities and having a rotated conformation in all the relevant redox active states at the same time. It is also now demonstrated that bioinspired di-iron electrocatalysts may have high activity for H_2 production.

References

1. Koroneos, C., Dompros, A., Roumbas, G., and Moussiopoulos, N. (2004) *Int. J. Hydrogen Energy*, **29**, 1443–1450.
2. Cookson, D.J. and Smith, B.E. (1992) *Energy Fuel*, **6**, 581–585.
3. Navarro, R.M., Peña, M.A., and Fierro, J.L.G. (2007) *Chem. Rev.*, **107**, 3952–3991.
4. Peña, M.A., Gómez, J.P., and Fierro, J.L.G. (1996) *Appl. Catal. Gen.*, **144**, 7–57.
5. Armor, J.N. (1999) *Appl. Catal. Gen.*, **176**, 159–176.
6. Mueller-Langer, F., Tzimas, E., Kaltschmitt, M., and Peteves, S. (2007) *Int. J. Hydrogen Energy*, **32**, 3797–3810.
7. Pilavachi, P.A., Chatzipanagi, A.I., and Spyropoulou, A.I. (2009) *Int. J. Hydrogen Energy*, **34**, 5294–5303.
8. Adams, M.W.W. and Stiefel, E.I. (1998) *Science*, **282**, 1842–1843.
9. Tard, C. and Pickett, C.J. (2009) *Chem. Rev.*, **109**, 2245–2274.
10. Das, D., Dutta, D., Nath, K., Meher Kotay, S., Das, A.K., and Veziroglu, T.N. (2006) *Curr. Sci.*, **90**, 1627–1637.
11. Adams, M.W.W. (1990) *Biochim. Biophys. Acta, Bioenerg.*, **1020**, 115–145.
12. Frey, M. (2002) *ChemBioChem*, **3**, 153–160.
13. Holm, R.H., Kennepohl, P., and Solomon, E.I. (1996) *Chem. Rev.*, **96**, 2239–2314.
14. Adams, M.W.W. and Mortenson, L.E. (1984) *J. Biol. Chem.*, **259**, 7045–7055.
15. Peters, J.W., Lanzilotta, W.N., Lemon, B.J., and Seefeldt, L.C. (1998) *Science*, **282**, 1853–1858.
16. Nicolet, Y., Piras, C., Legrand, P., Hatchikian, C.E., and Fontecilla-Camps, J.C. (1999) *Structure*, **7**, 13–23.
17. Pilet, E., Nicolet, Y., Mathevon, C., Douki, T., Fontecilla-Camps, J.C., and

Fontecave, M. (2009) *FEBS Lett.*, **583**, 506–511.
18. Peters, J.W., Szilagyi, R.K., Naumov, A., and Douglas, T. (2006) *FEBS Lett.*, **580**, 363–367.
19. Paschos, A., Glass, R.S., and Böck, A. (2001) *FEBS Lett.*, **488**, 9–12.
20. Reissmann, S., Hochleitner, E., Wang, H., Paschos, A., Lottspeich, F., Glass, R.S., and Böck, A. (2003) *Science*, **299**, 1067–1070.
21. Lenz, O., Zebger, I., Hamann, J., Hildebrandt, P., and Friedrich, B. (2007) *FEBS Lett.*, **581**, 3322–3326.
22. Watanabe, S., Matsumi, R., Arai, T., Atomi, H., Imanaka, T., and Miki, K. (2007) *Mol. Cell*, **27**, 29–40.
23. Roseboom, W., Blokesch, M., Böck, A., and Albracht, S.P.J. (2005) *FEBS Lett.*, **579**, 469–472.
24. Driesener, R.C., Challand, M.R., McGlynn, S.E., Shepard, E.M., Boyd, E.S., Broderick, J.B., Peters, J.W., and Roach, P.L. (2010) *Angew. Chem. Int. Ed.*, **49**, 1687–1690.
25. Blokesch, M., Albracht, S.P.J., Matzanke, B.F., Drapal, N.M., Jacobi, A., and Böck, A. (2004) *J. Mol. Biol.*, **344**, 155–167.
26. Happe, R.P., Roseboom, W., Pierik, A.J., Albracht, S.P.J., and Bagley, K.A. (1997) *Nature*, **385**, 126.
27. Albracht, S.J., Roseboom, W., and Hatchikian, E.C. (2006) *J. Biol. Inorg. Chem.*, **11**, 88–101.
28. (a) Wang, G., Benecky, M.J., Huynh, B.H., Cline, J.F., Adams, M.W.W., Mortenson, L.E., Hoffman, B.M., and Münck, E. (1984) *J. Biol. Chem.*, **259**, 14328–14331. (b) Rusnak, F.M., Adams, M.W.W., Mortenson, L.E., and Münck, E. (1987) *J. Biol. Chem.*, **262**, 38–41.
29. Popescu, C.V. and Münck, E. (1999) *J. Am. Chem. Soc.*, **121**, 7877–7884.
30. Pierik, A.J., Hagen, W.R., Dunham, W.R., and Sands, R.H. (1992) *Eur. J. Biochem.*, **206**, 705–719.
31. Münck, E. and Popescu, C. (2000) *Hyperfine Interact.*, **126**, 59–67.
32. Silakov, A., Reijerse, E.J., Albracht, S.P.J., Hatchikian, E.C., and Lubitz, W. (2007) *J. Am. Chem. Soc.*, **129**, 11447–11458.
33. Pereira, A.S., Tavares, P., Moura, I., Moura, J.J.G., and Huynh, B.H. (2001) *J. Am. Chem. Soc.*, **123**, 2771–2782.
34. Lubitz, W., Reijerse, E., and Van Gastel, M. (2007) *Chem. Rev.*, **107**, 4331–4365.
35. Adamska, A., Silakov, A., Lambertz, C., Rüdiger, O., Happe, T., Reijerse, E., and Lubitz, W. (2012) *Angew. Chem. Int. Ed.*, **51**, 11458–11462.
36. (a) Greco, C., Bruschi, M., De Gioia, L., and Ryde, U. (2007) *Inorg. Chem.*, **46**, 5911–5921. (b) Liu, Z.P. and Hu, P. (2002) *J. Am. Chem. Soc.*, **124**, 5175–5182. (c) Liu, Z.P. and Hu, P. (2002) *J. Chem. Phys.*, **117**, 8177–8180 (d) Fan, H.F. and Hall, M.B. (2001) *J. Am. Chem. Soc.*, **123**, 3828–3829.
37. (a) Berggren, G., Adamska, A., Lambertz, C., Simmons, T.R., Esselborn, J., Atta, M., Gambarelli, S., Mouesca, J.M., Reijerse, E., Lubitz, W., Happe, T., Artero, V., and Fontecave, M. (2013) *Nature*, **499**, 66–70. (b) Silakov, A., Wenk, B., Reijerse, E., and Lubitz, W. (2009) *Phys. Chem. Chem. Phys.*, **31**, 6592–6599.
38. Reihlen, H., Friedolsheim, A.V., and Oswald, W. (1928) *Justus Liebigs Ann. Chem.*, **465**, 72–96.
39. Hieber, W. and Spacu, P. (1937) *Z. Anorg. Allg. Chem.*, **233**, 353–364.
40. Hieber, W. and Scharfenberg, C. (1940) *Ber. Dtsch. Chem. Ges. A/B*, **73**, 1012–1021.
41. Hieber, W. and Gruber, J. (1958) *Z. Anorg. Allg. Chem.*, **296**, 91–103.
42. Hieber, W. and Beck, W. (1960) *Z. Anorg. Allg. Chem.*, **305**, 265–273.
43. Winter, A., Zsolnai, L., and Huttner, G. (1982) *Chem. Ber.*, **115**, 1286–1304.
44. King, R.B. (1962) *J. Am. Chem. Soc.*, **84**, 2460.
45. Dahl, L.F. and Wei, C.-H. (1963) *Inorg. Chem.*, **2**, 328–333.
46. Seyferth, D. and Henderson, R.S. (1979) *J. Am. Chem. Soc.*, **101**, 508–509.
47. Seyferth, D., Song, L.-C., and Henderson, R.S. (1981) *J. Am. Chem. Soc.*, **103**, 5103–5107.
48. Seyferth, D., Henderson, R.S., and Song, L.C. (1982) *Organometallics*, **1**, 125–133.

49. Seyferth, D., Womack, G.B., and Dewan, J.C. (1985) *Organometallics*, **4**, 398–400.
50. Cowie, M., DeKock, R.L., Wagenmaker, T.R., Seyferth, D., Henderson, R.S., and Gallagher, M.K. (1989) *Organometallics*, **8**, 119–132.
51. The reader is also directed towards the following reviews: (a) Wang, N., Wang, M., Chen, L., and Sun, L. (2013) *Dalton Trans.*, **42**, 12059–12071. (b) Darensbourg, M.Y. and Weigand, W. (2011) *Eur. J. Inorg. Chem.*, 994–1004. (c) Tschierlei, S., Ott, S., and Lomoth, R. (2011) *Energy Environ. Sci.*, **4**, 2340–2352. (d) Barton, B.E., Olsen, M.T., and Rauchfuss, T.B. (2010) *Curr. Opin. Biotechnol.*, **21**, 292–297. (e) Gloaguen, F. and Rauchfuss, T.B. (2009) *Chem. Soc. Rev.*, **38**, 100–108. (f) Ott, S. and Lomoth, R. (2009) *Dalton Trans.*, 9952–9959. (g) Wang, M., Chen, L., Li, X., and Sun, L. (2011) *Dalton Trans.*, **40**, 12793–12800. (h) Capon, J.-F., Gloaguen, F., Pétillon, F.Y., Schollhammer, P., and Talarmin, J. (2009) *Coord. Chem. Rev.*, **253**, 1476–1494. (i) Felton, G.A.N., Mebi, C.A., Petro, B.J., Vannucci, A.K., Evans, D.H., Glass, R.S., and Lichtenberger, D.L. (2009) *J. Organomet. Chem.*, **694**, 2681–2699. (j) Capon, J.F., Gloaguen, F., Pétillon, F.Y., Schollhammer, P., and Talarmin, J. (2008) *Eur. J. Inorg. Chem.*, 4671–4681. (k) Capon, J.-F., Gloaguen, F., Pétillon, F.Y., Schollhammer, P., and Talarmin, J. (2008) *C.R. Chim.*, **11**, 842–851. (l) Best, S. (2005) *Coord. Chem. Rev.*, **249**, 1536–1554. (m) Bruschi, M., Zampella, G., Fantucci, P., and De Gioia, L. (2005) *Coord. Chem. Rev.*, **249**, 1620–1640. (n) Capon, J.-F., Gloaguen, F., Schollhammer, P., and Talarmin, J. (2005) *Coord. Chem. Rev.*, **249**, 1664–1676. (o) Sun, L., Åkermark, B., and Ott, S. (2005) *Coord. Chem. Rev.*, **249**, 1653–1663. (p) Evans, D.J. and Pickett, C.J. (2003) *Chem. Soc. Rev.*, **32**, 268–275. (q) Georgakaki, I.P., Thomson, L.M., Lyon, E.J., Hall, M.B., and Darensbourg, M.Y. (2003) *Coord. Chem. Rev.*, **238-239**, 255–266. (r) Darensbourg, M.Y., Lyon, E.J., and Smee, J.J. (2000) *Coord. Chem. Rev.*, **206-207**, 533–561.
52. Gloaguen, F., Lawrence, J.D., Schmidt, M., Wilson, S.R., and Rauchfuss, T.B. (2001) *J. Am. Chem. Soc.*, **123**, 12518–12527.
53. Gloaguen, F., Lawrence, J.D., Rauchfuss, T.B., Bénard, M., and Rohmer, M.-M. (2002) *Inorg. Chem.*, **41**, 6573–6582.
54. Song, L.-C., Yang, Z.-Y., Hua, Y.-J., Wang, H.-T., Liu, Y., and Hu, Q.-M. (2007) *Organometallics*, **26**, 2106–2110.
55. Apfel, U.-P., Troegel, D., Halpin, Y., Tschierlei, S., Uhlemann, U., Goerls, H., Schmidt, M., Popp, J., Dunne, P., Venkatesan, M., Coey, M., Rudolph, M., Vos, J.G., Tacke, R., and Weigand, W. (2010) *Inorg. Chem.*, **49**, 10117–10132.
56. Harb, M.K., Windhager, J., Niksch, T., Görls, H., Sakamoto, T., Smith, E.R., Glass, R.S., Lichtenberger, D.L., Evans, D.H., El-khateeb, M., and Weigand, W. (2012) *Tetrahedron*, **68**, 10592–10599.
57. Gao, S., Fan, J., Sun, S., Peng, X., Zhao, X., and Hou, J. (2008) *Dalton Trans.*, 2128–2135.
58. Apfel, U.-P., Halpin, Y., Gottschaldt, M., Goerls, H., Vos, J.G., and Weigand, W. (2008) *Eur. J. Inorg. Chem.*, 5112–5118.
59. Harb, M.K., Niksch, T., Windhager, J., Görls, H., Holze, R., Lockett, L.T., Okumura, N., Evans, D.H., Glass, R.S., Lichtenberger, D.L., El-khateeb, M., and Weigand, W. (2009) *Organometallics*, **28**, 1039–1048.
60. Song, L.-C., Gai, B., Wang, H.-T., and Hu, Q.-M. (2009) *J. Inorg. Biochem.*, **103**, 805–812.
61. Windhager, J., Rudolph, M., Bräutigam, S., Görls, H., and Weigand, W. (2007) *Eur. J. Inorg. Chem.*, 2748–2760.
62. Windhager, J., Görls, H., Petzold, H., Mloston, G., Linti, G., and Weigand, W. (2007) *Eur. J. Inorg. Chem.*, 4462–4471.
63. Harb, M.K., Apfel, U.-P., Kuebel, J., Görls, H., Felton, G.A.N., Sakamoto, T., Evans, D.H., Glass, R.S., Lichtenberger, D.L., El-khateeb, M., and Weigand, W. (2009) *Organometallics*, **28**, 6666–6675.
64. Apfel, U.-P., Görls, H., Felton, G.A.N., Evans, D.H., Glass, R.S., Lichtenberger, D.L., and Weigand, W. (2012) *Helv. Chim. Acta*, **95**, 2168–2175.

65. Lawrence, J.D., Li, H., and Rauchfuss, T.B. (2001) *Chem. Commun.*, 1482–1483.
66. Lawrence, J.D., Li, H., Rauchfuss, T.B., Bénard, M., and Rohmer, M.-M. (2001) *Angew. Chem. Int. Ed.*, **40**, 1768–1771.
67. Ott, S., Kritikos, M., Åkermark, B., and Sun, L. (2003) *Angew. Chem. Int. Ed.*, **42**, 3285–3288.
68. Ott, S., Kritikos, M., Åkermark, B., Sun, L., and Lomoth, R. (2004) *Angew. Chem. Int. Ed.*, **43**, 1006–1009.
69. Ott, S., Borgström, M., Kritikos, M., Lomoth, R., Bergquist, J., Åkermark, B., Hammarström, L., and Sun, L. (2004) *Inorg. Chem.*, **43**, 4683–4692.
70. Song, L.-C., Yang, Z.-Y., Bian, H.-Z., and Hu, Q.-M. (2004) *Organometallics*, **23**, 3082–3084.
71. Li, H. and Rauchfuss, T.B. (2002) *J. Am. Chem. Soc.*, **124**, 726–727.
72. Schwartz, L., Eilers, G., Eriksson, L., Gogoll, A., Lomoth, R., and Ott, S. (2006) *Chem. Commun.*, 520–522.
73. Dong, W., Wang, M., Liu, X., Jin, K., Li, G., Wang, F., and Sun, L. (2006) *Chem. Commun.*, 305–307.
74. Löscher, S., Schwartz, L., Stein, M., Ott, S., and Haumann, M. (2007) *Inorg. Chem.*, **46**, 11094–11105.
75. Wang, Z., Liu, J.-H., He, C.-J., Jiang, S., Åkermark, B., and Sun, L.-C. (2007) *J. Organomet. Chem.*, **692**, 5501–5507.
76. Si, G., Wang, W.-G., Wang, H.-Y., Tung, C.-H., and Wu, L.-Z. (2008) *Inorg. Chem.*, **47**, 8101–8111.
77. Justice, A.K., Nilges, M.J., Rauchfuss, T.B., Wilson, S.R., De Gioia, L., and Zampella, G. (2008) *J. Am. Chem. Soc.*, **130**, 5293–5301.
78. Justice, A.K., De Gioia, L., Nilges, M.J., Rauchfuss, T.B., Wilson, S.R., and Zampella, G. (2008) *Inorg. Chem.*, **47**, 7405–7414.
79. Barton, B.E. and Rauchfuss, T.B. (2008) *Inorg. Chem.*, **47**, 2261–2263.
80. Song, L.-C., Li, C.-G., Gao, J., Yin, B.-S., Luo, X., Zhang, X.-G., Bao, H.-L., and Hu, Q.-M. (2008) *Inorg. Chem.*, **47**, 4545–4553.
81. Wang, N., Wang, M., Liu, T., Li, P., Zhang, T., Darensbourg, M.Y., and Sun, L. (2008) *Inorg. Chem.*, **47**, 6948–6955.
82. Tye, J.W., Lee, J., Wang, H.-W., Mejia-Rodriguez, R., Reibenspies, J.H., Hall, M.B., and Darensbourg, M.Y. (2005) *Inorg. Chem.*, **44**, 5550–5552.
83. Capon, J.-F., El Hassnaoui, S., Gloaguen, F., Schollhammer, P., and Talarmin, J. (2005) *Organometallics*, **24**, 2020–2022.
84. Duan, L., Wang, M., Li, P., Na, Y., Wang, N., and Sun, L. (2007) *Dalton Trans.*, 1277–1283.
85. Jiang, S., Liu, J., Shi, Y., Wang, Z., Åkermark, B., and Sun, L. (2007) *Polyhedron*, **26**, 1499–1504.
86. Liu, T. and Darensbourg, M.Y. (2007) *J. Am. Chem. Soc.*, **129**, 7008–7009.
87. Morvan, D., Capon, J.-F., Gloaguen, F., Le Goff, A., Marchivie, M., Michaud, F., Schollhammer, P., Talarmin, J., Yaouanc, J.-J., Pichon, R., and Kervarec, N. (2007) *Organometallics*, **26**, 2042–2052.
88. Thomas, C.M., Liu, T., Hall, M.B., and Darensbourg, M.Y. (2008) *Inorg. Chem.*, **47**, 7009–7024.
89. Thomas, C.M., Liu, T., Hall, M.B., and Darensbourg, M.Y. (2008) *Chem. Commun.*, 1563–1565.
90. Morvan, D., Capon, J.-F., Gloaguen, F., Pétillon, F.Y., Schollhammer, P., Talarmin, J., Yaouanc, J.-J., Michaud, F., and Kervarec, N. (2009) *J. Organomet. Chem.*, **694**, 2801–2807.
91. Song, L.-C., Luo, X., Wang, Y.-Z., Gai, B., and Hu, Q.-M. (2009) *J. Organomet. Chem.*, **694**, 103–112.
92. Song, L.-C., Gao, W., Luo, X., Wang, Z.-X., Sun, X.-J., and Song, H.-B. (2012) *Organometallics*, **31**, 3324–3332.
93. Le Cloirec, A., Best, S.P., Borg, S., Davies, S.C., Evans, D.J., Hughes, D.L., and Pickett, C.J. (1999) *Chem. Commun.*, 2285–2286.
94. Lyon, E.J., Georgakaki, I.P., Reibenspies, J.H., and Darensbourg, M.Y. (1999) *Angew. Chem. Int. Ed.*, **38**, 3178–3180.
95. Schmidt, M., Contakes, S.M., and Rauchfuss, T.B. (1999) *J. Am. Chem. Soc.*, **121**, 9736–9737.
96. Apfel, U.-P., Halpin, Y., Goerls, H., Vos, J.G., and Weigand, W. (2011) *Eur. J. Inorg. Chem.*, 581–588.
97. Mejia-Rodriguez, R., Chong, D., Reibenspies, J.H., Soriaga, M.P., and

Darensbourg, M.Y. (2004) *J. Am. Chem. Soc.*, **126**, 12004–12014.
98. Li, P., Wang, M., He, C., Li, G., Liu, X., Chen, C., Åkermark, B., and Sun, L. (2005) *Eur. J. Inorg. Chem.*, 2506–2513.
99. Lounissi, S., Capon, J.-F., Gloaguen, F., Matoussi, F., Pétillon, F.Y., Schollhammer, P., and Talarmin, J. (2011) *Chem. Commun.*, **47**, 878–880.
100. Tard, C., Liu, X.M., Ibrahim, S.K., Bruschi, M., De Gioia, L., Davies, S.C., Yang, X., Wang, L.S., Sawers, G., and Pickett, C.J. (2005) *Nature*, **433**, 610–613.
101. Razavet, M., Davies, S.C., Hughes, D.L., and Pickett, C.J. (2001) *Chem. Commun.*, 847–848.
102. Boyke, C.A., Rauchfuss, T.B., Wilson, S.R., Rohmer, M.-M., and Bénard, M. (2004) *J. Am. Chem. Soc.*, **126**, 15151–15160.
103. Van der Vlugt, J.I., Rauchfuss, T.B., Whaley, C.M., and Wilson, S.R. (2005) *J. Am. Chem. Soc.*, **127**, 16012–16013.
104. Van der Vlugt, J.I., Rauchfuss, T.B., and Wilson, S.R. (2006) *Chem. Eur. J.*, **12**, 90–98.
105. Razavet, M., Borg, S.J., George, S.J., Best, S.P., Fairhurst, S.A., and Pickett, C.J. (2002) *Chem. Commun.*, 700–701.
106. George, S.J., Cui, Z., Razavet, M., and Pickett, C.J. (2002) *Chem. Eur. J.*, **8**, 4037–4046.
107. Tard, C., Liu, X., Hughes, D.L., and Pickett, C.J. (2005) *Chem. Commun.*, 133–135.
108. Singleton, M.L., Bhuvanesh, N., Reibenspies, J.H., and Darensbourg, M.Y. (2008) *Angew. Chem. Int. Ed.*, **47**, 9492–9495.
109. Justice, A.K., Rauchfuss, T.B., and Wilson, S.R. (2007) *Angew. Chem. Int. Ed.*, **46**, 6152–6154.
110. Tye, J.W., Darensbourg, M.Y., and Hall, M.B. (2006) *Inorg. Chem.*, **45**, 1552–1559.
111. (a) Hsieh, C.-H., Erdem, Ö.F., Harman, S.D., Singleton, M.L., Reijerse, E., Lubitz, W., Popescu, C.V., Reibenspies, J.H., Brothers, S.M., Hall, M.B., and Darensbourg, M.Y. (2012) *J. Am. Chem. Soc.*, **134**, 13089–13202. (b) Olsen, M.T., Bruschi, M., De Gioia, L., Rauchfuss, T.B., and Wilson, S.R. (2008) *J. Am. Chem. Soc.*, **130**, 12021–12030. (c) Liu, Y.-C., Tu, L.-K., Yen, T.-H., Lee, G.-H., and Chiang, M.-H. (2011) *Dalton Trans.*, **40**, 2528–2541. (d) Trautwein, R., Almazahreh, L.R., Görls, H., and Weigand, W. (2013) *Z. Anorg. Allg. Chem.*, **639**, 1512–1519.
112. Munery, S., Capon, J.-F., De Gioia, L., Elleouet, C., Greco, C., Pétillon, F.Y., Schollhammer, P., Talarmin, J. Zampella, G. (2013) *Chem. Eur. J.*, **19**, 15458–15461, doi: 10.1002/chem.201303316
113. Gao, S., Fan, J., Sun, S., Song, F., Peng, X., Duan, Q., Jiang, D., and Liang, Q. (2012) *Dalton Trans.*, **41**, 12064–12074.
114. Felton, G.A.N., Vannucci, A.K., Chen, J., Lockett, L.T., Okumura, N., Petro, B.J., Zakai, U.I., Evans, D.H., Glass, R.S., and Lichtenberger, D.L. (2007) *J. Am. Chem. Soc.*, **129**, 12521–12530.
115. Capon, J.-F., Gloaguen, F., Schollhammer, P., and Talarmin, J. (2004) *J. Electroanal. Chem.*, **566**, 241–247.
116. Capon, J.-F., Gloaguen, F., Schollhammer, P., and Talarmin, J. (2006) *J. Electroanal. Chem.*, **595**, 47–52.
117. Gloaguen, F., Morvan, D., Capon, J.-F., Schollhammer, P., and Talarmin, J. (2007) *J. Electroanal. Chem.*, **603**, 15–20.
118. Chen, J., Vannucci, A.K., Mebi, C.A., Okumura, N., Borowski, S.C., Swenson, M., Lockett, L.T., Evans, D.H., Glass, R.S., and Lichtenberger, D.L. (2010) *Organometallics*, **29**, 5330–5340.
119. Goy, R., Apfel, U.-P., Elleouet, C., Escudero, D., Elstner, M., Görls, H., Dietzek, B., Talarmin, J., Schollhammer, P., González, L., and Weigand, W. (2013) *Eur. J. Inorg. Chem.*, 4466–4472.
120. Morneau, A., Geiger, W.E., Richmond, M.G., Don, M.-J., Watson, W.H., and Nagl, A. (2002) *Organometallics*, **21**, 1247–1256.
121. Cheah, M.H., Borg, S.J., and Best, S.P. (2007) *Inorg. Chem.*, **46**, 1741–1750.
122. Cheah, M.H., Borg, S.J., Bondin, M.I., and Best, S.P. (2004) *Inorg. Chem.*, **43**, 5635–5644.
123. Evans, D.H. (2008) *Chem. Rev.*, **108**, 2113–2144.

124. Geiger, W.E. (1985) *Prog. Inorg. Chem.*, **33**, 275–352.
125. Evans, D.H. and O'Connell, K.M. (1986) in *Electroanalytical Chemistry*, vol. 14 (ed. A.J. Bard), Marcel Dekker, New York, pp. 113–207.
126. Capon, J.-F., Ezzaher, S., Gloaguen, F., Pétillon, F.Y., Schollhammer, P., Talarmin, J., Davin, T.J., McGrady, J.E., and Muir, K.W. (2007) *New J. Chem.*, **31**, 2052–2064.
127. Wright, R.J., Zhang, W., Yang, X., Fasulo, M., and Tilley, T.D. (2012) *Dalton Trans.*, **41**, 73–82.
128. Liu, Y.-C., Chu, K.-T., Jhang, R.-L., Lee, G.-H., and Chiang, M.-H. (2013) *Chem. Commun.*, **49**, 4743–4745.
129. Schwartz, L., Eriksson, L., Lomoth, R., Teixidor, F., Vinas, C., and Ott, S. (2008) *Dalton Trans.*, 2379–2381.
130. Singh, P.S., Rudbeck, H.C., Huang, P., Ezzaher, S., Eriksson, L., Stein, M., Ott, S., and Lomoth, R. (2009) *Inorg. Chem.*, **48**, 10883–10885.
131. Charreteur, K., Kdider, M., Capon, J.-F., Gloaguen, F., Pétillon, F.Y., Schollhammer, P., and Talarmin, J. (2010) *Inorg. Chem.*, **49**, 2496–2501.
132. Orain, P.-Y., Capon, J.-F., Gloaguen, F., Schollhammer, P., and Talarmin, J. (2010) *Int. J. Hydrogen Energy*, **35**, 10797–10802.
133. Darchen, A., Mousser, H., and Patin, H. (1988) *J. Chem. Soc., Chem. Commun.*, 968–970.
134. Chong, D., Georgakaki, I.P., Mejia-Rodriguez, R., Sanabria-Chinchilla, J., Soriaga, M.P., and Darensbourg, M.Y. (2003) *Dalton Trans.*, 4158–4163.
135. Borg, S.J., Behrsing, T., Best, S.P., Razavet, M., Liu, X., and Pickett, C.J. (2004) *J. Am. Chem. Soc.*, **126**, 16988–16999.
136. Borg, S.J., Ibrahim, S.K., Pickett, C.J., and Best, S.P. (2008) *C.R. Chim.*, **11**, 852–860.
137. de Carcer, I.A., Di Pasquale, A., Rheingold, A.L., and Heinekey, D.M. (2006) *Inorg. Chem.*, **45**, 8000–8002.
138. Borg, S.J., Tye, J.W., Hall, M.B., and Best, S.P. (2007) *Inorg. Chem.*, **46**, 384–394.
139. Felton, G.A.N., Petro, B.J., Glass, R.S., Lichtenberger, D.L., and Evans, D.H. (2009) *J. Am. Chem. Soc.*, **131**, 11290–11291.
140. Greco, C., Zampella, G., Bertini, L., Bruschi, M., Fantucci, P., and De Gioia, L. (2007) *Inorg. Chem.*, **46**, 108–116.
141. Fontecilla-Camps, J.C., Amara, P., Cavazza, C., Nicolet, Y., and Volbeda, A. (2009) *Nature*, **460**, 814–822.
142. Bruschi, M., Greco, C., Kaukonen, M., Fantucci, P., Ryde, U., and De Gioia, L. (2009) *Angew. Chem. Int. Ed.*, **48**, 3503–3506.
143. Stanley, J.L., Heiden, Z.M., Rauchfuss, T.B., Wilson, S.R., De Gioia, L., and Zampella, G. (2008) *Organometallics*, **27**, 119–125.
144. Eilers, G., Schwartz, L., Stein, M., Zampella, G., De Gioia, L., Ott, S., and Lomoth, R. (2007) *Chem. Eur. J.*, **13**, 7075–7084.
145. Wang, F., Wang, M., Liu, X., Jin, K., Dong, W., and Sun, L. (2007) *Dalton Trans.*, 3812–3819.
146. Gao, W., Sun, J., Åkermark, T., Li, M., Eriksson, L., Sun, L., and Åkermark, B. (2010) *Chem. Eur. J.*, **16**, 2537–2546.
147. Carroll, M.E., Barton, B.E., Rauchfuss, T.B., and Carroll, P.J. (2012) *J. Am. Chem. Soc.*, **134**, 18843–18852.
148. Li, P., Wang, M., Chen, L., Liu, J., Zhao, Z., and Sun, L. (2009) *Dalton Trans.*, **38**, 1919–1926.
149. Jablonskyte, A., Wright, J.A., and Pickett, C.J. (2010) *Dalton Trans.*, **39**, 3026–3034.
150. Ezzaher, S., Capon, J.-F., Gloaguen, F., Pétillon, F.Y., Schollhammer, P., Talarmin, J., and Kervarec, N. (2009) *Inorg. Chem.*, **48**, 2–4.
151. Ezzaher, S., Gogoll, A., Bruhn, C., and Ott, S. (2010) *Chem. Commun.*, **46**, 5775–5777.
152. Matthews, S.L. and Heinekey, D.M. (2010) *Inorg. Chem.*, **49**, 9746–9748.
153. Zhao, X., Georgakaki, I.P., Miller, M.L., Yarbrough, J.C., and Darensbourg, M.Y. (2001) *J. Am. Chem. Soc.*, **123**, 9710–9711.
154. Zhao, X., Georgakaki, I.P., Miller, M.L., Mejia-Rodriguez, R., Chiang, C.-Y., and Darensbourg, M.Y. (2002) *Inorg. Chem.*, **41**, 3917–3928.

155. Zhao, X., Chiang, C.-Y., Miller, M.L., Rampersad, M.V., and Darensbourg, M.Y. (2003) *J. Am. Chem. Soc.*, **125**, 518–524.
156. Justice, A.K., Zampella, G., De Gioia, L., and Rauchfuss, T.B. (2007) *Chem. Commun.*, 2019–2021.
157. Justice, A.K., Zampella, G., De Gioia, L., Rauchfuss, T.B., Van der Vlugt, J.I., and Wilson, S.R. (2007) *Inorg. Chem.*, **46**, 1655–1664.
158. Adam, F.I., Hogarth, G., and Richards, I. (2007) *J. Organomet. Chem.*, **692**, 3957–3968.
159. Thomas, C.M., Rüdiger, O., Liu, T., Carson, C.E., Hall, M.B., and Darensbourg, M.Y. (2007) *Organometallics*, **26**, 3976–3984.
160. Barton, B.E., Zampella, G., Justice, A.K., De Gioia, L., Rauchfuss, T.B., and Wilson, S.R. (2010) *Dalton Trans.*, **39**, 3011–3019.
161. Xu, F., Tard, C., Wang, X., Ibrahim, S.K., Hughes, D.L., Zhong, W., Zeng, X., Luo, Q., Liu, X., and Pickett, C.J. (2008) *Chem. Commun.*, 606–608.
162. Fauvel, K., Mathieu, R., and Poilblanc, R. (1976) *Inorg. Chem.*, **15**, 976–978.
163. Orain, P.-Y., Capon, J.-F., Gloaguen, F., Pétillon, F.Y., Schollhammer, P., Talarmin, J., Zampella, G., De Gioia, L., and Roisnel, T. (2010) *Inorg. Chem.*, **49**, 5003–5008.
164. Gloaguen, F., Lawrence, J.D., and Rauchfuss, T.B. (2001) *J. Am. Chem. Soc.*, **123**, 9476–9477.
165. Ezzaher, S., Capon, J.-F., Gloaguen, F., Pétillon, F.Y., Schollhammer, P., Talarmin, J., Pichon, R., and Kervarec, N. (2007) *Inorg. Chem.*, **46**, 3426–3428.
166. Ezzaher, S., Capon, J.-F., Gloaguen, F., Kervarec, N., Pétillon, F.Y., Pichon, R., Schollhammer, P., and Talarmin, J. (2008) *C.R. Chim.*, **11**, 906–914.
167. Adam, F.I., Hogarth, G., Kabir, S.E., and Richards, I. (2008) *C.R. Chim.*, **11**, 890–905.
168. Ghosh, S., Hogarth, G., Hollingsworth, N., Holt, K.B., Richards, I., Richmond, M.G., Sanchez, B.E., and Unwin, D. (2013) *Dalton Trans.*, **42**, 6775–6792.
169. Orain, P.-Y., Capon, J.-F., Kervarec, N., Gloaguen, F., Pétillon, F.Y., Pichon, R., Schollhammer, P., and Talarmin, J. (2007) *Dalton Trans.*, 3754–3756.
170. Barton, B.E., Olsen, M.T., and Rauchfuss, T.B. (2008) *J. Am. Chem. Soc.*, **130**, 16834–16835.
171. Zaffaroni, R., Rauchfuss, T.B., Gray, D.L., De Gioia, L., and Zampella, G. (2012) *J. Am. Chem. Soc.*, **134**, 19260–19269.
172. Kramarz, K.W. and Norton, J.R. (1994) *Prog. Inorg. Chem.*, **42**, 1–65.
173. Jiang, S., Liu, J., and Sun, L. (2006) *Inorg. Chem. Commun.*, **9**, 290–292.
174. Jiang, S., Liu, J., Shi, Y., Åkermark, B., and Sun, L. (2007) *Dalton Trans.*, 896–902.
175. Wang, F., Wang, M., Liu, X., Jin, K., Dong, W., Li, G., Åkermark, B., and Sun, L. (2005) *Chem. Commun.*, 3221–3223.
176. Das, P., Capon, J.-F., Gloaguen, F., Pétillon, F.Y., Schollhammer, P., Talarmin, J., and Muir, K.W. (2004) *Inorg. Chem.*, **43**, 8203–8205.
177. Capon, J.-F., Ezzaher, S., Gloaguen, F., Pétillon, F.Y., Schollhammer, P., and Talarmin, J. (2008) *Chem. Eur. J.*, **14**, 1954–1964.
178. Si, Y., Ma, C., Hu, M., Chen, H., Chen, C., and Liu, Q. (2007) *New J. Chem.*, **31**, 1448–1454.
179. Song, L.-C., Wang, H.-T., Ge, J.-H., Mei, S.-Z., Gao, J., Wang, L.-X., Gai, B., Zhao, L.-Q., Yan, J., and Wang, Y.-Z. (2008) *Organometallics*, **27**, 1409–1416.
180. Erdem, O.F., Schwartz, L., Stein, M., Silakov, A., Kaur-Ghumaan, S., Huang, P., Ott, S., Reijerse, E., and Lubitz, W. (2011) *Angew. Chem. Int. Ed.*, **50**, 1439–1443.
181. Gao, W., Ekstrom, J., Liu, J., Chen, C., Eriksson, L., Weng, L., Åkermark, B., and Sun, L. (2007) *Inorg. Chem.*, **46**, 1981–1991.
182. Ezzaher, S., Capon, J.-F., Gloaguen, F., Pétillon, F.Y., Schollhammer, P., and Talarmin, J. (2007) *Inorg. Chem.*, **46**, 9863–9872.
183. Ezzaher, S., Orain, P.-Y., Capon, J.-F., Gloaguen, F., Pétillon, F.Y., Roisnel, T., Schollhammer, P., and Talarmin, J. (2008) *Chem. Commun.*, 2547–2549.

184. Olsen, M.T., Barton, B.E., and Rauchfuss, T.B. (2009) *Inorg. Chem.*, **48**, 7507–7509.
185. Nicolet, Y., De Lacey, A.L., Vernède, X., Fernandez, V.M., Hatchikian, E.C., and Fontecilla-Camps, J.C. (2001) *J. Am. Chem. Soc.*, **123**, 1596–1601.
186. Silakov, A., Wenk, B., Reijerse, E., and Lubitz, W. (2009) *Phys. Chem. Chem. Phys.*, **11**, 6592–6599.
187. Wang, N., Wang, M., Zhang, T., Li, P., Liu, J., and Sun, L. (2008) *Chem. Commun.*, 5800–5802.
188. Wang, N., Wang, M., Liu, J., Jin, K., Chen, L., and Sun, L. (2009) *Inorg. Chem.*, **48**, 11551–11558.
189. Wang, Y., Wang, M., Sun, L., and Ahlquist, M.S.G. (2012) *Chem. Commun.*, **48**, 4450–4452.
190. Lounissi, S., Zampella, G., Capon, J.-F., De Gioia, L., Matoussi, F., Mahfoudhi, S., Pétillon, F.Y., Schollhammer, P., and Talarmin, J. (2012) *Chem. Eur. J.*, **18**, 11123–11138.
191. DuBois, D.L. and Bullock, R.M. (2011) *Eur. J. Inorg. Chem.*, 1017–1027.
192. Rakowski DuBois, M. and DuBois, D.L. (2009) *Chem. Soc. Rev.*, **38**, 62–72.
193. Rakowski DuBois, M. and DuBois, D.L. (2009) *Acc. Chem. Res.*, **42**, 1974–1982.
194. Rakowski DuBois, M. and DuBois, D.L. (2008) *C.R. Chim.*, **11**, 805–817.
195. Helm, M.L., Stewart, M.P., Bullock, R.M., Rakowski DuBois, M., and DuBois, D.L. (2011) *Science*, **333**, 863–866.
196. Dey, S., Rana, A., Dey, S.G., and Dey, A. (2013) *ACS Catal.*, **3**, 429–436.
197. Quentel, F., Passard, G., and Gloaguen, F. (2012) *Chem. Eur. J.*, **18**, 13473–13479.
198. Quentel, F., Passard, G., and Gloaguen, F. (2012) *Energy Environ. Sci.*, **5**, 7757–7761.
199. Stubbert, B.D., Peters, J.C., and Gray, H.B. (2011) *J. Am. Chem. Soc.*, **133**, 18070–18073.
200. Bruschi, M., Greco, C., Fantucci, P., and De Gioia, L. (2008) *Inorg. Chem.*, **47**, 6056–6071.
201. Bruschi, M., Fantucci, P., and De Gioia, L. (2002) *Inorg. Chem.*, **41**, 1421–1429.
202. Zhou, T., Mo, Y., Liu, A., Zhou, Z., and Tsai, K.R. (2004) *Inorg. Chem.*, **43**, 923–930.
203. Na, Y., Wang, M., Jin, K., Zhang, R., and Sun, L. (2006) *J. Organomet. Chem.*, **691**, 5045–5051.
204. Feng, Y.-N., Xu, F.-F., Chen, R.-P., Wen, N., Li, Z.-H., and Du, S.-W. (2012) *J. Organomet. Chem.*, **717**, 211–217.
205. Liu, X., Ru, X., Li, Y., Zhang, K., and Chen, D. (2011) *Int. J. Hydrogen Energy*, **36**, 9612–9619.
206. Xu, E., Xiao, Z., Liu, H., Long, L., Li, L., and Liu, X. (2012) *RSC Adv.*, **2**, 10171–10174.
207. Vijaikanth, V., Capon, J.-F., Gloaguen, F., Schollhammer, P., and Talarmin, J. (2005) *Electrochem. Commun.*, **7**, 427–430.
208. Singleton, M.L., Crouthers, D.J., Duttweiler, R.P., Reibenspies, J.H., and Darensbourg, M.Y. (2011) *Inorg. Chem.*, **50**, 5015–5026.
209. Wang, H.-Y., Wang, W.-G., Si, G., Wang, F., Tung, C.-H., and Wu, L.-Z. (2010) *Langmuir*, **26**, 9766–9771.

5
The Third Hydrogenase

Callum Scullion and John A. Murphy

5.1
Introduction

Hydrogenases are a group of enzymes that catalyze the reversible cleavage of molecular hydrogen and are present within aerobic and anaerobic bacteria, archaea, and eukarya [1]. There are three distinct classes of hydrogenase, with two of these classes containing iron–sulfur clusters; [NiFe] hydrogenase contains nickel–iron–sulfur clusters, while [FeFe] hydrogenase contains iron–sulfur clusters only.

The third class of hydrogenase (which can be obtained from methanogenic archaea grown in nickel-limiting and red light conditions) [2] was originally believed to be metal-free, resulting in it being known as the *iron–sulfur-cluster free hydrogenase* or *H_2-forming methylenetetrahydromethanopterin dehydrogenase* (*Hmd*). However, in more recent years, the third hydrogenase has been found to contain an essential iron cofactor resulting in the new name of [Fe] hydrogenase. Although the iron cofactor within [Fe] hydrogenase contains a cysteine ligand, the chemistry at this iron center is quite distinct from that of the hydrogenases that contain iron–sulfur clusters.

[Fe] hydrogenase catalyzes the reversible reduction of N^5,N^{10}-methenyltetrahydromethanopterin ($CH{\equiv}H_4MPT^+$) **1** with H_2 to N^5,N^{10}-methylenetetrahydromethanopterin ($CH_2{=}H_4MPT$) **2** and a proton (Figure 5.1) [2]. This represents the reduction from the formic acid oxidation level to the formaldehyde oxidation level at the carbon center C_{14a}, which is originally derived from carbon dioxide within methanogenic archaea [1].

As this chapter discusses studies carried out on [Fe] hydrogenase before and after the discovery of the Fe cofactor, the abbreviation Hmd will be used for the enzyme for results before this major discovery.

Bioinspired Catalysis: Metal-Sulfur Complexes, First Edition.
Edited by Wolfgang Weigand and Philippe Schollhammer.

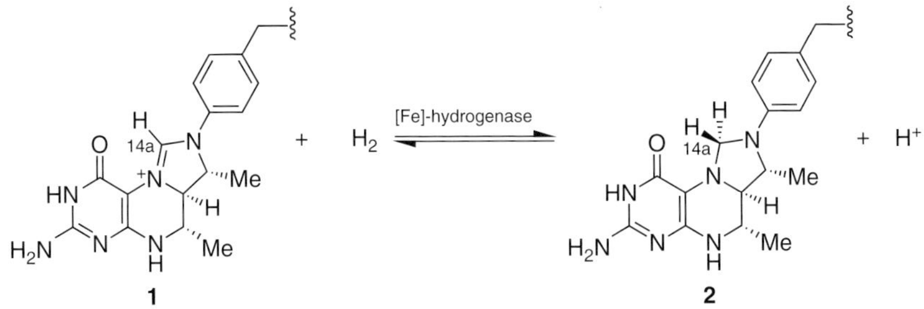

Figure 5.1 Reaction catalyzed by [Fe] hydrogenase.

5.2 Initial Studies of Hmd

Thauer and colleagues [2] first described the hydrogenase activity of Hmd in 1990 after isolating the pure enzyme from *Methanobacterium thermoautotrophicum.* In this study, Hmd was found to dehydrogenate $CH_2{=}H_4MPT$ **2** to form $CH{\equiv}H_4MPT^+$ **1** and H_2 as well as carry out the reduction of **1**–**2** with the liberation of a proton (Figure 5.1). In addition, it was found that the purified enzyme did not catalyze the reduction of coenzyme F_{420}, NAD^+, $NADP^+$, or methylviologen (it has been found that Hmd has a role in providing $CH_2{=}H_4MPT$ **2** for F_{420}-dependent methylenetetrahydromethanopterin dehydrogenase, which is needed for the reduction of coenzyme F_{420} in *M. thermoautotrophicum* grown under nickel-limiting conditions) [3]. While nickel was found in insufficient quantities to be active (<0.1–1 equiv of enzyme), 1 equiv of iron to 1 equiv of enzyme was identified; however, this finding was initially thought to be influenced by background contamination of iron rather than indicative of the presence of an active iron species within the enzyme [2]. This conclusion was perhaps influenced by the differences seen with [NiFe] hydrogenase and [FeFe] hydrogenase, which are redox active with viologens and have absorbances above 340 nm in UV/vis spectroscopy, which were not present with the Hmd isolated at this stage [4]. It was also established that the enzyme activity of [Fe] hydrogenase was rapidly lost under aerobic conditions and in the presence of dithiothreitol (DTT) or mercaptoethanol [2].

The effect of using standard inhibitors of transition-metal-containing enzymes in the presence of Hmd was examined in both redox directions. CO (5% and 50% in the gas phase), acetylene (5% and 50% in the gas phase), NO (1 and 10 mM), and azide (1 and 10 mM) were found to have no effect on enzyme activity [4]. Cyanide was found to have no effect at 0.1 mM but caused a 50% decrease in activity at 10 mM concentration [4].

To assess the conditions required for hydrogen cleavage in Hmd, tritium exchange experiments were carried out [4]. When pure Hmd alone was used, no exchange between $^3H_2$ and ^1H_2O was observed at a range of temperatures and pH values. However, when $CH{\equiv}H_4MPT^+$ **1** or $CH_2{=}H_4MPT$ **2** were added to the enzyme, exchange between $^3H_2$ and ^1H_2O was observed, with incorporation of tritium into the aqueous solution as well as into substrates $CH{\equiv}H_4MPT^+$ **1** and $CH_2{=}H_4MPT$ **2**. After the reaction of $CH{\equiv}H_4MPT^+$ **1** with $^3H_2$ was carried out, high-performance liquid chromatography (HPLC) separation of the products and radioactivity experiments showed that for every equivalent of $CH_2{=}H_4MPT$ **2** formed, 0.5 equiv of $^3H_2$ were incorporated. The substrate dependence for tritium exchange in Hmd is in contrast to [FeFe] hydrogenase and [NiFe] hydrogenase, which both actively carry out $^3H_2/^1H_2O$ exchange with no substrate present [4].

Several NMR studies have been carried out in order to gain information on the reversible process between $CH{\equiv}H_4MPT^+$ **1** and H_2.

In 1993, Schwörer *et al.* [5] examined the mechanism of dihydrogen formation in Hmd using deuterium labeling. First, the kinetic isotope effect was determined by comparing D_2 formation from $CD_2{=}H_4MPT$ **3** in D_2O and H_2 formation from $CH_2{=}H_4MPT$ **2** in H_2O (Figure 5.2), giving a value of ~4.7 for k_H/k_D [5]. The oxidation of $CD_2{=}H_4MPT$ **3** in H_2O was then examined, showing HD **6** and H_2 formation, while HD and D_2 were produced from $CH_2{=}H_4MPT$ **2** in D_2O (Figure 5.2). The production of H_2 from $CD_2{=}H_4MPT$ **3** in H_2O was accounted for by the reduction of $CD{\equiv}H_4MPT^+$ **4** by the HD produced, giving some amounts of $CHD{=}H_4MPT^+$ **5**, which could dehydrogenate to give H_2 (Figure 5.2). Indeed, $CHD{=}H_4MPT^+$ **5** was synthesized and subjected to the reaction conditions to form H_2 at a rate four times faster than HD. No exchange was observed between substrates **2** or **3** with gas or solvent without the presence of Hmd when blank experiments were run. When subjected to the same oxidative conditions with Hmd, but in the absence of $CH_2{=}H_4MPT$ **2** or $CD_2{=}H_4MPT$ **3**, no exchange between H_2 and D_2O or D_2 and H_2O was observed, complementing the previous tritium-labeling studies [4]. These results indicated that only one of the hydrogens from $CH_2{=}H_4MPT$ **2** was used in the formation of dihydrogen, with the

$$\underset{\mathbf{3}}{CD_2{=}H_4MPT} \xrightleftharpoons{D_2O} \underset{\mathbf{4}}{CD{\equiv}H_4MPT^+} + D_2$$

$$\underset{\mathbf{2}}{CH_2{=}H_4MPT} \xrightleftharpoons{H_2O} \underset{\mathbf{1}}{CH{\equiv}H_4MPT^+} + H_2$$

$$\underset{\mathbf{3}}{CD_2{=}H_4MPT} \xrightleftharpoons{H_2O} \underset{\mathbf{4}}{CD{\equiv}H_4MPT^+} + \underset{\mathbf{6}}{HD} + H_2$$

$$\underset{\mathbf{2}}{CH_2{=}H_4MPT} \xrightleftharpoons{D_2O} \underset{\mathbf{1}}{CH{\equiv}H_4MPT^+} + \underset{\mathbf{6}}{HD} + D_2$$

$$\underset{\mathbf{5}}{CHD{=}H_4MPT} \xrightleftharpoons{H_2O} \underset{\mathbf{4}}{CD{\equiv}H_4MPT^+} + \underset{\mathbf{6}}{HD} + H_2$$

Figure 5.2 Various reactions investigated by Schwörer *et al.* [5].

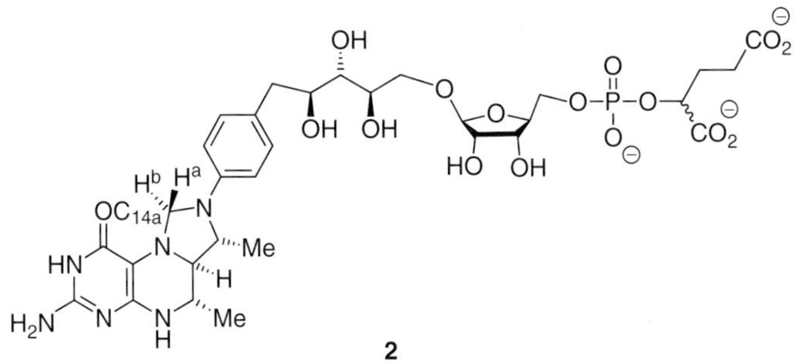

2

Figure 5.3 Structure of $CH_2{=}H_4MPT$ **2** showing magnetically distinct methylene protons H^a and H^b.

Table 5.1 Deuterium labeled experiments to determine stereoselectivity of Hmd process.

Entry number	Reaction	Major product
1	$^{13}CH{\equiv}H_4MPT^+ + D_2, D_2O$	$^{13}CH^aD^b{=}H_4MPT$
2	$^{13}CD{\equiv}H_4MPT^+ + H_2, H_2O$	$^{13}CD^aH^b{=}H_4MPT$
3	$^{13}CD{\equiv}H_4MPT^+ + H_2, D_2O$	$^{13}CD^aH^b{=}H_4MPT$
4	$^{13}CH{\equiv}H_4MPT^+ + D_2, H_2O$	$^{13}CH^aD^b{=}H_4MPT$
5	$^{13}CD{\equiv}H_4MPT^+ + D_2, H_2O$	$^{13}CD_2{=}H_4MPT$
6	$^{13}CH{\equiv}H_4MPT^+ + H_2, D_2O$	$^{13}CH_2{=}H_4MPT$

other hydrogen being obtained from solution. In forming dihydrogen, one atom is obtained from a proton in solution and the other from what is formally a hydride from $CH_2{=}H_4MPT$ **2** [4]; the pH dependence on the equilibrium of products confirms this, with alkaline pH ($pH > 7$) favoring the reduction of $CH{\equiv}H_4MPT^+$ **1** with H_2 and acidic pH ($pH < 7$) favoring the dehydrogenation of $CH_2{=}H_4MPT$ **2**.

Schleucher *et al.* [6] used two-dimensional NMR spectroscopy to determine the stereoselectivity of hydride transfer to $CH{\equiv}H_4MPT^+$ **1** catalyzed by Hmd. Owing to the complexity observed in the ^1H NMR of $CH_2{=}H_4MPT$ **2**, they simplified their analysis by using various deuterium-incorporating reactions and looking at the heteronuclear single-quantum coherence (HSQC) spectra of the products – this technique was further aided by ^{13}C labeling of the C_{14a} of the $CH{\equiv}H_4MPT^+$ **1** or $CD{\equiv}H_4MPT^+$ **4** starting materials. Owing to the chirality of $CH_2{=}H_4MPT$ **2**, the methylene protons (H^a and H^b shown in Figure 5.3) are chemically and magnetically distinct, with H^a observed at 4.9 ppm and H^b at 3.3 ppm in ^1H NMR. Table 5.1 gives a summary of the six reactions studied.

The results of these reactions showed that hydride is delivered stereoselectively to the *Re*-face of the amidine center in $CH{\equiv}H_4MPT^+$ **1** to take up the *pro-R* position (H^b) of C_{14a} in $CH_2{=}H_4MPT$ **2** [6].

The $CH{\equiv}H_4MPT^+$ **1**-dependent exchange of H and D is also well complemented by the observed conversion of *para*-H_2 to *ortho*-H_2 catalyzed by Hmd, which is also dependent on the presence of $CH{\equiv}H_4MPT^+$ **1** [7, 8].

5.3 Discovery that Hmd Contains a Bound Cofactor

In 2000, Buurman *et al.* [9] obtained Hmd enzyme from the heterologous expression of Hmd genes from *Methanopyrus kandleri* and *Methanococcus jannaschii* in *Escherichia coli*. The purified enzymes were found to be completely inactive. In the same study, Hmd purified from *Methanothermobacter marburgensis* was denatured in urea and the resulting mixture separated through a 10 kDa molecular mass cutoff membrane. The resulting ultrafiltrate was considered to contain components of too low molecular weight to contain the active enzyme. When the ultrafiltrate was added to the heterologously expressed Hmd, enzymatic activity (both hydrogenation and dehydrogenation) was observed, suggesting the presence of a bound cofactor in the wild enzyme [9]. HPLC of the ultrafiltrate followed by mass spectral analysis showed the active fractions to have apparent masses well below 10 kDa. UV/vis spectroscopy showed absorbance at 280 nm, which accounted for the loss in extinction coefficient at this value in the denatured enzyme [9]. In the same work, analysis by total reflection X-ray fluorescence spectrometry showed that the ultrafiltrate contained no nickel and only traces of iron and zinc (a result that was later explained by the conditions required for handling the enzyme) [10]. This further added to the opinion that Hmd did not contain a redox-active transition metal [9].

5.4 Discovery that Hmd is a Metalloenzyme

In 2004, Thauer and coworkers [10] made one of the most significant discoveries in the investigations into Hmd; they discovered that the 1 equiv of iron found in the sample of Hmd in 1990 [2] was due to an essential iron present within the enzyme.

Previously, Thauer *et al.* [2, 4] had shown that the Hmd that they had prepared had no absorption above 340 nm in UV/vis spectroscopy. However, the discovery that UV-A (320–400 nm) and blue light (400–500 nm) inactivate and bleach Hmd and its cofactor, suggested that the enzyme did contain UV/vis properties above 340 nm [10]. By preparing and isolating the Hmd enzyme in the dark at 4 °C (and storing anaerobically at −80 °C), it was possible to obtain a UV/vis spectrum of Hmd with absorbance maximum at ~360 nm [10]. This absorbance was seen to decrease the longer that the sample of enzyme was exposed to UV-A/blue light. Along with showing a decrease in activity when exposed to UV-A/blue light, iron was seen to be released by the enzyme, giving a strong indication that the iron was necessary for enzymatic activity [10].

Thauer and coworkers [10] also established that the enzyme is in fact inhibited by carbon monoxide, contrary to the previous investigations [4]. At a concentration of 100%, they observed 50% inhibition of the enzyme, a process that was found

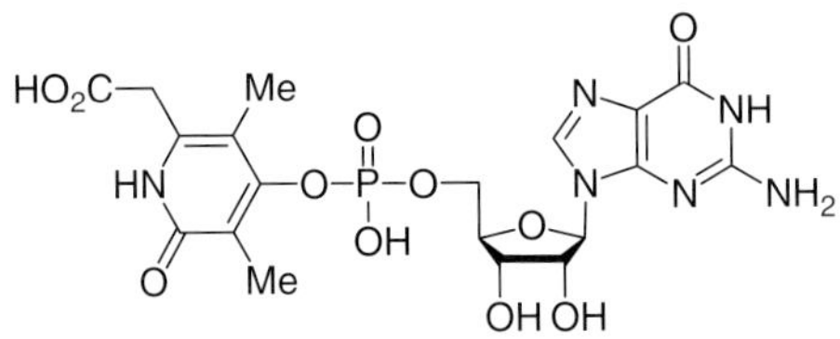

7

Figure 5.4 Light-inactivation product isolated from Hmd.

to be reversible when the same sample was used in the absence of carbon monoxide. Carbon monoxide inhibition provided a way of protecting the Hmd from light inactivation – a protection that was presumed to be due to the formation of an Fe–CO complex [10]. Again, cyanide inhibition was investigated and showed that the inhibition was most likely due to Hmd–cyanide complex formation rather than reaction with $CH{\equiv}H_4MPT^+$ **1** or $CD{\equiv}H_4MPT^+$ **2**, as originally proposed [4]. It should be noted here that although all three of the hydrogenase families are inhibited by CO, only Hmd is known to be inhibited by cyanide [10, 11].

Shima *et al.* [12] proposed a structure 7 (Figure 5.4) for the light-inactivation product from Hmd, most likely derived from the active cofactor. This species, separated by HPLC, had a mass of 542 Da (consistent with structure 7) and was proposed to contain a phosphodiester due to its reactivity with phosphodiesterase and the presence of a signal at −6.18 ppm in ^{31}P NMR. By isolating the light-inactivated product from Hmd produced in cells grown in $^{15}NH_4Cl$ rather than $^{14}NH_4Cl$, a mass of 548 Da rather than 542 Da was observed, which suggested that the product contained six nitrogens. Using various other mass spectroscopy, UV/vis and NMR techniques, it was concluded that structure 7 was the most likely structure for the light-inactivated compound.

Lyon *et al.* [11] reported further analysis of the products of light inactivation in Hmd. They found that for every mole of Hmd-enzyme-containing cofactor, ~2.4 ± 0.2 mol of CO was released; for Hmd apoenzyme, no CO was released. Analysis of the amount of Fe released also showed a value of 2.4 ± 0.2 mol of CO to Fe, suggesting the number of CO molecules in Hmd to be 2 or 3. The authors reported that the infrared analysis of the cofactor before light inactivation showed the presence of two bands of equal intensity at 2011 and 1944 cm^{-1}. It was suggested that this would account for the presence of an $Fe(CO)_2$ unit with an approximate 90° angle between the CO groups [11]. Considering the inhibition of Hmd by CO, the authors suggested that an $Fe(CO)_3$ compound was formed, indicating that an iron in the Fe^0, Fe^I, or Fe^{II} was present in Hmd [11]. The observed inhibition of Hmd by cyanide and the rarity of Fe^0-CN compounds were used to propose that the iron present in the enzyme was not present in the Fe^0 oxidation state. Finally, electron paramagnetic resonance (EPR) measurements were used to rule out an Fe^I center, and so it was proposed that the Hmd cofactor was an $Fe^{II}(CO)_2$ complex, which would be likely to be low-spin and six-coordinate [11].

It was also shown that when the Fe-containing cofactor was isolated, the CO bands in the infrared spectrum shifted from 2011 and 1944 cm^{-1} to 2031 and

$1972 cm^{-1}$. This was indicative of decreased electron density at the iron center, suggesting either an exchange of ligands or a change in the total number of ligands attached, that is, going from six-coordinate to five-coordinate [11].

Addition of $CH{\equiv}H_4MPT^+$ **1** or $CD{\equiv}H_4MPT^+$ **2** in argon to the Hmd enzyme showed slight changes in the infrared spectra recorded, indicative of binding close to or at the iron site in the enzyme [11]. The addition of H_2 increased the effects of $CH{\equiv}H_4MPT^+$ **1** on the IR spectrum of Hmd, possibly suggesting interaction between hydrogen and the iron center in the presence of **1**; however, no evidence of Fe–H or Fe–D stretches in the IR spectrum were found. This change in IR spectrum in argon or hydrogen was not observed without the presence of substrate **1** or **2** [11].

To gain further information about the electronic structure of iron in Hmd (and due to this center being EPR silent) [11], Shima *et al.* [13] studied the Mössbauer spectrum of ^{57}Fe-labeled Hmd prepared under dark conditions. The results showed a low isomer shift for Hmd and indicated the presence of a low-spin iron in a low oxidation state. Introduction of CO or cyanide inhibitors to the enzyme showed changes in isomer shift of the iron center, changes that the authors proposed indicated direct binding of the inhibitors to the iron. On the other hand, introduction of H_2 and/or $CH{\equiv}H_4MPT^+$ **1** to the enzyme had no significant effect in the spectrum, which shows that when Hmd interacts with its substrates or products, the electronic state of the iron site is not affected sufficiently to be detected by the Mössbauer-active nucleus [13].

Measurement of the magnetic properties suggested that the iron center in Hmd was diamagnetic with spin $S = 0$, consistent with Fe^{II} or Fe^0 [13]. On the basis of the infrared [11] and Mössbauer [13] studies together, it was proposed that an Fe^{II} was the more likely electronic state of the iron center. Additional information gained from the Mössbauer spectra of Hmd was a value of 1.14 ± 0.25 atoms of iron per protein monomer.

From these studies [10–13], it was concluded that the iron center in Hmd was an iron (II) guanylyl pyridone cofactor (FeGP cofactor) and that because of the effects of CO or cyanide on this center (causing changes in the properties of the iron and inhibiting enzymatic activity) it must be key to enzymatic activity. To gain further information on this cofactor, it was necessary to obtain information on the 3D structure of Hmd, which will henceforth be referred to as [*Fe*] *hydrogenase.*

5.5 Crystal Structure Studies of [Fe] Hydrogenase

In 2006 [14], the first X-ray crystal structures of [Fe] hydrogenase were reported, with the apoenzyme (without FeGP cofactor) isolated and crystallized from both *M. jannaschii* (Figure 5.5) and *M. kandleri*, giving structures of 1.75 and 2.4 Å resolution, respectively.

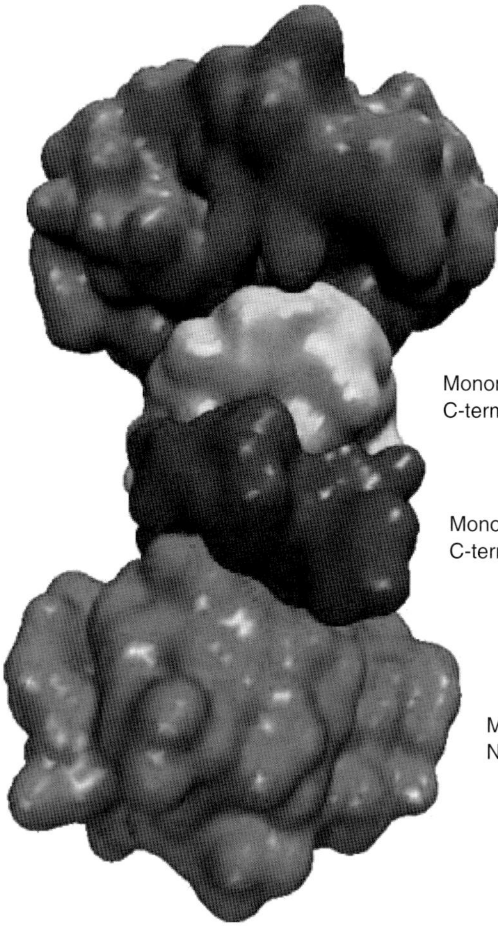

Figure 5.5 Crystal structure of homodimer of [Fe]-hydrogenase apoenzyme isolated from *M. jannaschii*; 2B0J in RCSB Protein Data Bank; quaternary structure file available from EMBL-EBI's Protein Data Bank in Europe (PDBe). (Image generated using VMD [15].)

[Fe] hydrogenase was found to be a homodimer with dimensions of ~90 Å × 50 Å × 40 Å, and with the architecture of one central and two identical peripheral globular units. The peripheral units corresponded to the N-terminal domains (red in Figure 5.5) of each subunit and were composed of Rossman fold-like α/β structure. The central globular unit corresponded to the C-terminal segments (green and blue in Figure 5.5) of both subunits, each containing four α-helices that form an intertwined intersubunit helix bundle [14]. Analysis of the relative arrangements of the globular units showed that [Fe] hydrogenase from *M. jannaschii* was present in a closed conformational state, while [Fe] hydrogenase from *M. kandleri* was in an open conformational state.

In the same study, activity measurements of mutant [Fe] hydrogenase were used to suggest that the iron-binding site was located in the pocket containing Cys176

[14]. Of the three conserved cysteine residues in [Fe] hydrogenase, only Cys176 was essential for activity. The authors mention the presence of an iron-ligating sulfur ligand in an unpublished X-ray absorption spectrum of the [Fe]-hydrogenase holoenzyme. Owing to the importance of Cys176, they suggest that this is the source of the ligation to iron [14]. In addition, in the case of *M. jannaschii*, [Fe] hydrogenase, a 13 Å long U-shaped electron density was observed in a region relatively close to the Cys176 residue (~6 Å away). This U-shape was assigned to poly(ethylene glycol) (PEG) present in the crystallization solution and this site was proposed (aided by some modeling) to be where $CH{\equiv}H_4MPT^+$ **1** binds [14].

In 2008, Shima *et al.* [15] obtained a 1.75 Å resolution X-ray crystal structure for the Hmd holoenzyme (the enzyme including the iron cofactor), isolated from *M. jannaschii*. By reconstituting [Fe]-hydrogenase apoenzyme with the labile FeGP cofactor under completely anaerobic and red-light conditions, the authors were able to form crystals of the enzyme. It was reported that the guanosine monophosphate moiety functions to anchor the FeGP cofactor to the enzyme [15]. In the active site, they found the iron center to be surrounded by a distorted square pyramidal or octahedral ligation shell dependent on the enzymatic state (Figure 5.6). They found that the nitrogen of the guanylylpyridinol was ligated to the iron, while the pyridinol and carboxylate substituents were not, although they noted that the latter substituent was partly disordered and so had to be interpreted with caution. The other species in the coordination sphere of iron were assigned as two CO molecules at 90° angles to each other, the sulfur from Cys176, a solvent water molecule, and an unknown ligand (shown as an oxygen in the X-ray structure) that could not be determined because of the high levels of disorder.

When the crystals were soaked in 3 mM cyanide, this led to an enhancement (1.6 times) in the electron density at the site of the unknown ligand, suggesting that it could be the site of reversible cyanide inhibition; the Fe–CN distance was determined as 1.8 Å. In addition, despite the water molecule being assigned trans to the nitrogen of the guanylylpyridinol, the oxygen atom was considered to be too far to be a ligand (2.8 Å for the holoenzyme and 2.7 Å for the cyanide-inhibited holoenzyme) [15]. This site was suggested to be the site of H_2 binding in the reaction with $CH{\equiv}H_4MPT^+$ **1** and also the possible site for the binding of extrinsic CO.

The authors proposed a mechanism for hydride delivery to $CH{\equiv}H_4MPT^+$ **1** based on these results [15]. They suggested that by binding H_2 in a manner side-on to the iron center, the pK_a of H_2 could be lowered (the pK_a of molecular H_2 being 35) [17] and that hydride could be delivered to “the strong hydride acceptor” $CH{\equiv}H_4MPT^+$ **1**, while the resulting proton could be accepted by a nearby proton acceptor. The candidate proton acceptors were suggested as the Cys176 ligand, the pyridinol ligand (nitrogen, oxygen, or carboxylate), or one of two histidines (His14 and His201) all within 6.5 Å of the iron. It was noted that the His14 → Ala mutation had a significant effect on the activity of the enzyme (while His201 → Ala mutation had little effect) and so it could be possible for this to be the proton acceptor [15].

In 2009 [18], the crystal structure of the FeGP cofactor within [Fe] hydrogenase was revised, based on the crystal structure of a mutated [Fe]-hydrogenase enzyme.

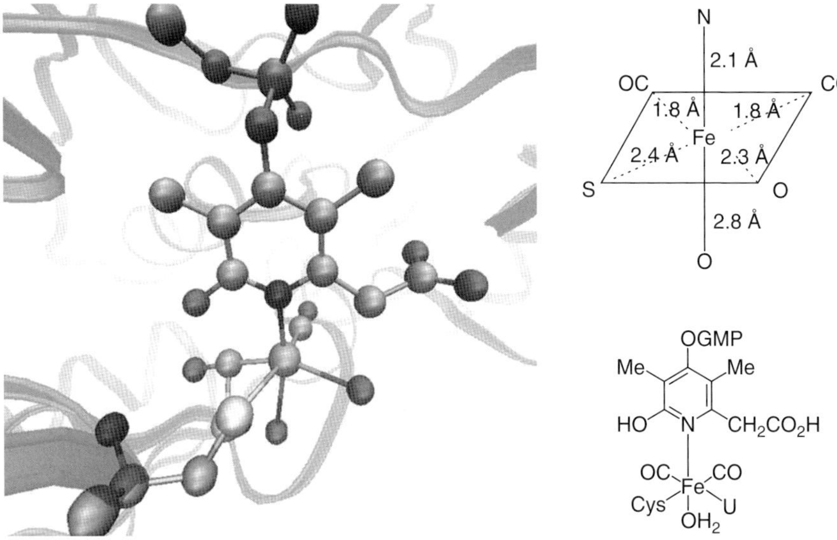

Figure 5.6 Crystal structure of FeGP cofactor binding site in [Fe]-hydrogenase holoenzyme isolated from *M. jannaschii*; 3DAG in RCSB Protein Data Bank. Fe center, two CO ligands, guanylyl pyridone ligand, and Cys176 shown as Corey–Pauling–Koltun (CPK) structures. (Image generated using VMD [16].)

The holoenzyme of Cys176 → Ala mutated [Fe] hydrogenase from *M. jannaschii* was crystallized in the presence of DTT to give a structure of 1.95 Å resolution (Figure 5.7). The structure showed an iron center octahedrally coordinated by one DTT-sulfur, one DTT-oxygen, two CO groups at 90° to each other (also shown by IR analysis), the 2-pyridinol's nitrogen, and the 2-pyridinol's 6-acylmethyl group in an acyl-iron ligation [18]. The DTT-sulfur had displaced the binding site of the Cys176 from the wild-type enzyme, while the DTT-oxygen had displaced the unknown ligand (Figures 5.6 and 5.7). The acyl-iron ligation showed that the previous "carboxylate" group assignment for the FeGP cofactor was also incorrect. This resulted in the pyridinyl ligand chelating the iron center, rotating it 180° from the previously determined structure. In addition, one of the CO ligands was determined to actually occupy the site previously suggested to be occupied by water (the water that was too distant to actually be a ligand). The acyl group oxygen was shown to link to the polypeptide chain by the formation of a hydrogen bond with the amide group of Ala176 (which would be Cys176 in the wild-type enzyme), while the 2-oxygen of the pyridinol ligand was shown to be in close proximity to the imidazole group of His14 and the 2- and 3-oxygens of DTT.

On the basis of the unprecedented finding of a biological acyl-iron complex, the structure of the wild-type [Fe]-hydrogenase holoenzyme was reinterpreted. This newly interpreted mode of ligation was found to agree better with the electron density observed not only in the holoenzyme but also the CO- and cyanide-inhibited structures. Hydrogen bonding was observed between the acyl group and the amide group of Cys176; however, no hydrogen bonding was observed between

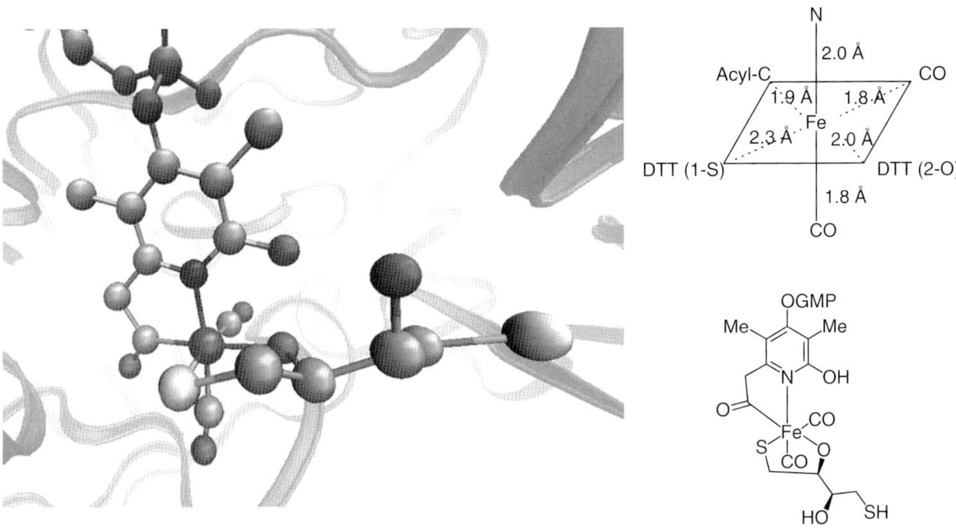

Figure 5.7 Crystal structure of FeGP cofactor binding site of DTT-treated, C176A-mutated [Fe]-hydrogenase holoenzyme isolated from *M. jannaschii*; 3F46 in RCSB Protein Data Bank. Fe center, two CO ligands, guanylyl pyridone ligand, and DTT shown as CPK structures. (Image generated using VMD [16].)

the pyridinol's hydroxyl group and His14, although an interaction with a solvent molecule anchored to Thr13 was observed.

To explain the reason for the original misinterpretation of the wild-type enzyme, the authors describe a "lack of imagination" concerning the possibility of an acyl ligand on the iron, which had resulted in a carboxylate facing toward the unpolar interior of protein. On the basis of these results, structure **8** (Figure 5.8) was proposed for the FeGP cofactor of [Fe] hydrogenase, where U is an unknown ligand. Further mass spectrometry and infrared spectroscopy evidence for structure **8** was recently reported by Shima *et al.* [19], who extracted the FeGP cofactor from [Fe] hydrogenase using acetic acid. The mass of an adduct with an acetate ligand binding in the Cys176 and "unknown" binding site was observed – previous attempts using 2-mercaptoethanol had proved unsuccessful for mass spectral analysis [19].

Recently, Shima and coworkers [20] carried out ^{13}C and ^{2}H labeling studies into the biosynthesis of FeGP cofactor **8** by feeding labeled acetate, pyruvate, and methionine to growing methanogenic archaea. The proposed pathway is shown (Scheme 5.1) and it was concluded that the ligands of the cofactor were derived from an acetate, two pyruvates, three carbon dioxides, and a methyl group from methionine [20].

In 2009, Hiromoto *et al.* [21] published the structure of a DTT-treated, C176A-mutated [Fe] hydrogenase with $CH{\equiv}H_4MPT^+$ **1** bound to it (Figure 5.9). The activity of this enzyme was inhibited by the formation of a DTT-chelate complex and

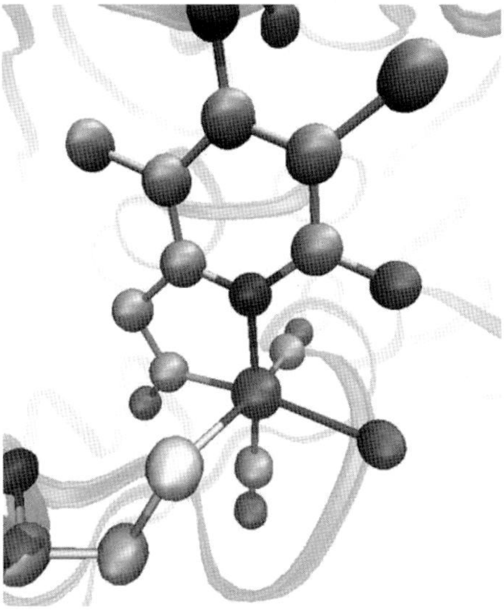

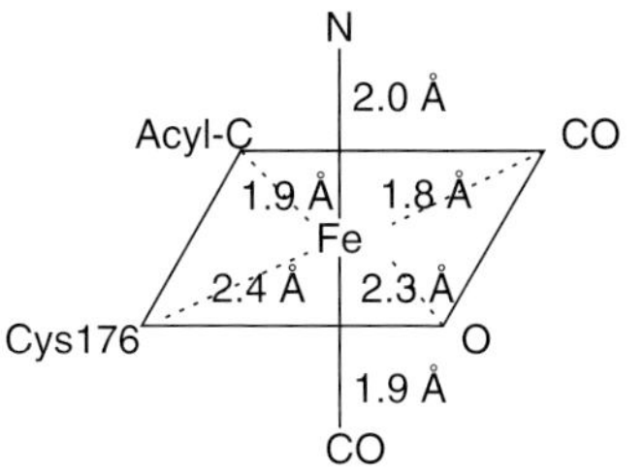

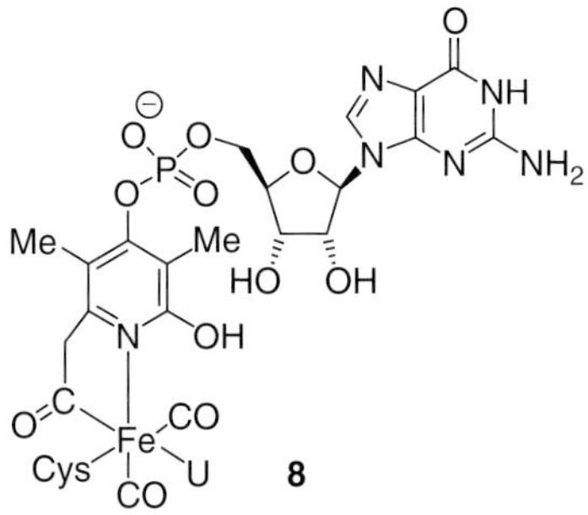

Figure 5.8 Reinterpreted crystal structure of the FeGP cofactor binding site of [Fe]-hydrogenase holoenzyme isolated from *M. jannaschii*; 3F47 in RCSB Protein Data Bank. Fe center, two CO ligands, guanylyl pyridone ligand, and DTT shown as CPK structures. (Image generated using VMD [16].)

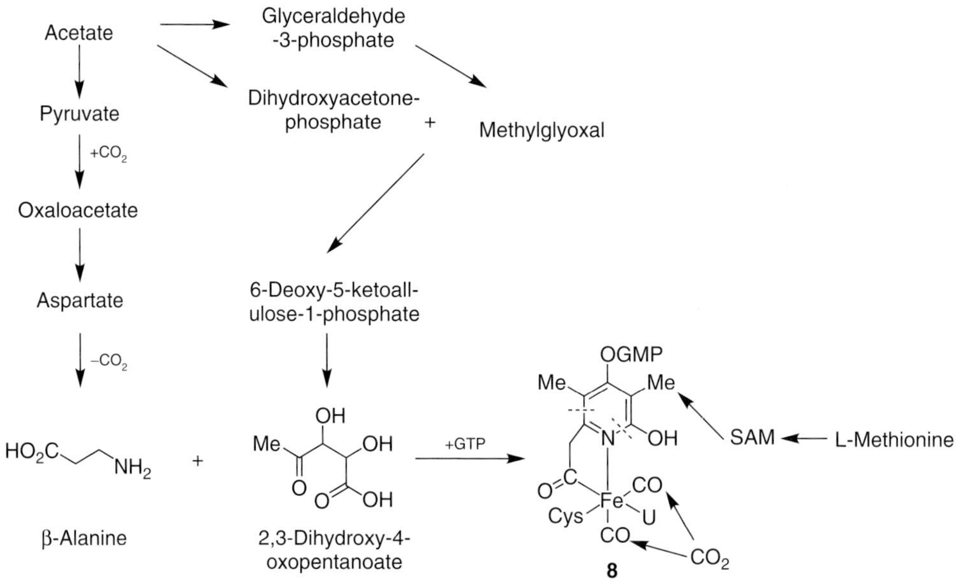

Scheme 5.1 Proposed biosynthesis of the FeGP cofactor.

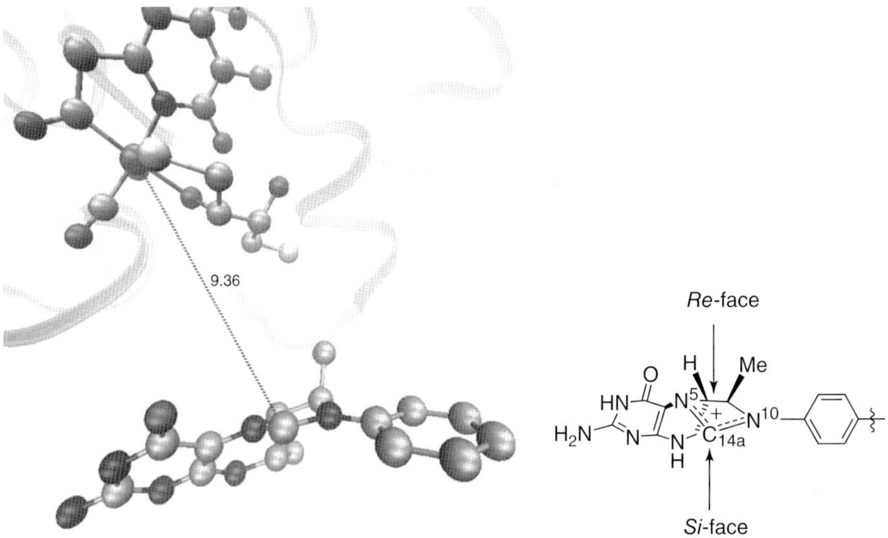

Figure 5.9 Crystal structure of $CH{\equiv}H_4MPT^+$ **1** binding in DTT-treated, C176A-mutated [Fe]-hydrogenase holoenzyme isolated from *M. jannaschii*; 3H65 in RCSB Protein Data Bank. Fe center, two CO ligands, guanylyl pyridone ligand, DTT, and $CH{\equiv}H_4MPT^+$ **1** shown as CPK structures. (Image generated using VMD [16].)

removal of Cys176, preventing the reaction of $CH{\equiv}H_4MPT^+$ **1**, which had previously hampered cocrystallization [21]. In the structure, the pterin, imidazolidine, and phenyl rings (the "head") of $CH{\equiv}H_4MPT^+$ **1** were present in the active cleft, with the remaining end (the "tail") of the molecule exposed to the bulk solvent and appearing disordered. As the conformation at N5 and N10 was planar in the observed X-ray structure of $CH{\equiv}H_4MPT^+$ **1** (Figure 5.9), whereas that in solution appeared to be tetrahedral (sp^3) at these two atoms (using NMR spectroscopy), the authors suggested that the crystal structure was not present in an activated conformation [21]. The crystal structure was deemed to be in an "open" form, leading to the conclusion that important interactions on the *Re*-face of substrate **1** were missing, allowing it to lie planar.

A distance of 9.3 Å lay between the iron center and C_{14a} of $CH{\equiv}H_4MPT^+$ **1** (Figure 5.9), which would be too long for transferring a hydride ion originating from an iron dihydrogen complex [21]. As such, the authors modeled closed forms of the apoenzyme with the FeGP cofactor **8** and substrate **1**, to explain the possible mechanism involved in the reaction of [Fe] hydrogenase. The models suggested that, in the closed form, the enzyme would be expected to position the iron center and C_{14a} of $CH{\equiv}H_4MPT^+$ **1** within 3 Å of each other. $CH{\equiv}H_4MPT^+$ **1** would be facing the "unknown" (U) site of the FeGP cofactor **8** and, therefore, it was reasoned that this site is where H_2 binds. These mechanistic investigations are described further in the next section, along with other mechanistic investigations.

5.6 Mechanistic Models of [Fe] Hydrogenase

5.6.1 Studies Before the Most Recent Assignment of the FeGP Cofactor

In 1995, before it had been established that [Fe] hydrogenase had a cofactor (and an essential iron), Berkessel and Thauer [22] presented an intriguing mechanistic proposal for the reaction of this enzyme based on superelectrophilic activation [23]. Their proposal was that protonation of $CH{\equiv}H_4MPT^+$ **1** in the enzyme active site would lead to formation of superelectrophilic dications **9** or **10** (Scheme 5.2). As Olah had shown that organic dications could abstract hydride from alkanes, the authors proposed that dications **9** or **10** might be active enough to abstract hydride from H_2. This proposal did not provide a role for the FeGP cofactor **8** that is now known to be present and essential to activity.

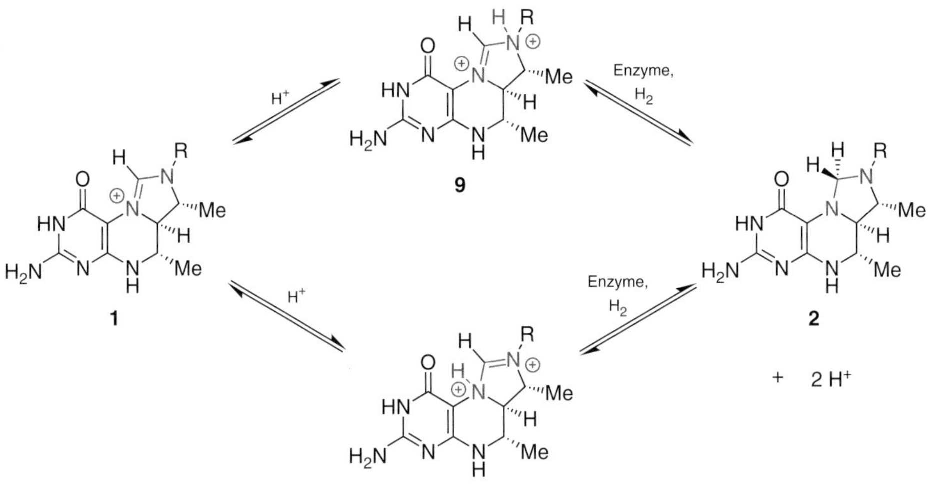

Scheme 5.2 Proposed superelectrophilic activation in [Fe] hydrogenase.

Inspired by the possibility of superelectrophilic activation in [Fe] hydrogenase, Corr *et al.* [24, 25] reported on the hydrogenation reaction of an amidine dication. The reduction of **11**–**14** (Scheme 5.3) in hydrogen was catalyzed by palladium on charcoal to give a 98% yield.

After the original (incorrect) crystal structure of the FeGP cofactor was suggested, Yang and Hall [26] carried out computational studies (density functional theory (DFT) calculations) on the possible role of the iron in activating hydrogen. They reported a trigger mechanism in which they used complex **16** (Scheme 5.4) as a model of the FeGP cofactor, where the 2-hydroxyl group has been moved to the 3-position in the pyridinol ligand that has also been used in the pyridone

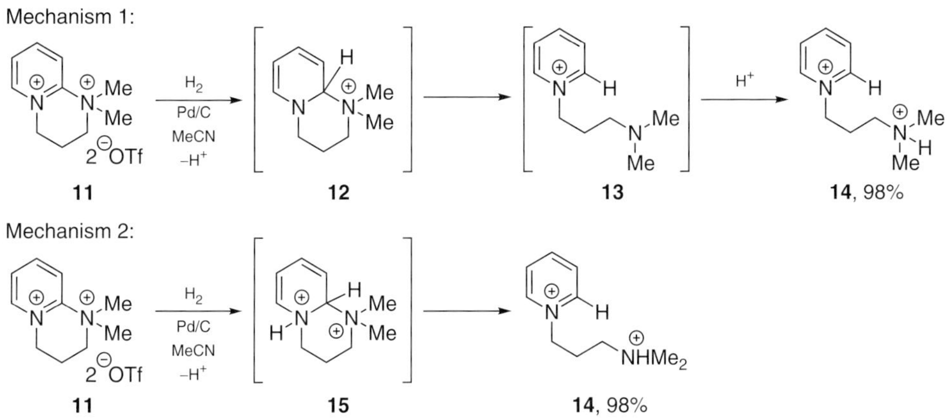

Scheme 5.3 Pd-catalyzed reaction of amidine dication **11** with H_2.

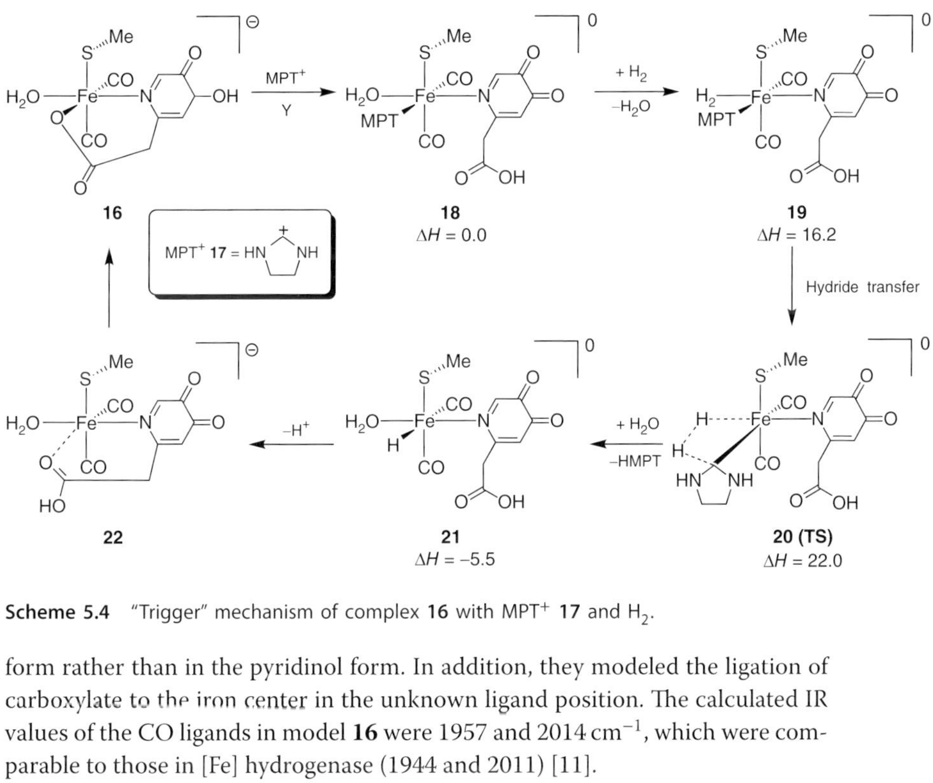

Scheme 5.4 "Trigger" mechanism of complex **16** with MPT^+ **17** and H_2.

form rather than in the pyridinol form. In addition, they modeled the ligation of carboxylate to the iron center in the unknown ligand position. The calculated IR values of the CO ligands in model **16** were 1957 and 2014 cm^{-1}, which were comparable to those in [Fe] hydrogenase (1944 and 2011) [11].

In the absence of MPT^+ **17** (the simplified model of $CH{\equiv}H_4MPT^+$ **1**), the splitting of H_2 was calculated to have a barrier of 39.7 kcal mol^{-1}. In the presence of MPT^+ **17**, the activation barrier was decreased to 22.0 kcal mol^{-1} (transition state **20**). Before reaching this transition state, the H_2/H_2O exchange to form structure **19** was calculated to occur at a value of 16.2 kcal mol^{-1}. This mechanism

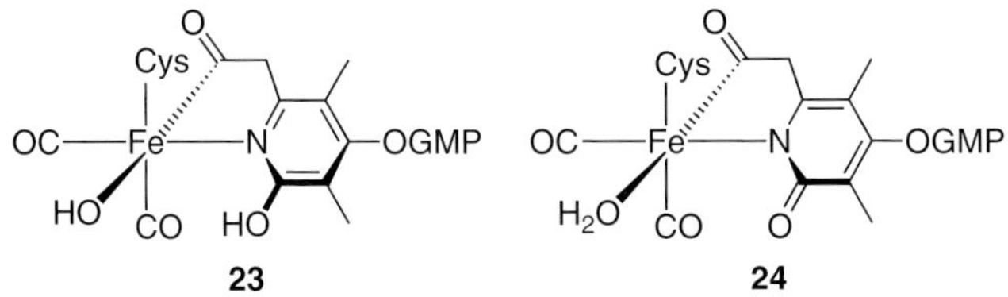

Figure 5.10 The proposed structure of the FeGP cofactor present in [Fe] hydrogenase suggested by Nakatani *et al.* [27].

was used to explain the dependence of H_2/H_2O exchange on the presence of MPT^+ **17**. The importance of the pyridone ligand was proposed to be due to its ability to "trigger" a change in its donor ability on the binding of MPT^+ **17**; this aided the polarization and cleavage of hydrogen and delivery of hydride to MPT^+ **17**.

5.6.2
Studies After the Most Recent Assignment of the FeGP Cofactor

As the original interpretation of the crystal structure [15] differed from the most recently proposed structure of the [Fe]-hydrogenase holoenzyme [18], Nakatani *et al.* [27] carried out DFT calculations on various model complexes of the active site to propose the mode of binding from the pyridinol ligand. On the basis of the comparison of calculated vibrational frequencies to those recorded experimentally for the [Fe]-hydrogenase enzyme, the authors suggested that the FeGP cofactor **8** was present as either complex **23** or complex **24** (Figure 5.10). Complex **23** consists of an acylpyridinol ligand, while complex **24** consists of an acylpyridone ligand. In addition, complex **23** has a hydroxide in the "unknown" coordination site, while complex **24** has a water molecule.

After the reinterpretation of the FeGP cofactor **8** by Hiromoto *et al.* [18], Yang and Hall [28] provided a new computational model for the possible mechanism in [Fe] hydrogenase. The authors based their DFT calculations on structure **25** (Figure 5.11) representing the FeGP cofactor **8**, where Cys176 is replaced by a methylthiolate ligand and the pyridinol ligand is simplified with the removal of 3,5-methyls and the guanosine monophosphate group. The site of the unknown ligand is occupied by a water molecule, which it was proposed can be exchanged with a molecule of water in an energetically neutral process aided by the favorable hydrogen bonding of the water molecule in bulk solution [28].

Scheme 5.5 shows the energy for exchange of hydrogen and water (taking complex **25** to start at 0 kcal mol^{-1} and without considering the favorable effect of solvation) to form complex **26** (8.8 kcal mol^{-1}). Through transition state **27** (with a free energy barrier of 6.6 kcal mol^{-1} above **26**), structure **28** can be formed, which is 3.4 kcal mol^{-1} more stable than **26**. In this structure, the dihydrogen has been cleaved, with one hydrogen atom bonded to the iron-bound-thiolate sulfur and the other hydrogen atom bonded to the iron center, stabilized by a dihydrogen bond to the 2-hydroxyl of the pyridinol ligand. This shows the potential importance of

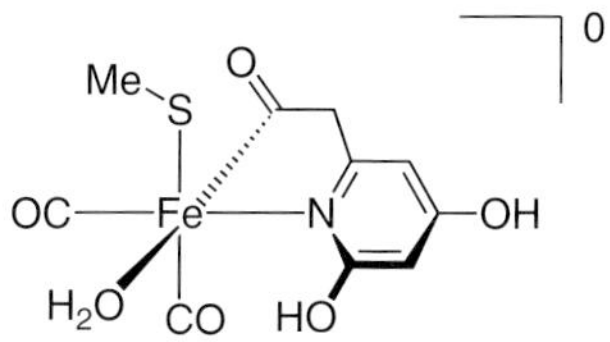

25

Figure 5.11 Yang and Hall's FeGP cofactor model **25**.

both the thiolate (Cys176) and the pyridinol hydroxyl group in the activation of hydrogen by [Fe] hydrogenase.

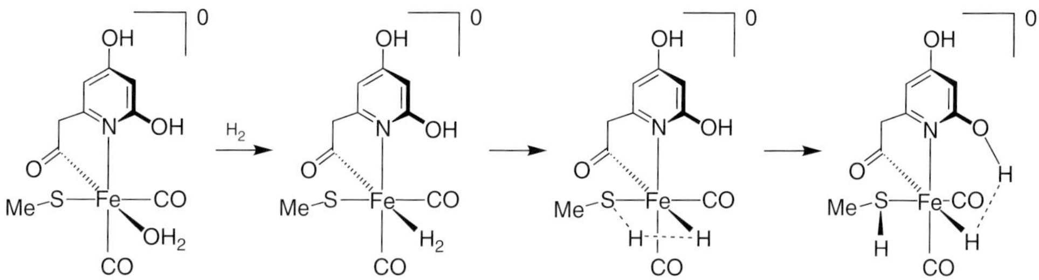

25 (0 kcal mol^{-1}) **26** (8.8 kcal mol^{-1}) **27 (TS)** (15.4 kcal mol^{-1}) **28** (5.4 kcal mol^{-1})

Scheme 5.5 Calculated energetics for the substitution of hydrogen with water in complex **25** and the subsequent formation of complex **28**.

The authors reported calculated values of 2007 and 1947 cm^{-1} for the vibrational frequencies of the two CO units in structure **25**, with structure **28** showing values of 2007 and 1949 cm^{-1} [28]. As these values showed very little difference and were very close to the infrared spectrum values found for the wild type [Fe]-hydrogenase enzyme (2011 and 1944 cm^{-1}) [11], it was suggested that the resting state of the enzyme could be represented by either structure. As the formation of **28** appeared to give a possible resting state, Yang and Hall [28] proceeded from this structure in their mechanistic calculations to provide a possible substrate-triggered hydride transfer mechanism.

The first pathway explored for hydrogen cleavage is represented in Scheme 5.6 and Figure 5.12. When Ph-MPT$^+$ **29** interacts with complex **28** to form complex **30**, a slight increase in energy is observed (0.4 kcal mol^{-1}) and a hydride can be transferred to Ph-MPT$^+$ **29** via transition state **31**, which lies at an energy value of 15.6 kcal mol^{-1} (the value for the total free energy value of the catalytic cycle). Simultaneous dissociation of Ph-HMPT and coordination of H_2 allows for the formation of complex **33**, which can proceed to reform complex **28** upon loss of a proton from two possible routes. The first route involves deprotonation of complex **33** on the pyridinol to give complex **34**, which was found to proceed at a higher energy than the route involving the deprotonation of the sulfur (i.e., via complex **26**).

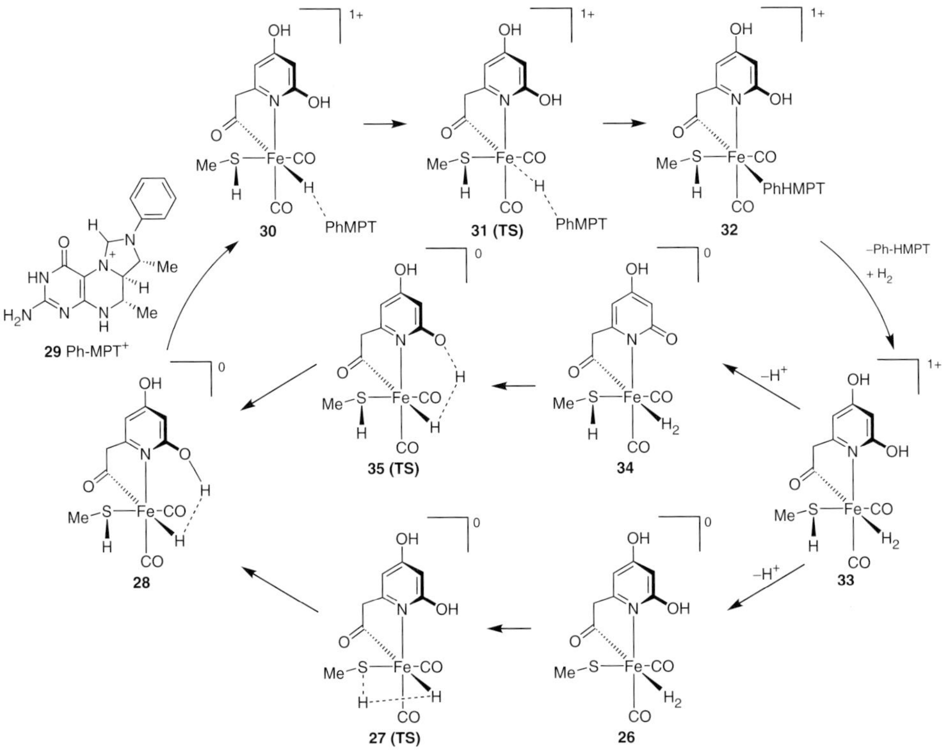

Scheme 5.6 Proposed mechanism for the reduction of Ph-MPT$^+$ **29** by complex **28**.

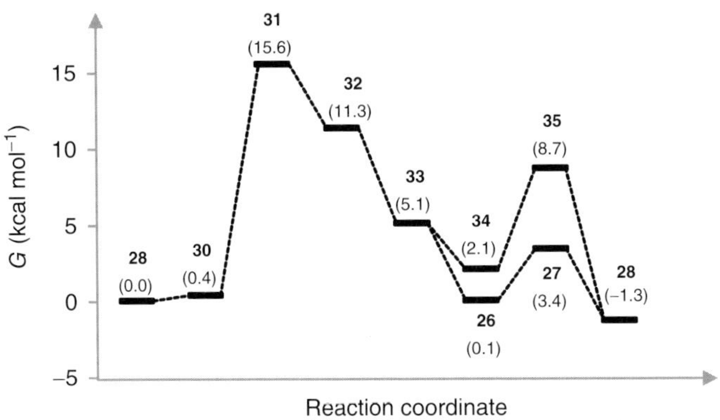

Figure 5.12 Calculated energy values for the mechanism shown in Scheme 5.6.

A second possible route to hydrogen transfer between the iron center and Ph-MPT$^+$ **29** was explored (see Scheme 5.7 and Figure 5.13) [28]. In this case, before the transfer of the hydride, complex **30** deprotonates to give complex **36** with an associated 16.3 kcal mol^{-1} increase in energy above complex **30**. A further 3.6 kcal mol^{-1} (to give transition state **37** at a total free energy of 20.3 kcal mol^{-1})

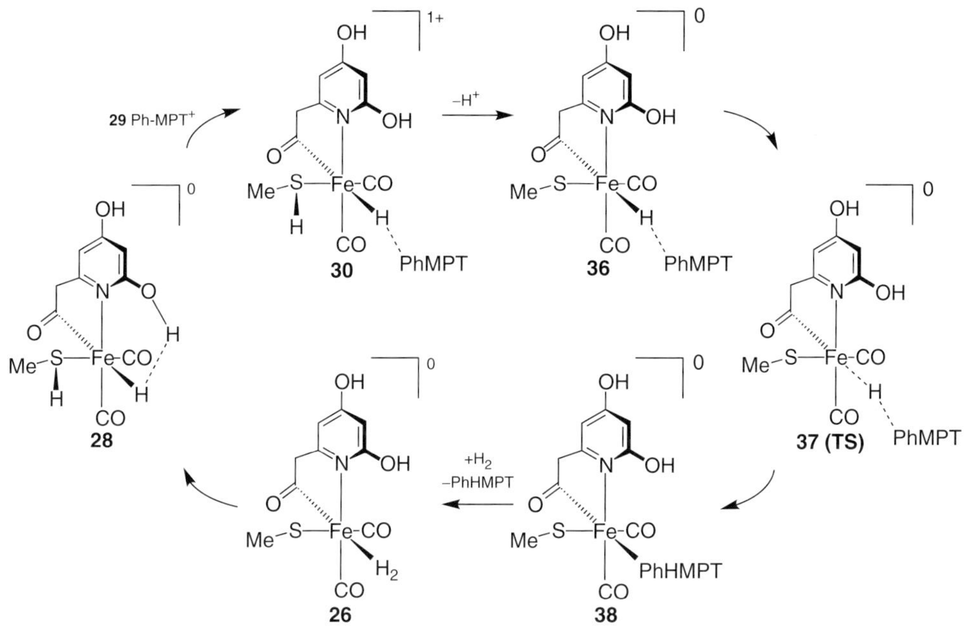

Scheme 5.7 A second proposed mechanism for the reduction of Ph-MPT$^+$ **29** by complex **28**.

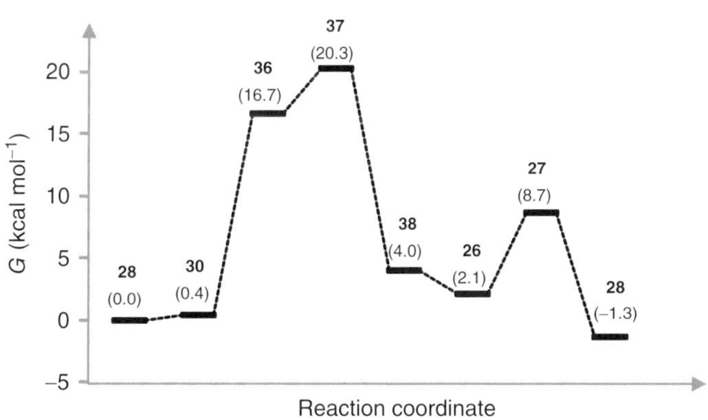

Figure 5.13 Calculated energy values for the mechanism shown in Scheme 5.7.

is then present for the transfer of hydride to the Ph-MPT and formation of complex **38**. Loss of Ph-HMPT and coordination of hydrogen then allows for the formation of complex **26**, which can go on to form complex **28** to allow the catalytic cycle to begin again.

In 2010, Dey [29] also published a DFT study of model FeGP cofactor complexes, in which complex **25** (Figure 5.14) was used as the resting complex. The substitution of the water ligand of complex **25** was investigated using CO, cyanide,

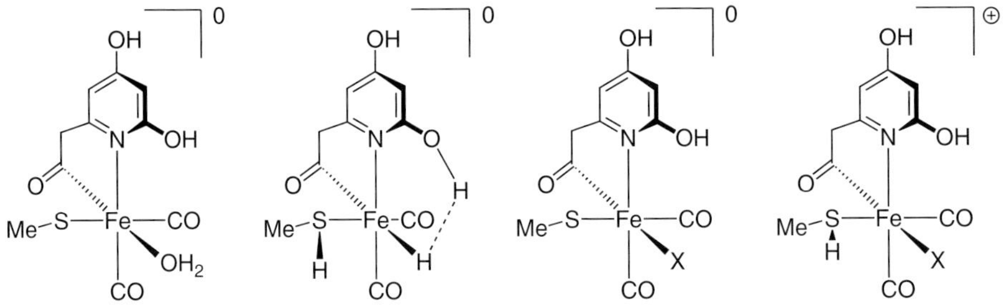

X = CO, CN⁻, H⁻, O₂, N₃⁻, or H₂

Figure 5.14 Complexes studied in the computational work of Dey [29].

hydride, oxygen, azide, and dihydrogen model complexes **39** (the corresponding thiol-bound complexes **40** were also investigated, as it was unclear if Cys176 was bound as a thiol or thiolate). In the case of $X = H_2$ for complex **39**, Dey reported the same structure as that reported by Yang and Hall [28] (complex **28**), in which heterolytic cleavage of dihydrogen (with the proton on the sulfur of the iron-bound thiolate and the hydride on the iron center) had been stabilized by a dihydrogen bond to the 2-hydroxy group of the pyridinol ligand. Removal of this dihydrogen bond by a 180° rotation of the 2-hydroxy group was associated with an 8 kcal mol^{-1} energy penalty. In this study, Dey's results agreed with Yang and Hall's findings [28] that substitution of water from complex **25** with H_2 was energetically unfavorable (without considering the energy gained from solvation). However, Dey [29] claimed that the space available between the iron center of the FeGP cofactor **8** and $CH{\equiv}H_4MPT^+$ **1** in the active (closed) form of [Fe] hydrogenase was too small to accommodate a water ligand. The author suggested that in the presence of substrate **1** and H_2, water would be dissociated from the FeGP cofactor and H_2 would be sufficiently small to bind in the free coordination site of the iron. This concept was also used to explain why CO and CN were not stronger inhibitors of [Fe] hydrogenase (both of these ligands were found to bind to the iron center more favorably than water in complex **39** where X = CO and CN^-), as both of these species are more sterically demanding than H_2.

As mentioned earlier, Hiromoto *et al.* [21] reported the crystal structure of a DTT-treated, (Cys176 to Ala176) mutated [Fe] hydrogenase in which $CH{\equiv}H_4MPT^+$ **1** was bound. In this study, they superimposed the structures of the apoenzyme, the holoenzyme, and the C176A-holoenzyme-substrate complex. This showed that the apoenzyme was in a "closed" form, while the holoenzyme was in an "open" form. The C176A-holoenzyme-substrate complex was shown to be closer to the "open" form, but had "closed" to some degree, with the slight rotation of the peripheral unit to the central cleft of the enzyme. The apoenzyme (which contains a PEG molecule in the $CH{\equiv}H_4MPT^+$ **1** binding position) and the C176A-holoenzyme-substrate complex contain a hydrogen bond between

Tyr267 and Ala282, while the holoenzyme contains a hydrogen bond between Tyr267 and Met252. In the apoenzyme and the C176A-holoenzyme-substrate complex, the amide of Met252 is hydrogen bonded to the PEG and the carbonyl of pterin-containing **1**, respectively. This is what prevents the Tyr267 hydrogen bonding to Met252 and causes a reorganization of the enzyme. Met252 and Ala255 as well as Tyr267 and Ala282 create new hydrogen bonds to each other and could be the cause of the peripheral unit rotating toward the central cleft. The more pronounced "closing" seen in the apoenzyme over the C176-holoenzyme-substrate complex is attributed to the presence of the DTT ligated onto the FeGP cofactor in the latter. It is suggested that this difference is enough to make the "closing" of the enzyme less energetically favorable.

The authors proposed a mechanism (Scheme 5.8) based on this "opening" and "closing" of the enzyme [21]. The first step involves binding of $CH{\equiv}H_4MPT^+$ **1** to the enzyme in the "open" form, which causes the enzyme to "close" and brings the $CH{\equiv}H_4MPT^+$ **1** binding site closer to the FeGP cofactor **8**. In the presence of hydrogen, H_2 binds to the "unknown" coordination site of (proposed) Fe^{II} center of the cofactor. The authors suggest that a side-on binding for dihydrogen then

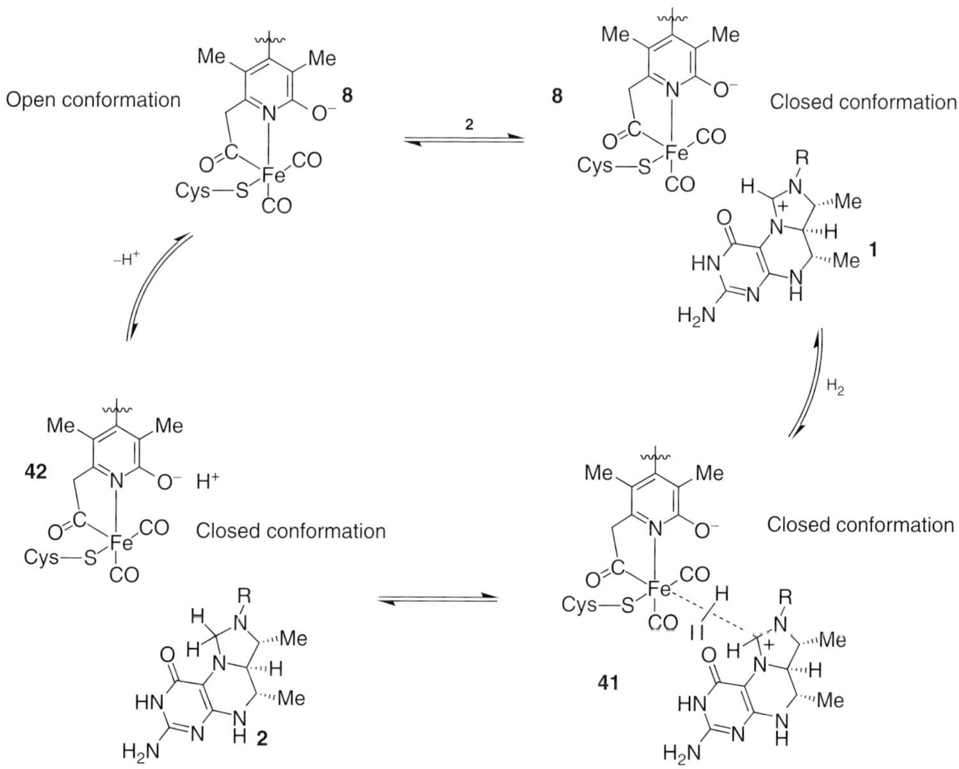

Scheme 5.8 Proposed "opening" and "closing" mechanism of [Fe] hydrogenase by Hiromoto *et al.* [21].

provides sufficient polarization of the molecule to allow hydride to be delivered to $CH{\equiv}H_4MPT^+$ **1** via complex **41**. At the same time, they suggest the proton could be accepted by the deprotonated pyridinol form complex **42** (equally, the sulfur of Cys176 was suggested as a possible proton acceptor). As the 2-hydroxy of the pyridinol is shown to be close to His14 (His14 to Ala14 mutation causes a big loss in activity), this might be a strong candidate for accepting a proton and then releasing it.

Further evidence for the conformational change of the [Fe]-hydrogenase enzyme upon the binding of $CH{\equiv}H_4MPT^+$ **1** and H_2 was obtained in the circular dichroism (CD) studies of Shima *et al.* [30]. In these studies, the CD signals of the enzyme were not significantly altered when $CH{\equiv}H_4MPT^+$ **1** was added to [Fe] hydrogenase under an atmosphere of N_2. However, when $CH{\equiv}H_4MPT^+$ **1** was added to the enzyme under an atmosphere of H_2, the CD signals were significantly altered. The authors described the change of signals in the 350–500 nm region as being indicative of changes in the environment of the FeGP cofactor **8** (in the presence of both substrate $CH{\equiv}H_4MPT^+$ **1** and H_2). As such, Shima and coworkers [30] concluded that these results support the idea that [Fe] hydrogenase rearranges to the "closed" form when substrate is bound.

5.6.3
Synthesized Model Complexes of the FeGP Cofactor

To date, several model complexes of the FeGP cofactor **8** have been synthesized; however, all of the compounds that have been described have had key differences to the natural complex.

In 2008, Royer *et al.* [31] reported the synthesis of rhodium and iridium complexes **43** and **44** as models for the active site of [Fe] hydrogenase. Indeed, complex **44** showed some hydrogenase-related reactivity by dehydrogenating alcohol **45** to acetophenone **46**, with a turnover number of 339 over 24 h and as a 50% solution in toluene. Although interesting for this dehydrogenation reactivity, the authors did not report any hydrogenation activity. In addition, the model did not contain the same metal as in the enzyme and once the crystal structure of the FeGP cofactor **8** was obtained [18], it was clear that the pyridinol ligated to the metal through the nitrogen and an acyl carbon rather than the nitrogen and a carboxylate oxygen (Scheme 5.9).

Sadique *et al.* [32] reported the synthesis of low-valent iron complex **47** (Figure 5.15) in which the center was found to be Fe^I, with vibrational carbonyl stretches at 1917 and 1996 cm^{-1}. This was the first example of a four-coordinate iron dicarbonyl complex and was synthesized before the reinterpreted crystal structure of the [Fe]-hydrogenase enzyme had been published [18], when the possibility of a four-coordinate cofactor was still being considered.

Complexes **48** and **49** (Figure 5.16) were prepared by Koch and coworkers [33]. These two complexes were reported to have infrared spectra that closely resembled the spectrum of the cyanide-inhibited [Fe]-hydrogenase enzyme and it was

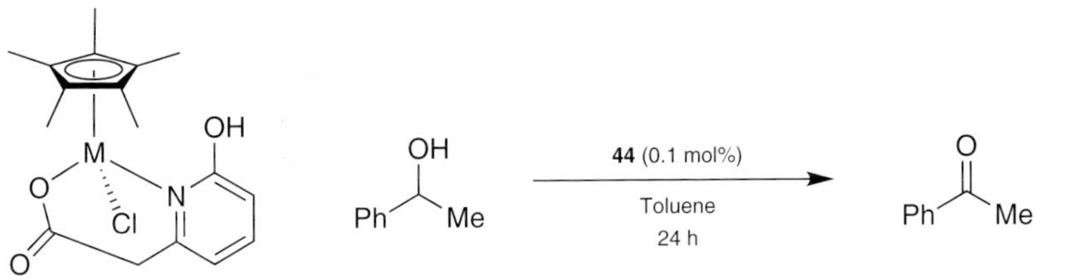

43, M = Rh
44, M = Ir

45

46, TON = 339

Scheme 5.9 Rhodium and iridium complexes **43** and **44** and the dehydrogenation of **45–46**.

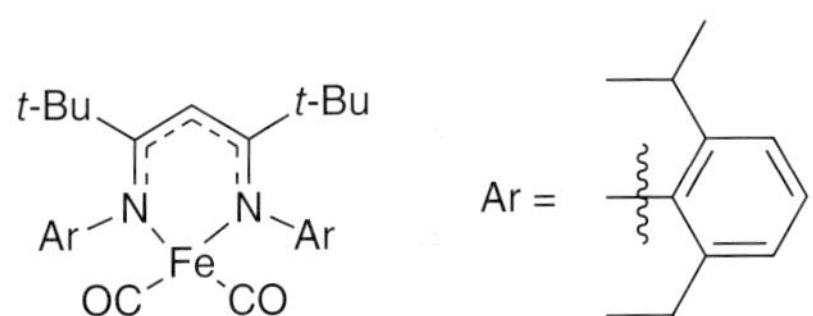

47

Figure 5.15 Four-coordinate Fe^{I} complex **47**.

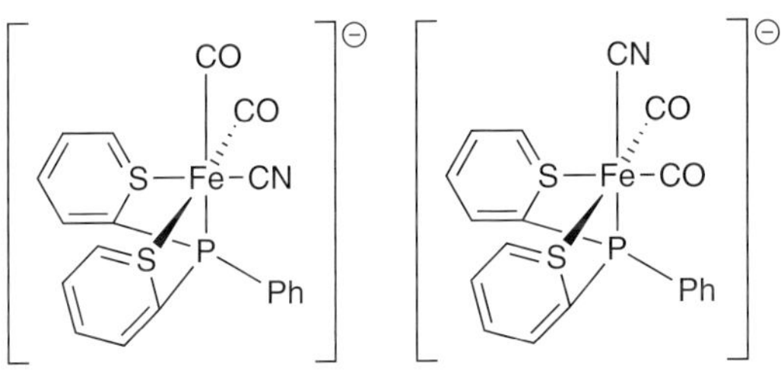

48

49

Figure 5.16 Complexes **48** and **49** prepared by Koch and coworkers [33].

suggested that the electron density at the iron center in the enzyme was similar to these complexes.

As part of an investigation into the ^{57}Fe nuclear resonance vibrational spectroscopy (NRVS) of the active site of [Fe] hydrogenase, Rauchfuss and coworkers [34, 35] prepared complex **50** (Figure 5.17). The comparison of the NRVS spectrum of this compound and the active site were used to propose a minimal ligand set for the enzyme cofactor consisting of two *cis*-carbonyls, a sulfur, and a light atom, suggested to be the nitrogen of the pyridinol cofactor.

Wang *et al.* [36] reported the synthesis of complex **51** (Figure 5.18), which had a tetradentate ligand composed of a pyridine, two thiolates, and a tertiary amine. This model was used to compare the Mössbauer and infrared spectra to those of [Fe] hydrogenase. The results were used to suggest that the iron center [Fe] hydrogenase was Fe^{II} and not Fe^{0} [36].

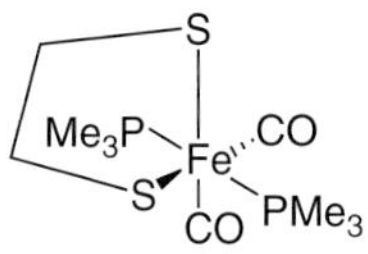

50

Figure 5.17 Complex **50** prepared by Rauchfuss and coworkers [34, 35].

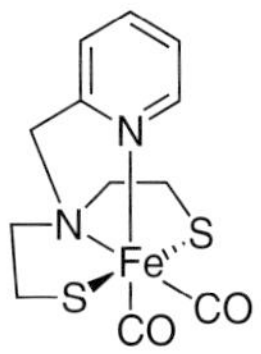

51

Figure 5.18 Complex **51** containing a tetradentate ligand.

In 2009, Hu and coworkers [37] synthesized complex **52**, which was also shown to be a precursor to compound **53** (Scheme 5.10). The carbonyl vibrational stretching values of these two complexes were 1987 and 2032 cm^{-1} and 1967 and 2021 cm^{-1}. The authors suggested that the complexes gave further evidence for the Fe^{II} state in the FeGP cofactor **8** of [Fe] hydrogenase. In both structures, the oxygen in the 2-position of the pyridine ligand was bound to the iron center, which had not been shown in the crystal structure of [Fe] hydrogenase [18].

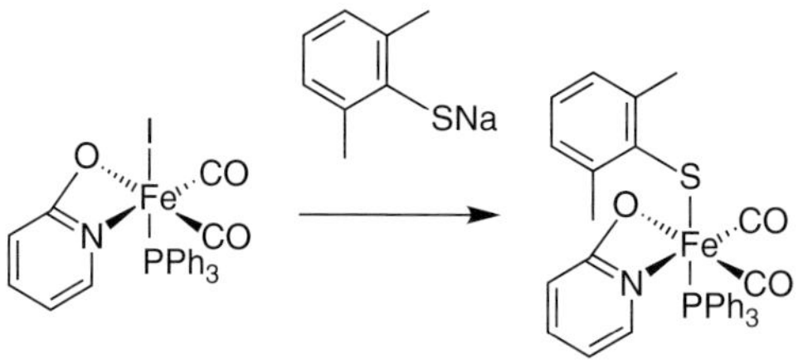

52 **53**

Scheme 5.10 Reaction of complex **52** to form complex **53**.

Darensbourg and coworkers [38] developed complexes **54** and **55** (Figure 5.19) as an analog of the FeGP cofactor **8**. In this complex, the oxygen in the 2-position of the pyridine ligand had been replaced by a sulfur and was shown to be bound as a thiolate to the metal center.

Later, Halder *et al.* [39, 40] synthesized complex **56** (Figure 5.20), which was the first FeGP mimic to contain a bond between Fe^{II} and a carbon ligand that was not a carbonyl. The lack of carbonyls and the fact that the carbon ligand was actually a carbene (of which there are none on the FeGP cofactor **8**) limited this system as a FeGP cofactor model.

In 2010, Darensbourg and coworkers [41] synthesized pentacoordinate complexes **57**, **58**, and **59** (Figure 5.21) as analogs of the [Fe] hydrogenase active site, citing the mechanistic proposal of Shima and coworkers that the "unknown" coordination site of the FeGP cofactor **8** could remain free before coordination

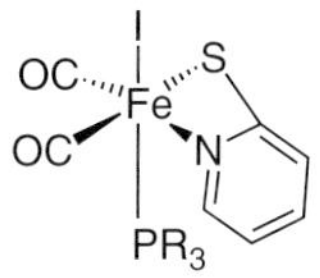

54, R = Cy
55, R = Ph

Figure 5.19 2-Thiopyridyl-containing complexes **54** and **55**.

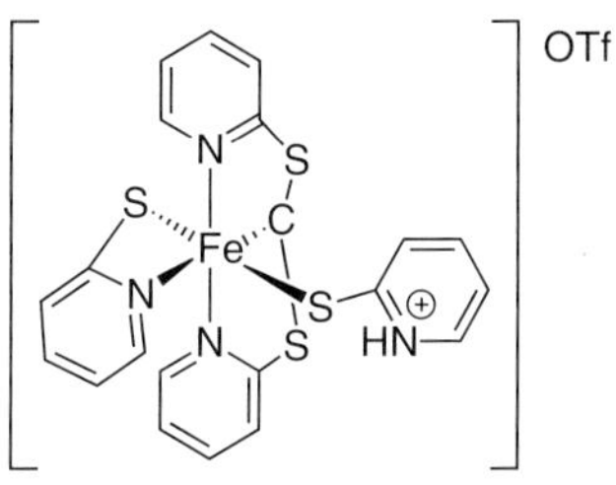

56

Figure 5.20 Complex **56** prepared by Halder *et al.* [39, 40].

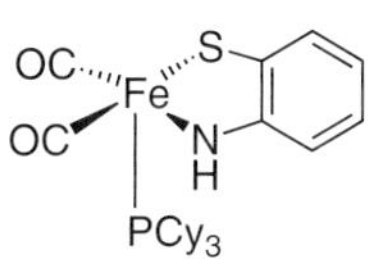

57

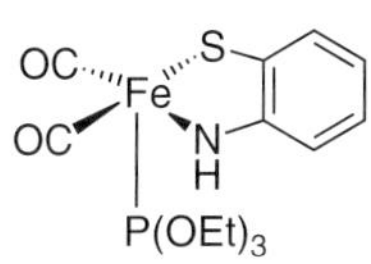

58

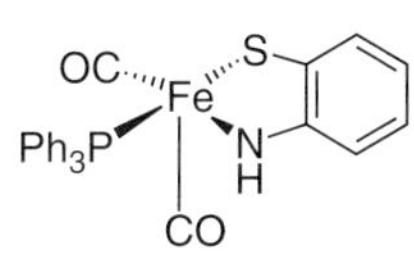

59

Figure 5.21 Pentacoordinate complexes **57–59** synthesized by Darensbourg and coworkers [41].

of hydrogen and delivery of hydride to $CH{\equiv}H_4MPT^+$ **1** [21]. No reactivity with hydrogen was discussed, but it was found that when complex **57** was protonated, it could accommodate an additional CO in the vacant site of iron. Under these conditions, extrinsic ^{13}CO was found to exchange with intrinsic CO [41], which differs from the behavior of [Fe] hydrogenase in which the intrinsic CO ligands are not exchanged [4].

Tanino *et al.* [42] synthesized complex **60** (Figure 5.22) as a model for the active site of [Fe] hydrogenase. The authors highlighted this complex for its *fac*$\{(CO)_3Fe\}$ group and suggested this compound could be used as a model for the CO-inhibited [Fe] hydrogenase. Differences from the FeGP were highlighted, such as the bromide ligand instead of an acyl ligand and the presence of π-stacking between the pyridine ring and one of the mesityl groups of the thiolate ligand [42].

Hu and coworkers [43] synthesized [Fe]-hydrogenase model complexes **61–64** (Figure 5.23) with acyl ligands and tight bite angle 2-thiopyridyl ligands. The authors reported that all of the carbonyls of complex **61** were exchangeable with ^{13}CO (including the acyl carbonyl), but no reactivity was observed with hydrogen.

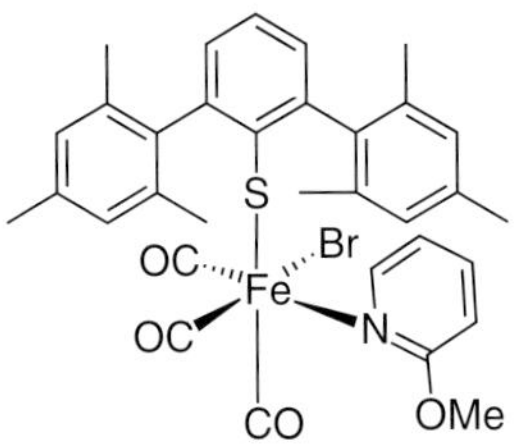

60

Figure 5.22 Model complex **60** with *fac*{$(CO)_3Fe$} group.

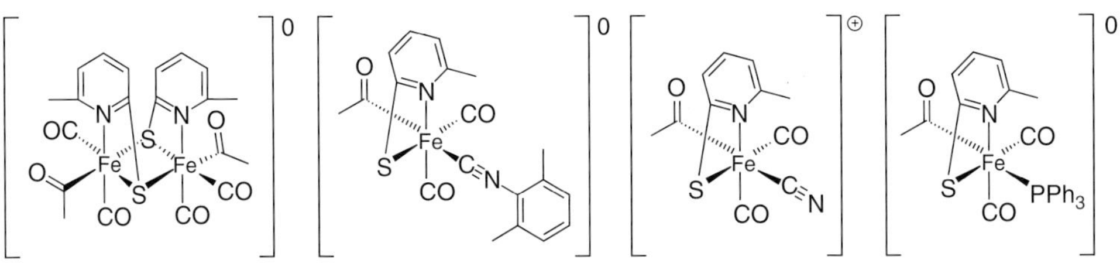

61 62 63 64

Figure 5.23 2-Thiopyridyl-containing complexes **61–64**.

Later, Pickett and coworkers [44] synthesized carbamoyl complex **65** (Scheme 5.11), of which they obtained a crystal structure that showed the same stereochemical arrangement as obtained for the FeGP cofactor **8** in [Fe] hydrogenase [18]. In the same journal issue, Hu and coworkers [45] synthesized compounds **66** and **68** (with the crystal structure of the latter disclosed), which was the first example of an [Fe]-hydrogenase model complex containing an acylmethylpyridinyl ligand, similar to the natural cofactor. In the case of compound

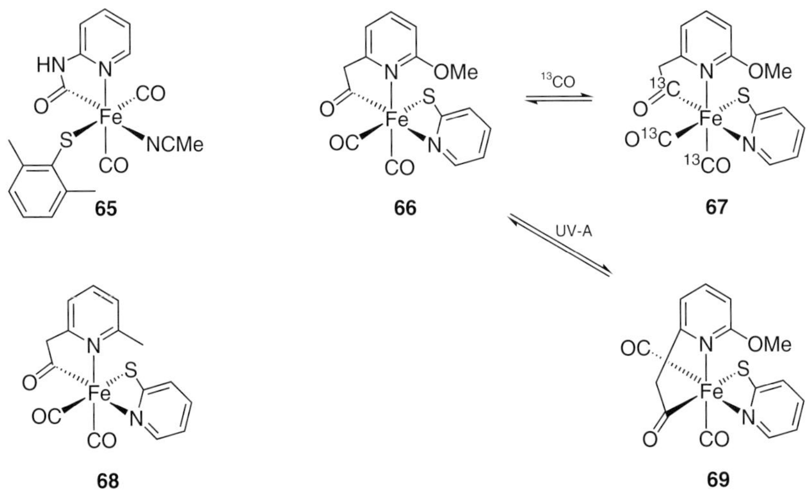

Scheme 5.11 Carbamoyl complex **65** and acylmethylpyridinyl complexes **66** and **68**.

66, when monitored by ^{13}C NMR, all of the carbonyls are observed to exchange with ^{13}CO to form compound **67**; a phenomenon not observed in [Fe] hydrogenase, in which the intrinsic carbonyls are not exchanged [4]. In the presence of UV-A light and absence of extrinsic CO, complex **66** isomerizes to complex **69** [45].

Royer *et al.* [46, 47] synthesized model complex **70** (Scheme 5.12), which they found to be indefinitely stable as a solid at room temperature, but that slowly decomposed as a solution kept at −10 °C. This complex was reacted with Et_4NCN to afford complex **71**, in which the cyanide ligand had been placed in the position trans to the acyl group, mimicking the stereochemical outcome of the reaction of cyanide with the FeGP cofactor **8** [15, 18], where the cyanide is placed in the "unknown" position trans to the acyl group of the pyridinol ligand. Complex **72** was additionally prepared by the protonation of complex **70**, in which the authors suggested that protonation of the thiolate sulfur or the acyl oxygen had occurred.

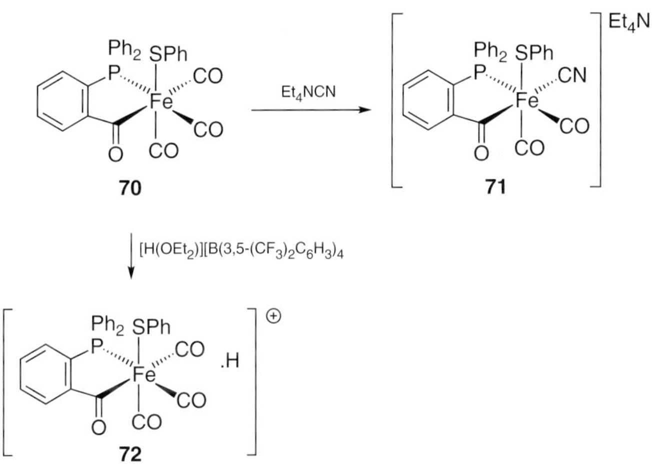

Scheme 5.12 Reactions of acyl-containing complex **70**.

In 2011, Hu and coworkers [48] isolated complex **73** (Scheme 5.13) and obtained a crystal structure. They found that this five-coordinate complex could accommodate a third carbonyl ligand in the presence of CO. With ^{13}C NMR and the use of ^{13}CO gas, the authors followed the incorporation of the sixth ligand trans to the acyl group to form complex **74** and then followed the subsequent exchange of the intrinsic carbonyl ligands to form complexes **75** and **76** (Scheme 5.13). The authors also reported that no H/D exchange occurred between H_2O, H_2, and D_2 in the presence of complex **73**, proposing that a mimic of $CH{\equiv}H_4MPT^+$ **1** may need to be present. Complex **73** was observed to be unreactive toward donor ligands such as water, acetonitrile, pyridine, and triethylamine, as well as with O_2. Owing to the fact that PPh_3 and 2-mercapto-6-methylpyridine were found to react with **73**, it was judged that the steric effect of the 2-methoxy of the pyridinyl ligand was not enough to account for the lack of reactivity with donor ligands. The authors suggested that the strong trans influence of the acyl ligand diminishes

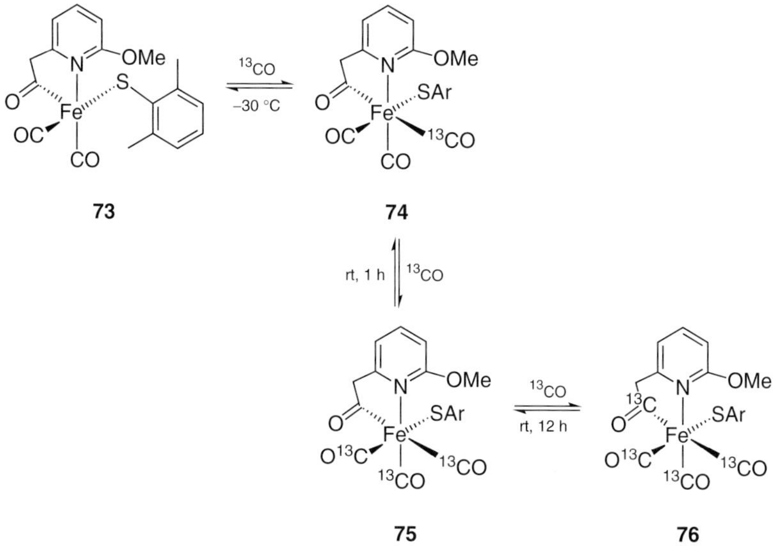

Scheme 5.13 Five-coordinate complex **73** and its reactivity with ^{13}CO.

the ligand-binding affinity at the vacant site and that the iron center was also too electron rich to bind σ-only donors and poor π acceptors.

More recently, Hu and coworkers [49] studied the protonation chemistry of complex 77 (Scheme 5.14). In this work, complex 77 was protonated in the presence of a donor ligand (MeCN or pyridine) to afford complexes **78** and **79**, where the thiolate is lost as a thiol. Complex 77 can then be regenerated from complexes **78** and **79** by treating them with sodium thiolate or the thiol in the presence of triethylamine (Scheme 5.14). The authors proposed that this reversible protonation process indicates that Cys176 in [Fe] hydrogenase is the immediate proton acceptor in the presence of H_2 when hydride is delivered to $CH{\equiv}H_4MPT^+$ **1**.

Hu and coworkers [50] also disclosed the infrared and Mössbauer analysis of complexes **52**, **55**, **61–64**, and **80–88** (Figure 5.24). The Mössbauer spectrum [13] of [Fe] hydrogenase had previously shown an isomeric shift of $\delta = +0.0\,\mathrm{mm\,s^{-1}}$ (+0.03 for extracted cofactor) and a positive quadrupole splitting of $\Delta E_Q = +0.65\,\mathrm{mm\,s^{-1}}$ (+0.43 for the extracted cofactor) for the diamagnetic iron center. Hu [50] reported the field-dependent Mössbauer study of complexes **61–64**, **80**, **83**, **85**, **87**, and **88**. The isomeric shifts of these complexes all lie in the range between 0.1 and $-0.02\,\mathrm{mm\,s^{-1}}$. Substitution of an iodide (e.g., in complex **80**) by a thiolate (e.g., in complexes **85** and **87**) showed an average of a $0.05\,\mathrm{mm\,s^{-1}}$ decrease in isomeric shift, while substitution of iodide (e.g., in complex **83**) by an acyl (e.g., in complex **64**) showed a decrease of $0.1\,\mathrm{mm\,s^{-1}}$ in isomeric shift. The isomeric shifts of complexes **62**, **63**, and **88** were on the lower end of the isomeric shift values (0.00, −0.02, and $0.01\,\mathrm{mm\,s^{-1}}$, respectively) because of the presence of π-acceptor ligands (CN^- and RNC). This observation correlated well with the Mössbauer studies of the [Fe]-hydrogenase enzyme [13].

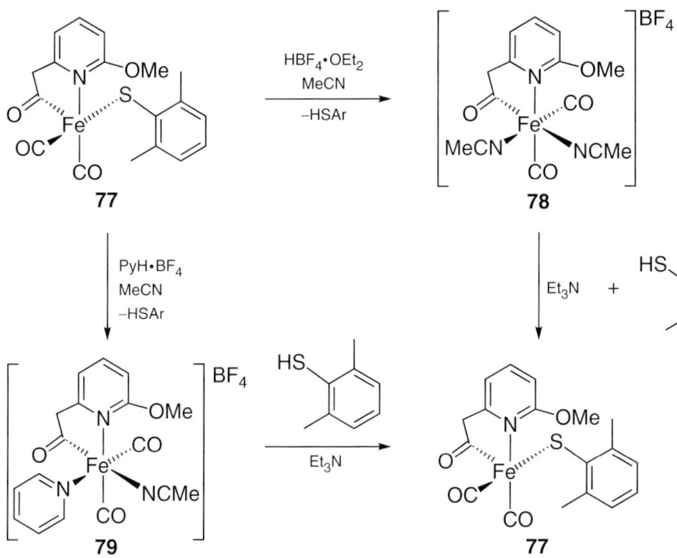

Scheme 5.14 Complex **77** and the reversible loss of the thiolate ligand upon protonation.

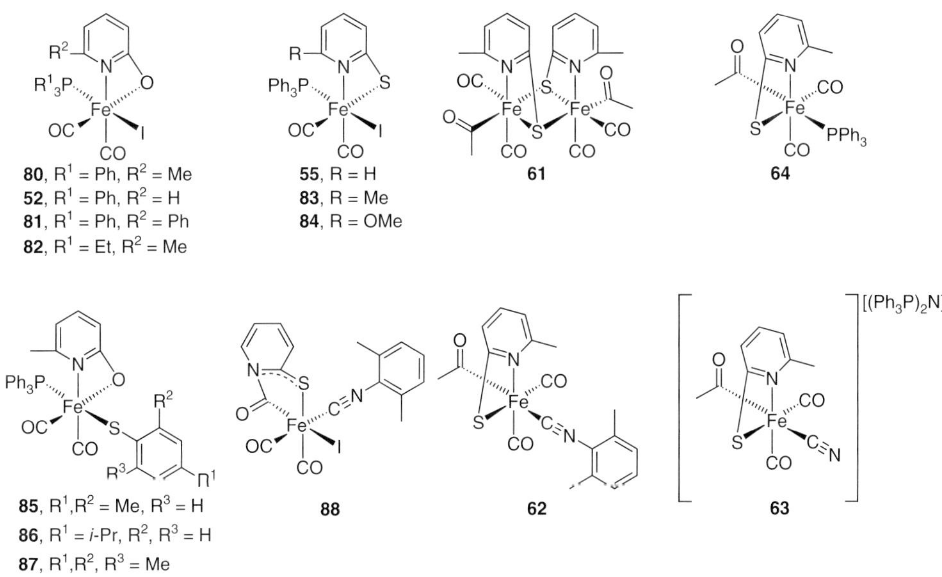

Figure 5.24 Various complexes synthesized by Hu and coworkers and analyzed by Mössbauer spectroscopy.

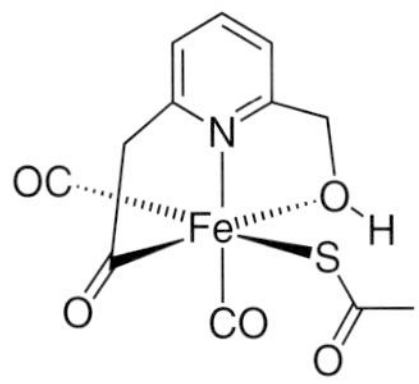

89

Figure 5.25 Acylmethylpyridyl-complex **89**, which contains a hydroxyl group.

Recently, Song and coworkers [51] reported structure **89** (Figure 5.25) as a biomimetic model of the active site of [Fe] hydrogenase. Other than the rhodium and iridium complexes of Royer *et al.* [31], this structure represented the first model complex of [Fe] hydrogenase with a hydroxyl-containing ligand. X-ray crystallography showed the distance between the iron and the hydroxyl oxygen was 2.128 Å (with an N–Fe–O angle of 78.13°), which is considerably shorter than the distance of 3.19 Å (with an N–Fe–O angle of 46.76°) between the hydroxyl oxygen and the iron in the [Fe]-hydrogenase enzyme [18]. As there is the methylene between the hydroxyl and the pyridine ligand (unlike the pyridinol ligand of the FeGP cofactor **8**), this allows the oxygen to fill a ligand position on the iron and affects the properties of the hydroxyl (e.g., pK_a). The authors suggest that the binding of the hydroxyl to the iron center may mimic the potential binding of water to the "unknown" ligand site in [Fe] hydrogenase [18].

As can be seen from this section, there is an abundance of interest in synthesizing model complexes of the [Fe]-hydrogenase active site. However, to date only one model complex has shown hydrogenase reactivity (dehydrogenase reactivity, to be precise) [31] and this is an iridium complex. With the increasing desire to convert carbon dioxide (industrially produced as well as atmospheric) into useful feedstock chemicals, understanding the way in which nature utilizes carbon dioxide is increasingly significant. As an enzyme involved in a step of the conversion of carbon dioxide to methane, [Fe] hydrogenase could be a vital tool in the production of industrially useful chemicals. Indeed, understanding and mimicking the chemistry carried out in [Fe] hydrogenase could aid the development of future industrially relevant processes and it is this potential that will hopefully see the synthesis of more model complexes of the [Fe]-hydrogenase active site.

References

1. Thauer, R.K., Kaster, A.-K., Goenrich, M., Schick, M., Hiromoto, T., and Shima, S. (2010) *Annu. Rev. Biochem.*, **79**, 507–536.
2. Zirngibl, C., Hedderich, R., and Thauer, R.K. (1990) *FEBS Lett.*, **261**, 112–116.
3. Afting, C., Hochheimer, A., and Thauer, R.K. (1998) *Arch. Microbiol.*, **169**, 206–210.
4. Zirngibl, C., Van Dongen, W., Schwörer, B., Von Bünau, R., Richter, M., Klein, A., and Thauer, R.K. (1992) *Eur. J. Biochem.*, **208**, 511–520.
5. Schwörer, B., Fernandez, V.M., Zirngibl, C., and Thauer, R.K. (1993) *Eur. J. Biochem.*, **212**, 255–261.
6. Schleucher, J., Griesinger, C., Schwörer, B., and Thauer, R.K. (1994) *Biochemistry*, **33**, 3986–3993.

7. Hartmann, G.C., Santamaria, E., Fernandez, V.M., and Thauer, R.K. (1996) *J. Biol. Inorg. Chem.*, **1**, 446–450.
8. Vogt, S., Lyon, E.J., Shima, S., and Thauer, R.K. (2008) *J. Biol. Inorg. Chem.*, **13**, 97–106.
9. Buurman, G., Shima, S., and Thauer, R.K. (2000) *FEBS Lett.*, **485**, 200–204.
10. Lyon, E.J., Shima, S., Buurman, G., Chowdhuri, S., Batschauer, A., Steinback, K., and Thauer, R.K. (2004) *Eur. J. Biochem.*, **271**, 195–204.
11. Lyon, E.J., Shima, S., Böcher, R., Thauer, R.K., Grevels, F.-W., Bill, E., Roseboom, W., and Albracht, S.P.J. (2004) *J. Am. Chem. Soc.*, **126**, 14239–14248.
12. Shima, S., Lyon, E.J., Sordel-Klippert, M., Kauβ, M., Kahnt, J., Thauer, R.K., Steinbach, K., Xie, X., Verier, L., and Griesing, C. (2004) *Angew. Chem. Int. Ed.*, **43**, 2547–2551.
13. Shima, S., Lyon, E.J., Thauer, R.K., Mienert, B., and Bill, E. (2005) *J. Am. Chem. Soc.*, **127**, 10430–10435.
14. Pilak, O., Mamat, B., Vogt, S., Hayemeier, C.H., Thauer, R.K., Shima, S., Vonrhein, C., Warkentin, E., and Ermler, U. (2006) *J. Mol. Biol.*, **358**, 798–809.
15. Shima, S., Pilak, O., Vogt, S., Schick, M., Stagni, M.S., Meyer-Klaucke, W., Warkentin, E., Thauer, R.K., and Ermler, U. (2008) *Science*, **321**, 572–575.
16. Humphrey, W., Dalke, A., and Schulten, K. (1996) *J. Mol. Graphics*, **14**, 33–38.
17. Buncel, E. and Menon, B. (1977) *J. Am. Chem. Soc.*, **99**, 4457–4461.
18. Hiromoto, T., Ataka, K., Pilak, O., Vogt, S., Stagni, M.S., Meyer-Klaucke, W., Wakentin, E., Thauer, R.K., Shima, S., and Ermler, U. (2009) *FEBS Lett.*, **583**, 585–590.
19. Shima, S., Schick, M., Kahnt, J., Ataka, K., Steinback, K., and Linne, U. (2012) *Dalton Trans.*, **41**, 767–771.
20. Schick, M., Xie, X., Ataka, K., Kahnt, J., Linne, U., and Shima, S. (2012) *J. Am. Chem. Soc.*, **134**, 3271–3280.
21. Hiromoto, T., Warkentin, E., Moll, J., Ermler, U., and Shima, S. (2009) *Angew. Chem. Int. Ed.*, **48**, 6457–6460.
22. Berkessel, A. and Thauer, R.K. (1995) *Angew. Chem., Int. Ed. Engl.*, **34**, 2247–2250.
23. Olah, G., Germain, A., Lin, H.C., and Forsyth, D.A. (1975) *J. Am. Chem. Soc.*, **97**, 2928–2929.
24. Corr, M.J., Gibson, K.F., Kennedy, A.R., and Murphy, J.A. (2009) *J. Am. Chem. Soc.*, **131**, 9174–9175.
25. Corr, M.J., Roydhouse, M.D., Gibson, K.F., Zhou, S.-Z., Kennedy, A.R., and Murphy, J.A. (2009) *J. Am. Chem. Soc.*, **131**, 17980–17985.
26. Yang, X. and Hall, M.B. (2008) *J. Am. Chem. Soc.*, **130**, 14036–14037.
27. Nakatani, N., Nakao, Y., Sato, H., and Sakaki, S. (2009) *Chem. Lett.*, **38**, 958–959.
28. Yang, X. and Hall, M.B. (2009) *J. Am. Chem. Soc.*, **131**, 10901–10908.
29. Dey, A. (2010) *J. Am. Chem. Soc.*, **132**, 13892–13901.
30. Shima, S., Vogt, S., Göbels, A., and Bill, E. (2010) *Angew. Chem. Int. Ed.*, **49**, 9917–9921.
31. Royer, A.M., Rauchfuss, T.B., and Wilson, S.R. (2008) *Inorg. Chem.*, **47**, 395–397.
32. Sadique, A.R., Brennessel, W.W., and Holland, P.L. (2008) *Inorg. Chem.*, **47**, 784–786.
33. Melarejo, D.Y., Chiarella, G.M., and Koch, S.A. (2007) Synthetic Analogs for the Iron Centers in Ni-Fe and the Iron-Sulfur Cluster Free Hydrogenase Enzymes, 234th ACS National Meeting, Boston, MA, August 19–23, 2007, Meeting Abstracts; *http://oasys2.confex.com/acs/234nm/techprogram/P1116341.HTM* (accessed 18 March 2014).
34. Guo, Y., Wang, H., Xiao, Y., Vogt, S., Thauer, R.K., Shima, S., Volkers, P.I., Rauchfuss, T.B., Pelmenshikov, V., Case, D.A., Alp, E.E., Sturhahn, W., Yoda, Y., and Cramer, S. (2008) *Inorg. Chem.*, **47**, 3969–3977.
35. Salomone-Stagni, M., Stellato, F., Whaley, C.M., Vogt, S., Morante, S., Shima, S., Rauchfuss, T.B., and Meyer-Klaucke, W. (2010) *Dalton Trans.*, **39**, 3057–3064.
36. Wang, X., Li, Z., Zeng, X., Luo, Q., Evans, D.J., Picket, C.J., and Liu, X. (2008) *Chem. Commun.*, **44**, 3555–3557.

37. Obrist, B.V., Chen, D., Ahrus, A., Schünemann, V., Scopelliti, R., and Hu, X. (2009) *Inorg. Chem.*, **48**, 3514–3516.
38. Li, B., Liu, T., Popescu, C.V., Bilko, A., and Darensbourg, M.Y. (2009) *Inorg. Chem.*, **48**, 11283–11289.
39. Halder, P., Dey, A., and Paine, T.K. (2009) *Inorg. Chem.*, **48**, 11501–11503.
40. Halder, P. and Paine, T.K. (2011) *Inorg. Chem.*, **50**, 708–710.
41. Liu, T., Li, B., Popescu, C.V., Bilko, A., Pérez, L.M., Hall, M.B., and Darensbourg, M.Y. (2010) *Chem. Eur. J.*, **16**, 3083–3089.
42. Tanino, S., Ohki, Y., and Tatsumi, K. (2010) *Chem. Asian J.*, **5**, 1962–1964.
43. Chen, D., Scopelliti, R., and Hu, X. (2010) *J. Am. Chem. Soc.*, **132**, 928–929.
44. Turrell, P.J., Wright, J.A., Peck, J.N.T., Oganesyan, V.S., and Pickett, C.J. (2010) *Angew. Chem. Int. Ed.*, **49**, 7508–7511.
45. Chen, D., Scopelliti, R., and Hu, X. (2010) *Angew. Chem. Int. Ed.*, **49**, 7512–7515.
46. Royer, A.M., Rauchfuss, T.B., and Gray, D.L. (2009) *Organometallics*, **28**, 3618–3620.
47. Royer, A.M., Salomone-Stagni, M., Rauchfuss, T.B., and Meyer-Klaucke, W. (2010) *J. Am. Chem. Soc.*, **132**, 16997–17003.
48. Chen, D., Scopelliti, R., and Hu, X. (2011) *Angew. Chem. Int. Ed.*, **50**, 5671–5673.
49. Chen, D., Scopelliti, R., and Hu, X. (2012) *Angew. Chem. Int. Ed.*, **51**, 1919–1921.
50. Chen, D., Ahrens-Botzong, A., Schünemann, V., Scopelliti, R., and Hu, X. (2011) *Inorg. Chem.*, **50**, 5249–5257.
51. Song, L.-C., Xie, Z.-J., Wang, M.-M., Zhao, G.-Y., and Song, H.-B. (2012) *Inorg. Chem.*, **51**, 7466–7468.

6
DFT Investigation of Models Related to the Active Site of Hydrogenases

Claudio Greco and Luca De Gioia

6.1
Introduction

The development of quantum-chemical theories and methods, coupled with the present efficiency of CPUs, has recently allowed the application of quantum mechanics (QM) computational tools to the investigation of large molecular systems, such as the active site of enzymes and related bioinspired synthetic compounds [1, 2]. Moreover, the development and critical evaluation of hybrid QM/MM (molecular mechanics) methods, where the reactive portion of a molecule is described at the QM level of theory, whereas its molecular environment, including the solvent molecules, is described using the MM theoretical description, has allowed the investigation of entire enzymes, and consequently to shed light on the interplay between the intrinsic chemical properties of the active site and the effects on its properties due to the molecular environment [3]. For all these reasons, QM and QM/MM calculations have been extensively used to dissect reaction pathways of hydrogenases and related models [4]. In fact, QM and QM/MM investigations not only allowed investigation of the catalytic mechanism of hydrogenases but they also led to a better comprehension of the atomic composition of the active site itself and of the ligands that are transiently bound to the catalytically active metal centers in specific states of the enzyme. In such a context, it has to be remarked that the elusive nature of H^+ and H_2 does not allow studying the regiochemistry of substrates binding by means of crystallographic techniques; quantum-chemical studies are instead well suited to investigate proton transfer paths and the process of H_2 binding and activation within these enzymes. Finally, computational methods proved important also to unveil the physiologically relevant redox states of metal ions within hydrogenases. In addition, QM methods have been also used to compute the spectroscopic properties of hydrogenases, with the aim of complementing and rationalizing experimental results targeted to the characterization of the structural and electronic properties of key enzyme forms. From a methodological perspective, most quantum-chemical studies of hydrogenases and related models have been carried out within the theoretical framework of the density functional theory

Bioinspired Catalysis: Metal-Sulfur Complexes, First Edition.
Edited by Wolfgang Weigand and Philippe Schollhammer.

(DFT), a choice that stems principally from the very good trade-off between the accuracy of DFT and its relatively low computational cost.

6.2 QM Studies of Hydrogenases

Three evolutionary distinct hydrogenase families are presently known: [FeFe], [NiFe], and [Fe] hydrogenases. Let us start with an overview of theoretical results obtained studying [FeFe]-hydrogenases. The first X-ray structure of a [FeFe] hydrogenase allowed disclosing the very peculiar structure of the iron–sulfur cluster (referred to as the *H-cluster*) found in the active site of these enzymes [5]. In parallel, Fourier transform infrared (FTIR) spectroscopy allowed to reveal the presence of CO and CN^- ligands in the coordination environment of the H-cluster [6, and references therein]. In particular, the H-cluster is formed by a classical [4Fe4S] cubane cluster (referred to as $[4Fe4S]_H$) bridged by a cysteine residue to a di-iron cluster, referred to as the $[2Fe]_H$ *subcluster* (Scheme 6.1). The iron atom of the $[2Fe]_H$ subcluster proximal to the $[4Fe4S]_H$ cluster (usually referred to as Fe_p) binds one CN^- and one CO ligand, and is bridged via one CO and a dithiolate ligand to the distal Fe atom (in the following referred to as Fe_d). The Fe_d atom of the $[2Fe]_H$ subcluster is further coordinated by one CO and one CN^- ligand. In addition, the Fe_d atom features a vacant coordination position, which is the most likely site where dihydrogen and/or protons bind. Different redox states of the $[2Fe]_H$ subcluster have been characterized spectroscopically. The oxidized and reduced forms of the enzyme are electron paramagnetic resonance (EPR) silent, and should correspond to Fe(II)Fe(II) and Fe(I)Fe(I) species, respectively (the latter is usually referred to as the H_{red} *state*). The partially oxidized form of [FeFe] hydrogenases (H_{ox}) is paramagnetic and corresponds to a Fe(I)Fe(II) redox state.

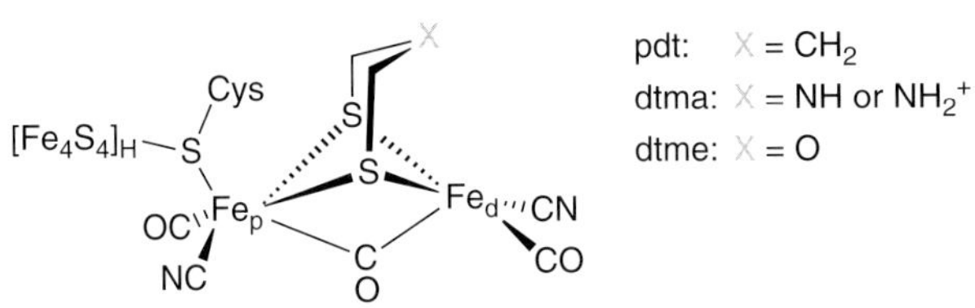

Scheme 6.1 Schematic structure of the H-cluster found in the active site of [FeFe] hydrogenases.

The publication of the first [FeFe]-hydrogenase crystal structures [5] immediately stimulated the interest of computational chemists. One of the first DFT investigations, performed by Cao and Hall [7], considered the presence of a propanedithiolate (pdt) as a chelating group bridging Fe_p and Fe_d (Scheme 6.1). DFT modeling allowed the authors to falsify the hypothesis that the active-ready, H_{ox} form of the enzyme could attain the Fe(II)Fe(III) redox state at the level of the $[2Fe]_H$ site [8]. In fact, computed IR frequencies for carbonyl ligands did not match the experimental values for any Fe(II)Fe(III) models, while theoretical spectra turned out to be compatible with the Fe(II)Fe(I) state. Moreover, a subsequent paper by Fiedler and Brunold [9], which presented computed EPR parameters for Fe(II)Fe(III) and Fe(II)Fe(I) models, reported results consistent with the given picture, as the experimental A-tensor for H_{ox} could be reasonably compared only to the theoretical counterpart for a Fe(II)Fe(I) $[2Fe]_H$ model. Cao and Hall also investigated the reactivity of the pdt-containing models toward H_2 by computing reaction energies and barriers for dihydrogen splitting and the concomitant proton transfer toward either the sulfur atoms of pdt, or the Fe_d-bound cyanide ligand (see Scheme 6.1 for atom labeling). The former path was found to be not unfeasible (reaction energy: 15 kcal mol^{-1}, with an activation energy of 17 kcal mol^{-1}), while proton transfer toward the cyanide turned out to be kinetically impeded (the activation energy in this case was computed to be as large as 38 kcal mol^{-1}). Notably, a very recent QM/MM study allowed to completely rule out the possibility of cyanide protonation at any stage of the [FeFe]-hydrogenases' catalytic cycle: in fact, theory showed that the protein matrix surrounding the active site is able to dramatically lower the basicity of iron-bound cyanides [10]. On the other hand, the alternative hypothesis of an involvement of the dithiolate sulfur atoms in the catalytic mechanism became less appealing, as soon as Hall, Hu, and their respective coworkers [11, 12] published novel DFT results on $[2Fe]_H$ models in which pdt was substituted by dithiomethylamine (dtma, see Scheme 6.1). Let us remark here that the actual nature of the chelating group in the protein active site could not be derived from crystallography, and pdt, dtma, or dithiomethylether (dtme, see Scheme 6.1) are all plausible candidates that were discussed in the [FeFe]-hydrogenases literature. In particular, DFT showed that the amine group of dtma could behave as a capable base to favor the splitting of a H_2 molecule bound to Fe_d: the barrier for intramolecular proton transfer turned out to be very low in this case, and the reaction is favorable also in terms of computed thermodynamic parameters [11, 13]. However, the presence of a dtma molecule in the enzyme active site remained rather speculative, until more recent quantum-chemical, and combined experimental/theoretical studies were published. The first of such works was presented by Lubitz and coworkers [14], who performed advanced EPR measurements on the enzyme, which allowed them to detect a ^{14}N nucleus with peculiar quadrupole coupling constants. The comparison between the latter and the theoretical quadrupole couplings obtained from DFT models of the $[2Fe]_H$ site with dtma were found to match, thus indicating the presence of a nitrogen atom in the pendant. A theoretical paper was then published by Ryde and coworkers [15, 16], who applied a newly developed QM/MM approach for the refinement of metalloproteins crystal structures. Ryde and collaborators found that dtma is the chelating group that better fits the electron density maps derived from high-resolution X-ray diffraction experiments on [FeFe] hydrogenases. Finally, Lubitz and coworkers published a second combined experimental/computational investigation on a Fe(II)Fe(I) biomimetic model of the $[2Fe]_H$ subcluster bearing a dithiolate as a bridging chelating group: such models underwent DFT analysis devoted to the computation of hyperfine couplings and quadrupole coupling constants. Synthesis and EPR characterization of the same compound allowed them to obtain the corresponding experimental values, which were very close not only to the theoretical ones but also to the experimental parameters typical of the H_{ox} form of the enzyme [17].

These convincing evidences on the nature of the pendant in the [FeFe] hydrogenases paved the way for theoretical analyses about the protonation states of the active site in the various experimentally known enzyme forms. Notably, the IR absorption of CO and CN ligands in the H-cluster is expected to be significantly influenced by the overall charge of the cluster and by regiochemistry of proton and dihydrogen binding to the Fe centers. In view of this, Reiher and coworkers [18] recently proposed an approach for accurate prediction of CO and CN groups vibrations, based on the calculation of IR frequencies and the subsequent application of specifically developed scaling factors. Such a methodological approach – originally pioneered by Hall, Darensbourg, and collaborators [19] – is based on the assumption that high-quality scaling factors can be developed in the case of [FeFe] hydrogenases, given the availability of a large number of experimental IR spectra of biomimetic models structurally related to the H-cluster. In particular, Reiher and collaborators [18] used the pure density functional BP86 in combination with a triple-ζ polarized basis set, which produced reliable molecular structures as well as harmonic vibrations for a large set of biomimetic models. Calculated CN and CO stretching frequencies were analyzed separately. Scaled vibrational frequencies were then applied to assign intermediates in the reaction cycle of [FeFe] hydrogenases. The infrared spectrum of the H_{ox} form turned out to be in very good agreement with the calculated spectrum of a Fe(I)Fe(II) model complex featuring a free coordination site at the distal Fe atom, as well as with the calculated spectra of complexes in which H_2 or H_2O are coordinated at this site. The spectrum of the reduced form of the enzyme (H_{red}) measured from the *Desulfovibrio desulfuricans* enzyme was found to be compatible with a mixture of a Fe(I)Fe(I) species with all terminal COs, and a Fe(I)Fe(I) species with protonated dtma ligand, while the spectrum of H_{red} recently measured from *Chlamydomonas reinhardtii* hydrogenase showed compatibility with a mixture of a Fe(I)Fe(I) species with a bridged CO, and a Fe(II)Fe(II) species with a terminal hydride bound to the Fe_d atom.

DFT modeling was also applied to propose plausible paths for the enzymatic process of H_2 evolution and oxidation. Several papers were published on such topic since 2001, some of which discussed the possibility of the formation of bridging hydrides as intermediates in the catalytic cycle [4, 20, 21]. However, the above-mentioned data, together with experimental results comparing the reactivity of μ- and terminal-hydride di-iron models, led to the current consensus that the distal iron atom is the sole center to be directly involved in the interaction with substrates (H^+ and H_2). In particular, DFT and QM/MM results indicate that proton transfer from the amine group of dtma to the vacant coordination site on Fe_d is highly favored in H_{red} [22, 23]. As shown in Scheme 6.2, the resulting

Fe(II)Fe(II) species with terminal hydride is proposed to undergo protonation of the amine group in the pendant.

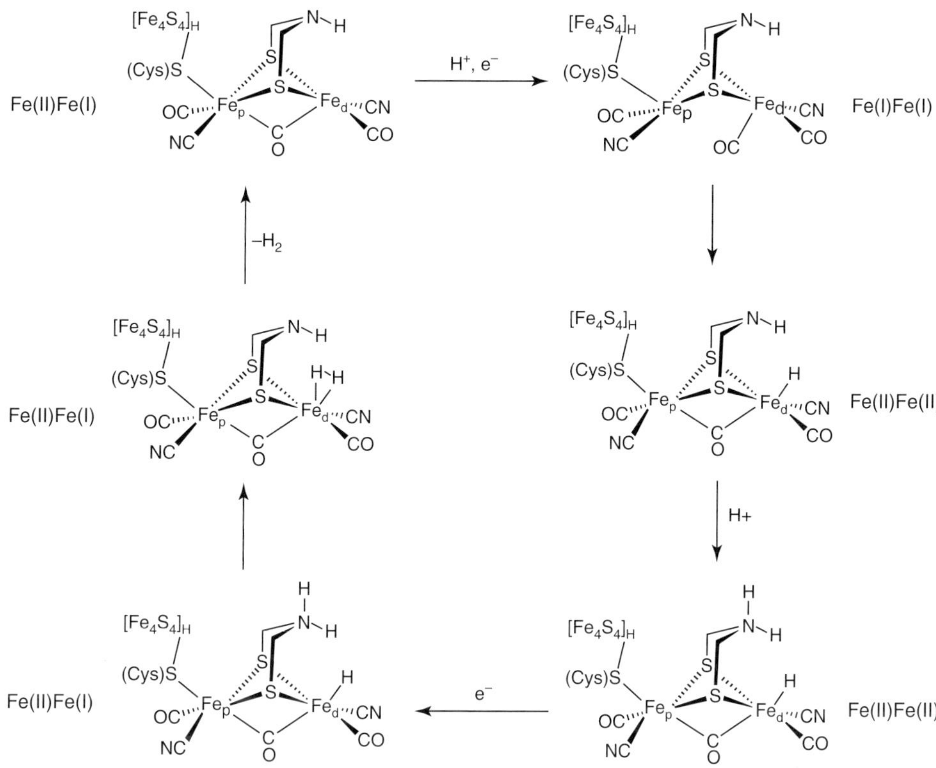

Scheme 6.2 The catalytic mechanism of [FeFe] hydrogenases, as deduced by recent QM and QM/MM calculations.

One-electron reduction of the active site following the latter proton transfer leads to a mixed-valence Fe(II)Fe(I) intermediate on which dihydrogen can easily form. H_2 detachment from Fe_d and reduction bring the H-cluster back to the initial Fe(I)Fe(I) redox state. The importance of dtma in the reversible process of H^+ supply from the bulk to Fe_d was also confirmed by recent QM/MM results on the acid–base reactivity of the protein matrix. In fact, Pachter and coworkers [24] used a combination of DFT and empirical molecular dynamics simulations to show that the most probable pathways for proton transfers in the enzymes from *D. desulfuricans* and *Clostridium pasteurianum* terminate in correspondence of a cysteine residue adjacent to the dtma amine group.

Extended QM treatment in QM/MM models proved important also to gain a better understanding of the redox processes taking place during catalysis. In fact, many of the known [FeFe]-hydrogenase variants feature the presence of ancillary Fe_2S_2 and/or Fe_4S_4 clusters that mediate electron transfer from and toward the

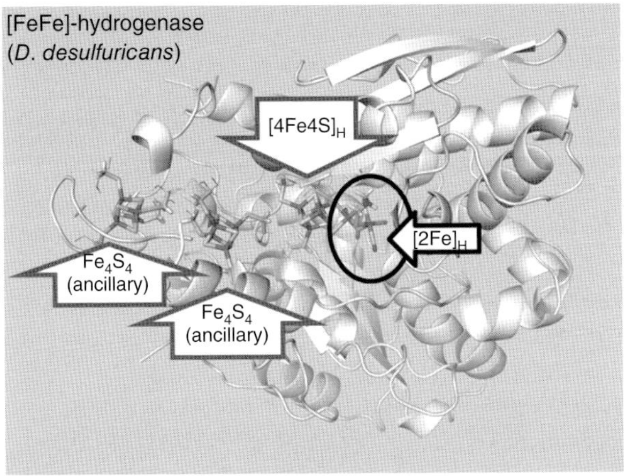

Figure 6.1 Structure of the [FeFe] hydrogenase from *D. desulfuricans*. The position of iron–sulfur clusters is shown with arrows.

H-cluster. In a recent contribution, the two ancillary Fe_4S_4 sites that flank the H-cluster in the *D. desulfuricans* enzyme (Figure 6.1) were included in the QM portion of a QM/MM model of H_{ox} [25].

It was shown that H_2 binding to Fe_d in such models leads to a significant reduction of the energy gap between the frontier molecular orbitals (FMOs) centered on the $[2Fe]_H$ subcluster and the FMOs of one of the ancillary Fe_4S_4 assemblies. Therefore, a hypothesis was formulated that H_2 binding to H_{ox} is able to promote reduction of the ancillary cluster at the expense of the $[2Fe]_H$ site; more specifically, coordination of dihydrogen to Fe_d was predicted to be concomitant with a Fe(II)Fe(I) → Fe(II)Fe(II) oxidation of the $[2Fe]_H$ subcluster, which is functional for H_2 activation.

Finally, DFT modeling turned out to be fundamental for a better understanding of inhibition of [FeFe] hydrogenases by molecular oxygen. In fact, exposure of the enzyme to O_2 rapidly leads to an irreversible damage of the H-cluster, a phenomenon intrinsically difficult to investigate experimentally. DFT modeling of the H-cluster and its surroundings demonstrated that O_2 is able to bind the vacant coordination site on Fe_d [26, 27]. This in turn triggers formation of reactive oxygen species such as O_2^- and HO_2^-, a premise for the development of the OOH radical and of H_2O_2. Both the latter species were shown to be able to compromise the structural integrity of the H-cluster by binding the Fe atoms of the $[Fe_4S_4]_H$ cluster or the S atoms of the cubane-coordinating cysteine residues, respectively.

Also, [NiFe] hydrogenases have been extensively studied using QM methods. The first X-ray structure of a [NiFe] hydrogenase was reported in 1995 and showed that in the active site a nickel ion is coordinated by four cysteine residues, while an iron atom is coordinated by two cysteine residues and three non-protein ligands (Scheme 6.3), which were assigned to one CO and two CN^- ligands by FTIR spectroscopic studies [28–30].

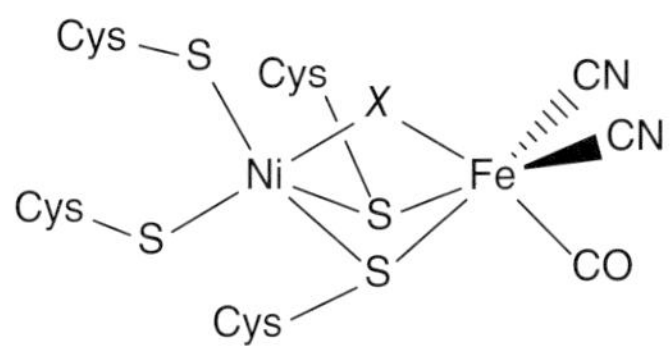

Scheme 6.3 Structure of the [NiFe]-hydrogenase active site as obtained by X-ray diffraction [28]. The chemical nature of the X atom could not be unambiguously deduced.

The more oxidized forms of [NiFe] hydrogenases are catalytically inactive and generally referred to as *Ni-A* and *Ni-B*. One-electron reduction of the Ni-B form leads to the formation of an EPR-silent and catalytically active species usually referred to as *Ni-SI*. Further reduction of the Ni-SI form yields a catalytically active paramagnetic species, referred to as *Ni-C*. Experimental and theoretical results converge in the description of Ni-C as a $Ni^{III}Fe^{II}$ species characterized by a hydride ligand bridging the two metal ions [31, 32]. The Ni-C form can be converted upon illumination to another paramagnetic state referred to as *Ni-L*, in a process that should imply the detachment of the bridging hydrogen atom as a proton, and the concomitant reduction of the metal–sulfur cluster from $Ni^{III}Fe^{II}$ to $Ni^{I}Fe^{II}$. One-electron reduction of the Ni-C form yields the most reduced state of the enzyme, which is referred to as *Ni-R*, and should correspond to a species featuring a hydride ligand simultaneously coordinated to Ni^{II} and Fe^{II} ions. When considering the electronic structure of the metal ions, the Fe^{II} ion has a low-spin configuration in all characterized redox states, whereas the spin state of the Ni ion in the Ni-SI state is still uncertain. In fact, Ni L-edge X-ray circular magnetic dichroism measurements favor a high-spin Ni^{II}, whereas FTIR spectra fit better with a low-spin state [33]. DFT calculations were initially used with the aim of shedding some light on this issue, but it turned out that reliable results about the spin state of Ni^{II} ions in models related to the Ni-SI state are difficult to obtain using DFT methods [34].

As for the catalytic mechanism, several DFT studies have complemented experimental efforts aimed at the characterization of the reactivity of the NiFe cluster [35–37]. The enzyme form that binds H_2 should correspond to the Ni-SI redox state, in which the Ni and Fe ions are four- and five-coordinated, respectively. In particular, initial DFT calculations showed that H_2 can bind to the Fe^{II} ion, whereas the corresponding adduct with Ni^{II} is unstable [36]. However, more recent DFT results, obtained using models of the active site in which the positions of the cysteine residues coordinated to Ni and Fe were constrained to the corresponding positions observed in the crystal, indicated that both Ni-H_2 and Fe-H_2 adducts are plausible intermediate species in the catalytic mechanism [38].

Another controversial issue about the catalytic mechanism of [NiFe] hydrogenases concerns the redox state of the enzyme form where H_2 cleavage takes place. To address this issue, theoretical methods have been extensively used. However, results from QM studies cannot be considered fully conclusive, because it was shown that they can be affected by differences in the chemical models adopted in the quantum-chemical calculations. In particular, H_2 activation is predicted to take place on a $Ni^{III}Fe^{II}$ species when small models of the active site are used, whereas results obtained using larger models suggest that H_2 activation and cleavage take place on a $Ni^{II}Fe^{II}$ species [38].

According to most DFT investigations, either the S atom of one cysteine terminally coordinated to Ni or a water molecule could play the role of proton acceptor and trigger the heterolytic cleavage of H_2 [37, 20] (Scheme 6.4). This acid–base reaction would produce a Ni^{II}-H^--Fe^{II} species, which should correspond to the Ni-R form. One-electron oxidation of Ni-R would lead to a Ni^{III}-H^--Fe^{II} species, which should correspond to the Ni-C form of the enzyme. Loss of a proton and an electron from Ni-C would close the catalytic cycle. However, a recent theoretical investigation suggested that H_2 heterolytic cleavage takes place only in the first catalytic cycle, leading to the transient formation of a species where nickel has oxidation state Ni(I). In the following catalytic cycles, H_2 would be cleaved according to an oxidative addition step [39].

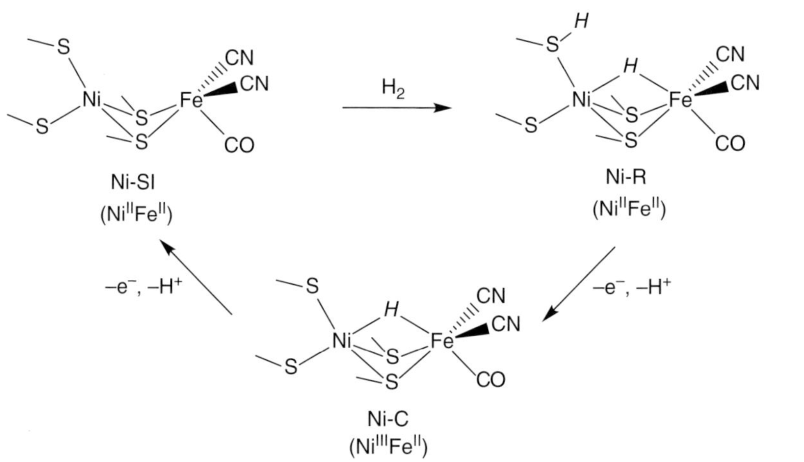

Scheme 6.4 Key steps in the catalytic mechanism of [NiFe] hydrogenases, as deduced by several QM studies.

A third family of hydrogenases, referred to as *[Fe] hydrogenases*, which contain in the active site a guanylyl-pyridinol cofactor (referred to as *FeGP*), was recently characterized by X-ray diffraction [40]. [Fe] hydrogenases catalyze the reversible reduction of N^5,N^{10}-methenyl-tetrahydromethanopterin ($CH{\equiv}H_4MPT^+$) by H_2, to form N^5,N^{10}-methylene-tetrahydromethanopterin ($CH_2{=}H_4MPT$) and a proton (Scheme 6.5), a reaction that is a key step in the reduction of CO_2 to CH_4 in the metabolism of methanogenic archaea. Therefore, [Fe] hydrogenases differ from [NiFe] hydrogenases and [FeFe] hydrogenases because they catalyze the transfer of a hydride to an organic molecule, and also because H_2 activation takes place only when the substrate is present in the active site.

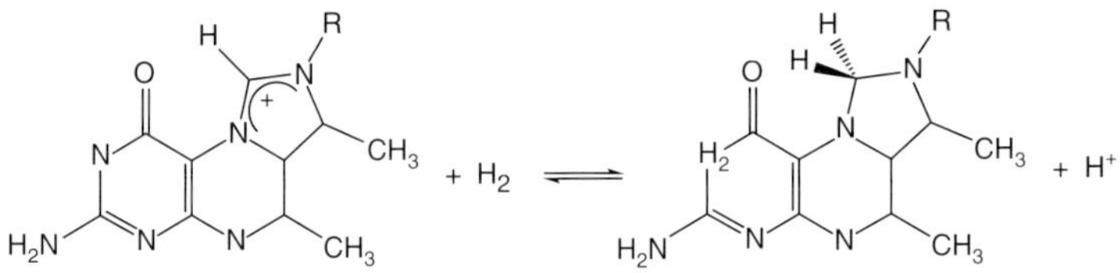

Scheme 6.5 The reaction catalyzed by [Fe] hydrogenases.

Also, for [Fe] hydrogenases, the experimental characterization of the three-dimensional structure stimulated theoretical studies aimed at shedding light on plausible catalytic reaction pathways. The first DFT study about the reactivity of the [Fe]-hydrogenase active site [41] highlighted that H_2 cleavage is kinetically hindered on an iron ion coordinated by the nitrogen atom of a pyridinol ring, a carboxylate group, two CO molecules, and a methylthiolate ligand. However, it was found that the presence of a $CH{\equiv}H_4MPT^+$ group close to the metal cofactor reduces significantly the reaction energy barrier. In particular, a key role in the catalytic mechanism would be played by the pyridinol group, which is a strong donor and therefore can avoid H_2 heterolytic cleavage in the absence of $CH{\equiv}H_4MPT^+$. The presence of $CH{\equiv}H_4MPT^+$ in the enzyme active site would decrease the donor strength of the heterocyclic moiety, and consequently favor the transfer of a H^- to $CH{\equiv}H_4MPT^+$. Notably, after a recent revision of the structure of the FeGP cofactor of [Fe] hydrogenases, another DFT study of the catalytic mechanism was reported, suggesting that H_2 activation implies H_2 coordination to the Fe^{II} ion, followed by heterolytic cleavage [42]. According to this study, the role of the base group could be played either by the cysteinate ligand or by the OH group of the pyridinol group (Scheme 6.6). The metal-hydride species produced in this way would then react with $CH{\equiv}H_4MPT^+$, forming $CH_2{=}H_4MPT$.

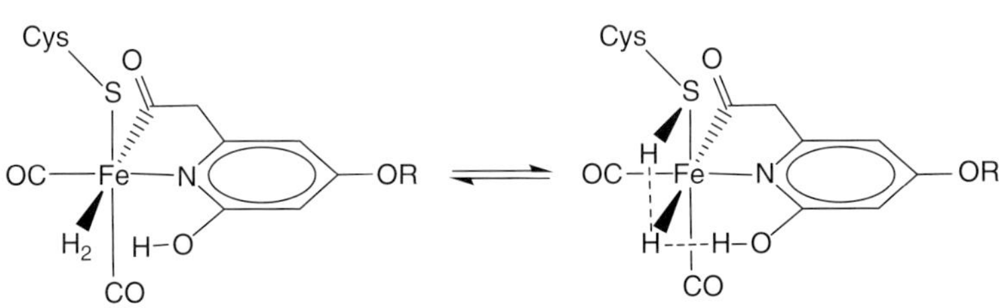

Scheme 6.6 H_2 activation in the active site of [Fe] hydrogenases, as deduced by DFT studies.

6.3 QM Studies of Synthetic Complexes Related to the Active Site of Hydrogenases

The remarkable catalytic properties of hydrogenases, as well as the biologically unusual structural features of the metal–sulfur cofactors found in their active sites, have attracted the attention of several organometallic chemists and have stirred efforts to design, synthesize, and characterize hundreds of biomimetic or bioinspired compounds resembling, to various extents, the structure of the active sites of hydrogenases [43–46].

From a general chemical perspective, it is important to remark that, even if the reaction that should be catalyzed by synthetic models related to the active site of hydrogenases is apparently very simple,

$$2H^+ + 2e^- \rightleftarrows H_2$$

it contains both the key aspects of chemistry: that is, redox and acid–base reaction steps. In fact, a catalysts for H_2 production must undergo both reduction and proton-binding reactions (in the following, we will arbitrarily consider protons and electrons as the reactants and H_2 as the product, essentially because very few bioinspired or biomimetic catalysts able to oxidize H_2 are presently known). Therefore, both the redox and the acid–base properties of biomimetic and bioinspired synthetic models are crucial and, as expected, have been the subject of several DFT investigations.

When considering the different families of hydrogenases, it should be noted that biomimetic models of [FeFe] hydrogenases are by far the most extensively studied, essentially due to the relatively easy synthetic accessibility of coordination compounds deriving by the $[Fe_2(Xdt)(CO)_6]$ scaffold (where Xdt is usually edt, pdt, or adt). For this reason, the overview presented in the following will be focused on QM investigations of [FeFe]-hydrogenase synthetic models.

6.3.1
DFT Studies about Structural and Redox Properties of Synthetic Complexes Related to the Active Site of [FeFe] Hydrogenases

As discussed when describing [FeFe] hydrogenases, different redox states of the $[2Fe]_H$ cluster have been characterized spectroscopically. However, some ambiguities about the structural features of the $[2Fe]_H$ subcluster in the H_{red} redox state still exist, and therefore several biomimetic complexes have been synthesized and characterized with the aim of shedding light on this issue. More specifically, even if several simple synthetic di-iron dithiolate complexes can be considered as suitable models of the composition of the active site of [FeFe] hydrogenases, it is noteworthy that all Fe^IFe^I complexes synthesized to date fail to reproduce the orientation of the diatomic ligands that is observed in the enzyme active site (rotated vs unrotated geometries; Scheme 6.7). Because this structural difference could be correlated with the observation that synthetic di-iron complexes are generally poor catalysts when compared to the enzyme, DFT has been extensively used with the aim of rationally designing synthetic complexes resembling the structural features of the enzyme active site. In fact, experimental and computational studies converge to the conclusion that the peculiar structure of the $[2Fe]_H$ subcluster in the H_{red} state could be due to a combination of electronic effects caused by iron ligands and specific interactions between the cyanide nitrogens and the peptide chain. However, the position of hydrogen bond donors observed in the protein is extremely difficult to reproduce in small synthetic analogs, and therefore DFT has been used to evaluate how changes in the first coordination sphere of synthetic compounds can affect the structural features of the complex [47].

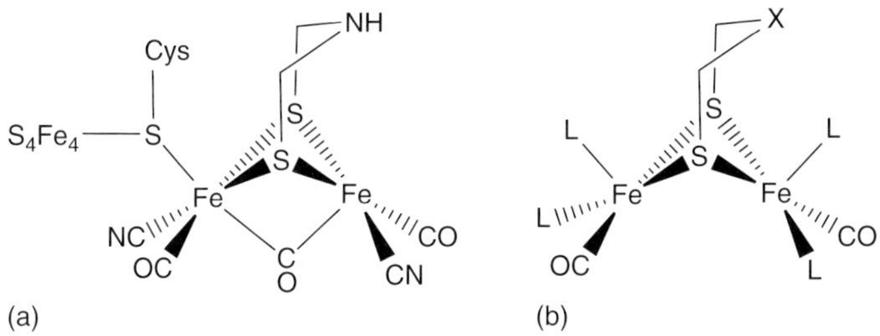

Scheme 6.7 Comparison of the structure of the active site in [FeFe] hydrogenases (a, referred to in the following as the *rotated form*) and related synthetic Fe^IFe^I complexes (b, referred to as the *unrotated form*).

It turned out that the most promising "ingredients" to obtain Fe^IFe^I complexes resembling the structure of the $[2Fe]_H$ subcluster include the insertion into the scaffold of borane or amine functional groups, the addition of a tertiary butyl group to the central methylene unit of the pdt chelating ligand, and the substitution of some CO ligands with better σ-donors, to give asymmetric complexes. In particular, DFT calculations suggest that the combination of a sterically demanding dithiolate bridge with the asymmetric presence of strong donor ligands could be a viable strategy to obtain synthetic di-iron complexes with structural properties similar to the $[2Fe]_H$ active site.

When considering rotated and unrotated isomers, it is noteworthy that several experimental and theoretical studies have established that the one-electron oxidation of Fe^IFe^I dithiolate analogs of the [FeFe]-hydrogenase active site induces isomerization from the unrotated to the rotated form. However, it is often not clear whether the structure change follows the electron transfer (EC mechanism) or it is concerted with it (a so called quasi-reversible process). As an example, Schollhammer and collaborators [48] recently investigated electrochemically and theoretically the oxidatively induced geometry change in $[Fe_2(CO)_4(dppe)(\mu\text{-}pdt)]$ (dppe = $Ph_2PCH_2CH_2PPh_2$). However, the experimental study of the oxidation of $[Fe_2(CO)_4(dppe)(\mu\text{-}pdt)]$ did not allow to discriminate between an EC or a quasi-reversible mechanism. DFT calculations showed that an EC mechanism for the one-electron oxidation of $[Fe_2(CO)_4(dppe)(\mu\text{-}pdt)]$ is more likely, as a $[Fe_2(CO)_4(dppe)(\mu\text{-}pdt)]^+$ structure characterized by an unrotated geometry resembling the neutral parent compound corresponds to an energy minimum. Then, the $[Fe_2(CO)_4(dppe)(\mu\text{-}pdt)]^+$ unrotated isomer would rapidly rearrange to the thermodynamically more stable isomer characterized by rotated geometry.

DFT was also used to complement experimental efforts aimed at the design of biomimetic Fe_6S_6 compounds. As an example, Pickett and coworkers [49], obtained and characterized $[Fe_4S_4(L)_3\{Fe_2(CH_3C(CH_2S)_3)(CO)_5\}]^{2-}$ (L = 1,3,5-tris (4,6-dimethyl-3-mercaptophenylthio)-2,4,6-tris-(*p*-tolyl-thio)-benzene), which closely resembles the chemical composition of the H-cluster. However, compared to the H-cluster, in the binuclear portion of the synthetic organometallic cluster, the ligands are disposed in an unrotated conformation. The DFT investigation of such biomimetic Fe_6S_6 compound showed that the reduced form of the complex can be formally described as an Fe(I)Fe(I)[2Fe(II)2Fe(III)] species, and that the rotated conformation is significantly higher in energy relative to the unrotated isomer. More remarkably, the DFT investigation of the one-electron oxidized complex showed that, again compared to the H-cluster, the Fe_4S_4 moiety is directly involved in the redox process, while the binuclear subcluster remains in the Fe(I)Fe(I) redox state, highlighting a key difference between the synthetic assembly and the enzyme, because in H_{red} the binuclear portion of the H-cluster switches from Fe(I)Fe(I) to Fe(I)Fe(II) upon one-electron oxidation of the enzyme.

QM calculations can also help shed light on the electronic properties that control redox potentials, and therefore can be particularly useful when designing molecules with specific redox properties. Prompted by these observations, with the aim of evaluating the ability of DFT in the quantitative predictions of redox potentials, as well as to investigate the effects of the electron-donor properties of ligands, Roy *et al.* [50] have recently computed the redox potential of a series of Fe(I)Fe(I) complexes. It turned out that the fit between calculated and experimental values is very good (R^2 values of 0.94 and 0.88 for the oxidation and reduction potentials, respectively) when using the BP86 functional, whereas a systematic error was observed when using B3LYP. It was also found that the nature of the chelating ligand does not affect appreciably the reduction potential, even though a large shift occurs in complexes where the chelating ligand changes from aliphatic to aromatic (0.18 V shift toward more positive potentials). This trend was rationalized considering that the lowest unoccupied molecular orbital (LUMO) of the Fe^IFe^I hexacarbonyl complexes contains little dithiolate bridging character. On the other hand, the oxidation potential for azadithiolate-containing complexes is shifted by about 30 mV when compared to μ-dithiobis(tricarbonyl-iron) derivatives. The authors wondered also if the calculated one-electron redox potentials could correlate with the HOMO (highest occupied molecular orbital) and LUMO energies of the Fe^IFe^I complexes, as well as with the singly occupied molecular orbital (SOMO) of the oxidized/reduced species. Remarkably, DFT results showed a good agreement between the orbital energies and the redox potentials, suggesting that the computational effort required for the calculation of cationic, neutral, and anionic species could be unnecessary. In other words, the reduction potential of a biomimetic [FeFe]-hydrogenase compound could be obtained just from a single geometry optimization calculation, with an error within 0.13 and 0.07 V for oxidation and reduction, respectively.

6.3.2
DFT Studies about the Reactivity of Synthetic Models Related to the Active Site of [FeFe] Hydrogenases

Several DFT investigations have been carried out with the aim of disclosing key factors affecting the reactivity of biomimetic and bioinspired compounds. In this context, it is important to realize that the order of reduction and protonation steps can be different when considering biomimetic complexes and [FeFe] hydrogenases. In fact, even if these reaction steps can be strongly coupled, experimental and computational results led to describe the process of H_2 formation into the enzyme as taking place according to the following steps (Scheme 6.8):

1) Protonation of the Fe(I)Fe(I) form of the $[2Fe]_H$ subcluster, to form an Fe(II)Fe(II)-H^- species;
2) One-electron reduction of the Fe(II)Fe(II)-H^- species to give a Fe(II)Fe(I)-H^- form;
3) Further protonation of the Fe(II)Fe(I)-H^- form, to give a labile Fe(II)Fe(I)-H_2 species, which can easily release H_2;
4) One-electron reduction to close the catalytic cycle.

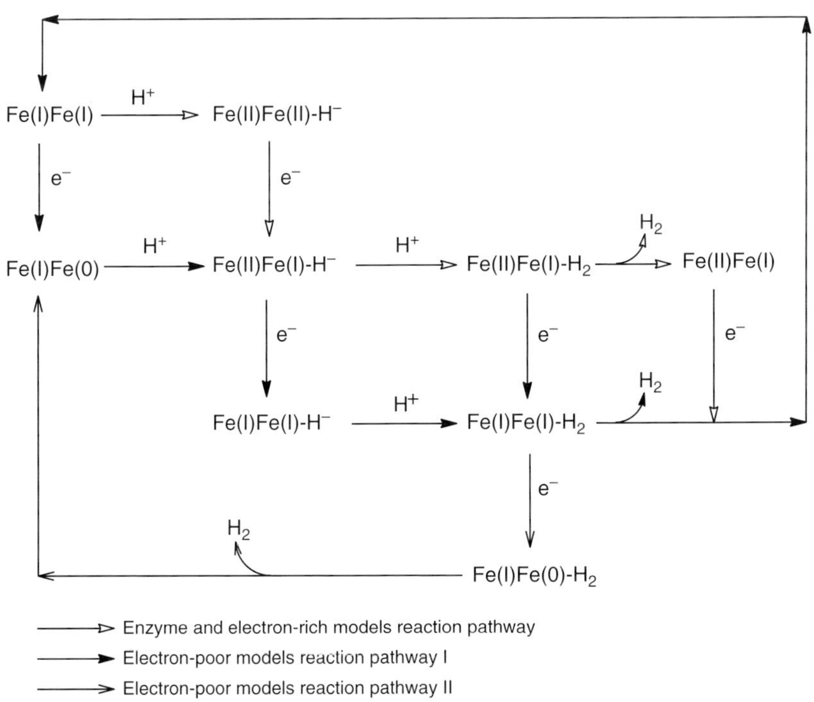

Scheme 6.8 Possible reaction pathways for H_2 formation in [FeFe] hydrogenases and related biomimetic compounds.

On the other hand, in several synthetic models related to the active site of [FeFe] hydrogenases, one-electron reduction is the first step in the process of electrocatalytic proton reduction. In fact, CO-rich compounds such as $Fe_2(SR)_2(CO)_6$ cannot be easily protonated, and their involvement in catalytic processes often implies formation of redox states that do not have correspondence in the enzymatic system (Scheme 6.8):

1) One-electron reduction of the Fe(I)Fe(I) species to give a Fe(I)Fe(0) form;
2) Protonation of the Fe(I)Fe(0) form, to give an Fe(II)Fe(I)-H^- species;
3) Further one-electron reduction and protonation steps to close the catalytic cycle.

In particular, cyclic voltammetry experiments showed that even the very simple $Fe_2(S_2C_3H_6)(CO)_6$ complex, which is the scaffold upon which more sophisticated biomimetic models of [FeFe]-hydrogenases are built, can catalyze dihydrogen evolution. However, $Fe_2(S_2C_3H_6)(CO)_6$ is very electron poor and consequently difficult to protonate, and therefore must be reduced to the formal Fe(I)Fe(0) redox state before protonation. Indeed, with such a complex it was possible experimentally to distinguish between two hydrogen-evolving processes (reaction pathways I and II, Scheme 6.8), which take place at different potentials and are characterized by different reaction rates. Prompted by these observations, DFT was used to dissect the mechanism leading to dihydrogen formation and to disclose the structural and electronic properties of key intermediate species formed in the catalytic cycle [51]. According to computational results, one-electron reduction and protonation of $Fe_2(S_2C_3H_6)(CO)_6$ yields a species characterized by a hydride ligand bridging the two iron ions. Further reduction and protonation yields a species that is characterized by both μ-H and μ-CO groups (Scheme 6.9). However, release of H_2 from such a species was computed to be kinetically hindered, in agreement with experimental data.

When CO ligands are replaced by more basic donors, the redox and acid base behavior becomes more similar to the enzymatic system, that is, initial protonation of the Fe(I)Fe(I) species is followed by one-electron reduction. However, also when dealing with such more biomimetic complexes, important differences between the natural and synthetic systems have been observed. In particular, in almost all hydride derivatives of di-iron dithiolates, the hydride ligand bridges the two metals. Indeed, some di-iron complexes with terminal-hydride ligands have been characterized. As an example, the terminal-hydride complexes characterized by the general structure $[HFe_2(pdt)(CO)_4(chel)]^+$ (chel, chelating ligand) are often observable at low temperature by NMR spectroscopy, but they isomerize to bridging hydride complexes upon warming. Notably, the stability of terminal-hydride complexes is enhanced in sterically crowded or very electron-rich di-iron dithiolato carbonyls such as $Fe_2(pdt)(CO)_2(dppv)_2$ (dppv = *cis*-$C_2H_2(PPh_2)_2$), which undergoes protonation to give a terminal hydride with a half-life of several minutes at room temperature. Remarkably, the terminal-hydride form of the complex $Fe_2(pdt)(CO)_2(dppv)_2$ was shown to reduce at a potential about 100 mV less negative than that required to reduce the corresponding bridging hydride complex.

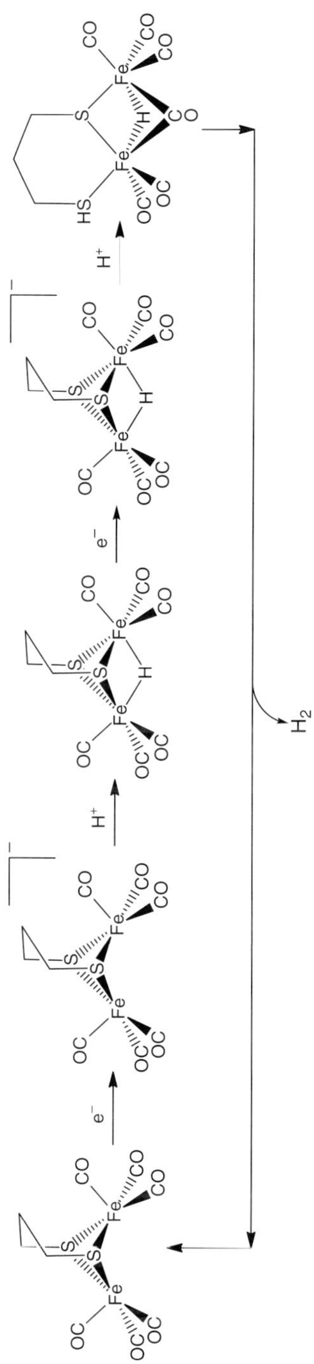

Scheme 6.9 Mechanism of H_2 formation catalyzed by $[Fe_2(S_2C_3H_6)(CO)_6]$.

These and related observations stimulated DFT studies aimed at disclosing factors affecting both the regiochemistry of protonation of di-iron biomimetic complexes and the mechanism of isomerization from terminal to bridging hydride species.

6.3.3 DFT Studies about Regiochemistry of Protonation of Synthetic Complexes Related to the Active Site of [FeFe] Hydrogenases

As discussed, in biomimetic models of the [FeFe]-hydrogenase active site, the regiochemistry of protonation has significant implications for catalysis as terminal hydrides are more susceptible to protonolysis and more readily reduced than the corresponding μ-hydride isomers. Therefore, in light of the differing reactivity for the different binding modes, it is crucial to understand the factors leading to the preferential formation of terminal-hydride species. In this context, it is important to note that the presence in the [FeFe]-hydrogenase active site of an adt chelating ligand that could facilitate proton transfer from the bulk to the di-iron subcluster (and vice versa), has prompted the design and synthesis of several complexes characterized by the presence of an amine-containing chelating group. A complex of this family that has been thoroughly characterized, also by DFT calculations, is $[Fe_2(\mu\text{-}SCH_2N(CH_2Ph)CH_2S)(CO)_4(PMe_3)_2]$ [52]. In particular, upon the addition of various acids, three differently protonated derivatives can be formed, resulting from protonation at the nitrogen atom of the adt moiety, protonation of the FeFe bond, and protonation of both the adt and the Fe–Fe bond. Interestingly, extended X-ray absorption fine structure (EXAFS) measurements suggested that structural changes were already induced by protonation of only the adt nitrogen, which is quite distant from the two iron atoms. This conclusion was corroborated by DFT calculations, which showed that the protonation of adt locks the complex in a conformation where the two PMe_3 groups are in basal positions, with a concomitant elongation of the FeFe bond and rotation of CO ligands. Moreover, it was suggested that these changes could facilitate protonation of the Fe–Fe bond to yield the bridging hydride. In addition, it was also proposed that, in the isomer featuring only adt protonation, the oxidation state of iron could be described as in between Fe(I) and Fe(II) and that Fe–Fe bond protonation could be coupled to only partial iron and sulfur oxidation.

The effect of the presence of an amine-containing pendant in biomimetic models of [FeFe] hydrogenases has also been recently explored using both experimental and computational methods by Schollhammer and collaborators [53], who synthesized complexes containing the chelating PNPR ligand ($PNPR = \{R'_2PCH_2\}_2NR$), which was successfully used by DuBois and DuBois [54, 55] for the design of functional models of [NiFe]-hydrogenase complexes. A similar study was also reported by Wang *et al.* [56]. It turned out that the presence of the PNPR ligand increases significantly the rate of protonation and the consequent formation of a bridging hydride, with respect to the ordinary dppe analog. However, although the reduction potential of species featuring a μ-hydride is positively shifted by further N-protonation of the pendant, the overall proton reduction process was found to be slow. In this work, DFT calculations allowed the identification of the energy-costly reaction pathways for proton migration and H_2 release.

As discussed previously, one of the reasons for the very different catalytic activity of the enzyme when compared to synthetic catalysts is thought to stem from the different regiochemistry of proton binding to Fe(I)Fe(I) species. In fact, available experimental and theoretical data suggest that, in the enzyme, protons bind in a terminal manner to the distal iron center of the $[2Fe]_H$ cluster, whereas proton binding to Fe(I)Fe(I) organometallic complexes eventually leads to μ-H species, which usually are less reactive in the H_2 formation reaction (Scheme 6.10).

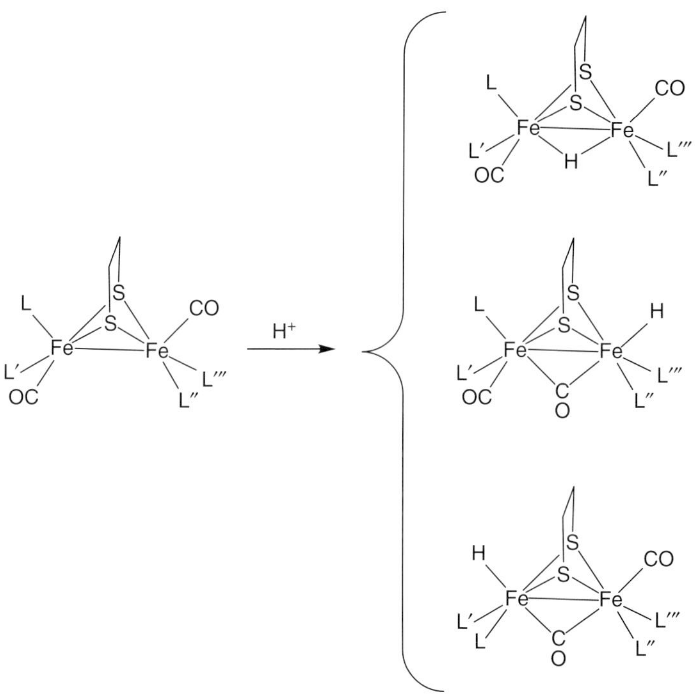

Scheme 6.10 Regiochemistry of iron protonation in Fe_2S_2 biomimetic complexes. L in biomimetic models generally corresponds to CO or phosphine groups.

With the aim of better understanding factors affecting protonation regiochemistry, a DFT investigation of the reactivity with H^+ of a series of organometallic complex resembling the structure of the $[2Fe]_H$ subcluster was recently reported [57]. According to this study, protonation at the Fe–Fe bond was found to be thermodynamically favored relative to terminal iron protonation, irrespective of the ligand stereo-electronic properties. In particular, the DFT investigation of asymmetric models showed that, from a thermodynamics perspective, protonation generally takes place preferentially on the less electron-rich iron atom, that is, the iron atom coordinated by the largest number of CO ligands.

This trend is consistent with experimental data and was rationalized considering that, on the one hand protonation of the Fe atom coordinated by the largest number of phosphine ligands is favored because this iron atom is the electron richest, but on the other hand it is disfavored because in the reaction product the CO ligands (which compete among themselves for back-donation from the iron atoms) are all coordinated to the same metal center. Instead, protonation on the iron atom coordinated by a few (or none) phosphines is disfavored because it is less electron-rich. However, in the reaction product, the CO ligands would be more symmetrically distributed, decreasing the competition for back-donation. The two effects are almost balanced when small phosphines are present. However, when the phosphine ligand coordinated is dppv, steric factors also play a role, and therefore preferential protonation at the less electron-rich iron atom is generally observed. DFT results also suggested that, in asymmetric species, even from a kinetic perspective, terminal protonation at the less electron-rich Fe atom is favored because the formation of the μ-CO transition state is favored by back-donation from the "nonrotating" Fe atom to the μ-CO group.

DFT results highlighted also the importance of the presence of at least one electron donor ligand to allow thermodynamically facile terminal protonation. The latter observation agrees well with a recent report showing that in the synthetic complex $[(CO)(dppe)Fe(\mu SCH_2N(i\text{Pr})CH_2S)Fe(CO)_3]$ protonation takes place exclusively at the N atom.

6.3.4 DFT Studies about the Isomerization of Synthetic Complexes Related to the Active Site of [FeFe] Hydrogenases

It is well known that di-iron synthetic complexes featuring a terminal hydride spontaneously and quickly interconvert to the corresponding less reactive μ-H forms [58, 59]. Therefore, the isomerization from terminal- to μ-hydrides is one of the limiting factors toward obtaining efficient di-iron catalysts for proton reduction. In light of these observations, the detailed characterization of the mechanism of isomerization from terminal- to μ-H species becomes one of the key steps for the rational design of synthetic complexes featuring kinetically persistent (and reactive) terminal-hydride forms. In this scenario, DFT was used to compare plausible isomerization pathways in biomimetic complexes resembling the [FeFe]-hydrogenase bimetallic cluster [60, 61]. It turned out that isomerization processes commonly referred to as *Bailar* (*trigonal*) and *Ray–Dutt* (*rhombic*) twists can be both operative in di-iron hydride complexes. In particular, a terminal- to μ-hydride isomerization proceeding according to Bailar twist steps implies an initial 120° rotation along the FeFe axes of the $Fe(L)_2H$ group (Scheme 6.11), yielding a basal-hydride intermediate, followed by a second 120° rotation of the three ligands along an axis perpendicular to the FeFe bond. Correspondingly, an isomerization pathway characterized only by Ray–Dutt steps also implies the intermediate formation of a basal-hydride species, but in this case the simultaneous movement of four ligands takes place in the two steps (Scheme 6.11).

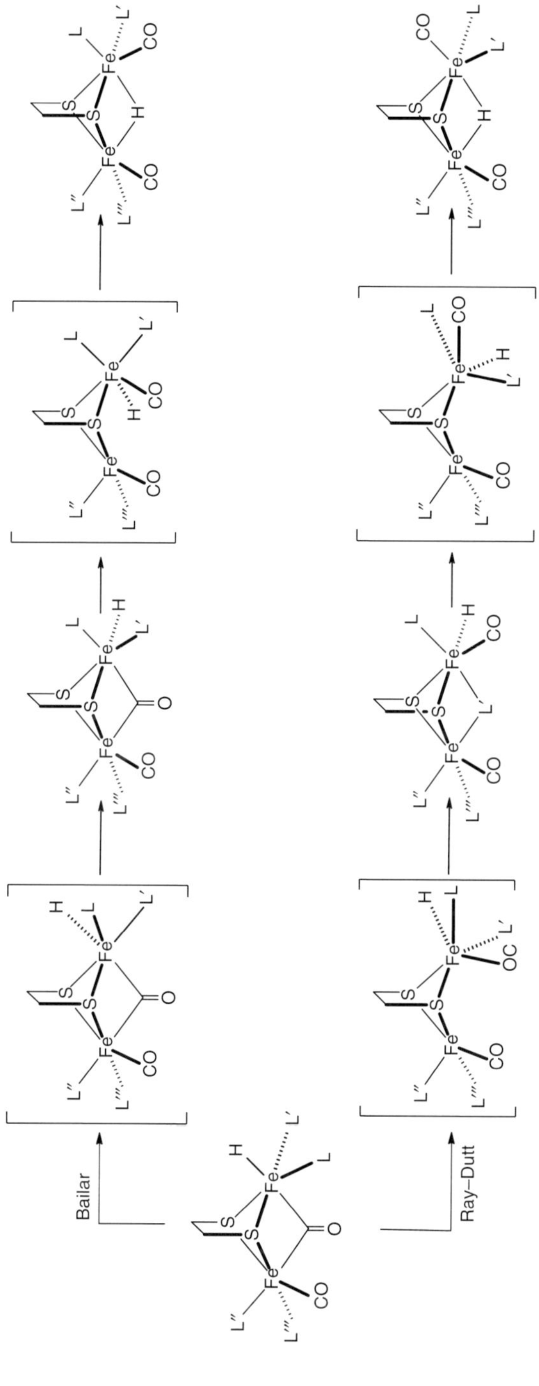

Scheme 6.11 Isomerization of terminal- to μ-hydrides in models of the [FeFe]-hydrogenase bimetallic cofactor. L corresponds to CO or phosphine ligands.

According to DFT results, the most favored isomerization pathways, in terms of activation energies, are generally those proceeding through Ray–Dutt processes. However, it is worth noting that an isomerization mechanism consisting only of Ray–Dutt steps cannot be always operative. In fact, when two phosphine ligands are coordinated to the isomerizing iron center, the Ray–Dutt mechanism is unfeasible because it would imply the transient formation of high-energy intermediates characterized by a μ-phosphine ligand.

6.4 Conclusions

In this chapter, we have presented a noncomprehensive overview of QM studies of molecular models related to the active site of hydrogenases, with the aim of showing how computational studies have contributed to the elucidation of key properties of hydrogenases, as well as of related synthetic compounds. Indeed, the large number of recently published studies containing combined computational and experimental results well illustrates the value of using such complementary approaches to unravel key structure–function relationships. However, several issues about the chemistry of hydrogenases remain elusive and are expected to be the focus of ongoing and forthcoming studies. As an example, the recently characterized structure of *Ralstonia eutropha* [NiFe] hydrogenases [62], featuring a very peculiar iron–sulfur cluster that has been proposed to be fundamental to catalyze the reduction of O_2 to H_2O and consequently protect the active site from oxidative damage, discloses the possibility of designing robust oxygen-tolerant enzymes [63]. When considering biomimetic models, recent results have highlighted the importance of coupled proton–electron transfer for efficient H_2 formation or oxidation [64], and computational studies are expected to play an important role in the design of novel and more efficient synthetic catalysts [65].

References

1. Rokob, T.A., Srnec, M., and Rulisek, L. (2012) Theoretical calculations of physico-chemical and spectroscopic properties of bioinorganic systems: current limits and perspectives. *Dalton Trans.*, **41**, 5754–5768.
2. Acevedo, O. and Jorgensen, W.L. (2010) Advances in quantum and molecular mechanical (QM/MM) simulations for organic and enzymatic reactions. *Acc. Chem. Res.*, **43**, 142–151.
3. Ranaghan, K.E. and Mulholland, A.J. (2010) Investigations of enzyme-catalysed reactions with combined quantum mechanics/molecular mechanics (QM/MM) methods. *Int. Rev. Phys. Chem.*, **29**, 65–133.
4. Siegbahn, P.E.M., Tye, J.W., Hall, M.B. *et al.* (2007) Computational studies of [NiFe] and [FeFe] hydrogenases. *Chem. Rev. (Columbus)*, **107**, 4414.
5. Peters, J.W., Lanzilotta, W.N., Lemon, B.J., and Seefeldt, L.C. (1998) X-ray crystal structure of the Fe-only hydrogenase (Cpl) from Clostridium pasteurianum to 1.8 angstrom resolution. *Science*, **282**, 1853–1858.
6. Nicolet, Y., De Lacey, A.L., Vernede, X., Fernandez, V.M., Hatchikian, E.C., and Fontecilla-Camps, J.C. (2001) Crystallographic and FTIR spectroscopic

evidence of changes in Fe coordination upon reduction of the active site of the Fe-only hydrogenase from Desulfovibrio desulfuricans. *J. Am. Chem. Soc.*, **123**, 1596–1601.

7. Cao, Z. and Hall, M.B. (2001) Modeling the active sites in metalloenzymes. 3. Density functional calculations on models for [Fe]-hydrogenase: structures and vibrational frequencies of the observed redox forms and the reaction mechanism at the Diiron Active Center. *J. Am. Chem. Soc.*, **123**, 3734–3742.
8. Popescu, C.V. and Munck, E. (1999) Electronic structure of the H cluster in [Fe]-hydrogenases. *J. Am. Chem. Soc.*, **121**, 7877–7884.
9. Fiedler, A.T. and Brunold, T.C. (2005) Computational studies of the H-cluster of Fe-only hydrogenases: geometric, electronic, and magnetic properties and their dependence on the [Fe4S4] cubane. *Inorg. Chem.*, **44**, 9322–9334.
10. Greco, C., Bruschi, M., Fantucci, P., Ryde, U., and De Gioia, L. (2011) Isocyanide in biochemistry? A theoretical investigation of the electronic effects and energetics of cyanide ligand protonation in [FeFe]-hydrogenases. *Chem. Eur. J.*, **17**, 1954–1965.
11. Fan, H.J. and Hall, M.B. (2001) A capable bridging ligand for Fe-only hydrogenase: density functional calculations of a low-energy route for heterolytic cleavage and formation of dihydrogen. *J. Am. Chem. Soc.*, **123**, 3828–3829.
12. Liu, Z.P. and Hu, P. (2002) A density functional theory study on the active center of Fe-only hydrogenase: characterization and electronic structure of the redox states. *J. Am. Chem. Soc.*, **124**, 5175–5182.
13. Zampella, G., Greco, C., Fantucci, P., and De Gioia, L. (2006) Proton reduction and dihydrogen oxidation on models of the [2Fe] H cluster of [Fe] hydrogenases. A density functional theory investigation. *Inorg. Chem.*, **45**, 4109–4118.
14. Silakov, A., Wenk, B., Reijerse, E., and Lubitz, W. (2009) (14)N HYSCORE investigation of the H-cluster of [FeFe] hydrogenase: evidence for a nitrogen in the dithiol bridge. *Phys. Chem. Chem. Phys.*, **11**, 6592–6599.
15. Ryde, U., Greco, C., and De Gioia, L. (2010) Quantum refinement of [FeFe] hydrogenase indicates a dithiomethylamine ligand. *J. Am. Chem. Soc.*, **132**, 4512–4513.
16. Ryde, U. and Nilsson, K. (2003) Quantum refinement – a combination of quantum chemistry and protein crystallography. *J. Mol. Struct. THEOCHEM*, **632**, 259–275.
17. Erdem, Ö.F., Schwartz, L., Stein, M., Silakov, A., Kaur-Ghumaan, S., Huang, P., Ott, S., Reijerse, E.J., and Lubitz, W. (2011) A model of the [FeFe] hydrogenase active site with a biologically relevant azadithiolate bridge: a spectroscopic and theoretical investigation. *Angew. Chem. Int. Ed.*, **50**, 1439–1443.
18. Yu, L., Greco, C., Bruschi, M., Ryde, U., De Gioia, L., and Reiher, M. (2011) Targeting intermediates of [FeFe]-hydrogenase by CO and CN vibrational signatures. *Inorg. Chem.*, **50**, 3888–3900.
19. Tye, J.W., Darensbourg, M.Y., and Hall, M.B. (2008) Refining the active site structure of iron–iron hydrogenase using computational infrared spectroscopy. *Inorg. Chem.*, **47**, 2380–2388.
20. Bruschi, M., Zampella, G., Fantucci, P., and De Gioia, L. (2005) DFT investigations of models related to the active site of [NiFe] and [Fe] hydrogenases. *Coord. Chem. Rev.*, **249**, 1620–1640.
21. Bertini, L., Bruschi, M., De Gioia, L., Fantucci, P., Greco, C., and Zampella, G. (2007) *Atomistic Approaches in Modern Biology*, Springer, pp. 1–46.
22. Greco, C., Bruschi, M., De Gioia, L., and Ryde, U. (2007) A QM/MM investigation of the activation and catalytic mechanism of Fe-only hydrogenases. *Inorg. Chem.*, **46**, 5911–5921.
23. Bruschi, M., Greco, C., Kaukonen, M., Fantucci, P., Ryde, U., and De Gioia, L. (2009) Influence of the [2Fe] H subcluster environment on the properties of key intermediates in the catalytic cycle of [FeFe] hydrogenases: hints for the ratio nal design of synthetic catalysts. *Angew. Chem. Int. Ed.*, **48**, 3503–3506.
24. Hong, G., Cornish, A.J., Hegg, E.L., and Pachter, R. (2011) On understanding

proton transfer to the biocatalytic [Fe–Fe](H) sub-cluster in [Fe–Fe] H(2)ases: QM/MM MD simulations. *Biochim. Biophys. Acta, Bioenerg.*, **1807**, 510–517.
25. Greco, C., Bruschi, M., Fantucci, P., Ryde, U., and De Gioia, L. (2011) Mechanistic and physiological implications of the interplay among iron-sulfur clusters in [FeFe]-hydrogenases. A QM/MM perspective. *J. Am. Chem. Soc.*, **133**, 18742–18749.
26. Stiebritz, M.T. and Reiher, M. (2009) Theoretical study of dioxygen induced inhibition of [FeFe]-hydrogenase. *Inorg. Chem.*, **48**, 7127–7140. Erratum in: *Inorg Chem.* (2010) **49**, 8645.
27. Stiebritz, M.T. and Reiher, M. (2012) Hydrogenases and oxygen. *Chem. Sci.*, **3**, 1739–1751.
28. Volbeda, A., Charon, M., Piras, C., Hatchikian, E., Frey, M., and Fontecillacamps, J. (1995) Crystal-structure of the nickel-iron hydrogenase from desulfovibrio-gigas. *Nature*, **373**, 580–587.
29. Volbeda, A. and Fontecilla-Camps, J.C. (2005) Structure–function relationships of nickel–iron sites in hydrogenase and a comparison with the active sites of other nickel–iron enzymes. *Coord. Chem. Rev.*, **249**, 1609–1619.
30. De Lacey, A.L., Hatchikian, E.C., Volbeda, A., Frey, M., Fontecilla-Camps, J.C., and Fernandez, V.M. (1997) Infrared-spectroelectrochemical characterization of the [NiFe] hydrogenase of desulfovibrio gigas. *J. Am. Chem. Soc.*, **119**, 7181–7189.
31. Pandelia, M.-E., Ogata, H., and Lubitz, W. (2010) Intermediates in the catalytic cycle of [NiFe] hydrogenase: functional spectroscopy of the active site. *ChemPhysChem*, **11**, 1127–1140.
32. Ogata, H., Lubitz, W., and Higuchi, Y. (2009) [NiFe] hydrogenases: structural and spectroscopic studies of the reaction mechanism. *Dalton Trans.*, 7577–7587.
33. Gu, W., Jacquamet, L., Patil, D.S., Wang, H.X., Evans, D.J., Smith, M.C., Millar, M., Koch, S., Eichhorn, D.M., Latimer, M. *et al.* (2003) Refinement of the nickel site structure in *Desulfovibrio gigas* hydrogenase using range-extended EXAFS spectroscopy. *J. Inorg. Biochem.*, **93**, 41–51.
34. Bruschi, M., Gioia, L., Zampella, G., Reiher, M., Fantucci, P., and Stein, M. (2004) A theoretical study of spin states in $Ni-S_4$ complexes and models of the [NiFe] hydrogenase active site. *J. Biol. Inorg. Chem.*, **9**, 873–884.
35. Stein, M. and Lubitz, W. (2002) Quantum chemical calculations of [NiFe] hydrogenase. *Curr. Opin. Chem. Biol.*, **6**, 243–249.
36. Niu, S. and Hall, M.B. (2001) Modeling the active sites in metalloenzymes 5. The heterolytic bond cleavage of H(2) in the [NiFe] hydrogenase of desulfovibrio gigas by a nucleophilic addition mechanism. *Inorg. Chem.*, **40**, 6201–6203.
37. De Gioia, L., Fantucci, P., Guigliarelli, B., and Bertrand, P. (1999) Ni–Fe hydrogenases: a density functional theory study of active site models. *Inorg. Chem.*, **38**, 2658–2662.
38. Wu, H. and Hall, M.B. (2008) Density functional theory on the larger active site models for [NiFe] hydrogenases: two-state reactivity? *C.R. Chim.*, **11**, 790–804.
39. Lill, S.O.N. and Siegbahn, P.E.M. (2009) An autocatalytic mechanism for NiFe-hydrogenase: reduction to Ni (I) followed by oxidative addition. *Biochemistry*, **48**, 1056–1066.
40. Shima, S., Pilak, O., Vogt, S., Schick, M., Stagni, M.S., Meyer-Klaucke, W., Warkentin, E., Thauer, R.K., and Ermler, U. (2008) The crystal structure of [Fe]-hydrogenase reveals the geometry of the active site. *Science*, **321**, 572–575.
41. Yang, X. and Hall, M.B. (2008) Trigger mechanism for the catalytic hydrogen activation by monoiron (iron- sulfur cluster-free) hydrogenase. *J. Am. Chem. Soc.*, **130**, 14036–14037.
42. Yang, X. and Hall, M.B. (2009) Monoiron hydrogenase catalysis: hydrogen activation with the formation of a dihydrogen, Fe–H(delta-)…H(delta+)-O, bond and methenyl-H4MPT + triggered hydride transfer. *J. Am. Chem. Soc.*, **131**, 10901–10908.
43. Evans, D.J. and Pickett, C.J. (2003) Chemistry and the hydrogenases. *Chem. Soc. Rev.*, **32**, 268–275.

44. Tard, C. and Pickett, C.J. (2009) Structural and functional analogues of the active sites of the [Fe]-,[NiFe]-, and [FeFe]-hydrogenases. *Chem. Rev.*, **109**, 2245–2274.
45. Georgakaki, I.P., Thomson, L.M., Lyon, E.J., Hall, M.B., and Darensbourg, M.Y. (2003) Fundamental properties of small molecule models of Fe-only hydrogenase: computations relative to the definition of an entatic state in the active site. *Coord. Chem. Rev.*, **238**, 255–266.
46. Rauchfuss, T.B. (2004) Research on soluble metal sulfides: from polysulfido complexes to functional models for the hydrogenases. *Inorg. Chem.*, **43**, 14–26.
47. Tye, J.W., Darensbourg, M.Y., and Hall, M.B. (2006) De novo design of synthetic Di-iron(I) complexes as structural models of the reduced form of iron–iron hydrogenase. *Inorg. Chem.*, **45**, 1552–1559.
48. Chouffai, D., Zampella, G., Capon, J.-F., De Gioia, L., Gloaguen, F., Pétillon, F.Y., Schollhammer, P., and Talarmin, J. (2011) Oxidatively induced reactivity of $[Fe_2(CO)_4(\kappa_2\text{-dppe})(\mu\text{-pdt})]$: an electrochemical and theoretical study of the structure change and ligand binding processes. *Inorg. Chem.*, **50**, 12575–12585.
49. Tard, C., Liu, X., Ibrahim, S.K., Bruschi, M., De Gioia, L., Davies, S.C., Yang, X., Wang, L.S., Sawers, G., and Pickett, C.J. (2005) Synthesis of the H-cluster framework of iron-only hydrogenase. *Nature*, **433**, 610–613.
50. Roy, L.E., Batista, E.R., and Hay, P.J. (2008) Theoretical studies on the redox potentials of Fe dinuclear complexes as models for hydrogenase. *Inorg. Chem.*, **47**, 9228–9237.
51. Greco, C., Zampella, G., Bertini, L., Bruschi, M., Fantucci, P., and De Gioia, L. (2007) Insights into the mechanism of electrocatalytic hydrogen evolution mediated by $Fe_2(S_2C_3H_6)(CO)_6$: the simplest functional model of the Fe-hydrogenase active site. *Inorg. Chem.*, **46**, 108–116.
52. Löscher, S., Schwartz, L., Stein, M., Ott, S., and Haumann, M. (2007) Facilitated hydride binding in an Fe–Fe hydrogenase active-site biomimic revealed by X-ray absorption spectroscopy and DFT calculations. *Inorg. Chem.*, **46**, 11094–11105.
53. Lounissi, S., Zampella, G., Capon, J.-F., De Gioia, L., Matoussi, F., Mahfoudhi, S., Petillon, F.Y., Schollhammer, P., and Talarmin, J. (2012) Electrochemical and theoretical investigations of the role of the appended base on the reduction of protons by [Fe(2) (CO)(4) (kappa(2) -PNP(R))(mu-S(CH(2))(3) S] (PNP(R) = {Ph(2) PCH(2) }(2) NR, R = Me, Ph). *Chemistry (Weinheim an Der Bergstrasse)*, **18**, 11123–11138.
54. Yang, J.Y., Bullock, R.M., Shaw, W.J., Twamley, B., Fraze, K., DuBois, M.R., and DuBois, D.L. (2009) Mechanistic insights into catalytic H2 oxidation by Ni complexes containing a diphosphine ligand with a positioned amine base. *J. Am. Chem. Soc.*, **131**, 5935–5945.
55. O'Hagan, M., Shaw, W.J., Raugei, S., Chen, S., Yang, J.Y., Kilgore, U.J., DuBois, D.L., and Bullock, R.M. (2011) Moving protons with pendant amines: proton mobility in a nickel catalyst for oxidation of hydrogen. *J. Am. Chem. Soc.*, **133**, 14301–14312.
56. Wang, Y., Wang, M., Sun, L., and Ahlquist, M.S.G. (2012) Pendant amine bases speed up proton transfers to metals by splitting the barriers. *Chem. Commun.*, **48**, 4450–4452.
57. Zampella, G., Fantucci, P., and De Gioia, L. (2009) Unveiling how stereoelectronic factors affect kinetics and thermodynamics of protonation regiochemistry in [FeFe] hydrogenase synthetic models: a DFT investigation. *J. Am. Chem. Soc.*, **131**, 10909–10917.
58. Capon, J.F., Gloaguen, F., Pétillon, F.Y., Schollhammer, P., and Talarmin, J. (2009) Electron and proton transfers at diiron dithiolate sites relevant to the catalysis of proton reduction by the [FeFe]-hydrogenases. *Coord. Chem. Rev.*, **253**, 1476–1494.
59. Zaffaroni, R., Rauchfuss, T.B., Gray, D.L., De Gioia, L., and Zampella, G. (2012) Terminal vs bridging hydrides of diiron dithiolates: protonation of $Fe_2(dithiolate)(CO)_2(PMe_3)_4$. *J. Am. Chem. Soc.*, **134**, 19260–19269.

60. Zampella, G., Fantucci, P., and De Gioia, L. (2010) DFT characterization of the reaction pathways for terminal-to μ-hydride isomerisation in synthetic models of the [FeFe]-hydrogenase active site. *Chem. Commun.*, **46**, 8824–8826.
61. Liu, C., Peck, J.N.T., Wright, J.A., Pickett, C.J., and Hall, M.B. (2011) Density functional calculations on protonation of the [FeFe]-hydrogenase model complex $Fe_2(\mu\text{-pdt})(CO)_4(PMe_3)_2$ and subsequent isomerization pathways. *Eur. J. Inorg. Chem.*, **2011**, 1080–1093.
62. Fritsch, J., Scheerer, P., Frielingsdorf, S., Kroschinsky, S., Friedrich, B., Lenz, O., and Spahn, C.M.T. (2011) The crystal structure of an oxygen-tolerant hydrogenase uncovers a novel iron-sulphur centre. *Nature*, **479**, 249–252.
63. Volbeda, A., Amara, P., Darnault, C., Mouesca, J.-M., Parkin, A., Roessler, M.M., Armstrong, F.A., and Fontecilla-Camps, J.C. (2012) X-ray crystallographic and computational studies of the O-2-tolerant [NiFe]-hydrogenase 1 from Escherichia coli. *Proc. Natl. Acad. Sci. U.S.A.*, **109**, 5305–5310.
64. Camara, J.M. and Rauchfuss, T.B. (2012) Combining acid-base, redox and substrate binding functionalities to give a complete model for the [FeFe]-hydrogenase. *Nat. Chem.*, **4**, 26–30.
65. Greco, C. and De Gioia, L. (2011) A theoretical study on the enhancement of functionally relevant electron transfers in biomimetic models of [FeFe]-hydrogenases. *Inorg. Chem.*, **50**, 6987–6995.

7
Mechanistic Aspects of Biological Hydrogen Evolution and Uptake

Joseph A. Wright and Christopher J. Pickett

7.1
Introduction

This chapter is focused on the structural and mechanistic chemistry related to the active sites of the biological hydrogen-evolving or uptake enzymes. These are the [FeFe], [NiFe], and [Fe] hydrogenases, and the nitrogenases. There is a reciprocal relationship between the biochemistry and the chemistry of model systems: advances in each area inform and help the development of the other, besides driving new research. Understanding how the metallosulfur machinery in the natural systems function at the molecular level is not just a "need-to-know" pursuit for potential technological applications – such as replacing precious metals in fuel or producer cells. In the wider context, it advances our knowledge of the fundamental chemistry of dihydrogen and hydrides at transition metal centers, how coligands control electronic properties, and how second coordination sphere or neighboring groups influence catalytic behavior.

7.2
[FeFe] Hydrogenases

7.2.1
Overview of the Catalytic Cycle

The [FeFe] hydrogenases reduce protons rapidly to dihydrogen at low overpotentials with respect to the thermodynamic potential for the $H^+/1/2H_2$ couple, about −413 mV versus the standard hydrogen electrode (SHE) at pH 7 [1]. The catalysis is reversible, but in their natural physiological role they are generally (but not exclusively) engaged in proton reduction [1]. This complements the [NiFe] hydrogenases, which are biased toward a hydrogen uptake role.

The core geometry of the active site of the [FeFe] hydrogenase comprises a "butterfly" di-iron dithiolate unit in which the two sulfur atoms are linked by a $-CH_2NHCH_2-$ chain [2, 3]. Each iron atom is also coordinated by one CN^- and

Bioinspired Catalysis: Metal-Sulfur Complexes, First Edition.
Edited by Wolfgang Weigand and Philippe Schollhammer.

one CO ligand. The "proximal" iron is ligated by a cysteinyl sulfur, which bridges to the {4Fe4S} electron-transfer unit. In the resting state a (semi)bridging CO ligand links the two iron atoms leaving a vacancy on the distal iron atom (Scheme 7.1) [4].

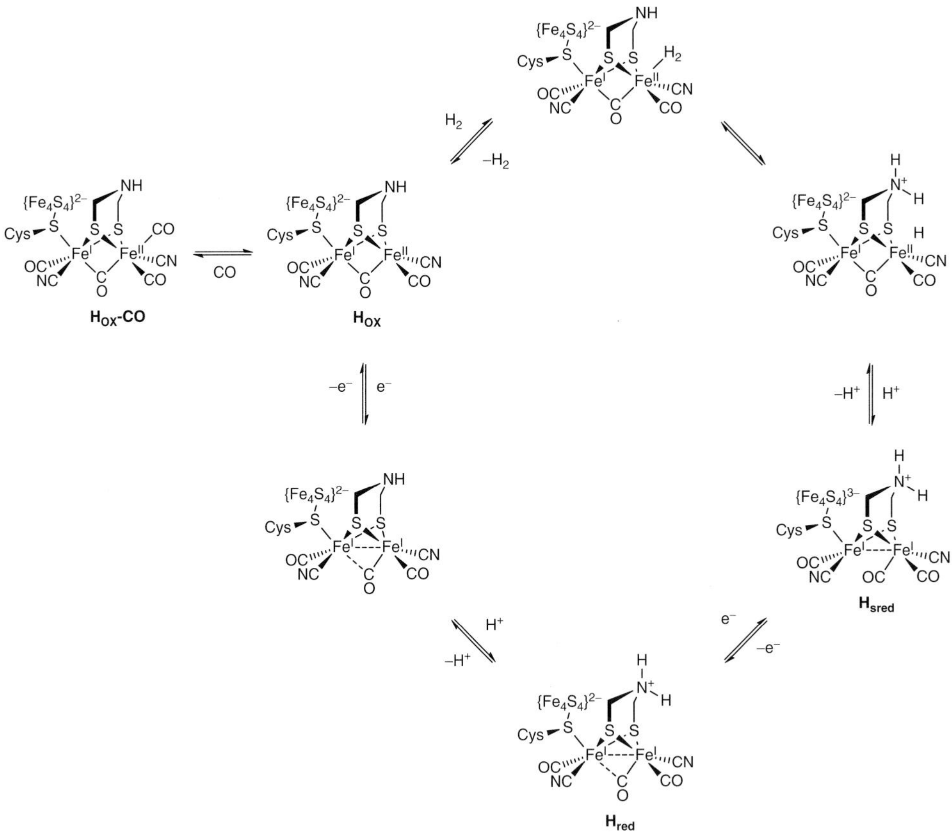

Scheme 7.1 Catalytic cycle for [FeFe] hydrogenase [4].

Synthetic models of the structures presented in the biological cycle (Scheme 7.1) have been numerous since the first crystallographic characterizations of the H-cluster appeared in late 1998 [2, 3], and have been extensively reviewed [1, 5–15]. The current understanding of the biological cycle for hydrogen evolution/uptake by [FeFe] hydrogenase based on chemical precedent, spectroscopy, and biochemistry is represented in Scheme 7.1 [4]. This scheme includes assignments of oxidation states for the subsite and represents a consensus view of the catalytic cycle. While not all intermediates proposed have been detected, this cycle nevertheless provides a basis for discussion of the chemistry of model systems. An outstanding example of the interplay between the synthetic organometallic chemistry of the subsites, spectroscopy, and biology has been the demonstration that synthetic cofactors can be introduced into maturase and apo-hydrogenase proteins to give semisynthetic hydrogenases with high activity [16, 17].

The [FeFe] hydrogenases are extremely sensitive to oxygen damage. Understanding of the breakdown of model systems by oxygen has been reviewed [18]; while the model systems typically react with oxygen at the bridging thiolates, generation of superoxide, which attacks the iron–sulfur cluster, is most probably the cause of inactivation in the enzyme: it has been suggested that binding O_2 at the distal iron precedes the release of superoxide [19].

Key issues relevant to our understanding of the mechanism of the biological process are as follows: (i) the nature of the bridgehead ligand, (ii) mixed valency and the rotated state, (iii) the role of bridging or terminal hydrides in the hydrogen evolution pathway, (iv) coupling between proton and electron transfer within the H-cluster, and (v) how biology attains fast turnover kinetics at low overpotentials.

7.2.2 The Nature of the Bridgehead Atom

As is clear from Scheme 7.1, the amine bridgehead ligand is thought to play a key role in the hydrogen evolution/uptake cycle, both in guiding the proton to the distal iron site and in facilitating proton-hydride coupling. The X-ray crystallographic data do not distinguish bridgehead CH_2, O, or NH groups [2, 3]; nevertheless Peters and Fontecilla-Camps [20] were the first to propose a mechanism involving the reversible (bridging ammonium–terminal hydride) hydrogen evolution pathway. Proof that the bridgehead group is a protonatable amine rests with ENDOR (electron nuclear double resonance)/HYSCORE (hyperfine sublevel correlation) studies on both enzyme [21] and model systems [22], which show that the ^{15}N quadrupolar couplings observed are only consistent with an NH group at the center of the dithiolate bridge. This assignment is supported by quantum refinement of the X-ray data for *Clostridium pasteurianum*, which gives the best agreement between the two techniques when the bridging group is NH [23].

Dramatic reinforcement of the essential role of the bridgehead amine has come from studies of the activation of apo-HydA1 by the scaffold protein HydF in which synthetic subsites (Figure 7.1; **1a**, **1b**, and **1c**) [16, 17]. While Fourier transform infrared (FT-IR) established that all of the synthetic di-iron units could be transferred from the scaffold protein to apo-HydA1, only **1a** led to the reconstitution of an active H-cluster capable of hydrogen production at rates comparable with a native hydrogenase [16]. Remarkably, direct incorporation of these synthetic subsites into apo-HydA1 was achieved by incubation without an intermediary scaffold protein HydF.

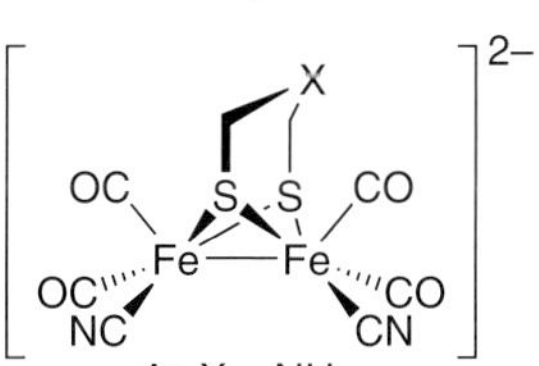

1a X = NH
1b X = CH_2
1c X = O

Figure 7.1 Synthetic subsites employed by Fontecave and coworkers [16, 17].

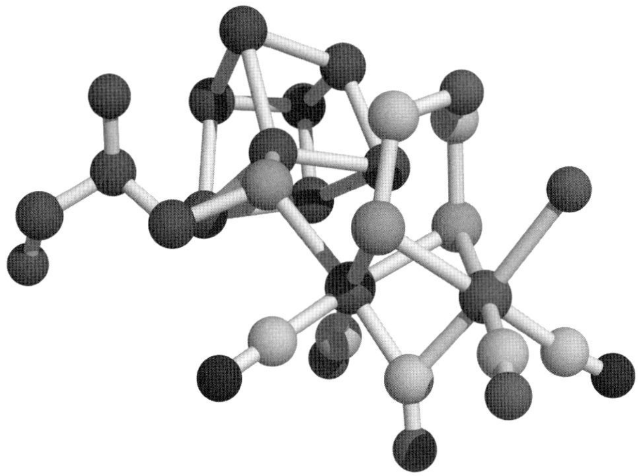

Figure 7.2 Structure of H_{ox} from *Clostridium pasteurianum* (central atom of thiolate bridge modeled as oxygen) [24].

7.2.3
Structural Features of the Resting State (H_{ox}) and Reduced State (H_{red}) of the Active Site

The mixed valency $\mathbf{H_{ox}}$ state can be considered as the resting state of the enzyme. Structures of the enzyme were obtained using enzymes from *C. pasteurianum* [2] and *Desulfovibrio desulfuricans* [3]. While both were produced under nominally reducing conditions, the consensus view is now that the *C. pasteurianum* structure is best regarded as representing $\mathbf{H_{ox}}$, while that from *D. desulfuricans* is in state $\mathbf{H_{red}}$ [6, 13]. More recently, a structure of the enzyme from *C. pasteurianum*, again in the $\mathbf{H_{ox}}$ form, has been reported at a resolution of 1.39 Å [24].

The structure of $\mathbf{H_{ox}}$ from *C. pasteurianum* is shown in Figure 7.2 [24] and that of $\mathbf{H_{red}}$ from *D. desulfuricans* is shown in Figure 7.3 [3]. With the caution that we are comparing sites from two different organisms, the structural consequences of the formal one-electron reduction of $\mathbf{H_{ox}}$ to $\mathbf{H_{red}}$ can be considered. Key features of $\mathbf{H_{ox}}$ are (i) a bridging CO ligand, (ii) the "rotated" state at the distal iron, (iii) a weakly bound water molecule at the distal iron, and (iv) hydrogen-bonded cyanide ligands. Reduction conserves the rotated state but (v) the CO ligand is no longer bound in the fully bridging position: it is considerably displaced toward a terminal bonding mode. In addition, (vi) there is no evidence for a bound water molecule at the distal iron.

7.2.4
Relationship between Structural and Spectroscopic Properties of H_{ox}, H_{red}, and H_{sred}

The infrared spectra of $\mathbf{H_{ox}}$, as determined in various organisms, indicate a high degree of consistency in the frequencies of the terminal CO, the bridging CO, and CN bands, as shown in Table 7.1 [25–30]. The bridging CO frequencies, for

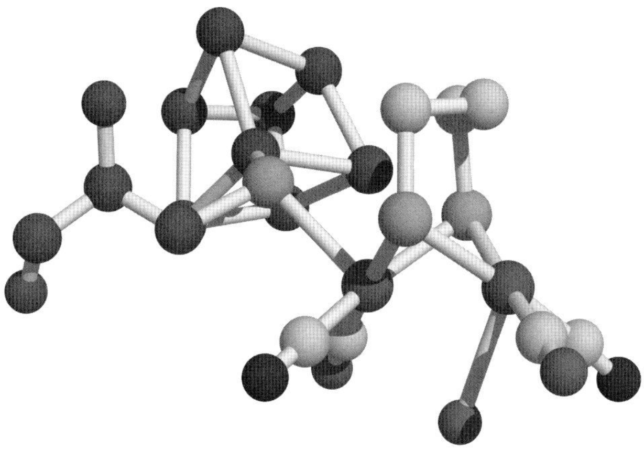

Figure 7.3 Structure of H_{red} from *Desulfovibrio desulfuricans* (central atom of thiolate bridge modeled as carbon, rotated CO modeled as an oxygen atom) [3].

example, differ by <2 cm^{-1}. The weak CN bands and terminal bands also occur at very similar frequencies (Table 7.1). This is fully consistent with the high degree of sequence homology in the respective enzymes around the active site. *A priori*, one might expect that frequencies associated with the other turnover states that have been detected, $\mathbf{H_{red}}$ and $\mathbf{H_{sred}}$ (super reduced), should show very similar vibrational signatures.

Against this background spectroscopy for $\mathbf{H_{ox}}$, we have reevaluated all the available FT-IR data for the reduction of [FeFe] hydrogenases [25–30]. This leads to the following observations. Reduction of $\mathbf{H_{ox}}$ in various spectroscopic and spectroelectrochemical experiments yields infrared spectra for $\mathbf{H_{red}}$ and $\mathbf{H_{sred}}$ at closely corresponding frequencies irrespective of the organism. Such reductions invariably lead to a mixture of redox states, $\mathbf{H_{ox}}$, $\mathbf{H_{red}}$, and $\mathbf{H_{sred}}$, with a distribution dependent on the strength of the reductant or reducing potential. For example, reduction with dithionite gives $\mathbf{H_{red}}$ and $\mathbf{H_{sred}}$, while incubation under dihydrogen gives a mixture of all three states.

The key feature of the reduction of $\mathbf{H_{ox}}$ to $\mathbf{H_{red}}$ is the switch of the bridging CO ligand to a terminal binding mode: three terminal CO bands are observed near 1894, 1916, and 1965 cm^{-1}. The shift of the lowest frequency band by over 90 cm^{-1} cannot be interpreted other than as a switch from bridging to terminal CO bonding mode (see subsequent text). It has been suggested that there is a $\mathbf{H_{red}}$ state in which a bridging CO is retained and that this shows an organism-dependent variation in the structure of a state of the H-cluster [29]. However, our reexamination of the frequencies reported for this system suggests the band assigned to $\mathbf{H_{red}}$ at 1793 cm^{-1} is better attributed to $\mathbf{H_{ox}}$, that at 1935 cm^{-1} to either $\mathbf{H_{ox}}$ and $\mathbf{H_{sred}}$ or a mixture of these, while those at 1891 and 1919 cm^{-1} can be assigned to $\mathbf{H_{red}}$. This interpretation is strongly supported by other data for the same organism, which shows the same set of bands as for $\mathbf{H_{red}}$ in *D. desulfuricans*, *Desulfovibrio vulgaris*, and *Megasphaera elsdenii* [30]. Simulated IR spectra from density

Table 7.1 Infrared stretching frequencies for CO groups in [FeFe] hydrogenases.

Organism	Wavenumber (cm^{-1})		
	H_{ox}	H_{red}	H_{sred}
Desulfovibrio desulfuricans [28]	1802	1894	1883
	1940	1916	1932
	1965	1965	1955
Chlamydomonas reinhardtii [29]	1800	1793	1882
	1940	1891	1919
	1964	1935	1954
Chlamydomonas reinhardtii [30]	1802	1891	1883
	1940	1916	1933
	1964	1963	1954
Desulfovibrio vulgaris [25]	(1812)	1894	—
	1940	1916	—
	1965	1964	—
Desulfovibrio vulgaris [26]	~1800	1894	—
	1940	1916	—
	1965	1965	—
Megasphaera elsdenii [25]	N.R.	1892	—
	1938	N.R.	—
	1965	1965	—
Clostridium pasteurianum [27]	1802	—	—
	1948	—	—
	1971	—	—

Gray highlighted frequencies are reassigned in text. N.R. = Not reported

functional theory (DFT) calculations on simplified models for $\mathbf{H_{red}}$ are consistent with the presence of both unprotonated and protonated bridgehead species [31].

Infrared data suggest one-electron reduction of $\mathbf{H_{red}}$ to give $\mathbf{H_{sred}}$ is possible. Consistent values are available for two organisms [28–30]. The shift in infrared frequencies to lower values on reduction are relatively small, and this is in accord with the process involving reduction of the $\{4Fe4S\}^{2+}$ unit to the $\{4Fe4S\}^{+}$ state as established by electron paramagnetic resonance (EPR) [4]. The $\mathbf{H_{sred}}$ state does not possess a bridging CO; the structure of the subsite is essentially unchanged from that in $\mathbf{H_{red}}$.

Carbon monoxide reversibly inhibits hydrogen evolution by [FeFe] hydrogenase (Scheme 7.1). Structural studies reveal that $\mathbf{H_{ox}}$ binds carbon monoxide at the distal iron site [32, 33], and FT-IR studies of the enzyme in the presence and absence of CO clearly show the interconversion of $\mathbf{H_{ox}}$ and $\mathbf{H_{ox}}$**-CO** (CO-inhibited) [28, 29]. There is no spectroscopic evidence that carbon monoxide binds to $\mathbf{H_{red}}$ at the vacant apical site on the distal iron. This is understandable if the Fe(I) centers are electronically closed shell with extensive Fe–Fe bonding. In the single crystallographically characterized $\mathbf{H_{red}}$ state,

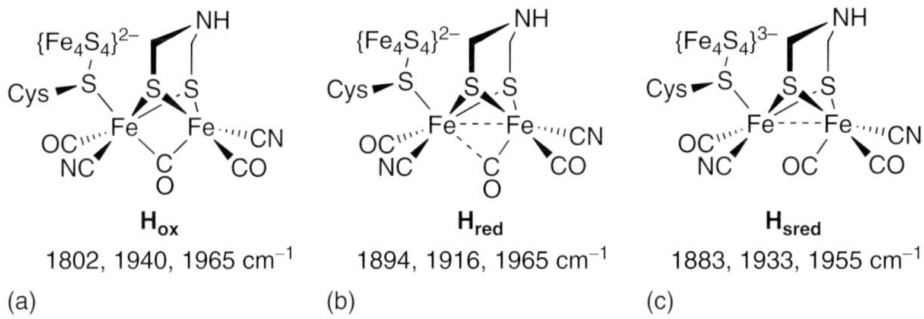

Figure 7.4 Consensus values for CO stretching frequencies in (a) $\mathbf{H_{ox}}$, (b) $\mathbf{H_{red}}$, and (c) $\mathbf{H_{sred}}$.

the Fe–Fe bond distance is 2.6 Å [3], which is in the metal–metal bonding regime.

In summary, structural and spectroscopic data are consistent with the interconversion of the $\mathbf{H_{ox}}$, $\mathbf{H_{red}}$, and $\mathbf{H_{sred}}$ states of the H-cluster as shown in Figure 7.4, which directly relate to part of the catalytic cycle represented in Scheme 7.1. There is no unequivocal evidence for these states possessing terminal hydride ligation or for bridging CO in the $\mathbf{H_{red}}$ or $\mathbf{H_{sred}}$ states. Protonation of the nitrogen bridgehead in the spectroscopically characterized reduced states is plausible but not established.

7.2.5
The Rotated State and Mixed Valency: Synthetic Systems

Studies of synthetic systems have informed our understanding of the interrelationship of the structures of Fe(I)Fe(I) and the mixed-valence Fe(I)Fe(II) redox levels of the subsite. Early work opened up the chemistry of synthetic Fe(I)Fe(I) systems, providing structures with key elements of the natural system in the $\mathbf{H_{red}}$ state [34–42]. In the enzyme, X-ray crystallography of the $\mathbf{H_{red}}$ state shows that the distal iron has a rotated state configuration (Figures 7.3 and 7.5, **2**), whereas all of the earlier models possessed the (*pseudo*)symmetric arrangement (Figure 7.5, **3**), with more or less free turnstile interchange of the terminal ligands around each metal center [43, 44]. The introduction of bulky groups at the bridgehead in synthetic models has recently been shown to provide solid-state structures that approach the conformation found in the enzyme, but which remain dynamic in solution (Figure 7.6) [45]. Undoubtedly, within the protein matrix, hydrogen bonding interactions of the cyanide groups stabilize the rotated state (Figure 7.7) [3].

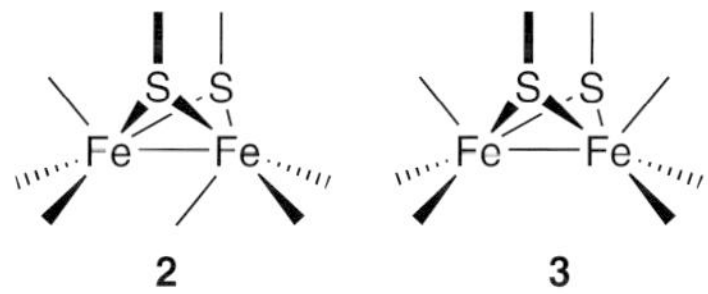

Figure 7.5 Rotated and symmetric conformations at the two iron centers.

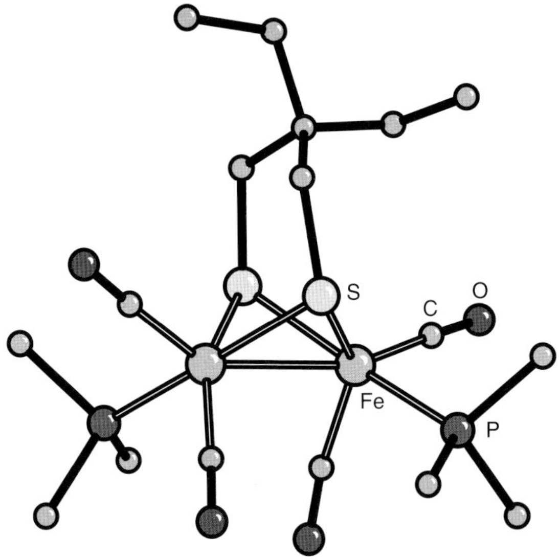

Figure 7.6 Structure of 2Fe2S model compound bearing bulky bridgehead substituents [45].

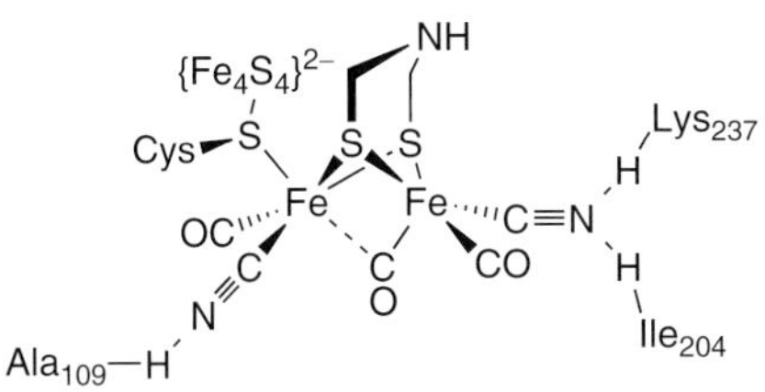

Figure 7.7 Hydrogen bonding from protein to subsite in structure from *D. desulfuricans* [3].

Early spectroelectrochemical and stopped-flow experiments showed that a mixed-valence Fe(I)Fe(II) system could be generated from a stable Fe(I)Fe(I) precursor possessing the key structural elements of the natural subsite in the CO-inhibited form, $\mathbf{H_{ox}}$**-CO** (Scheme 7.2) [46]. Although stable mixed-valence structures with cyanide coligands remain elusive, considerable advances have been made in the characterization of di-iron units possessing abiological coligands (Figure 7.8) [45, 47–51]. A key aspect of this chemistry is the rearrangement upon oxidation of one of the square-based pyramids to the rotated state exposing a coordination site. Stabilization of this type of a structure is favored by strongly electron-donating carbene [47] or rigid phosphine ligands [48, 50, 51]. It has also been shown that increasing the bulk at the bridgehead can favor the formation of a rotated state upon one-electron oxidation of Fe(I)Fe(I) precursors [45, 49]. As is evident in the structure of the $\mathbf{H_{red}}$ state [3], in the protein pocket of the natural system hydrogen bonding interactions maintain the rotated state configuration.

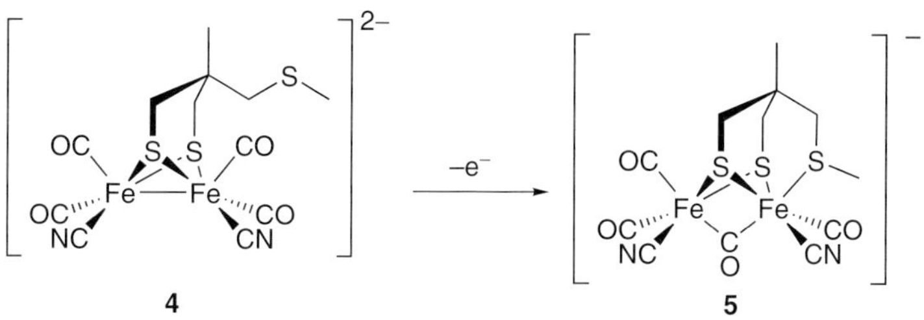

Scheme 7.2 Mixed-valence Fe(I)Fe(II) system [46].

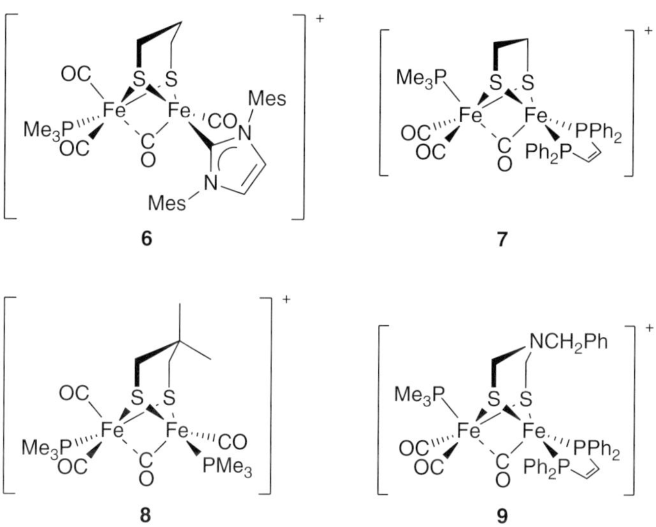

Figure 7.8 Mixed-valence Fe(I)Fe(II) systems with (semi)bridging CO groups and vacancy at the distal iron (Mes = 2,4,6-$Me_3C_6H_2$) [45, 47–51].

To date, structurally characterized mixed-valence systems have been limited to species with CH_2 or NBn bridgehead dithiolate moieties (Figure 7.8); mixed-valence systems with a bridging NH ligand and vacancy at the distal iron have not been structurally characterized.

While the $\mathbf{H_{ox}}$ state of the H-cluster shows a delocalized spin density, the complex **7**, and the NH-bridgehead analog of **9** are formulated as mixed-valent Fe(II)Fe(I) species with the spin predominantly found in a d_{z^2}-like orbital on the basis of EPR and Mössbauer data [52]. This underscores the difficulty in reproducing the precise electronic environment of the native system.

In summary, it has been shown that with a suitable choice of coligands, mixed-valence Fe(I)Fe(II) systems with the rotated state configuration can be synthesized, which are structurally akin to $\mathbf{H_{ox}}$. A structure at the Fe(I)Fe(I) level with a semibridging CO ligand and distortion toward the rotated state has been synthesized by using a bridgehead ligand, which approaches the conformation of $\mathbf{H_{red}}$.

7.2.6
Hydrides

Hydride species have been widely postulated as intermediates in the enzyme cycle (Scheme 7.1) but so far have eluded detection by FT-IR, EPR/ENDOR, or other methods. Studies on synthetic models of the natural subsite have provided crystallographic and spectroscopic data for bridging and terminal hydride species and mechanistic information on factors related to their formation. Synthetic di-iron subsite analogs have been extensively studied as electrocatalysts for hydrogen evolution and a role for hydrides in these systems is undoubted.

Protonation of the metal–metal bond in Fe(I)Fe(I) dithiolate systems can give bridging hydrides, provided the coligands confer sufficient basicity on the di-iron

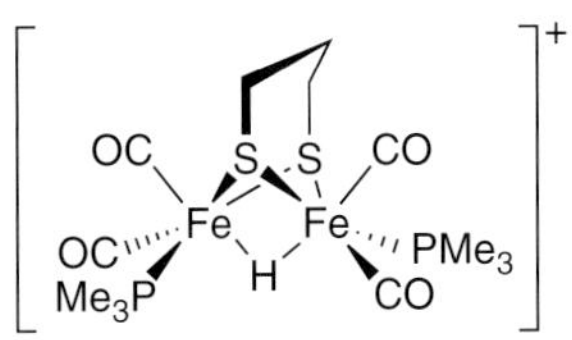

10

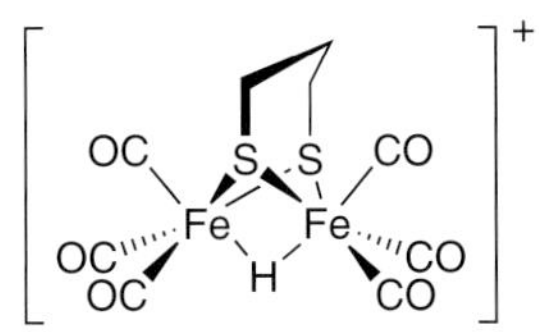

11

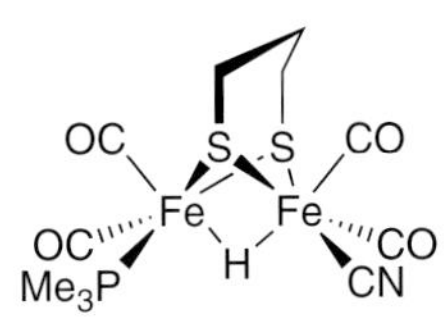

12

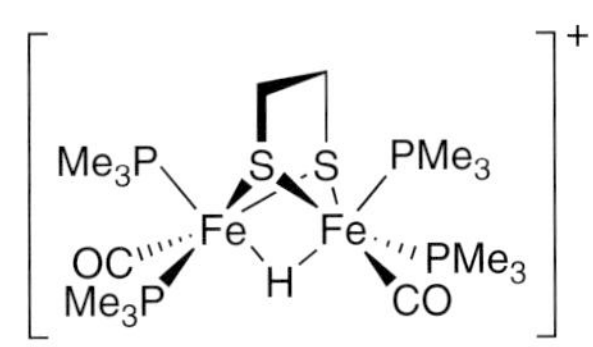

13

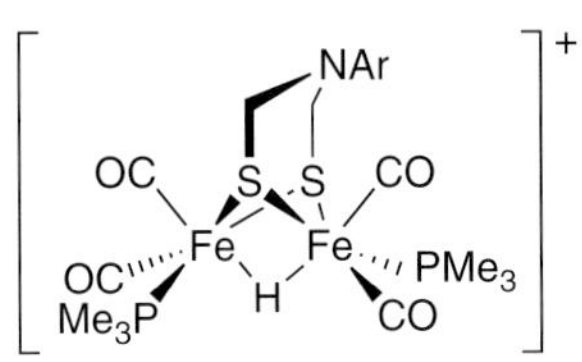

14

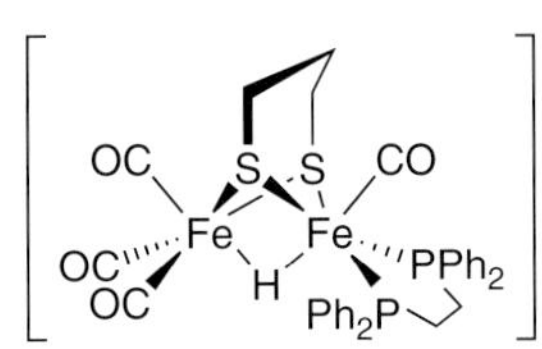

15

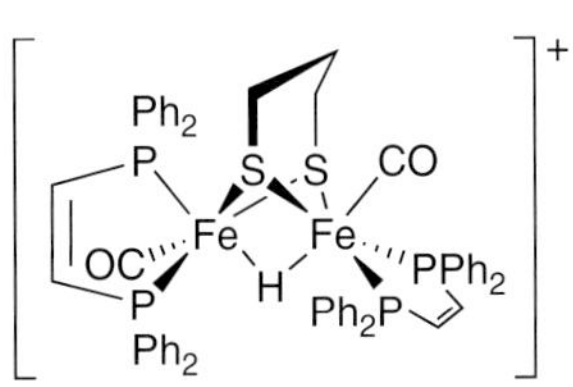

16

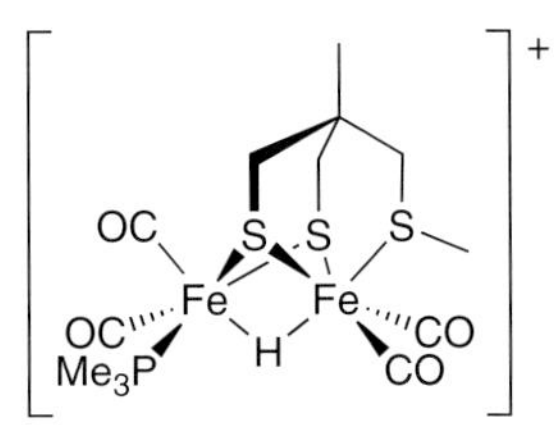

17

Figure 7.9 Examples of structurally characterized bridging hydrides (Ar = p-$C_6H_4NO_2$) [40, 53, 55–60].

unit or the acid is sufficiently strong [53]. There are now 23 structurally characterized bridging hydride systems [54]; some representative examples are shown in Figure 7.9 [40, 53, 55–60], including a recent example with a {2Fe3S} core [60].

Rauchfuss and coworkers [56] showed that a terminal hydride could be produced indirectly by hydride attack on a Fe(II)Fe(II) dithiolate cation, providing both the first example of a crystallographically characterized terminal hydride and key NMR spectroscopic information on the different signatures for bridging and terminal systems with the same {2Fe2S} cores. Schollhammer and coworkers [58] provided the first evidence for the *direct* formation of terminal hydrides by *protonation* of di-iron dithiolates. To date, three di-iron dithiolate species bearing terminal hydrides have been structurally characterized (Figure 7.10) [56, 61, 62]. Importantly, compound **19** represents the first crystallographically characterized terminal hydride in which the amine bridge is also protonated.

In earlier discussions of the Schollhammer system, it was also thought that the generation of a terminal hydride was favored by dissymmetric substitution at the two iron atoms [58], as is postulated for the subsite in the enzyme; counterintuitively, this is the *less* electron-rich distal iron (Scheme 7.1). Dissymmetric substitution will confer different electron densities on the two metal centers and polarize the Fe–Fe metal bond. It has been argued that protonation in model systems occurs more readily at the less electron-rich iron atom because this center can more easily adopt the rotated state configuration [63]. There is the caveat that later work has demonstrated that asymmetric substitution is not a prerequisite for terminal hydride formation (Figure 7.10) [61, 62].

It was originally proposed that the formation of terminal hydrides by protonation of di-iron units was the kinetically favored step and such species were intermediates on the pathway to the thermodynamically favored bridging hydrides.

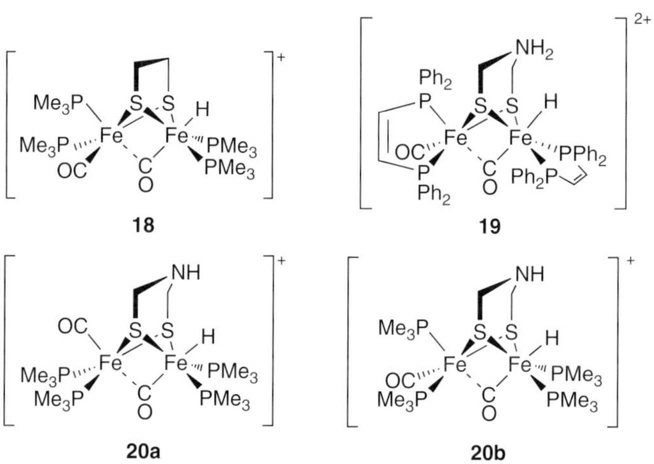

Figure 7.10 Structurally characterized terminal hydride systems (**20** crystallizes as a mixture of two isomeric forms) [56, 61, 62].

However, it has been shown that for an electron-rich symmetric complex, protonation at −70 °C gives the bridging hydride directly *before* detection of the terminal species [62]. At higher temperatures, both terminal and bridging hydrides are observed, ultimately resulting in the conversion of all of the terminal species to the bridging form at room temperature. The half-life for this conversion is ~2.5 h for protonation of compound **21** (Figure 7.11). At −90 °C, there is some evidence that protonation occurs initially at sulfur. Subsequent intramolecular migration gives the terminal species, while an intermolecular process, second order in complex, is suggested to lead to the bridged species. These observations are summarized in Scheme 7.3.

Stopped-flow kinetic studies on a range of di-iron units at 21 °C have provided information on the primary rates of protonation of the metal–metal bond and rates for subsequent rearrangement to the most stable isomer [64–66]. Terminal

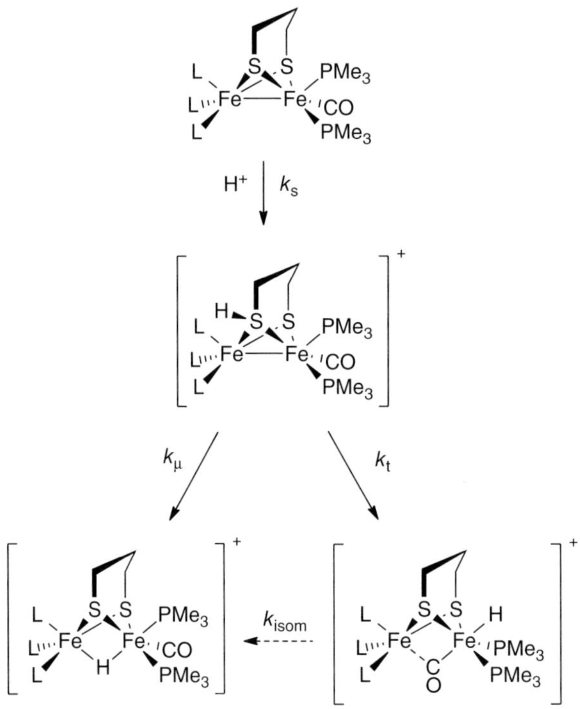

Scheme 7.3 Protonation mechanism proposed by Rauchfuss [62] (L = CO, PMe_3).

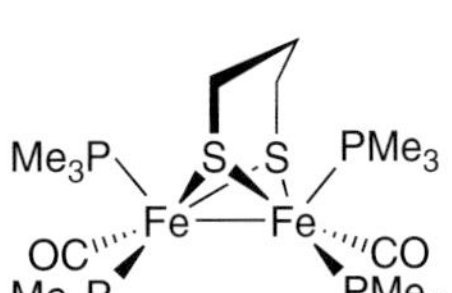

21

Figure 7.11 Highly electron-rich di-iron complex exhibiting both terminal and bridging protonation [62].

hydride intermediates have *not* been detected, consistent with two competitive rather than sequential protonation pathways and in accord with what is observed for **21** [62]. However, DFT calculations have suggested that the intimate details of such reactions may involve attack from the "underside" of the Fe–Fe bond with formation of a short-lived hydride terminal intermediate [67]. Such an intermediate is geometrically distinct from isolated or otherwise spectroscopically characterized terminal hydrides insofar as the latter involve protonation at Fe in the rotated state arrangement (see preceding text). Specifically, DFT calculations on the protonation of the [FeFe] hydrogenase model complex **22** (Scheme 7.4) show that di-iron bridging hydride species are more stable than iron terminal-hydride, sulfur-hydride, or formyl isomers. Consistent with experimental observation, the *transoid* basal/basal forms are more stable than other μ-H isomers. With an ether as the proton carrier, $[Et_2OH]^+$, the favored reaction pathways appear to involve weak coordination to CO (not sulfur) followed by transfer of the proton from ether to an iron terminal site rather than directly to the bridging site. These kinetically favored terminal-hydride species isomerize through a low-energy Ray–Dutt twist to produce the apical/basal bridging hydride isomer. This isomer rearranges over somewhat higher barrier Bailar twists to the *cisoid* and *transoid* basal/basal isomers. The proposed overall pathway is shown in Scheme 7.4.

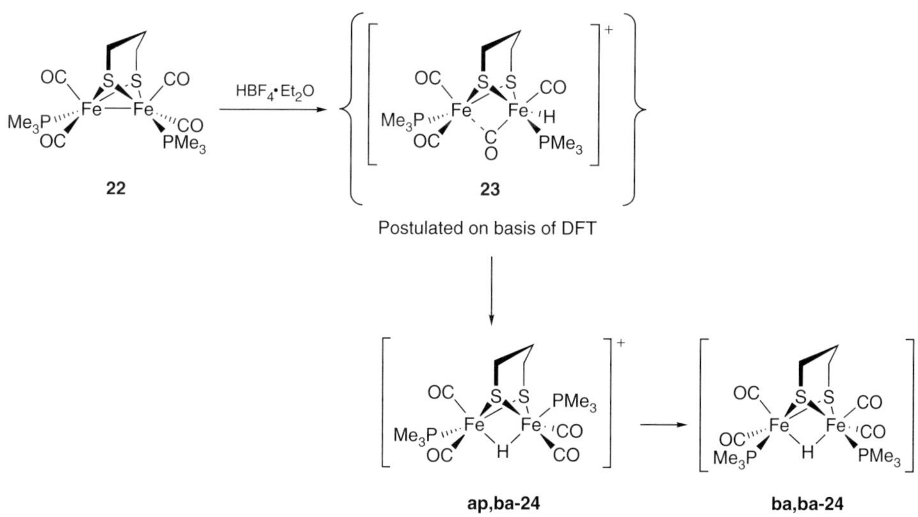

Scheme 7.4 Proposed mechanism for formation of a μ-hydride species [67].

The structure of the di-iron unit dramatically controls the overall rate of protonation. For example, the complex **25** protonates to give the bridging hydride at a rate that is 10^6 faster than protonation of **26** (Figure 7.12) [68]. Figure 7.13 shows the relationship between the oxidation potential of the parent complex and the relative activation energy for its protonation to give a bridging hydride. There is an

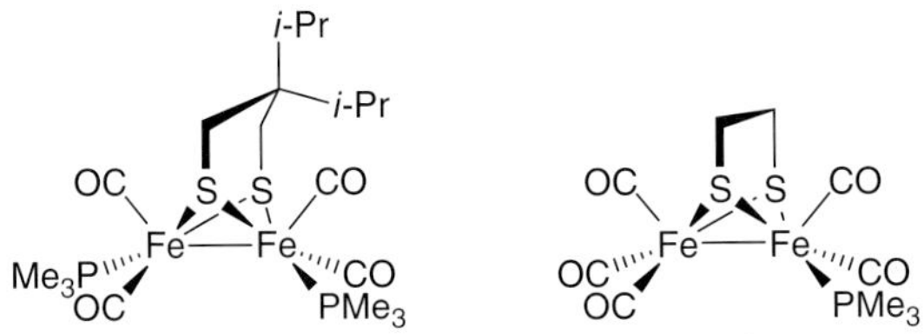

Figure 7.12 Relative rates of bridging hydride formation [68].

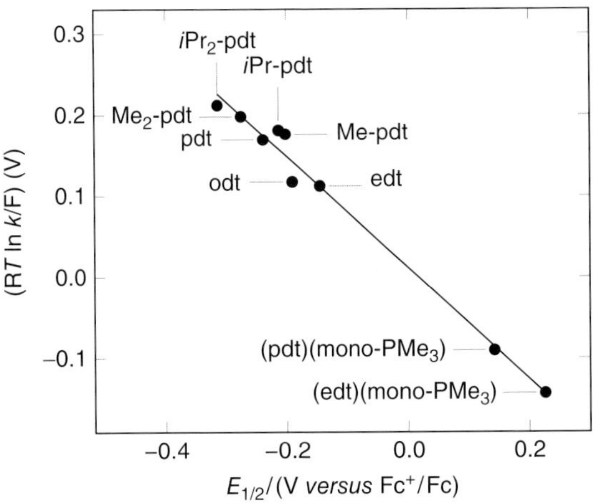

Figure 7.13 Correlation between oxidation potential and rate of protonation for $Fe_2(xdt)(CO)_{(6-n)}(PMe_3)_n$ by $HBF_4 \cdot Et_2O$ in MeCN (edt, ethane-1,2-dithiolate; pdt, propane-1,3-dithiolate; odt, 2-oxopropane-1,3-dithiolate; $n = 2$ unless specified) [68].

essentially linear correlation with a Brønsted coefficient of about 0.7. This is consistent with the relative energy of the HOMO (highest occupied molecular orbital) in these complexes having a ground-state effect that dominates the magnitude of the activation energy barrier of the process. The oxidation potential of a general base usually tightly correlates with pK_a. As the HOMO in the complexes has a predominantly metal–metal bond character and is also the site of protonation, it is reasonable that the basicity of this bond largely determines protonation rates.

7.2.7 Hydrides and Electrocatalysis of Hydrogen Evolution

As discussed, in enzyme turnover hydrides are most probably extant as transition states or intermediates in hydrogen evolution or uptake pathways. While such species have not been detected for the [FeFe] hydrogenase, there is unambiguous spectroscopic evidence for μ-hydride formation in the [NiFe] hydrogenase (see subsequent text) and in nitrogenase chemistry. Hydrides are plausible intermediates in electrocatalysis by synthetic di-iron species and such processes are covered in detail in Chapter 4 of this book. Here, we wish to consider the evidence for terminal and/or bridging hydrides in such electrocatalytic systems together with the role or otherwise of protonation at the bridgehead nitrogen.

Early work on di-iron dithiolate systems focused on electrocatalysis by hexacarbonyl systems, for example, $Fe_2(pdt)(CO)_6$, and detailed mechanistic studies led to the conclusion that opening of one of the sulfurs of the ligand and/or the loss of a CO was the key aspect of the electrocatalysis [10, 69]. Electron transfer probably activates the di-iron unit by generation of a site at which the substrate proton can bind in a terminal manner. Other work on tetranuclear thiolate systems provided evidence for a terminal to bridging CO rearrangement following two-electron transfer, again exposing a site for proton binding in a terminal position and fast catalysis [70]. Importantly, this electrocatalysis probably involves Fe(I)Fe(I) rather than the Fe(I)Fe(0) or Fe(0)Fe(0) levels of the hexacarbonyl systems. As discussed, if the di-iron unit is sufficiently electron rich, protonation at the Fe(I)Fe(I) level can occur to give terminal hydrides without the need for reduction to Fe(I)Fe(0) or Fe(0)Fe(0) states. This has more relevance to the natural system, the Fe(I)Fe(I) level corresponding to the $\mathbf{H_{red}}$ state.

A bridgehead NH and other systems in which a base is appended to a di-iron unit can also be thought of as "guiding" the proton to a basic iron site. Competition between bridging and terminal sites of protonation has been discussed: where a protonatable dithiolate bridgehead group is present (O, NH, or NR), terminal protonation rates are enhanced *vis-à-vis* bridging protonation [71]. Similarly, a protonatable diphosphine ligand disposed toward the basal face of the di-iron unit has been shown to enhance the rate of protonation at the Fe–Fe bond (Figure 7.14) [72, 73].

Protonation at bridging or terminal sites in Fe(I)Fe(I) species formally generates Fe(II)Fe(II) hydrides: these are electroactive and are plausible intermediates in an electrocatalytic cycle [14]. Terminal hydrides have been shown to be easier to reduce by about 200 mV than their isomeric bridging hydride counterparts [59]. Spectroelectrochemical studies have provided data for paramagnetic bridging hydrides establishing that a mixed-valence Fe(II)Fe(I) level can be attained

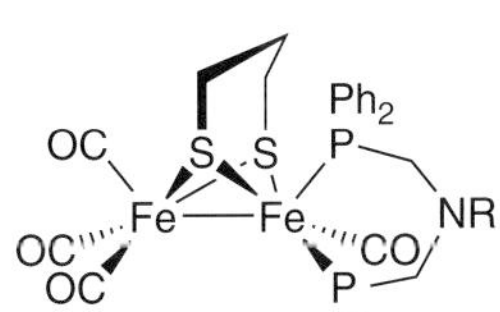

27

Figure 7.14 Model system featuring a protonatable phosphine ligand (R = Me [72], *n*-Pr [73]). The crystal structure for the case R = *n*-Pr shows the phosphines in the apical–basal arrangement [73], but this is dynamic as demonstrated by the fact that the structure for the doubly protonated derivative has a basal–basal phosphine ligand geometry.

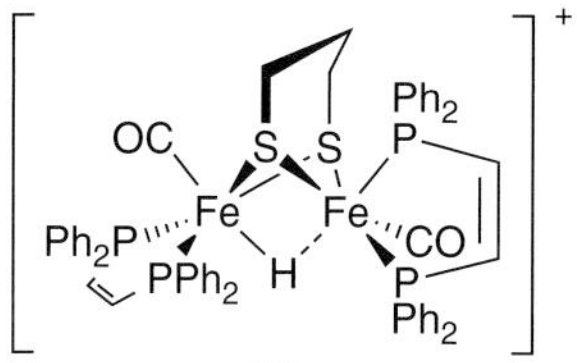

28

Figure 7.15 Crystallographically characterized paramagnetic hydride [75].

without geometric rearrangement [74]. Subsequently, the paramagnetic hydride **28** was crystallographically characterized (Figure 7.15) [75].

It has been generally thought that in an electrocatalytic cycle electron transfer to give an Fe(II)Fe(I) terminal or bridging hydride would enhance basicity (hydricity), enabling protic attack and hydrogen evolution (Scheme 7.5). However, deuterium labeling studies of **28** have shown that protonation of the Fe(II)Fe(I) species evolves dihydrogen, but *the bridging hydride ligand is retained.* This has led to the suggestion that in electrocatalytic cycles involving a μ-H, the latter is merely a spectator ligand (Scheme 7.5) [75]. The bridging hydride systems can be viewed as electron-poor catalysts akin to the hexacarbonyl species, which reduce at similar potentials. As discussed, electrocatalysis by, for example, $Fe_2(pdt)(CO)_6$ involves thiolate or CO decoordination; it may well be the case that catalysis involving spectator μ-H ligands also necessitates ligand loss to generate terminal hydride protonation sites. Interestingly, Fe–μ-H–Fe species have been invoked in hydrogen evolution pathways catalyzed by the FeMo-cofactor in nitrogenase [76].

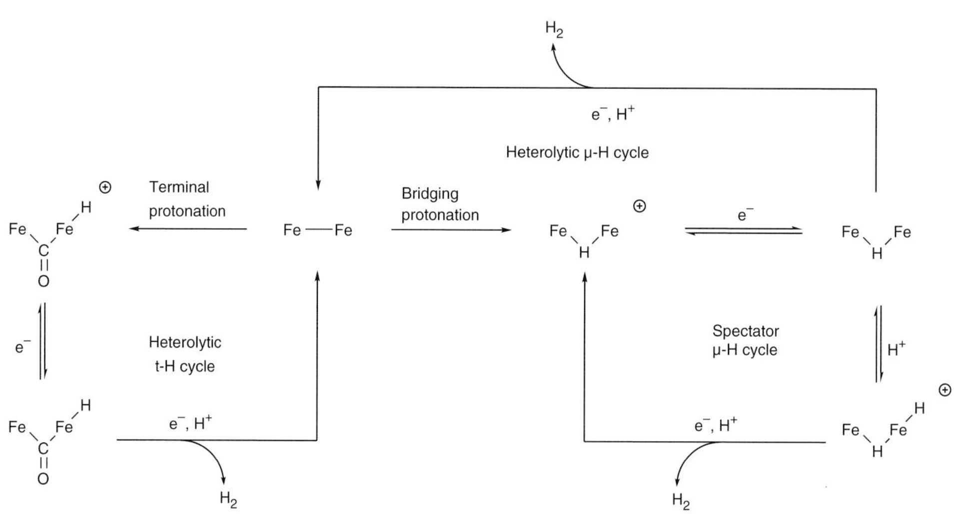

Scheme 7.5 Mechanistic cycles for H_2 release by a generalized di-iron center (t-H = terminal hydride; μ-H = bridging hydride).

A protonatable NH-bridgehead group can greatly enhance the rate of hydrogen evolution from a synthetic di-iron system possessing a coordination environment that favors terminal protonation [61]. Under similar conditions, the azo-propanedithiolate complex **19** electrocatalyzes hydrogen evolution about 10 000 times faster than its propane dithiolate counterpart. This is attributed to the presence of the bridgehead amine relaying protons for hydrogen evolution to the terminal site. The comparison is made under a protic regime, where the catalysis is independent of acid concentration. The magnitude of k_{cat} for the systems with amine bridgehead atoms is increased by a factor of 250 on switching from μ-H to t-H. Thus, the two species must cycle without a common intermediate, and therefore the μ-H cannot go via the unprotonated state (Scheme 7.5, heterolytic μ-H cycle). For those systems reported using the same acid, at the limit of the slowest system the acid-dependent kinetics are remarkably similar with k_{obs} at 10 equiv $ClCH_2COOH$ of about 3.5. We suggest that this is consistent with the rate-limiting protonation of a comparable terminal site.

7.2.8 Dihydrogen Oxidation

In contrast to proton reduction, oxidation of dihydrogen has proved to be rather more difficult to achieve using synthetic models. The first demonstration of activation of dihydrogen by a model system was reported in 2009. The paramagnetic mixed-valence complex **29**, which has a bridgehead amine group, was found to cleave dihydrogen to give the diamagnetic bridging hydride **31** via a terminal hydride intermediate (Scheme 7.6) [77]. The H_2 cleavage reaction also provided the first example of introducing a hydride at a di-iron dithiolate from molecular hydrogen. Importantly, the related O and CH_2 bridgehead dithiolate species were unreactive under the same conditions, stressing the importance

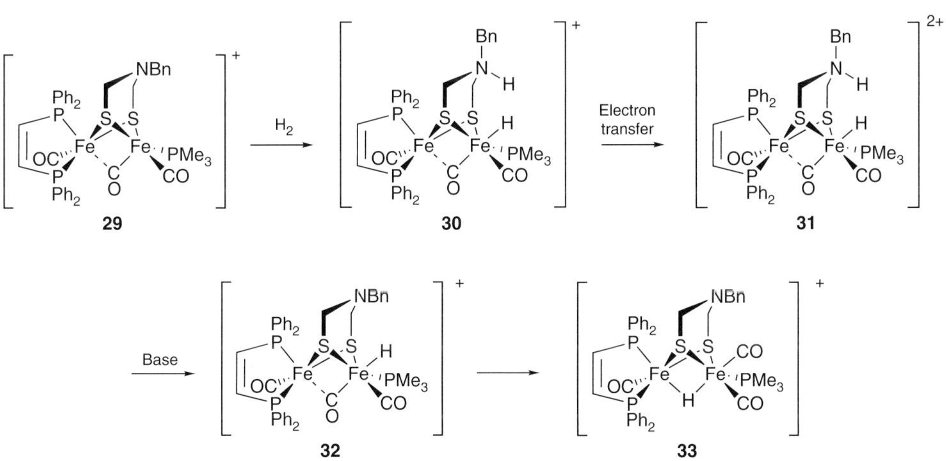

Scheme 7.6 Oxidation of dihydrogen by a di-iron system [77].

of an internal base in assisting the heterolytic cleavage. This system requires significant hydrogen pressure (18 000 psi) to achieve formation of a hydride product. In addition, the stoichiometry suggests that **29** must also function as a sacrificial electron acceptor. In the natural system, the {4Fe4S} cluster provides the "electron sink" to support the redox chemistry occurring at the {2Fe2S} subsite, concertedly removing an electron from bound dihydrogen or paramagnetic (35-electron) terminal intermediate. This was exploited by Camara and Rauchfuss, who combined the proven ability of **29** to activate dihydrogen with an excess of the oxidant $Cp^*{}_2Fe^+$ ($Cp^* = \eta^5$-pentamethylcyclopentadienyl) as surrogate for the cubane electron sink. This allowed activation of dihydrogen under much milder conditions (2 atm) [51]. Further elegant work has shown that by combining a decamethylferrocenium/decamethylferrocene relay and the di-iron subsite mimic in a single molecule, *catalytic* oxidation of dihydrogen can be achieved in the presence of the hindered base P(*o*-tolyl)$_3$ and an excess oxidant $[(\eta^5\text{-}C_5H_5)_2Fe][B\{C_6H_3(CF_3)_2\}_4]$ (Scheme 7.7) [78]. This catalysis is extremely slow: the rate is $0.0001\ s^{-1}$ compared to $28\,000\ s^{-1}$ for the natural system [79]. However, in demonstrating that the combination of a basic bridgehead group, a di-iron unit with a proximal electron-relay oxidizes dihydrogen, the system offers both an insight into the natural system and will undoubtedly guide development of new catalytic systems.

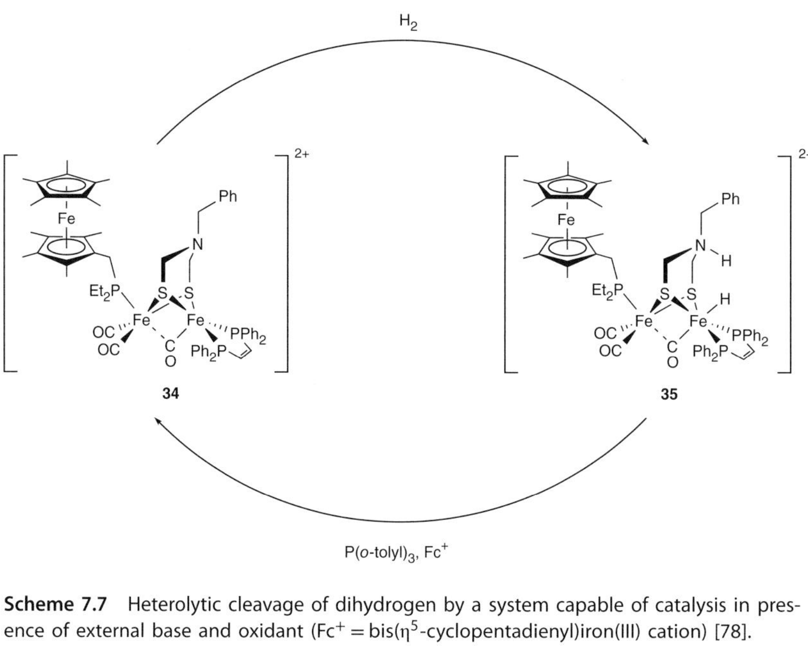

Scheme 7.7 Heterolytic cleavage of dihydrogen by a system capable of catalysis in presence of external base and oxidant (Fc^+ = bis(η^5-cyclopentadienyl)iron(III) cation) [78].

Theoretical calculations related to this system have suggested that H_2 binding and heterolytic cleavage take place concertedly with substrate-binding triggering electron transfer to the acceptor ferrocene moiety [80]. Formation of an iron–dihydrogen complex as a reaction intermediate was therefore excluded. Neither dihydrogen nor indeed hydride states have been detected in the natural system, but this is perhaps not surprising given the fast turnover.

While these systems operate through terminal hydride/appended base interactions, it has now been shown that a basic group proximal to the metal–metal bond facilitates heterolytic dihydrogen cleavage by Fe(I)–Fe(II) mixed-valent species (Scheme 7.8) [81]. The products at the end of the reaction (about 4 h) are *bridging* hydrides **37** and **38** together with the bridgehead base protonated species **39**. In the presence of ferrocenium as an oxidant and the exogenous weak base P(*o*-tolyl)$_3$, the system turns over slowly, less than one turnover per hour, at a rate comparable to that for **34**. The presumption is that the reaction mechanism does not proceed via a terminal hydride intermediate; because bridging hydrides are the thermodynamic sink, some caution in this interpretation is needed.

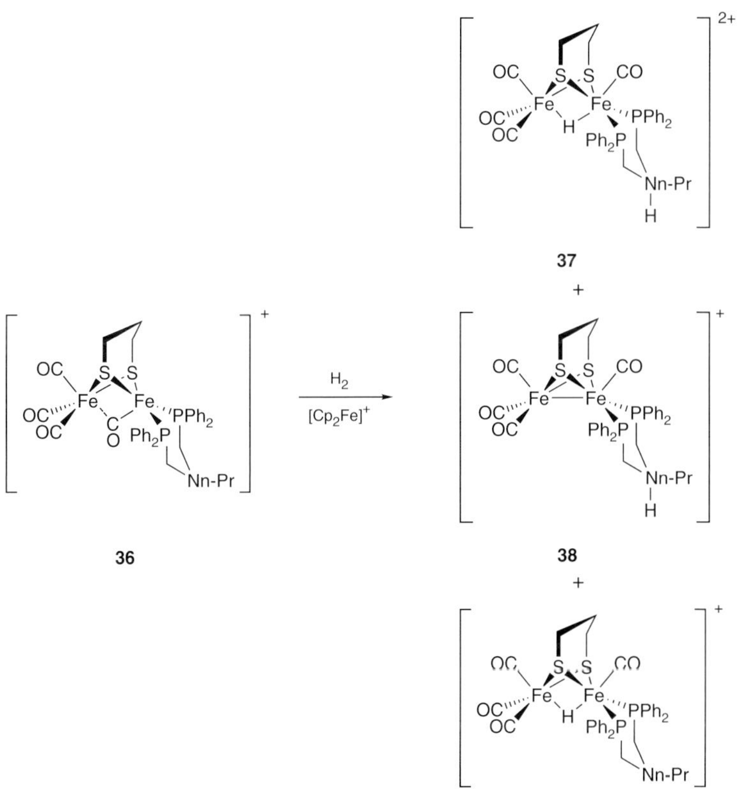

Scheme 7.8 Activation of dihydrogen by a basic ligand system [81].

7.2.9 Final Comments

While much progress has been made over the past 15 years in understanding the basic proton, electron, and electrocatalytic chemistry of synthetic subsite analogs, there are still significant challenges in progress toward competent synthetic (electro)catalysts based on di-iron subsite chemistry. Matching both the kinetics and turnover potential of the natural system has not been achieved. The role of the protein matrix in tuning the chemistry of the H-cluster for fast low overpotential catalysis is not well understood, beyond the undoubted importance of hydrogen bonding to cyanide ligands in stabilizing the rotated state together with "directed" proton channeling to the distal iron site. It is in the area of rational design of second coordination sphere interactions around synthetic subsite or H-cluster analogs that one might expect significant future advances. Seminal work on introducing synthetic subsites directly to an apoenzyme provides a bridge between organometallic chemistry, control of the second coordination sphere environment, and enzyme chemistry [16, 17]. It opens up the way for construction of hybrid systems and a means of probing how the protein matrix controls catalysis.

7.3 [NiFe] Hydrogenases

7.3.1 Overview of the Catalytic Cycle

The [NiFe] hydrogenases reversibly oxidize dihydrogen to protons. As with the [FeFe] hydrogenases, the enzymes can operate bidirectionally at low overpotentials for hydrogen oxidation or reduction, but their physiological role is in hydrogen uptake [1].

The core structure of the [NiFe] hydrogenase is a bimetallic system bridged by two cysteinyl ligands, coordinated at iron by two CN^- and one CO ligands and at nickel by two additional cysteine ligands, which are retained in the catalytic cycle [11, 82]. The iron atom has a square pyramidal or octahedral geometry, depending on the state of the enzyme, while the nickel center has an approximately disphenoidal ("seesaw") or distorted trigonal pyramidal geometry. Carbon monoxide, an inhibitor of the enzyme, binds at the nickel atom [83].

A representation of the catalytic cycle and pathways for converting inactive to active enzyme states of [NiFe] hydrogenase is given in Scheme 7.9 [11, 82, 84]. This cycle encompasses the spectroscopically and/or crystallographically identified state. The EPR-active **Ni-A** state has been most extensively characterized by spectroscopic and crystallographic methods [11]. This aerobically-isolated form of the enzyme is inactive and is unlikely to be a physiologically relevant form. However, it can be activated slowly under reducing conditions. It is thought that this

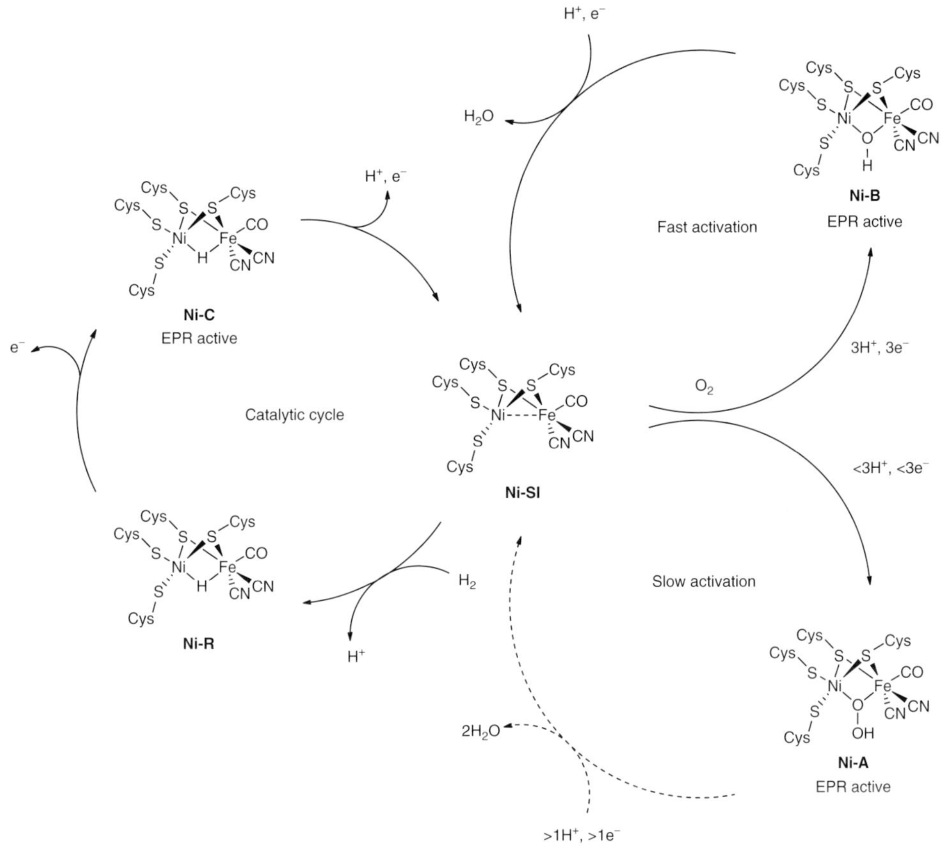

Scheme 7.9 Overview of the known species in the catalytic and activation cycles of [NiFe] hydrogenase [82, 84].

state has a bridging hydroperoxide ligand that links Ni(III) and Fe(II) [85]. The slow activation is attributed to the necessity of removing this ligand to enter a catalytically competent, EPR-silent, Ni(II)Fe(II) level **Ni-SI**. Isolation of the enzyme under anaerobic and reducing conditions followed by controlled oxidation at low temperature has allowed crystallographic characterization of the **Ni-B** form of the enzyme, which is rapidly activated under reducing conditions to give **Ni-SI** [85]. The structures of the EPR-active **Ni-B** state indicate that the Ni(III) and Fe(II) centers are also bridged by an oxygen ligand, but in this case as a hydroxide. Notably, the infrared data attributed to these two states is essentially indistinguishable but the EPR differs [11].

The literature on the redox/protonation states of the [NiFe] hydrogenase and how this relates to changes in the FT-IR spectra of the CO and CN^- bands has much complexity. For example, there are three putative **Ni-R** and two **Ni-SI** substrates [6, 11]. FT-IR frequencies for those species in the cycle(s) for which we

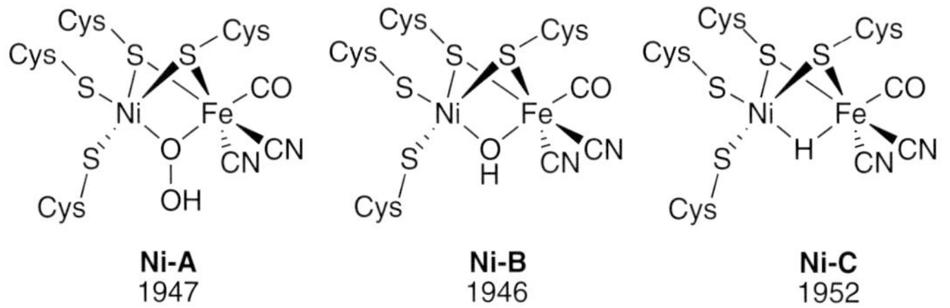

Figure 7.16 Infrared CO stretching frequencies (per centimetre) for catalytically relevant EPR-active species [86].

have some confidence in assigning are given in Figure 7.16 (values are for [NiFe] hydrogenase from *Desulfovibrio gigas*) [86].

X-ray crystallographic data for reduced states of the enzyme, for example, EPR-active **Ni-C** or EPR-silent **Ni-SI**, are available although assignment to a single state can be ambiguous [87]. What is clear is that these reduced (active) states show no evidence for an exogenous ligand bridging the metal centers. However, spectroscopic evidence for the **Ni-C** state unambiguously shows the presence of a bridging hydride [88, 89], which of course would not be resolved by protein crystallography.

7.3.2 Structural Models of Ni-A, Ni-B, and Ni-SI States

Mononuclear nickel and iron systems have been extensively studied in the context of the core geometry of these metals in [NiFe] hydrogenase, and this been reviewed [11]. Here, we focus on dinuclear structures that primarily relate to turnover. Early examples of dithiolate-bridged assemblies were reported by Sellmann [90] and Evans [91] (Figure 7.17).

Structural models have been synthesized which reproduce key features of the active site of the enzyme: sulfur-ligating nickel, bridging thiolate ligands, and CO and CN coordinating iron (Figure 7.18) [92, 93]. These are arguably closest in structure to CO-inhibited forms of the enzyme, but notably bear the additional CO ligand at iron rather than at nickel, as is observed in the natural system [82].

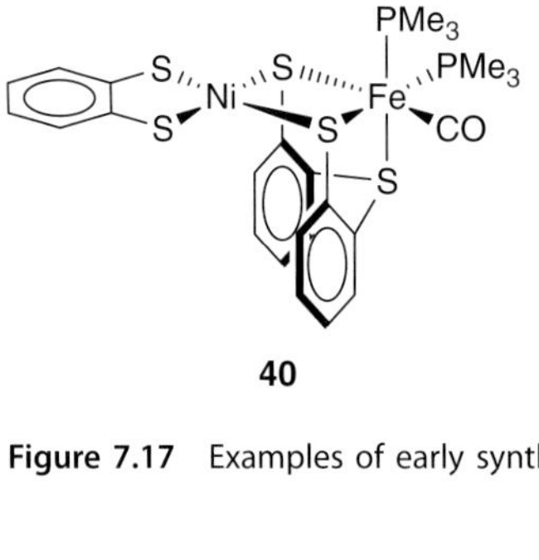

40

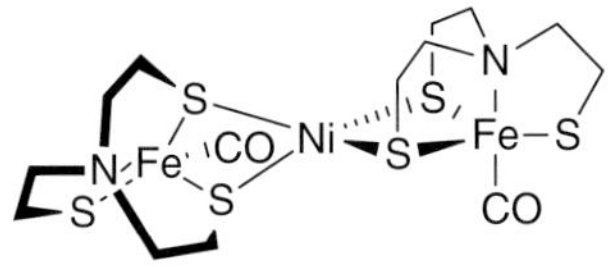

41

Figure 7.17 Examples of early synthetic dithiolate-bridged bimetallic assemblies [90, 91].

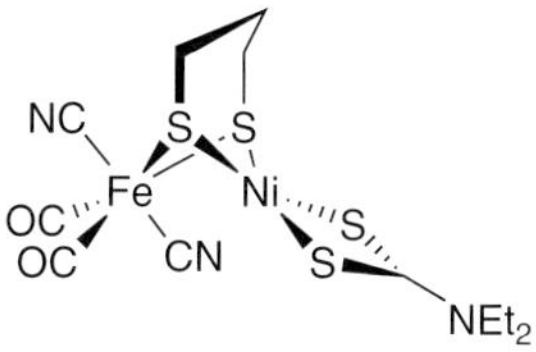

42

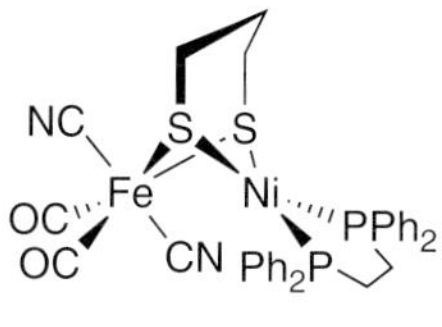

43

Figure 7.18 Structural models for the [NiFe] hydrogenase [92, 93].

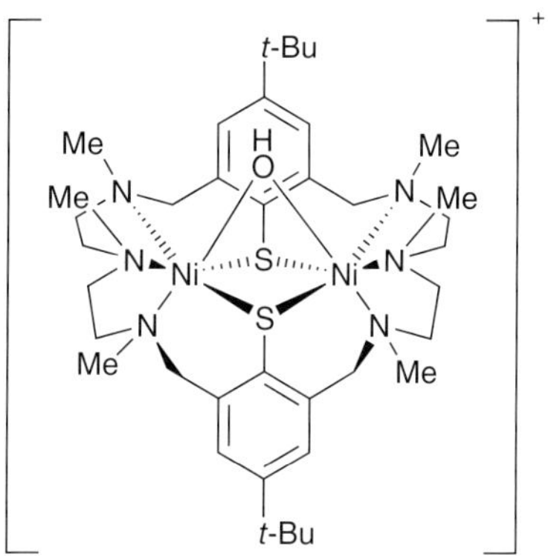

44

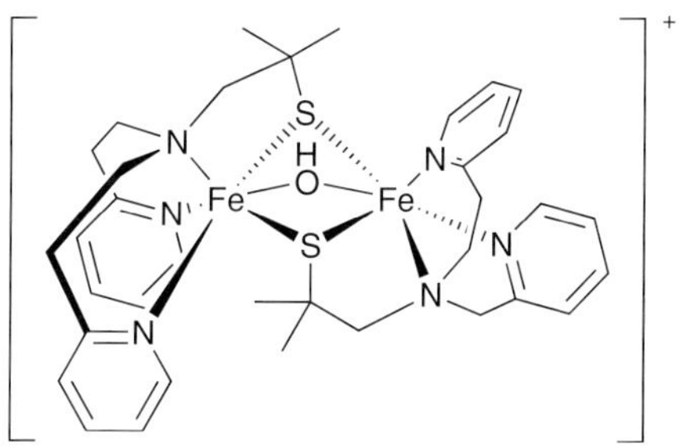

45

Figure 7.19 Nickel and iron "butterfly" dithiolate complexes containing μ-OH groups [94, 95].

There are no examples of nickel–iron systems with bridging dithiolates that also possess a bridging oxygen atom as in the peroxy-bridged **Ni-A** form or the hydroxyl-bridged **Ni-B** form of the enzyme. Indeed, in the Cambridge Structural Database, there are only two examples of dinuclear first-row transition metal complexes possessing a κ^2-O peroxy bridge.[1)] These are dicopper and dicobalt assemblies, and neither possess terminal or bridging thiolate ligands nor CO. The bridging hydroxy motif is common in first-row transition metal complexes: there are some six structures that have a butterfly dithiolate geometry and μ-OH.[2)] The most relevant structures are for the Ni–Ni and Fe–Fe species shown in Figure 7.19 [94, 95]; neither of these complexes shows chemistry pertinent to the natural system.

7.3.3
Hydride Chemistry Related to Ni-C/Ni-R: Functional Models

The first evidence that dihydrogen can be cleaved at a dinuclear dithiolate-bridged nickel system was demonstrated by Ogo [96] using the Ni/Ru assembly **46** shown

1) Cambridge Structural Database reference codes HPEDCO and MAMTEA.
2) Cambridge Structural Database reference codes VORMAQ, RAKLIY, MOSMOW, NAZREM, ODAGAC, and ODAFUV.

in Scheme 7.10. Somewhat related to this, dihydrogen evolution at a Ni/Ru assembly has been shown, but this operates at a high overpotential [97].

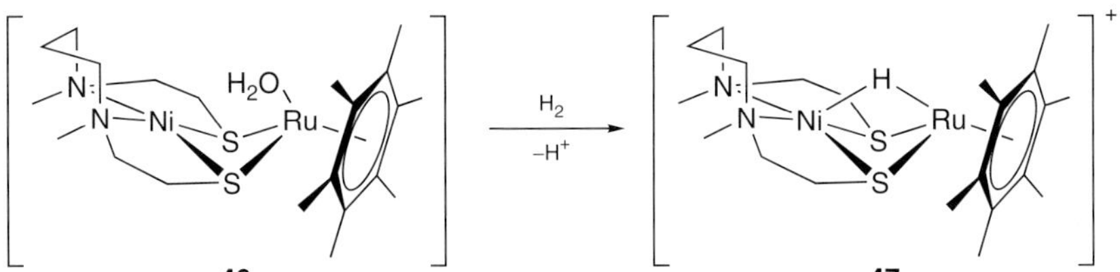

46 47

Scheme 7.10 Activation of dihydrogen by a Ni/Ru complex [96].

Much chemistry of Ni/Fe assemblies related to **Ni-C** has been based on tertiary phosphine systems rather than dithiolate ligation at nickel or cyanide ligation at iron. This mirrors the chemistry of the [FeFe] hydrogenase models, where tertiary phosphine ligation dominates the model chemistry.

The first crystallographically characterized Ni/Fe complex (**48**) with a μ-H ligand as found in **Ni-C** and **Ni-R** is shown in Figure 7.20, which was synthesized by protonation of a tricarbonyl iron/nickel precursor [98]. The formal oxidation states of the metal centers are Ni(II) and Fe(II), corresponding more closely to the **Ni-R** state. This and related bridging hydride complexes electrocatalyze proton reduction at high overpotentials, but show no propensity for dihydrogen oxidation [99, 100]. There are no examples of paramagnetic Ni/Fe bridging hydride species corresponding to **Ni-C**; oxidation of **48** is reported to be irreversible [99], in contrast to the enzyme system which shows reversibility [6].

The first example of the cleavage of dihydrogen at a dinuclear Ni/Fe assembly to give a hydride complex was also demonstrated by Ogo and coworkers [101] (Scheme 7.11). The rate of protonation of the hydride complex **50** by strong acid (to give hydrogen with regeneration of **49**) is slow, and the system is not catalytic.

Dihydrogen activation using a Ni/Fe assembly arguably more closely related to the natural system has been reported by Manor and Rauchfuss [102]. The cyanide ligands in complex **43** form adducts with the Lewis acid tris(pentafluorophenyl)borane. This makes a CO ligand susceptible to oxidation and removal by an amine oxide to provide a vacant site at which H_2 is

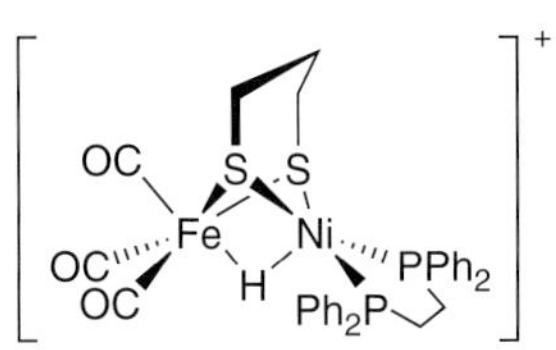

48

Figure 7.20 First hydride-bridged Ni/Fe model [98].

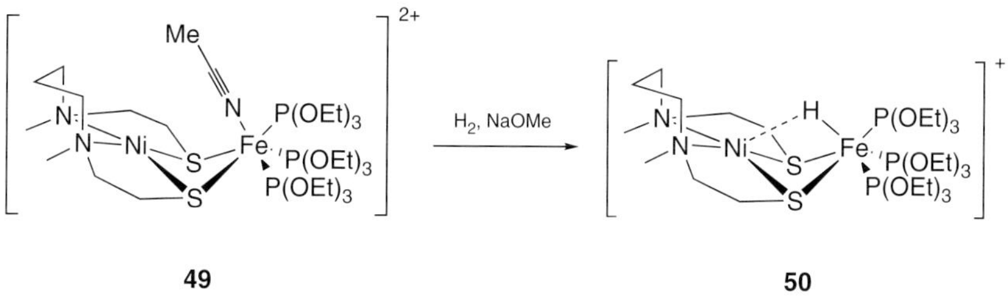

Scheme 7.11 Stoichiometric activation of H_2 by a Ni/Fe complex [101].

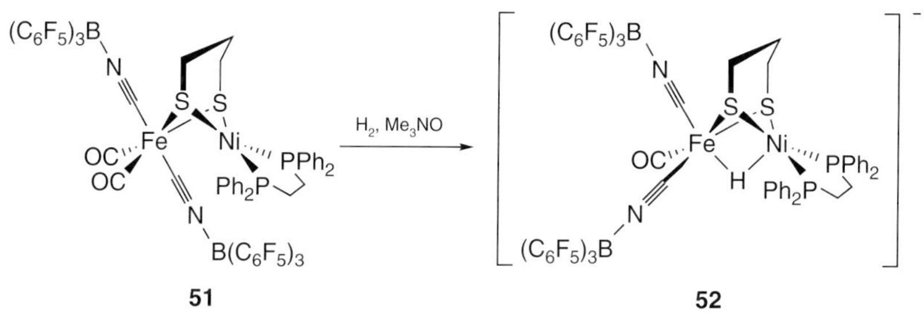

Scheme 7.12 Formation of bridging Ni/Fe complex [102].

heterogeneously cleaved to give a bridging hydride (**52**, Scheme 7.12) [102]. In contrast to cationic hydrides such as **48**, the anionic hydride **52** is reversibly oxidized at modest potential (−0.08 V vs Fc^+/Fc). This gives a mixed-valence redox state nominally corresponding to **Ni-C**. The insensitivity of the redox potential to the nature of the diphosphine is argued to indicate that oxidation takes place at the iron atom rather than at nickel. Cyclic voltammetry under dihydrogen in the presence and absence of the sterically hindered base 1,8-diazabicyclo[5.4.0]undec-7-ene (DBU) suggests that the system electrocatalytically oxidizes dihydrogen; controls under argon would further support this. That hydrogen oxidation is apparently observed on the short timescale of the voltammetric experiment (seconds) contrasts with the very slow oxidations of dihydrogen reported for the di-iron complexes **34** and **36**.

In seminal work, fast electrocatalytic oxidation of dihydrogen at a single nickel center with amine-functionalized diphosphine ligands has been demonstrated by DuBois and coworkers [103, 104] (Figure 7.21). This mononuclear system cannot function via a bridging hydride pathway without the (unlikely) involvement of binuclear intermediates. In addition, the coordination environment of nickel is abiological. Nevertheless, the role of second coordination sphere amines in promoting fast catalysis clearly relates to the biological catalysis of hydrogen oxidation by the hydrogenases (see Chapter 4).

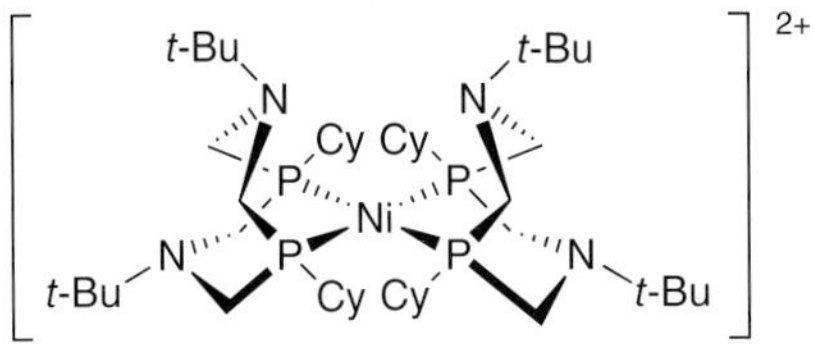

53

Figure 7.21 Mononickel complex capable of activation of dihydrogen [103].

7.3.4 Final Comments

Much progress has been made in the area of Ni/Fe model chemistry since 2005. Undoubtedly, improved systems for hydrogen oxidation will evolve and structural/electronic models for the natural states, particularly **Ni-C**, are likely to emerge. As with di-iron models, there remains the problem of low rates/turnover frequencies at low overpotentials. For example, the fastest oxidation system is based on the mononuclear nickel complex **53**, which has a turnover frequency $65\,s^{-1}$ at an overpotential of about 300 mV under 1 atm of dihydrogen [104].

7.4 [Fe] Hydrogenase

7.4.1 Overview

The [Fe] hydrogenase catalyzes the stereospecific delivery of a hydride to a single substrate, N^5,N^{10}-methenyltetrahydromethanopterin (methenyl-H_2MPT) (Scheme 7.13) [105–107], and in contrast to the [FeFe] and [NiFe] hydrogenase families does not undergo a redox process at the metal as part of the catalytic cycle. The active site of the enzyme contains a biologically unique iron-acyl unit

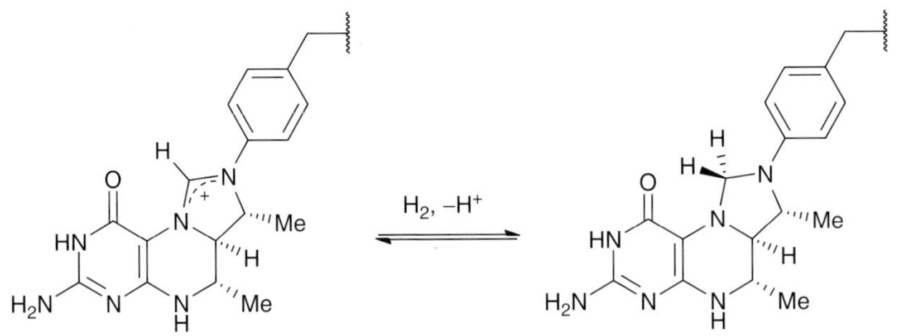

Scheme 7.13 Hydride transfer catalyzed by the [Fe] hydrogenase [105, 106].

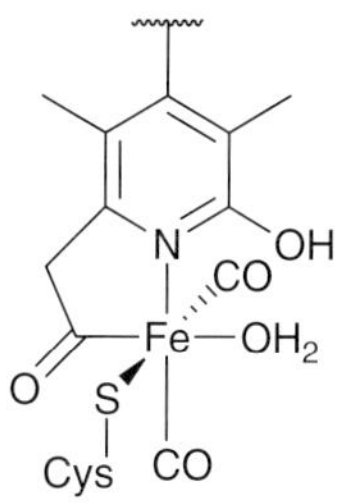

Figure 7.22 Active site of the [Fe] hydrogenase [108].

ligated by a hydroxypyridine group (Figure 7.22) [108]. In the active form of the enzyme, the iron atom bears two CO ligands: reversible CO inhibition occurs at high carbon monoxide concentrations, giving a species with three CO ligands bound to iron [106]; see also Chapter 5.

7.4.2
Biological Mechanism

Labeling studies show that the enzyme does not catalyze the exchange of tritium with the unlabeled proton in water unless the substrate is present, and that the kinetic isotope effect observed is only consistent with a hydride transfer to the substrate [105, 109, 110]. Two-dimensional NMR experiments have established that the hydride is delivered stereospecifically (pro-R face) to the substrate during catalysis, and that this is derived directly from molecular hydrogen [111, 112].

para-Hydrogen studies have shown that heterolytic cleavage of dihydrogen by the enzyme is dependent on the presence of the substrate [113]. This experiment utilizes the fact that dihydrogen exists in two spin states, with the antiparallel (*para*) state giving an enhanced NMR signal: this is only lost slowly unless the H–H bond is cleaved [5]. In the presence of either the substrate or the enzyme alone, *para*-hydrogen enhancement is unaffected. However, the enhancement is rapidly depleted when the two are combined. When the same experiment was carried out in D_2O, both D_2 and HD were formed with little H_2 production, which was interpreted as indicating rapid loss of a proton from the enzyme active site following dihydrogen cleavage. These results can be rationalized by considering a side-on binding mode of H_2 to the metal center, rather than with hydrogen dissociation, k_{off}, approximately equal to that of H^+/D^+ exchange, $k_{exchange}$ (Scheme 7.14).

Taken together, the data from isotopic labeling and NMR spectroscopy shows that the iron center cannot bear an exchangeable protonated site: this would be able to scramble isotopic labels in the absence of substrate. There also cannot be an intermediate in which dihydrogen is cleaved to give two bound but nonexchangeable protons: this would lead to a loss of the *para*-hydrogen signal in the absence of a substrate. Two possible mechanisms are therefore consistent with the given data and with a catalytic role for the metal center. First, "push–pull" activation by both the metal and substrate may be involved (see subsequent text). Second, the binding site for H_2 may only be exposed on docking by the substrate to the enzyme and the subsequent rearrangement.

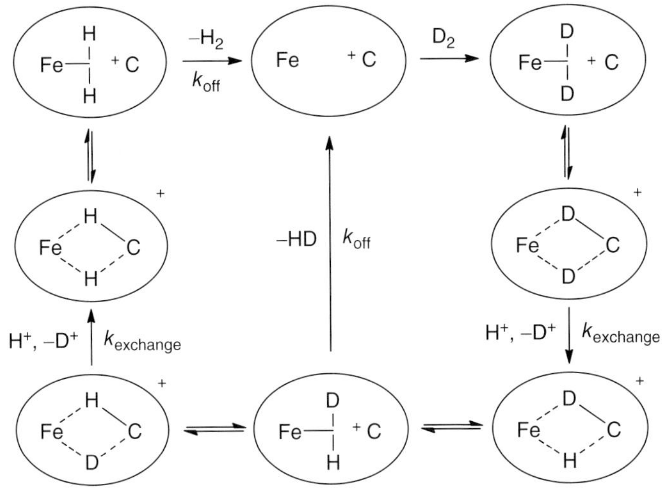

Scheme 7.14 Kinetic scheme for heterolytic cleavage of H_2 by the [Fe] hydrogenase based on *para*-hydrogen data [113].

There is good evidence to show that while substrate binding is linked to a conformational change, this is not linked to blocking the H_2 activation site on the metal.

Shima and coworkers have investigated conformational changes in the enzyme on substrate binding using mutant [114] and circular dichromism studies [115]. They found that in the absence of the substrate the enzyme is "open," with the metal center accessible. They propose that binding of the substrate alters the conformation to a "closed" state, bringing the metal and substrate into proximity (Scheme 7.15). This leads to a mechanism invoking side-on dihydrogen binding and activation by both iron and the substrate. This is a development of the earlier proposal by Berkessel and Thauer [116], who proposed the carbocation acted as a hydride acceptor before it was known that the enzyme active site contained a metal center.

DFT modeling by Yang and Hall [117] and by Reiher and coworkers [118] support the concept of "push–pull" activation of the substrate (Scheme 7.16). In the Hall mechanism, binding of H_2 by the metal proceeds before the substrate is "close." A strong hydrogen bond to the pyridinol OH means that exchange of the bound hydride does not occur until "activated" by the proximity of the substrate, which disrupts the hydrogen bonding. Notably, there are two paths of similar energy in the Hall mechanism, one of which requires protonation/deprotonation of the cysteinyl sulfur; Hu and coworkers [119] recently observed reversible protonation of sulfur in a model system. Dey [120] has used DFT methods to examine the influence of the ligand system on the ability of the [Fe] hydrogenase to bind small molecules, and concurs with Yang and Hall's conclusion that H_2 binding is only energetically favorable in the presence of the substrate.

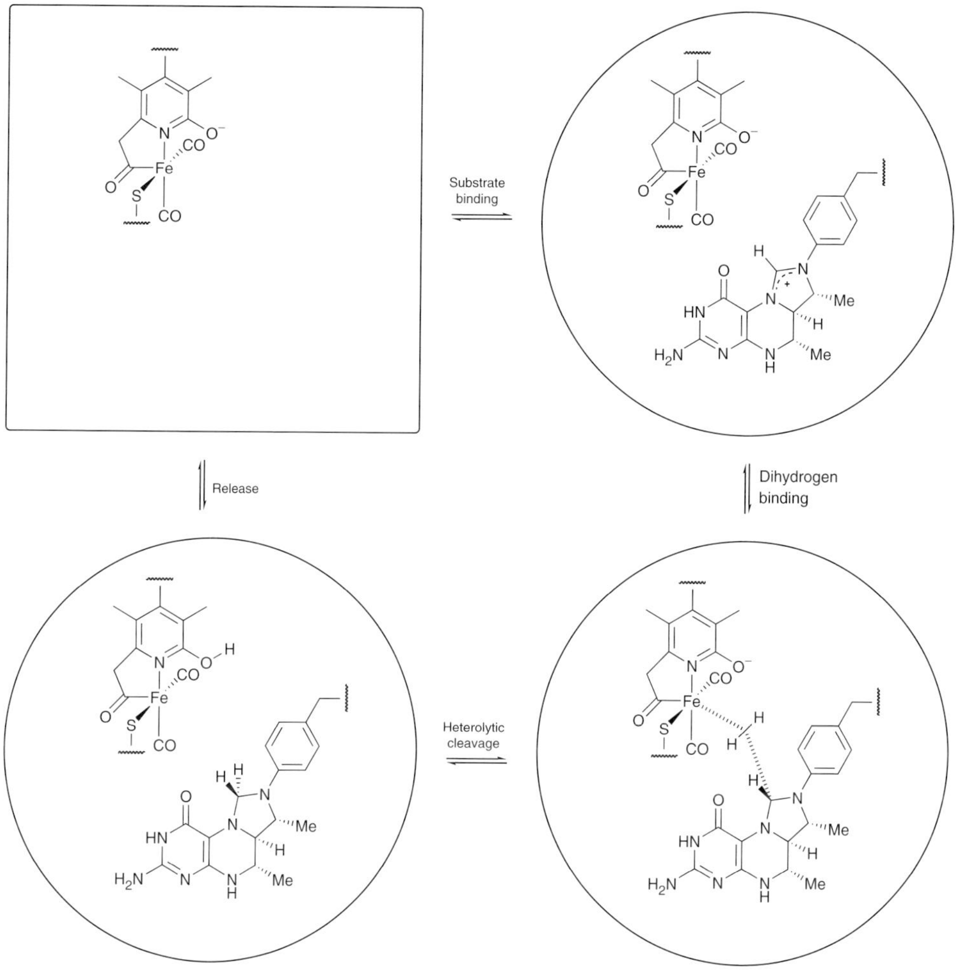

Scheme 7.15 Substrate-dependent conformal change and cleavage mechanism proposed by Shima. The circles represent a "closed" enzyme conformation; the square represents an "open" conformation [114].

7.4.3 Model Studies

Reproducing the structure of the active site has been tackled by a number of groups [121, 122], and several good structural models are now available (Figure 7.23) [123–125]. These systems unambiguously support the assignment of the oxidation state of the active site as Fe(II) [126]. While functionality at the 2-position on the pyridine ring has been introduced in model compounds, this has so far been limited to OMe [124, 125], OPMB (PMB, 4-methoxybenzyl) [127], and NH_2 [128], with no successful synthesis of a mononuclear species with

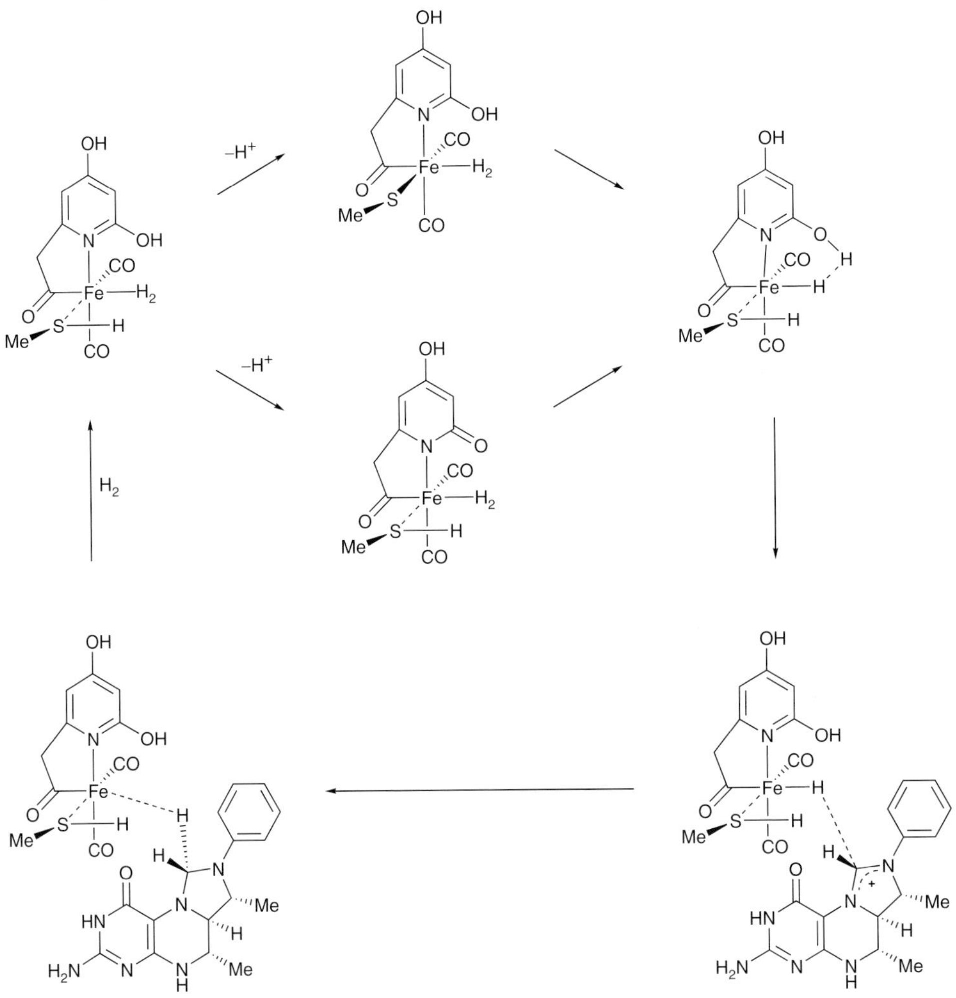

Scheme 7.16 Mechanism for dihydrogen activation by the [Fe] hydrogenase as proposed by Yang and Hall [117].

an OH group as occurs in the biological system. This is a clear limitation in the model chemistry with respect to studies of functionality where a ligand bound neighboring group is likely to play a critical role in dihydrogen cleavage, comparable to the azadithiolate in [FeFe] hydrogenase. Certain functionality of model systems directly related to the enzyme site has been shown insofar as the synthetic Fe-centers **54** and **56** reversibly bind one additional CO ligand [122, 125]. Evidence for the mechanism of reaction of the [Fe] hydrogenase is therefore primarily derived from studies with the enzyme itself, supported by computational investigations, and still awaits well-defined prototypically chemistry.

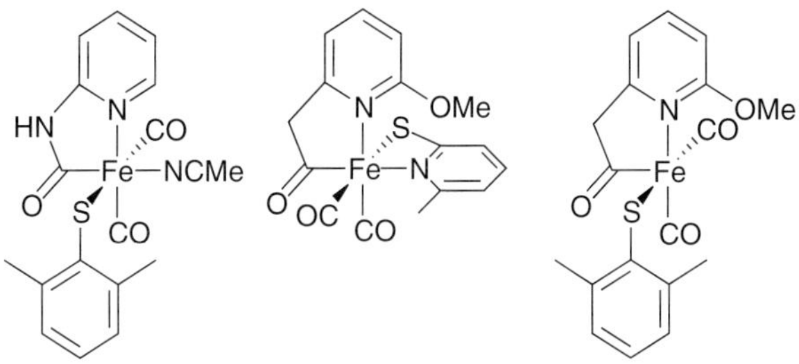

54 **55** **56**

Figure 7.23 Models for the [Fe] hydrogenase [123–125].

7.4.4
Final Comments

While good structural models for the active site of the [Fe] hydrogenase are now available, this has not thus far been translated into studies showing catalytic activity with synthetic systems, nor evidence for formation of hydride intermediates. A key challenge in this area is the need to combine active site mimetics with H_2 reduction of substrate or substrate analog molecules. Such investigations may open the way to using monoiron complexes as hydride delivery agents in synthetic organic chemistry.

7.5
Nitrogenase

7.5.1
Overview

Chapter 9 discusses the fundamental chemistry of substrate reduction by nitrogenases. Hydrogen evolution is an integral part of nitrogen fixation. In the absence of dinitrogen as a substrate, nitrogenase catalyzes the reduction of protons to dihydrogen [76, 129]. In the presence of dinitrogen, even at 50 atm, hydrogen evolution is not suppressed. It is generally recognized that the nitrogen fixation process involves the obligatory release of one H_2 molecule for every molecule of N_2 reduced to ammonia. Hydrides are implicated in the turnover of the enzyme and models have emerged which suggest that the addition of an electron to the Fe_6S_9Mo core is accompanied by proton transfer and the formation of Fe–H [76]. This section focuses on the chemistry of metal hydride systems pertinent to nitrogenase.

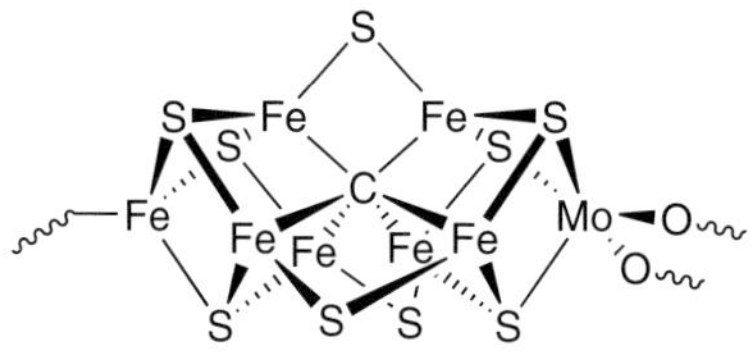

Figure 7.24 The FeMo-cofactor (FeMoco) of nitrogenase [130–133].

7.5.2
Hydrogen Evolution by Mo-Nitrogenase

The structure of the active site of nitrogenase, FeMoco [130–133], at which protons and dinitrogen are reduced, is represented in Figure 7.24. In the widely accepted kinetic model of nitrogen fixation, the enzyme states associated with turnover are represented in the Lowe–Thornley model shown in Scheme 7.17. Hydrogen is evolved in the absence of substrate dinitrogen at the $\mathbf{E_2}$ level. In the presence of dinitrogen, reducing equivalents are directed toward ammonia formation after attainment of the $\mathbf{E_4}$ level, where H_2 is concomitantly released with dinitrogen binding.

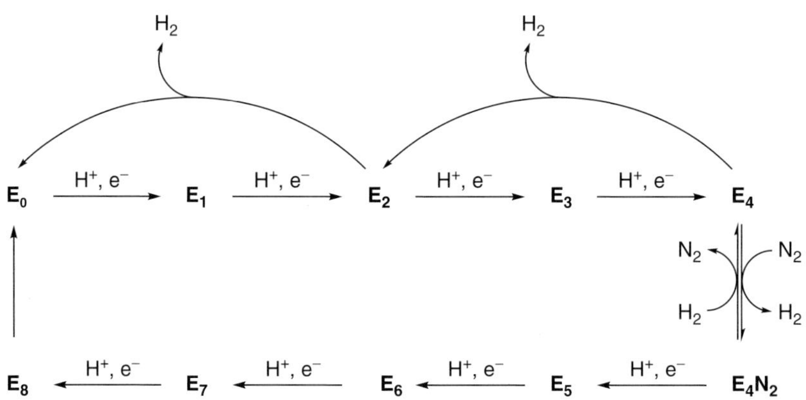

Scheme 7.17 Simplified Lowe–Thornley mechanism [129].

Direct spectroscopic evidence for hydride formation at FeMoco has been provided by detailed EPR and ENDOR studies using isotopic labeling [134]. Initial studies have shown that FeMoco of the nitrogenase α-70^{Ile} molybdenum–iron (MoFe) protein variant accumulates a novel $S = 1/2$ state that can be trapped during the reduction of protons to H_2. ENDOR measurements showed the presence of two hydrides with large isotropic hyperfine couplings, $A_{\text{iso}} \approx 23$ MHz, indicating they are bound to the cofactor. The data disfavors any model that involves protonation of sulfides, and thus suggest that the intermediate instead contains two chemically equivalent bound hydrides; it appeared unlikely that these were terminal monohydrides. Further work provided evidence that the hydride bridged two iron atoms, that is, μ-H at two symmetrically disposed sites at the $\mathbf{E_4}$ level [135]. Mo–H–Fe bridging in this state was excluded on the basis of the observed EPR parameters.

7.5.3 Paramagnetic Bridging Fe/Fe Hydrides

The one-electron reduction of the bridging hydride di-iron complex **10** (Figure 7.9) gave a $S = 1/2$ signal showing strong hyperfine coupling to the hydride, which changed appropriately on deuterium labeling [74]. Complementary FT-IR and DFT studies established that one-electron reduction led to the retention of geometry in the mixed-valence [Fe(1.5)–μ-H–Fe(1.5)] species. Later work with a related system allowed isolation and crystallographic characterization of a 35-electron paramagnetic hydride (**28**, Figure 7.15) [75].

An aspect of the Fe_6-core of FeMoco is the four-coordinate geometry of the iron sites, which lie between tetrahedral and trigonal pyramidal. A model complex **57** (Figure 7.25) has been shown to have a $S = 1/2$ mixed-valence Fe(II)/Fe(III) electronic structure [136], in which the unpaired electron is shared equally between the two iron centers as is the case in the one-electron reduction of complex **10** [74]. It is argued that the structural and electronic parameters for this complex lend support to the proposal that FeMoco contains two [Fe–μ-H–Fe] fragments in the catalytic intermediate that has accumulated four reducing equivalents ($\mathbf{E_4}$).

Paramagnetic terminal hydrides or, indeed, integer spin hydrides cannot be ruled out as intermediates in the catalytic cycle; however, none have yet been detected. The spectroscopic EPR signature for certain low-valent terminal hydrides of the type shown in Figure 7.26 have been described [137].

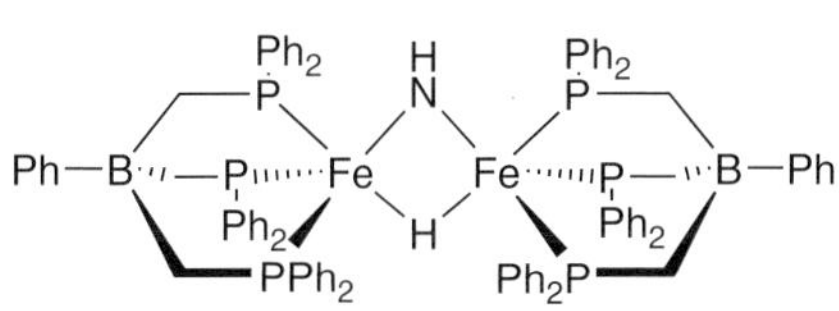

57

Figure 7.25 Mixed-valence bridging hydride model for the binding of hydride by FeMoco [136].

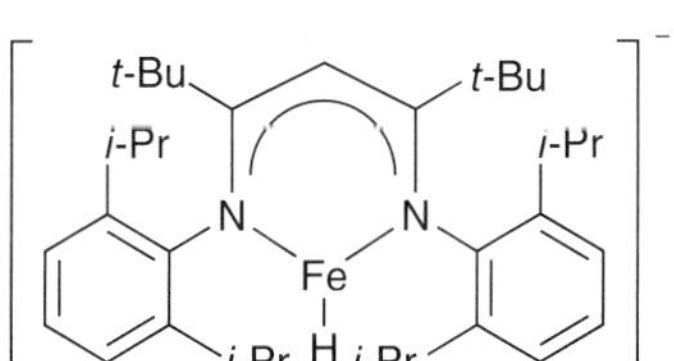

58

Figure 7.26 Low-valent iron terminal hydride complex [137].

7.5.4 Final Comments

There have been many biochemical studies of hydrogen evolution where isotopic labeling has shown dinitrogen-dependent H_2/D_2 exchange and the conversion $H_2 + 2D^+ + 2e^- \leftarrow 2HD$ [138]. Whether the natural system involves iron dihydrides or iron-dihydrogen intermediates or whether hydrogen evolution results from protic attack on a bridging hydride remains to be determined. There is much background chemistry on both dihydrogen complexes [139] and on the reversible displacement of dihydrogen from these by dinitrogen.

References

1. Vincent, K.A., Parkin, A., and Armstrong, F.A. (2007) *Chem. Rev.*, **107**, 4366–4413.
2. Peters, J.W., Lanzilotta, W.N., Lemon, B.J., and Seefeldt, L.C. (1998) *Science*, **282**, 1853–1858.
3. Nicolet, Y., Piras, C., Legrand, P., Hatchikian, C.E., and Fontecilla-Camps, J.C. (1999) *Structure*, **7**, 13–23.
4. Adamska, A., Silakov, A., Lambertz, C., Rüdiger, O., Happe, T., Reijerse, E., and Lubitz, W. (2012) *Angew. Chem. Int. Ed.*, **51**, 11458–11462.
5. Vignais, P.M. and Billoud, B. (2007) *Chem. Rev.*, **107**, 4206–4272.
6. Fontecilla-Camps, J.C., Volbeda, A., Cavazza, C., and Nicolet, Y. (2007) *Chem. Rev.*, **107**, 4273–4303.
7. De Lacey, A.L., Fernández, V.M., Rousset, M., and Cammack, R. (2007) *Chem. Rev.*, **107**, 4304–4330.
8. Lubitz, W., Reijerse, E., and van Gastel, M. (2007) *Chem. Rev.*, **107**, 4331–4365.
9. Siegbahn, P.E.M., Tye, J.W., and Hall, M.B. (2007) *Chem. Rev.*, **107**, 4414–4435.
10. Gloaguen, F. and Rauchfuss, T.B. (2009) *Chem. Soc. Rev.*, **38**, 100–108.
11. Tard, C. and Pickett, C.J. (2009) *Chem. Rev.*, **109**, 2245–2274.
12. Gordon, J.C. and Kubas, G.J. (2010) *Organometallics*, **29**, 4682–4701.
13. Mulder, D.W., Shepard, E.M., Meuser, J.E., Joshi, N., King, P.W., Posewitz, M.C., Broderick, J.B., and Peters, J.W. (2011) *Structure*, **19**, 1038–1052.
14. Tschierlei, S., Ott, S., and Lomoth, R. (2011) *Energy Environ. Sci.*, **4**, 2340–2352.
15. Wang, N., Wang, M., Chen, L., and Sun, L. (2013) *Dalton Trans.*, **42**, 12059–12071.
16. Berggren, G., Adamska, A., Lambertz, C., Simmons, T.R., Esselborn, J., Atta, M., Gambarelli, S., Mouesca, J.-M., Reijerse, E., Lubitz, W., Happe, T., Artero, V., and Fontecave, M. (2013) *Nature*, **499**, 66–69.
17. Esselborn, J., Lambertz, C., Adamska-Venkatesh, A., Simmons, T., Berggren, G., Noth, J., Siebel, J., Hemschemeier, A., Artero, V., Reijerse, E., Fontecave, M., Lubitz, W., and Happe, T. (2013) *Nat. Chem. Biol.*, **9**, 607–609.
18. Darensbourg, M.Y. and Weigand, W. (2011) *Eur. J. Inorg. Chem.*, **2011**, 994–1004.
19. Stripp, S.T., Goldet, G., Brandmayr, C., Sanganas, O., Vincent, K.A., Haumann, M., Armstrong, F.A., and Happe, T. (2009) *Proc. Natl. Acad. Sci. U.S.A.*, **106**, 17331–17336.
20. Nicolet, Y., Lemon, B.J., Fontecilla-Camps, J.C., and Peters, J.W. (2000) *Trends Biochem. Sci.*, **25**, 138–143.
21. Silakov, A., Wenk, B., Reijerse, E., and Lubitz, W. (2009) *Phys. Chem. Chem. Phys.*, **11**, 6592–6599.
22. Erdem, Ö.F., Schwartz, L., Stein, M., Silakov, A., Kaur-Ghymaan, S., Huang, P., Ott, S., Reijerse, E.J., and Lubitz, W. (2011) *Angew. Chem. Int. Ed.*, **50**, 1439–1443.
23. Ryde, U., Greco, C., and De Gioia, L. (2010) *J. Am. Chem. Soc.*, **132**, 4512–4513.

24. Pandey, A.S., Harris, T.V., Giles, L.J., Peters, J.W., and Szilagyi, R.K. (2008) *J. Am. Chem. Soc.*, **130**, 4533–4540.
25. Van Der Spek, T.M., Arendsen, A.F., Happe, R.P., Yun, S., Bagley, K.A., Stufkens, D.J., Hagen, W.R., and Albracht, S.P.J. (1996) *Eur. J. Biochem.*, **237**, 629–634.
26. Pierik, A.J., Hulstein, M., Hagen, W.R., and Albracht, S.P.J. (1998) *Eur. J. Biochem.*, **258**, 572–578.
27. Chen, Z., Lemon, B.J., Huang, S., Swartz, D.J., Peters, J.W., and Bagley, K.A. (2002) *Biochemistry*, **41**, 2036–2043.
28. Roseboom, W., Lacey, A.L.D., Fernandez, V.M., Hatchikian, E.C., and Albracht, S.P.J. (2006) *J. Biol. Inorg. Chem.*, **11**, 102–118.
29. Silakov, A., Kamp, C., Reijerse, E., Happe, T., and Lubitz, W. (2009) *Biochemistry*, **48**, 7780–7786.
30. Mulder, D.W., Ratzloff, M.W., Shepard, E.M., Byer, A.S., Noone, S.M., Peters, J.W., Broderick, J.B., and King, P.W. (2013) *J. Am. Chem. Soc.*, **135**, 6921–6929.
31. Yu, L., Greco, C., Bruschi, M., Ryde, U., Gioia, L.D., and Reiher, M. (2011) *Inorg. Chem.*, **50**, 3888–3900.
32. Lemon, B.J. and Peters, J.W. (1999) *Biochemistry*, **38**, 12969–12973.
33. Bennett, B., Lemon, B.J., and Peters, J.W. (2000) *Biochemistry*, **39**, 7455–7460.
34. Le Cloirec, A., Best, S.P., Borg, S., Davies, S.C., Evans, D.J., Hughes, D.L., and Pickett, C.J. (1999) *Chem. Commun.*, **35**, 2285–2286.
35. Lyon, E.J., Georgakaki, I.P., Reibenspies, J.H., and Darensbourg, M.Y. (1999) *Angew. Chem. Int. Ed.*, **38**, 3178–3180.
36. Schmidt, M., Contakes, S.M., and Rauchfuss, T.B. (1999) *J. Am. Chem. Soc.*, **121**, 9736–9737.
37. Gloaguen, F., Lawrence, J.D., Schmidt, M., Wilson, S.R., and Rauchfuss, T.B. (2001) *J. Am. Chem. Soc.*, **123**, 12518–12527.
38. Lawrence, J.D., Li, H., Rauchfuss, T.B., Bérnard, M., and Rohmer, M.-M. (2001) *Angew. Chem. Int. Ed.*, **40**, 1768–1771.
39. Razavet, M., Davies, S.C., Hughes, D.L., and Pickett, C.J. (2001) *Chem. Commun.*, **37**, 847–848.
40. Zhao, X., Georgakaki, I.P., Miller, M.L., Yarbrough, J.C., and Darensbourg, M.Y. (2001) *J. Am. Chem. Soc.*, **123**, 9710–9711.
41. George, S.J., Cui, Z., Razavet, M., and Pickett, C.J. (2002) *Chem. Eur. J.*, **8**, 4037–4046.
42. Li, H. and Rauchfuss, T.B. (2002) *J. Am. Chem. Soc.*, **124**, 726–727.
43. Lyon, E.J., Georgakaki, I.P., Reibenspies, J.H., and Darensbourg, M.Y. (2001) *J. Am. Chem. Soc.*, **123**, 3268–3278.
44. Zhao, X., Georgakaki, I.P., Miller, M.L., Mejia-Rodriguez, R., Chiang, C.-Y., and Darensbourg, M.Y. (2002) *Inorg. Chem.*, **41**, 3917–3928.
45. Hsieh, C.-H., Erdem, Ö.F., Harman, S.D., Singleton, M.L., Reijerse, E., Lubitz, W., Popescu, C.V., Reibenspies, J.H., Brothers, S.M., Hall, M.B., and Darensbourg, M.Y. (2012) *J. Am. Chem. Soc.*, **134**, 13089–13102.
46. Razavet, M., Borg, S.J., George, S.J., Best, S.P., Fairhurst, S.A., and Pickett, C.J. (2002) *Chem. Commun.*, **36**, 700–701.
47. Liu, T. and Darensbourg, M.Y. (2007) *J. Am. Chem. Soc.*, **129**, 7008–7009.
48. Justice, A.K., Rauchfuss, T.B., and Wilson, S.R. (2007) *Angew. Chem. Int. Ed.*, **46**, 6152–6154.
49. Singleton, M.L., Bhuvanesh, N., Reibenspies, J.H., and Darensbourg, M.Y. (2008) *Angew. Chem. Int. Ed.*, **47**, 9492–9495.
50. Justice, A.K., De Gioia, L., Nilges, M.J., Rauchfuss, T.B., Wilson, S.R., and Zampella, G. (2008) *Inorg. Chem.*, **47**, 7405–7414.
51. Camara, J.M. and Rauchfuss, T.B. (2011) *J. Am. Chem. Soc.*, **133**, 8098–8101.
52. Silakov, A., Olsen, M.T., Sproules, S., Reijerse, E.J., Rauchfuss, T.B., and Lubitz, W. (2012) *Inorg. Chem.*, **51**, 8617–8628.
53. Matthews, S.L. and Heinekey, D.M. (2010) *Inorg. Chem.*, **49**, 9746–9748.
54. A search of the Cambridge Structure Database: Allen, F.H. (2002) *Acta Crystallogr., Sect. B*, **58**, 380–388.

v 5.34 (released November 2012) reveals 23 monocationic bridging hydrides featuring unique {2Fe2S} cores.
55. Gloaguen, F., Lawrence, J.D., Rauchfuss, T.B., Bérnard, M., and Rohmer, M.-M. (2002) *Inorg. Chem.*, **41**, 6573–6582.
56. van der Vlugt, J.I., Rauchfuss, T.B., Whaley, C.M., and Wilson, S.R. (2005) *J. Am. Chem. Soc.*, **127**, 16012–16013.
57. Dong, W., Wang, M., Liu, X., Jin, K., Li, G., Wang, F., and Sun, L. (2006) *Chem. Commun.*, **40**, 305–307.
58. Ezzaher, S., Capon, J.-F., Gloaguen, F., Pétillon, F.Y., Schollhammer, P., and Talarmin, J. (2007) *Inorg. Chem.*, **46**, 3426–3428.
59. Barton, B.E. and Rauchfuss, T.B. (2008) *Inorg. Chem.*, **47**, 2261–2263.
60. A. Jablonskytė, J. A. Wright, S. A. Fairhurst, L. R. Webster and C. J. Pickett, submitted.
61. Carroll, M.E., Barton, B.E., Rauchfuss, T.B., and Carroll, P.J. (2012) *J. Am. Chem. Soc.*, **134**, 18843–18852.
62. Zaffaroni, R., Rauchfuss, T.B., Gray, D.L., De Gioia, L., and Zampella, G. (2012) *J. Am. Chem. Soc.*, **134**, 19260–19269.
63. Barton, B.E., Zampella, G., Justice, A.K., De Gioia, L., Rauchfuss, T.B., and Wilson, S.R. (2010) *Dalton Trans.*, **39**, 3011–3019.
64. Wright, J.A. and Pickett, C.J. (2009) *Chem. Commun.*, **45**, 5719–5721.
65. Jablonskytė, A., Wright, J.A., and Pickett, C.J. (2010) *Dalton Trans.*, **39**, 3026–3034.
66. Jablonskytė, A., Wright, J.A., and Pickett, C.J. (2011) *Eur. J. Inorg. Chem.*, **2011**, 1033–1037.
67. Liu, C., Peck, J.N.T., Wright, J.A., Pickett, C.J., and Hall, M.B. (2011) *Eur. J. Inorg. Chem.*, **2011**, 1080–1093.
68. A. Jablonskytė, L. R. Webster, J. A. Wright and C. J. Pickett, submitted.
69. Borg, S., Behrsing, T., Best, S.P., Razavet, M., Liu, X., and Pickett, C.J. (2004) *J. Am. Chem. Soc.*, **126**, 16988–16999.
70. Cheah, M.H., Tard, C., Borg, S.J., Liu, X., Ibrahim, S.K., Pickett, C.J., and Best, S.P. (2007) *J. Am. Chem. Soc.*, **129**, 11085–11092.
71. Barton, B.E., Olsen, M.T., and Rauchfuss, T.B. (2008) *J. Am. Chem. Soc.*, **130**, 16834–16835.
72. Ezzaher, S., Capon, J.-F., Gloaguen, F., Pétillon, F.Y., Schollhammer, P., Talarmin, J., and Kervarec, N. (2009) *Inorg. Chem.*, **48**, 2–4.
73. Wang, N., Wang, M., Zhang, T., Li, P., Liu, J., and Sun, L. (2008) *Chem. Commun.*, **44**, 5800–5802.
74. Jablonskytė, A., Wright, J.A., Fairhurst, S.A., Peck, J.N.T., Ibrahim, S.K., Oganesyan, V.S., and Pickett, C.J. (2011) *J. Am. Chem. Soc.*, **133**, 18606–18609.
75. Wang, W., Nilges, M.J., Rauchfuss, T.B., and Stein, M. (2013) *J. Am. Chem. Soc.*, **135**, 3633–3639.
76. Hoffman, B.M., Lukoyanov, D., Dean, D.R., and Seefeldt, L.C. (2013) *Acc. Chem. Res.*, **46**, 587–595.
77. Olsen, M.T., Barton, B.E., and Rauchfuss, T.B. (2009) *Inorg. Chem.*, **48**, 7507–7509.
78. Camara, J.M. and Rauchfuss, T.B. (2012) *Nat. Chem.*, **4**, 26–30.
79. Frey, M. (2002) *ChemBioChem*, **3**, 153–160.
80. Greco, C. (2013) *Inorg. Chem.*, **52**, 1901–1908.
81. Wang, N., Wang, M., Wang, Y., Zheng, D., Han, H., Ahlquist, M.S.G., and Sun, L. (2013) *J. Am. Chem. Soc.*, **135**, 13688–13691.
82. Shafaat, H.S., Rüdiger, O., Ogata, H., and Lubitz, W. (2013) *Biochim. Biophys. Acta, Bioenerg.*, **1827**, 986–1002.
83. Ogata, H., Mizoguchi, Y., Mizuno, N., Miki, K., Adachi, S.-I., Yasuoka, N., Yagi, T., Yamauchi, O., Hirota, S., and Higuchi, Y. (2002) *J. Am. Chem. Soc.*, **124**, 11628–11635.
84. Goris, T., Wait, A.F., Saggu, M., Fritsch, J., Heidary, N., Stein, M., Zebger, I., Lendzian, F., Armstrong, F.A., Friedrich, B., and Lenz, O. (2011) *Nat. Chem. Biol.*, **7**, 310–318.
85. Volbeda, A., Martin, L., Cavazza, C., Matho, M., Faber, B.W., Roseboom, W., Albracht, S.P.J., Garcin, E., Rousset, M., and Fontecilla-Camps, J.C. (2005) *J. Biol. Inorg. Chem.*, **10**, 239–245.
86. de Lacey, A.L., Hatchikian, E.C., Volbeda, A., Frey, M.,

Fontecilla-Camps, J.C., and Fernandez, V.M. (1997) *J. Am. Chem. Soc.*, **119**, 7181–7189.
87. Volbeda, A., Amara, P., Darnaulta, C., Mouesca, J.-M., Parkin, A., Roessler, M.M., Armstrong, F.A., and Fontecilla-Camps, J.C. (2012) *Proc. Natl. Acad. Sci. U.S.A.*, **109**, 5305–5310.
88. Brecht, M., van Gastel, M., Buhrke, T., Friedrich, B., and Lubitz, W. (2003) *J. Am. Chem. Soc.*, **125**, 13075–13083.
89. Foerster, S., van Gastel, M., Brecht, M., and Lubitz, W. (2005) *J. Biol. Inorg. Chem.*, **10**, 51–62.
90. Sellmann, D., Geipel, F., Lauderbach, F., and Heinemann, F.W. (2002) *Angew. Chem. Int. Ed.*, **41**, 632–634.
91. Smith, M.C., Barclay, J.E., Cramer, S.P., Davies, S.C., Gu, W.-W., Hughes, D.L., Longhurst, S., and Evans, D.J. (2002) *J. Chem. Soc., Dalton Trans.*, **31**, 2641–2647.
92. Li, Z., Ohki, Y., and Tatsumi, K. (2005) *J. Am. Chem. Soc.*, **127**, 8950–8951.
93. Jiang, J., Maruani, M., Solaimanzadeh, J., Lo, W., Koch, S.A., and Millar, M. (2009) *Inorg. Chem.*, **48**, 6359–6361.
94. Kersting, B. and Steinfeld, G. (2001) *Chem. Commun.*, **37**, 1376–1377.
95. Krishnamurthy, D., Sarjeant, A.N., Goldberg, D.P., Caneschi, A., Totti, F., Zakharov, L.N., and Rheingold, A.L. (2005) *Chem. Eur. J.*, **11**, 7328–7341.
96. Ogo, S., Kabe, R., Uehara, K., Kure, B., Nishimura, T., Menon, S.C., Harada, R., Fukuzumi, S., Yoshiki Higuchi, T.O., Tamada, T., and Kuroki, R. (2007) *Science*, **316**, 585–587.
97. Oudart, Y., Artero, V., Pécaut, J., and Fontecave, M. (2006) *Inorg. Chem.*, **45**, 4334–4336.
98. Barton, B.E., Whaley, C.M., Rauchfuss, T.B., and Gray, D.L. (2009) *J. Am. Chem. Soc.*, **131**, 6942–6943.
99. Barton, B.E. and Rauchfuss, T.B. (2010) *J. Am. Chem. Soc.*, **132**, 14877–14885.
100. Carroll, M.E., Barton, B.E., Gray, D.L., Mack, A.E., and Rauchfuss, T.B. (2011) *Inorg. Chem.*, **50**, 9554–9563.
101. Ogo, S., Ichikawa, K., Kishima, T., Matsumoto, T., Nakai, H., Kusaka, K., and Ohhara, T. (2013) *Science*, **339**, 682–684.
102. Manor, B.C. and Rauchfuss, T.B. (2013) *J. Am. Chem. Soc.*, **135**, 11895–11900.
103. Yang, J.Y., Chen, S., Dougherty, W.G., Kassel, W.S., Bullock, R.M., DuBois, D.L., Raugei, S., Rousseau, R., Dupuis, M., and DuBois, M.R. (2010) *Chem. Commun.*, **46**, 8618–8620.
104. Shaw, W.J., Helm, M.L., and DuBois, D.L. (2013) *Biochim. Biophys. Acta, Bioenerg.*, **1827**, 1123–1139.
105. Zirngibl, C., Van Dongen, W., Schwörer, B., Von Bünau, R., Richter, M., Klein, A., and Thauer, R.K. (1992) *Eur. J. Biochem.*, **208**, 511–520.
106. Shima, S. and Thauer, R.K. (2007) *Chem. Rec.*, **7**, 37–46.
107. Corr, M.J. and Murphy, J.A. (2011) *Chem. Soc. Rev.*, **40**, 2279–2292.
108. Hiromoto, T., Ataka, K., Pilak, O., Stagni, M.S., Meyer-Klaucke, W., Warkentin, E., Thauer, R.K., Shima, S., and Ermler, U. (2009) *FEBS Lett.*, **583**, 585–590.
109. Schwörer, B., Fernandez, V.M., Zirngibl, C., and Thauer, R.K. (1993) *Eur. J. Biochem.*, **212**, 255–261.
110. Vogt, S., Lyon, E.J., Shima, S., and Thauer, R.K. (2008) *J. Biol. Inorg. Chem.*, **13**, 97–106.
111. Schleucher, J., Griesinger, C., Schwörer, B., and Thauer, R.K. (1994) *Biochemistry*, **33**, 3986–3993.
112. Schleucher, J., Schwörer, B., Thauer, R.K., and Griesinger, C. (1995) *J. Am. Chem. Soc.*, **117**, 2941–2942.
113. Hartmann, G.C., Santamaria, E., Fernández, V.M., and Thauer, R.K. (1996) *J. Biol. Inorg. Chem.*, **1**, 445–450.
114. Hiromoto, T., Warkentin, E., Moll, J., Ermler, U., and Shima, S. (2009) *Angew. Chem. Int. Ed.*, **48**, 6457–6460.
115. Shima, S., Vogt, S., Göbels, A., and Bill, E. (2010) *Angew. Chem. Int. Ed.*, **49**, 9917–9921.
116. Berkessel, A. and Thauer, R.K. (1995) *Angew. Chem. Int. Ed. Engl.*, **34**, 2247–2250.
117. Yang, X. and Hall, M.B. (2009) *J. Am. Chem. Soc.*, **131**, 10901–10908.
118. Finkelmann, A.R., Stiebritz, M.T., and Reiher, M. (2013) *J. Phys. Chem. B*, **117**, 4806–4817.

119. Chen, D., Scopelliti, R., and Hu, X. (2012) *Angew. Chem. Int. Ed.*, **51**, 1919–1921.
120. Dey, A. (2010) *J. Am. Chem. Soc.*, **132**, 13892–13901.
121. Wright, J.A., Turrell, P.J., and Pickett, C.J. (2010) *Organometallics*, **29**, 6146–6156.
122. Schultz, K.M., Chen, D., and Hu, X. (2013) *Chem. Asian J.*, **8**, 1068–1075.
123. Turrell, P.J., Wright, J.A., Peck, J.N.T., Oganesyan, V.S., and Pickett, C.J. (2010) *Angew. Chem. Int. Ed.*, **49**, 7508–7511.
124. Chen, D., Scopelliti, R., and Hu, X. (2010) *Angew. Chem. Int. Ed.*, **49**, 7512–7515.
125. Chen, D., Scopelliti, R., and Hu, X. (2011) *Angew. Chem. Int. Ed.*, **50**, 5671–5673.
126. Wang, X., Li, Z., Luo, Q., Evans, D.J., Pickett, C.J., and Liu, X. (2008) *Chem. Commun.*, **44**, 3555–3557.
127. Song, L.-C., Zhao, G.-Y., Xie, Z.J., and Zhang, J.-W. (2013) *Organometallics*, **32**, 2509–2512.
128. Turrell, P.J., Hill, A.D., Ibrahim, S.K., Wright, J.A., and Pickett, C.J. (2013) *Dalton Trans.*, **42**, 8140–8146.
129. Seefeldt, L.C., Hoffman, B.M., and Dean, D.R. (2009) *Annu. Rev. Biochem.*, **78**, 701–722.
130. Kim, J. and Rees, D.C. (1992) *Nature*, **360**, 553–560.
131. Kim, J. and Rees, D.C. (1992) *Science*, **257**, 1677–1682.
132. Lancaster, K.M., Roemelt, M., Ettenhuber, P., Hu, Y., Ribbe, M.W., Neese, F., Bergmann, U., and DeBeer, S. (2011) *Science*, **334**, 974–977.
133. Spatzal, T., Aksoyoglu, M., Zhang, L., Andrade, S.L.A., Schleicher, E., Weber, S., Rees, D.C., and Einsle, O. (2011) *Science*, **334**, 940.
134. Igarashi, R.Y., Laryukhin, M., Dos Santos, P.C., Lee, H.-I., Dean, D.R., Seefeldt, L.C., and Hoffman, B.M. (2005) *J. Am. Chem. Soc.*, **127**, 6231–6241.
135. Lukoyanov, D., Yang, Z.-Y., Dean, D.R., Seefeldt, L.C., and Hoffman, B.M. (2010) *J. Am. Chem. Soc.*, **132**, 2526–2527.
136. Kinney, R.A., Saouma, C.T., Peters, J.C., and Hoffman, B.M. (2012) *J. Am. Chem. Soc.*, **134**, 12637–12647.
137. Chiang, K.P., Scarborough, C.C., Horitani, M., Lees, N.S., Ding, K., Dugan, T.R., Brennessel, W.W., Bill, E., Hoffman, B.M., and Holland, P.L. (2012) *Angew. Chem. Int. Ed.*, **51**, 3658–3662.
138. Burgess, B.K. and Lowe, D.J. (1996) *Chem. Rev.*, **96**, 2983–3012.
139. Kubas, G.J. (2007) *Chem. Rev.*, **107**, 4152–4205.

Part III
Nitrogen Fixation

Bioinspired Catalysis: Metal-Sulfur Complexes, First Edition.
Edited by Wolfgang Weigand and Philippe Schollhammer.

8
Structures and Functions of the Active Sites of Nitrogenases

Chi Chung Lee, Jared A. Wiig, Yilin Hu, and Markus W. Ribbe

8.1
Introduction

Nitrogenases are a family of complex metalloenzymes that catalyze the biological nitrogen fixation, a reaction that converts atmospheric dinitrogen (N_2) to ammonia (NH_3) [1–16]. This process is carried out exclusively by bacteria and archaea, and it is estimated to reduce (or fix) ~90 million tons of nitrogen per year [8, 10, 17, 18]. Thus, it represents a key step for fixed nitrogen to enter the global food chain. Furthermore, unlike the industrial Haber–Bosch process, which reduces N_2 to NH_3 under high temperature and pressure, biological nitrogen fixation by nitrogenases relies on the chemical energy generated through the hydrolysis of ATP [8, 12, 19]. Therefore, it embodies the formidable chemistry to break the inert N≡N triple bond, one of the most challenging reactions in terms of the activation barrier (~950 kJ mol^{-1}), under ambient conditions [20, 21]. Because of its biological and chemical significance, nitrogenase has remained one of the focal points for decades of bioinorganic chemistry research.

There are four classes of nitrogenases: the molybdenum (Mo) nitrogenase, the vanadium (V) nitrogenase, the iron (Fe)-only nitrogenase, and the superoxide dismutase (SOD)-dependent nitrogenase from *Streptomyces thermoautotrophicus* [5, 8, 10, 22, 23]. All four nitrogenases are complex metalloenzymes that contain metalloclusters essential for their catalytic functions; however, while the first three nitrogenases are highly homologous to each other, the fourth is not related to them. Containing Mo at its active site, the Mo nitrogenase is the best characterized among the nitrogenases identified to date. Recent studies, on the other hand, have furthered our understanding of the V nitrogenase, which has some distinct structural and catalytic features from its Mo counterpart. Comparative studies of Mo- and V nitrogenases, therefore, provide important insights into the structure and function of this fascinating enzyme family.

The objective of this article is to review the similarities and distinctions between the active metal clusters of Mo- and V nitrogenases. Following an introduction of the basic properties of Mo nitrogenase in Section 8.2, the catalytic mechanisms of Mo nitrogenase are discussed in Section 8.3. The distinctive features of

Bioinspired Catalysis: Metal-Sulfur Complexes, First Edition.
Edited by Wolfgang Weigand and Philippe Schollhammer.

V nitrogenase are surveyed in Section 8.4, with an emphasis on the comparison of this nitrogenase with its Mo counterpart; and the catalytic properties of the solvent-extracted metal clusters of nitrogenases are covered in Section 8.5, with a focus on the ability of these clusters to catalyze ATP-independent substrate reduction.

8.2 Properties of Mo Nitrogenase

Mo nitrogenase is the best characterized form of nitrogenase, and it is commonly found in diazotrophic microorganisms, such as *Azotobacter vinelandii, Klebsiella pneumoniae,* and *Clostridium pasteurianum* [9, 10, 24]. The reaction catalyzed by Mo nitrogenase is generally depicted as follows [8, 11, 12]:

$$N_2 + 8H^+ + 8e^- + 16ATP \rightarrow 2NH_3 + H_2 + 16ADP + 16P_i$$

It should be noted that while the stoichiometry of the reaction catalyze by the Mo nitrogenase is well established, the distribution of electrons toward the formation of H_2 and NH_3 is rather different when N_2 reduction is catalyzed by the V nitrogenase (see Section 8.4.3).

Mo nitrogenase consists of two redox-active protein components. One of them is referred to as the *iron (Fe) protein,* which contains a ferrodoxin-type $[Fe_4S_4]$ cluster; whereas the other is termed the *molybdenum–iron (MoFe) protein,* which houses two pairs of unique metal centers: the P-cluster ($[Fe_8S_7]$) and the iron–molybdenum cofactor, or FeMo-co ($[MoFe_7S_9C$-homocitrate$]$). During catalysis, these two proteins complex with each other, permitting electrons to be transferred from the reduced $[Fe_4S_4]$ cluster in the Fe protein, via the P-cluster, to the FeMo-co in MoFe protein concomitant with the hydrolysis of ATP [3–6, 12, 13]. Substrate reduction eventually takes place at the FeMo-co site once it accumulates sufficient electrons. As such, the structural and chemical properties of the component proteins of nitrogenase and their associated metal clusters are crucial for the catalytic features of this enzyme and will serve as the focus of the subsequent discussion.

8.2.1 Properties of Fe Protein and its Associated Cluster

8.2.1.1 The Fe Protein Structure

The Fe protein of *A. vinelandii* is a γ_2-homodimer of ~60 kDa, and its subunits are encoded by *nifH* [1–4, 6–9]. This protein contains a subunit-bridging $[Fe_4S_4]$ cluster and a Walker A motif (residues 9–16: $^{9}\underline{G}KGGI\underline{GKS}^{16}$; consensus residues underlined) for the binding of nucleotides (MgATP, MgADP, and their analogs) within each subunit [4, 25–28]. The MgADP-bound and nucleotide-free structures have been solved for the *A. vinelandii* Fe protein (Figure 8.1) [29, 30]. In both structures, the $[Fe_4S_4]$ cluster is located at the interface of the two monomeric

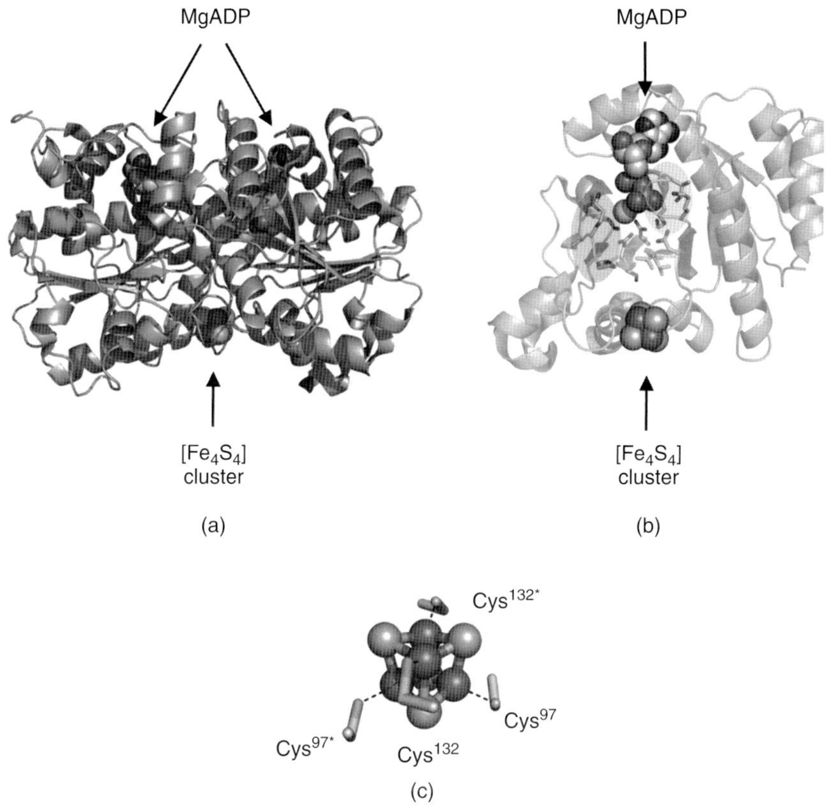

Figure 8.1 (a) Crystal structures of the MgADP-bound Fe protein from *A. vinelandii*, (b) its three structural domains, and (c) its $[Fe_4S_4]$ cluster. The subunits of Fe protein are indicated in grayscale, and the cluster is shown in ball-and-stick presentation. The P-loop formed by the Walker A motif, the Switch I region, and the Switch II region are highlighted in gray on the right, left, and bottom, respectively. Atoms are colored as follows: C, gray; O, red; N, blue; Mg, green; Fe, orange; S, yellow. The cysteinyl ligands from the two subunits are indicated as Cys and Cys*, respectively. PYMOL was used to generate this figure (protein data bank (PDB) entry 1FP6).

subunits near the protein surface. Interestingly, the nucleotide-binding site is 15 Å away from the $[Fe_4S_4]$ cluster, indicative of protein conformational changes that connect the events at the nucleotide sites to the $[Fe_4S_4]$ cluster (Figure 8.1) [30]. Mutagenic analyses suggest that such protein conformational changes may occur upon the binding of MgATP to the Fe protein and likely involve the Walker A motif, the Switch I (residues 38–44) region, the Switch II (residues 125–129) region, and a conserved region around the residue Ala^{157} [31–34]. This argument is supported by the observation of a loop that wraps around the phosphate groups of the nucleotide in the MgADP-bound conformation of Fe protein [30]. Formed by the Walker A motif, this loop has the hydroxyl group of Ser^{16} coordinating the Mg^{2+} ion and the invariant Lys^{10}, thereby forming a salt bridge with the β-phosphate of the nucleotide. On the basis of this observation, it has been proposed that (i) Switch I acts as direct link that communicates between the nucleotide-binding site and the Fe protein/MoFe protein docking interface and (ii) Switch II mediates signal transduction between the nucleotide-binding site and the $[Fe_4S_4]$ cluster [4].

8.2.1.2 The $[Fe_4S_4]$ Cluster

The $[Fe_4S_4]$ cluster is symmetrically ligated by two cysteine residues, $Cys^{\alpha97}$ and $Cys^{\alpha132}$, from each subunit of the Fe protein [3, 10]. It has a +1 oxidization state when the protein is isolated in excess dithionite [35]. In this state, the cluster displays a mixed spin state of $S = 1/2$ and $S = 3/2$ and the distribution of spin states is thought to be influenced by a number of factors, such as temperature and solvents [35–41]. The $[Fe_4S_4]^{1+}$ cluster can be oxidized by one electron, either enzymatically by its redox partner (i.e., MoFe protein), or chemically by oxidizing dyes (e.g., indigodisulfonate (IDS)) [35]. The $[Fe_4S_4]^{1+/2+}$ couple is regarded as physiologically relevant, and the midpoint potential of *A. vinelandii* Fe protein is measured at ∼−310 mV versus normal hydrogen electrode (NHE) in solutions [42]. Interestingly, this potential is decreased by ∼100 mV upon binding of MgATP, which is likely a result of the conformation change of the Fe protein that is induced by nucleotide binding [4, 42]. The $[Fe_4S_4]^{1+}$ cluster can be further reduced to $[Fe_4S_4]^0$ by strong reductants, such as Ti(III) citrate, which has a solution potential of ∼−800 mV versus NHE [43]. Although the identification of the all-ferrous $[Fe_4S_4]^0$ cluster points to the possibility of Fe protein in transferring two electrons via a $[Fe_4S_4]^{0/2+}$ redox pair, the physiological relevance of the $[Fe_4S_4]^0$ state is not clear at this point [44].

8.2.2 Properties of MoFe Protein and its Associated Clusters

8.2.2.1 The MoFe Protein

The MoFe protein of *A. vinelandii* is a $\alpha_2\beta_2$-heterotetramer of ∼220 kDa, and its α- and β-subunits are encoded by *nifD* and *nifK*, respectively [7]. This protein consists of a pair of αβ-dimers that are related by a twofold rotational axis (Figure 8.2). The α- and β-subunits are both divided into three structural domains: αI, αII, αIII and βI, βII, βIII, respectively, and all these domains consist of α-helical- and parallel β-sheet-type polypeptide folds [3, 45–47]. The P-cluster is located at the interfaces of αI- and βI-domains, whereas the FeMo-co is buried within a cavity between the αI-, αII-, and αIII-subunits, 14 Å away from the P-cluster [45–52].

8.2.2.2 The P-Cluster

The P-cluster is coordinated to the MoFe protein by six cysteine ligands, three from the α-subunit ($Cys^{\alpha62}$, $Cys^{\alpha88}$, $Cys^{\alpha154}$) and three from the β-subunit ($Cys^{\beta70}$, $Cys^{\beta95}$, $Cys^{\beta153}$) (Figure 8.2). In the presence of excess dithionite, the P-cluster has a structure of two $[Fe_4S_4]$ subcubanes sharing a μ_6-sulfide [46, 47, 49].

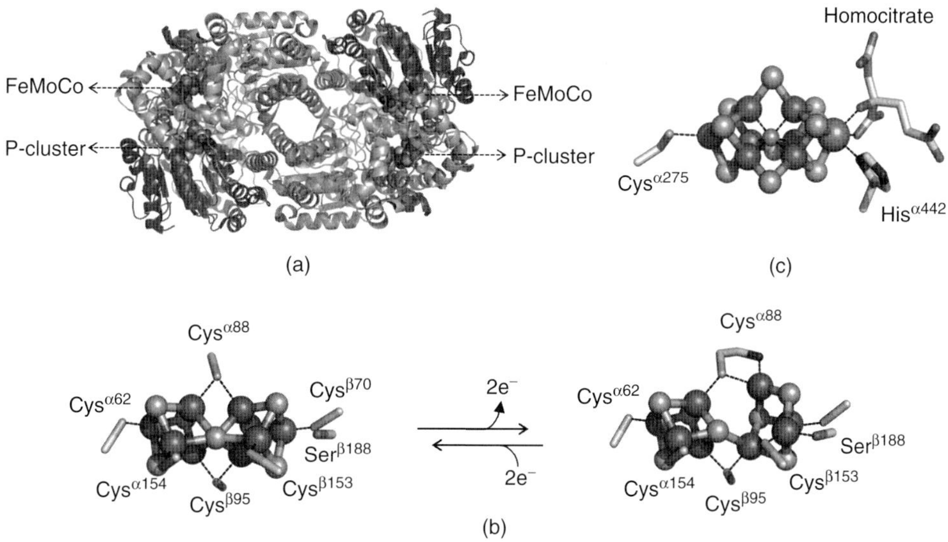

Figure 8.2 (a) Crystal structures of the MoFe protein from *A. vinelandii*, (b) its FeMo-co center, and (c) its P-cluster in the P^N (left) and P^{OX} (right) state. The α- and β-subunits are shaded dark and light gray, respectively, and the clusters are shown in ball-and-stick presentations. Atoms are shown in grayscale. The interstitial C^{4-} atom in FeMo-co is shaded dark gray. PYMOL was used to generate this figure (PDB entry 1M1N).

The dithionite-reduced P-cluster (termed the *P^N state*) is believed to be all-ferrous despite indications of some ferric characteristics of its Fe atoms [4, 53, 54]. While the P^N cluster cannot be further reduced, it can undergo a two-electron oxidation step in the presence of IDS [55]. The IDS-oxidized P-cluster (termed the *P^{OX} state*) has half of its structure, that is, the subcubane coordinated by $Cys^{\beta70}$, $Cys^{\beta95}$, and $Cys^{\beta153}$, in a more open conformation in which the ligation of two Fe atoms to the μ_6-sulfide is replaced by coordination of these Fe atoms by two protein ligands: the Oγ atom from the side chain of $Ser^{\beta188}$ and the N atom from the backbone amide of $Cys^{\alpha188}$ (Figure 8.2c) [10, 49]. Thus, it is clear that the interconversion between the P^N and P^{OX} states results in a change of the core structure and ligation environment of the cluster. However, the reason behind such a dramatic structural rearrangement of the P-cluster upon oxidation is unclear particularly when the Franck–Condon principle, which states that efficient electron transfer occurs when there is a minimum change of the nuclear configuration, is considered [56].

Both the P^N and P^{OX} states have been well characterized by electron paramagnetic resonance (EPR) and Mössbauer spectroscopy. While P^N cluster is mostly diamagnetic, the P^{OX} cluster exhibits an $S = 3$ or 4 integer spin state, which can be best recognized by a parallel-mode EPR signal at $g = 11.8$ [54, 57]. Furthermore, there is an intermediate oxidation level between the P^N and P^{ox} redox states. This oxidation state (termed the *P^{1+} state*) can be generated upon an one-electron oxidation of P^N, and it displays an interesting, mixed spin state of $S = 1/2 : 5/2$ [58].

Although the exact physiological relevance of each of the redox states of the P-cluster is unclear, it is generally assumed that they are involved in the electron-transfer processes during catalysis.

8.2.2.3 The FeMo-co

The [$MoFe_7S_9C$] core of FeMo-co can be viewed as [Fe_4S_3] and [$MoFe_3S_3$] subcubanes that are bridged by three μ_2-sulfides and one μ_6-carbide [45–48, 51, 59, 60]. While the Mo–Fe–S core composition of FeMo-co was established nearly two decades ago, the identity of the interstitial carbide was only demonstrated recently by X-ray emission spectroscopy (XES), electron spin echo envelope modulation (ESEEM), and high-resolution X-ray crystallography [59, 60]. In addition to its metal core, the FeMo-co also has a homocitrate entity attached to the Mo atom, which is believed to be important for the incorporation of FeMo-co to the MoFe protein. FeMo-co is located within the α-subunit of the MoFe protein, and it is coordinated by only two residues: a $His^{\alpha 442}$ at the Mo end and a $Cys^{\alpha 275}$ at the opposite Fe terminal (Figure 8.2). FeMo-co is known to undergo one-electron reduction/oxidation processes [4]. The protein-bound FeMo-co exhibits a well-characterized $S = 3/2$ EPR signal at $g = 4.7$, 3.7, and 2.0 in the presence of excess dithionite. This signal disappears when MoFe protein undergoes a one-electron reduction by Fe protein, leading to the hypothesis that FeMo-co receives electrons from the Fe protein for the subsequent reduction of substrates. The FeMo-co can also be extracted as an intact entity into organic solvents, such as *N*-methylformamide (NMF) [4]. Upon extraction, the FeMo-co exhibits an $S = 3/2$ EPR signal at $g = 5.94$, 4.66, and 3.50 that is similar to, yet broader in shape than the signal of its protein-bound counterpart [4]. The isolated FeMo-co can be used to reconstitute the FeMo-co-deficient MoFe protein and, perhaps more interestingly, it can catalyze the reduction of certain substrates on its own when a strong reductant is supplied (see Section 8.5).

8.3 Catalysis by Mo Nitrogenase

Structural insights into the catalytic mechanism of Mo nitrogenase were obtained from an MgADP·AlF_4^--stabilized complex between the Fe protein and the MoFe protein [45]. The crystal structure of this complex is believed to represent a "transition state" of catalysis by Mo nitrogenase upon MgATP hydrolysis, and it provides compelling evidence that the electrons are transferred from the Fe protein, via the P-cluster, to the FeMo-co, where electrons are passed on to the bound substrate for its subsequent reduction (Figure 8.3). Apart from this structural work, a general mechanism of nitrogenase catalysis was proposed by Thorneley and Lowe [61] ~30 years ago, whereas an alternative reaction pathway of N_2 reduction was suggested by recent spectroscopic work, along with the plausible binding modes of reaction intermediates and the alternative substrates of nitrogenase.

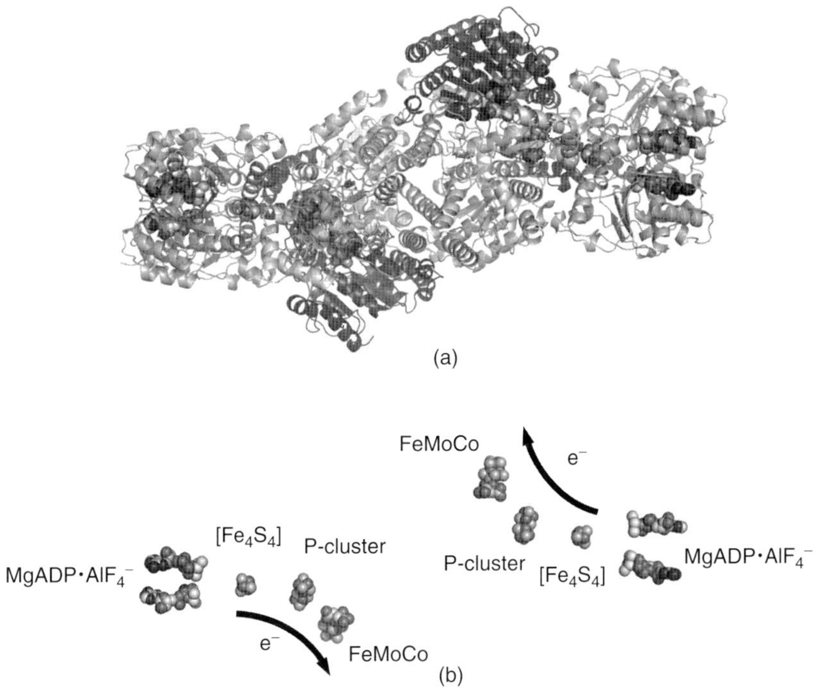

Figure 8.3 (a) Crystal structure of the ADP·AlF_4^--stabilized Fe protein/MoFe protein complex and (b) the relative positions of components involved in the electron flow. The two subunits of Fe protein are shaded dark gray, and the α- and β-subunits of MoFe protein are shaded dark and light gray, respectively. All clusters and ADP·AlF_4^- are shown as space-filling models. Atoms are shown in grayscale. PYMOL was used to create the figure (PDB entry 1M34).

8.3.1 The Thorneley–Lowe Model

8.3.1.1 The Fe Protein Cycle

The first part of the Thorneley–Lowe model is known as the *Fe protein cycle*, and it describes the events that occur during the electron transfer between Fe protein and MoFe protein (Figure 8.4) [4, 8, 10, 61]. The Fe protein cycle starts with the reduction of Fe protein, which renders its $[Fe_4S_4]$ cluster in the +1 oxidation state. Subsequently, the reduced Fe protein binds two MgATP molecules and undergoes a conformational change, which allows it to complex with the MoFe protein. The formation of Fe protein/MoFe protein complex then facilities the interprotein, one-electron transfer from the $[Fe_4S_4]^{1+}$ cluster of Fe protein to the P-cluster of MoFe protein, which occurs concomitant with the hydrolysis of two MgATP molecules. Finally, the complex dissociates upon the release of the inorganic phosphates, and the re-reduction of the $[Fe_4S_4]^{1+}$ cluster is accompanied by the release MgADP, which allows the protein to start the next round of electron transfer. While the structure of the MgATP-bound Fe protein is not available,

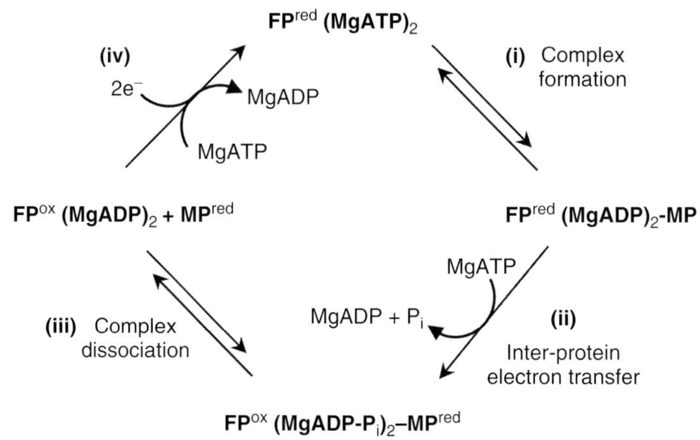

Figure 8.4 The Fe protein cycle of Mo nitrogenase. This cycle describes (i) the complex formation of the MgATP-bound Fe protein (FP) with the MoFe protein (MP); (ii) the ATP hydrolysis-dependent electron transfer from FP to MP; (iii) the dissociation of complex upon the release of the inorganic phosphate (P_i); and (iv) the re-reduction of the FP and the replacement of MgADP with MgATP. Subscripts: red, reduced; ox, oxidized.

X-ray scattering studies indicate that the binding of MgATP indeed induces a conformational change of the Fe protein, which is reflected by a decrease of protein gyration radius by about 2 Å [62]. This observation aligns well with the kinetics of Fe chelation, which suggests that the $[Fe_4S_4]$ cluster is more surface exposed upon MgATP binding [63–65].

8.3.1.2 **The MoFe Protein Cycle**

The second part of the Thorneley–Lowe model is known as the *MoFe protein cycle*, and it depicts the process of proton and electron transfer by MoFe protein to the substrate, N_2, during turnover (Figure 8.5) [4, 8, 10, 61]. The MoFe protein cycle consists of eight distinct steps, each representing one of the eight electron/proton transfer steps that are needed to complete a turnover of one N_2 molecule to two NH_3 molecules (i.e., $N_2 + 8H^+ + 8e^- \rightarrow 2NH_3 + H_2$) [10]. In this model, the resting state of one functional αβ-half of the MoFe protein, which contains one P-cluster and one FeMo-co, is designated as M_0. M_0 can be converted to various turnover states, M_n, where n represents the number of proton/electron pairs added on M_0. The binding of substrates to MoFe protein occurs at a more reduced state than M_0, and the exact stage of binding depends on the nature of the substrates. In the case of N_2, it likely binds at M_3, where MoFe protein has already accumulated three protons and three electrons concomitant with the release of one H_2. After that, Mo nitrogenase would undergo five or six steps of proton/electron addition to yield two molecules of NH_3. On the basis of the Thorneley–Lowe model, these steps are carried out through an asymmetric reduction of N_2, where one N atom is fully reduced to NH_3 before the other N atom is reduced to the same product

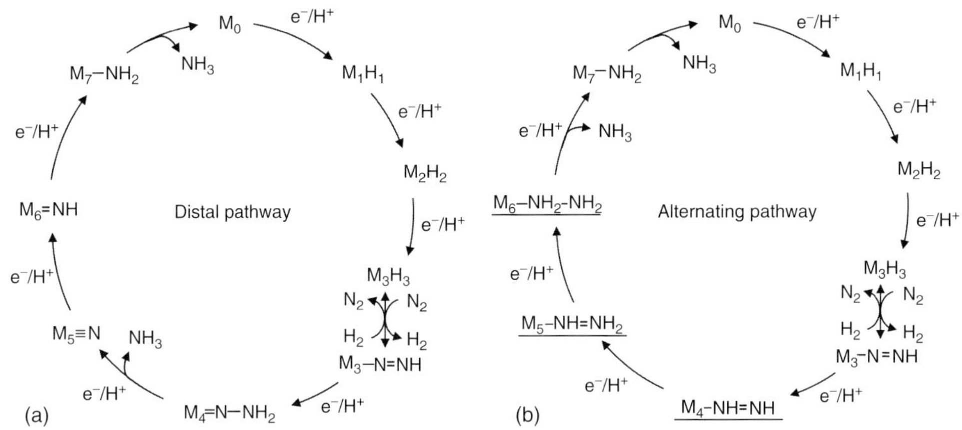

Figure 8.5 The MoFe protein cycle of Mo nitrogenase, featuring the distal (a) and alternating (b) pathway of N_2 reduction. (a) The distal pathway, proposed by Thorneley and Lowe, describes one plausible sequence of events during N_2 reduction, in which the distal N is reduced first and released as NH_3 before the M_5 stage, followed by the subsequent reduction of the proximal N and the release of a second NH_3 at the M_7 stage. (b) The alternating pathway, based on recent spectroscopic studies, shares most of the steps with the distal pathway, but differs at stages M_4, M_5, and M_6 (underlined). It features the alternating addition of electrons and protons at both the distal and proximal N atoms. Consequently, the two NH_3 molecules are released at the M_6 and M_7 stages, respectively, late in the sequence. In both pathways, M represents one functional unit of the MoFe protein, which consists of one αβ-dimer and one pair of FeMo-co/P-cluster. Subscripts 0–7 indicate the number of electrons transferred to M from the Fe protein via the Fe protein cycle (see Figure 8.4).

(Figure 8.5a) [61, 66]. Thus, this model predicts the formation of a hydrazido intermediate (M=N−NH_2) and an imido intermediate (M=NH) at stage M_4 and M_6, respectively, as well as the stepwise release of two NH_3 molecules at stages M_5 and M_7. Following the release of the second NH_3, the MoFe protein is cycled back to M_0 and ready for another round of electron/proton addition and substrate reduction. In the absence of substrates, however, M_2 can be cycled back to M_0 through the release of H_2.

Such a N_2 reduction pathway (designated the distal pathway) aligns well with the reaction scheme of NH_3 formation by mononuclear Mo metal complexes; however, it has not been rigorously investigated in the context of nitrogenase [61, 67]. One line of experimental evidence that supports this mechanism came from the detection of hydrazine formation upon acid and base quenching of the pre-steady-state Mo nitrogenase during N_2 turnover [68]. On the basis of the analogy between the chemistry of Mo nitrogenase and the dinitrogen complexes of Mo and W, hydrazine is thought to represent the reduced form of the M=N−NH_2 intermediate, which presumably binds to the M_4 state of MoFe protein [4, 68]. However, the observation of hydrazine formation does not rule out the appearance of other intermediates, such as M−NH=NH or M−NH_2−NH_2, in this process, although the formation of these intermediates is not favored in this reaction pathway.

8.3.2
Recent Development

8.3.2.1 Alternative Pathway of N_2 Reduction

In the past decade, our knowledge of the reaction mechanism of N_2 reduction was expanded on the basis of the data acquired from advanced spectroscopic studies. Electron nuclear double resonance (ENDOR) and ESEEM spectroscopy have been applied to the studies of the turnover state of wild-type Mo nitrogenase and its variant forms [69–74]. The outcome of these experiments has led to the proposal of a different N_2 reduction pathway (Figure 8.5b) [70]. This model is modified from the Thorneley–Lowe model and shares the first three steps with the latter (Figure 8.5a); however, it proposes that protons and electrons are added to the two N atoms of N_2 in an alternating manner (designated the alternating pathway) (Figure 8.5b). On the basis of this scheme, the release of the two NH_3 molecules would occur at stages M_6 and M_7 of the MoFe protein cycle, while a diazene intermediate (M–NH=NH) and hydrazine intermediate ($M-NH_2-NH_2$) would be bound on the FeMo-co at stage M_4 and M_6, respectively. Support for such a reaction pathway stems primarily from the capture of a common intermediate species by using hydrazine (N_2H_4), diazene (N_2H_2), and N_2 as substrates, suggesting that the reduction of N_2H_4 and N_2H_2 share the same intermediates of N_2 reduction [69–71]. However, structural characterization of this common intermediate by ENDOR reveals a $-NH_x$ species, which likely represents M_7 [70, 72]. Thus, this result alone does not distinguish between the two proposed reaction pathways. As such, further effort is required to address this point; in particular, the capture and characterization of the N_2 reduction intermediates bound at the M_4, M_5, or M_6 stages will likely provide key insights into this issue.

8.3.2.2 Plausible Structures of N_2 Reduction Intermediates

Structural characterization of the FeMo-co-bound intermediates of N_2 reduction using ENDOR or ESEEM spectroscopy was hampered by a number of obstacles. First, there are many possible intermediate species that cannot be conclusively differentiated by ENDOR spectroscopy, particularly considering the existence of both alternating and distal pathways of N_2 reduction. The complexity of fitting the correct model to the spectroscopic data is illustrated in Figure 8.6, where the eight possible structures of N_2 reduction intermediates at the M_5 and M_6 stages are shown [73]. Second, there is insufficient knowledge about the number of electrons and protons that are added at each trapped intermediate state, which further complicates the interpretation of spectroscopic data [74]. To address this issue, advanced kinetic methods of nitrogenase turnover were developed to independently determine the reduction level (i.e., M_n) of Mo nitrogenase during the catalytic cycle. Recently, such a protocol was employed to characterize a previously identified hydride intermediate, which was determined to be representing the M_4 stage of the N_2 reduction cycle [74]. Combined ENDOR spectroscopy and kinetic analysis are thus expected to be used to narrow down the possible intermediate structures. Third, there is a limitation for the placement of nuclear-spin

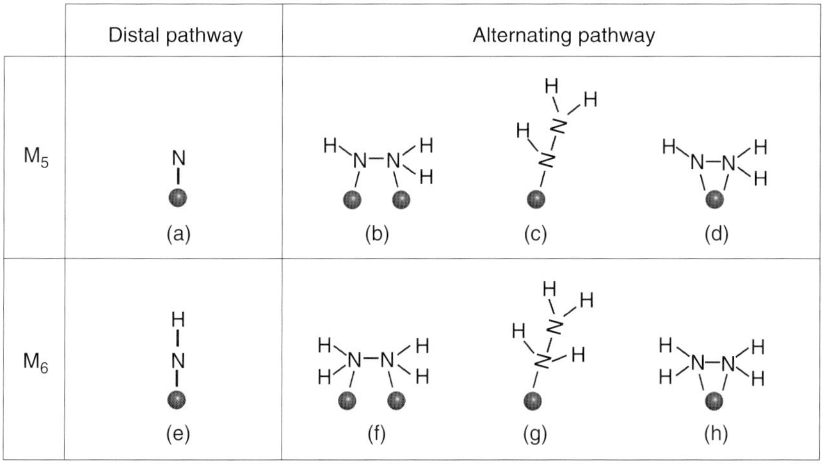

Figure 8.6 The eight possible binding modes of N_2 reduction intermediates at M_5 and M_6 stages. Proposed structural models of MoFe-protein-bound intermediates of N_2 reduction at the M_5 (a–d) and M_6 (e–h) stages for both distal (a, e) and alternating (b–d, f–h) pathways based on kinetic, ESEEM and ENDOR analyses.

probes on the substrate-derived intermediates owing to the symmetric nature of N_2, N_2H_2, and N_2H_4 molecules [73]. Deuterium labeling of N for ^{1,2}H ENDOR spectroscopy is also not an option, because N–H are solvent exchangeable. To circumvent this issue, alternative substrates of nitrogenases, such as methyldiazene (NH=N–CH_3), was used in the place of the nitrogenous substrates [75]. In this case, the N atoms, as well as the H atoms that are associated with them, can be selectively labeled and differentiated in the ENDOR analysis because of the presence of the methyl group. Other examples of spectroscopic characterization using various alternative substrates, as well as the catalytic features of these substrates, are discussed subsequently.

8.3.2.3 Alternative Substrates of Nitrogenase

Nitrogenase is known to reduce a variety of small molecules aside from its physiological substrate N_2. The functional groups of these alternative substrates are typically doubly or triply bonded C, N, or O moieties [4, 8, 66, 76–78]. There are two advantages to using these alternative substrates as structural probes of nitrogenase mechanism. First, many of these substrates, including cyanide (HCN/CN^-) and carbon monoxide (CO), bind to MoFe protein or NMF-isolated FeMo-co at a relatively more oxidized level (i.e., the M_2 stage or a stage before M_2) [4, 79–81]. Considering that the early turnover states, such as stages M_0 through M_2, can be generated with relative ease, robust trapping and observation of the binding of these substrates to MoFe protein or isolated FeMo-co can be achieved. Second, because some of these substrates, such as CO or propargyl alcohol (HC≡CHCH_2OH), have slower reaction kinetics, reaction intermediates that are relatively stable can be generated and characterized by spectroscopic

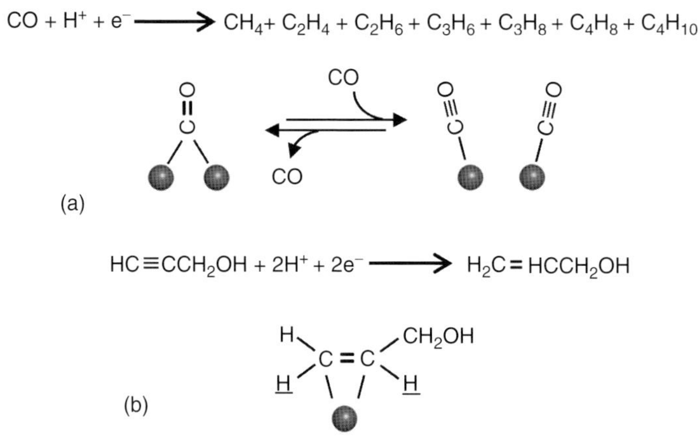

Figure 8.7 Proposed binding modes of alternative substrates of Mo nitrogenase. (a) The reaction of CO reduction and the proposed structural models of MoFe-protein-bound CO at low (left) and high (right) CO concentrations. Note that the exact stoichiometry of CO reduction is not clear. (b) The reaction of two-electron reduction of propargyl alcohol and the proposed structure of the reaction intermediate. This intermediate consists of an allyl alcohol bound side-on to a single Fe ion through its alkene group. Solvent-derived hydrogen atoms are underlined.

means [73, 78]. The ENDOR-based structural studies of the MoFe-protein-bound HC≡CHCH$_2$OH and CO are summarized subsequently, whereas the electrochemical, spectroscopic, and catalytic investigations of the binding of CN$^-$ and CO to isolated FeMo-co are discussed in Section 8.5.

The two-electron reduction of HC≡C–CH$_2$OH to allyl alcohol (H$_2$C=CH–CH$_2$OH) provides a simple, yet effective, model for the structural characterization of nitrogenase reaction intermediates [73, 82]. Facilitated by the slow turnover rates, an $S = 1/2$ species can be easily generated from HC≡C–CH$_2$OH and subjected to spectroscopic analysis. Furthermore, continuous wave (CW) and pulsed ^{13}C ENDOR can be employed if the protein samples are labeled with ^{13}C and ^{2}H and turned over in H$_2$O and D$_2$O buffers. Indeed, data obtained from these experiments suggest that the trapped species contains all three C atoms of the substrate and is bound to the MoFe protein through one or two of these carbon atoms. ^{1,2}H ENDOR further demonstrated the presence of three coupled protons: one substrate-derived and two solvent-derived [73, 82]. The observation of the coupling of two solvent-derived protons indicates that this species likely represents the reaction product with two added protons, whereas the relative strength of the hyperfine tensors of each coupled proton provides important insights into the structure of this enzyme-bound species. Together with the structural information of other model compounds, these data suggest that the functional alkene group of the allyl alcohol product likely binds side-on to one of the Fe atoms of FeMo-co (Figure 8.7). This model is also supported by other independent density functional theory (DFT) calculations [83].

CO is one of the first nitrogenase inhibitors/substrates subjected to spectroscopic analysis [84–87]. The initial interest in CO-related studies stemmed from the fact that CO is isoelectric to N_2 and, therefore, may serve as a good analog of N_2 for studies of the interactions between the substrate and the active site of nitrogenase [4]. Furthermore, it was established that multiple EPR-active species could be generated upon incubation of varying amounts of CO with Mo nitrogenase under turnover conditions. Specifically, a so-called lo-CO EPR signal could be generated in the presence of low CO concentrations, whereas a "hi-CO" EPR signal could be generated in the presence of high CO concentrations [84, 85]. Both "lo-CO" and "hi-CO" states of MoFe protein exhibited $S = 1/2$ signals, and the two states were shown to be interconvertible through photolysis studies. ^{57}Fe, ^{13}C ENDOR and ESEEM spectroscopy revealed that the "lo-CO" signal corresponded to a form of MoFe protein with one CO molecule bridged by two Fe atoms across the "waist" of FeMo-co, whereas the "hi-CO" signal corresponded to a form of MoFe protein with two CO molecules terminally bound to two Fe atoms of FeMo-co (Figure 8.7a,b) [84, 85]. Interestingly, it was shown recently that Mo nitrogenase could catalyze the reductive coupling of CO into hydrocarbon products, such as ethylene (C_2H_4), ethane (C_2H_6), propylene (C_3H_6), and propane (C_3H_8), in the presence of high concentrations of CO [78]. While the mechanism of this intriguing reaction is not well understood, the existence of the "hi-CO" state suggests the capacity of FeMo-co to bind and process two substrates/intermediates at the same time. The details of this exciting discovery are discussed further in the following section.

8.4 Unique Features of V Nitrogenase

The V nitrogenase shares a good degree of homology in sequence and function with its Mo counterpart. It is also a binary system that consists of an iron (Fe) protein component and a vanadium iron (VFe) protein component [5, 78, 88]. Moreover, the catalysis of V nitrogenase, similar to that of Mo nitrogenase, requires the action of both protein components and ATP-dependent association/dissociation between them that permits the transfer of electrons from the Fe protein to the active cofactor site (FeVco) of VFe protein [5, 78, 88]. On the other hand, despite the overall similarities between the two nitrogenases, V nitrogenase displays some unique structural and catalytic features that are clearly distinct from its Mo counterpart (see subsequent text).

8.4.1 Structural Features of Fe Protein and its Associated Cluster

8.4.1.1 The Fe Protein

The Fe protein of V nitrogenase (encoded by *vnfH*) from *A. vinelandii* shares ~91% of sequence identity with its counterpart in Mo nitrogenase (encoded by *nifH*)

[78]. Similar to NifH, VnfH is a ~60 kDa homodimer that contains four iron atoms and four acid-labile sulfur atoms per protein [89, 90]. The two cysteinyl ligands of the $[Fe_4S_4]$ cluster, as well as the consensus Walker A motif, are conserved in the sequences of VnfH and NifH [91]. In the dithionite-reduced state, both VnfH and NifH display the same, mixed $S = 3/2 : 1/2$ EPR signals, although the $S = 3/2$ and $S = 1/2$ signals of VnfH are 10% and 27%, respectively, more intense than those of NifH [89, 90, 92]. In addition, VnfH displays a midpoint potential comparable to that of NifH in the MgADP-bound state, and it also utilizes the $[Fe_4S_4]^{2+/1+}$ couple for the one-electron transfer event during catalysis [93].

8.4.1.2 The $[Fe_4S_4]$ Cluster

Surprisingly, despite the significant degree of homology between VnfH and NifH, a recent Fe K-edge XAS/EXAFS study shows that there are subtle differences between the redox properties and geometric arrangements of the $[Fe_4S_4]$ clusters of these two Fe proteins [92]. The rising edge of XAS data shows that the iron atoms in the $[Fe_4S_4]$ cluster of VnfH are more reduced than those in NifH. In addition, the $[Fe_4S_4]$ cluster in VnfH is best modeled by two identical $[Fe_2S_2]$ rhomboids that are stacked and offset by 90° from each other, whereas the $[Fe_4S_4]$ cluster in NifH can be modeled alternatively by two unequal $[Fe_2S_2]$ rhomboids that are bent to a greater degree out of plane relative to each other (Figure 8.8). It is unclear whether or how these subtle differences affect the catalytic capacities of the V- and Mo nitrogenases. However, given the differences between the P-clusters of VFe and MoFe proteins, as well as the fact that the $[Fe_4S_4]$ clusters of VnfH and NifH directly interact with the P-clusters in these proteins, it can be reasoned that the clusters of Fe proteins are fine-tuned by nature to work best with their respective P-cluster "partners" during catalysis [92].

8.4.2 Structural Features of VFe Protein and its Associated Clusters

8.4.2.1 The VFe Protein

The VFe protein consists of three subunits: α, β, and δ, which are encoded by *vnfD*, *vnfK*, and *vnfG*, respectively. While the δ subunit is unique to VFe protein, the α- and β-subunits of the VFe protein share ~33% and ~32% of sequence identity, respectively, with their counterparts in the MoFe protein [78]. In *Azotobacter chroococcum*, VFe protein was shown to be a $\alpha_2\beta_2\delta_2$-hexameric of ~240 kDa; in *A. vinelandii*, however, the same protein was identified recently as a $\alpha_2\beta_2\delta_4$-octamer of ~270 kDa [5, 94]. Metal and spectroscopic analyses established the presence of P-cluster and FeVco at a ~1 : 1 ratio in VFe protein [5, 94]. Furthermore, the significant degree of sequence conservation between MoFe and VFe proteins, particularly with regard to the ligands of P-cluster and FeMo-co or the residues surrounding these clusters, strongly suggests the presence of analogous metal centers in these proteins.

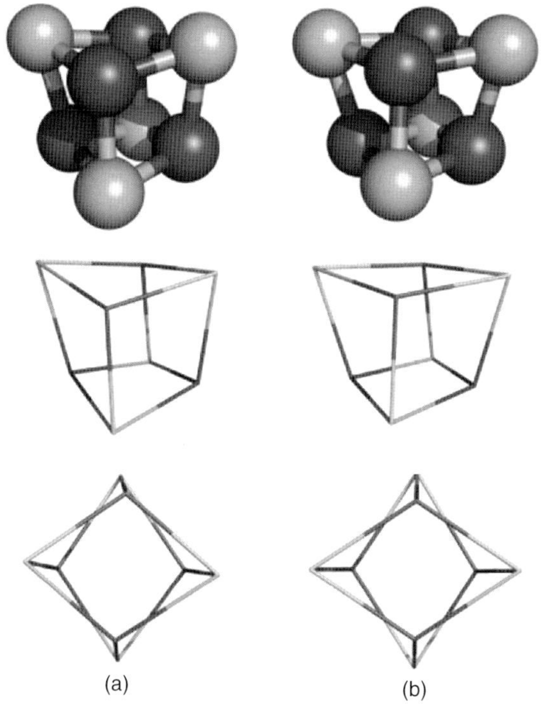

Figure 8.8 EXAFS-derived models of $[Fe_4S_4]$ clusters in NifH (a) and VnffH (b). The $[Fe_4S_4]$ clusters are shown in ball-and-stick (top) and line (middle, bottom) presentations. Both clusters are rotated to show the overlapping of the $[Fe_2S_2]$ rhomboids (a and b, bottom). (a) In the case of NifH, the two rhomboids are unequal and bent to a greater degree out of plane relative to each other, (b) whereas in VnfH, two identical $[Fe_2S_2]$ rhomboids are stacked and offset by 90°. Atoms are colored as follows: Fe, dark gray; S, light gray. PYMOL was used to generate this figure, with coordinates of the crystal structure of NifH $[Fe_4S_4]$ cluster (PDB entry 1NIP) modified on the basis of the EXAFS fits.

8.4.2.2 **The P-Cluster of VFe Protein**

While earlier spectroscopic characterization of the *A. chroococcum* VFe protein has pointed to a structural/functional resemblance between the P-clusters in VFe and MoFe proteins, recent characterization of the *A. vinelandii* VFe protein revealed differences between these P-cluster species [94]. First, the IDS-oxidized P-cluster of VFe protein does not display a signal at $g = 11.8$, which is associated with the P^{OX} state of MoFe protein (see Section 8.2.2.2 above). The dithionite-reduced P-cluster of VFe protein, on the other hand, displays a strong $S = 1/2$ signal, which is absent from the spectrum of the P^N-cluster of MoFe protein [94]. The intensity of this signal correlates with the activity of the VFe protein and, therefore, is catalytically relevant. Further, VFe protein displays an $S = 5/2$ signal in the dithionite-reduced state, which is attributed to the P-cluster based on results of the temperature-dependent EPR analysis [94, 95]. This signal bears a resemblance to an analogous $S = 5/2$ signal, which is associated with the so-called

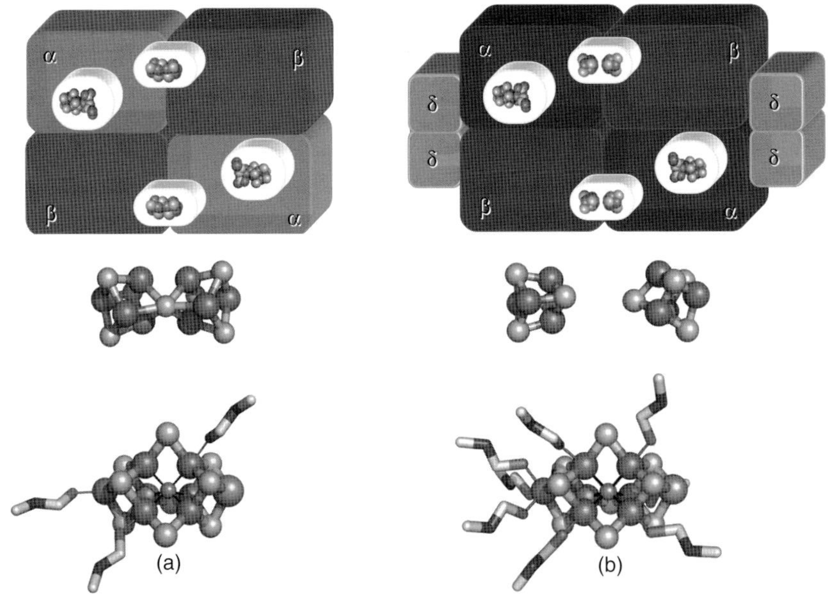

Figure 8.9 Schematic representation of MoFe protein (a, top) and VFe protein (b, top), EXAFS-derived models of P-clusters in MoFe (a, middle) and VFe (b, middle) protein, and EXAFS-derived models of NMF-extracted FeMo-co (a, bottom) and FeVco (b, bottom). The MoFe protein is an $\alpha_2\beta_2$ heterotetramer (a, top); whereas the VFe protein is an $\alpha_2\beta_2\delta_4$ heterooctamer (b, top). The P-cluster of MoFe protein has an $[Fe_8S_7]$ topology (a, middle), whereas P-cluster of VFe protein consists of paired $[Fe_4S_4]$ clusters (b, middle). The isolated FeVco (b, bottom) resembles the isolated FeMo-co (a, bottom) in core structure, although it is more extended in structure and bound with more NMF molecules. Clusters are shown in ball-and-stick presentations; NMF molecules are shown as sticks. Atoms are shown in grayscale. PYMOL was used to generate this figure, with coordinates of the P-cluster and FeMo-co derived from the MoFe protein crystal structure (PDB entry 1M1N) modified on the basis of the EXAFS fits.

P^{1+} state, an intermediate redox state between the P^N and the P^{OX} state of the P-cluster of MoFe protein [58]. Consistent with the different EPR properties of the P-clusters in VFe protein, Fe K-edge EXAFS analysis of a cofactor-deficient form of VFe protein suggests that its P-clusters consist of paired $[Fe_4S_4]$-like clusters rather than the "standard" $[Fe_8S_7]$ clusters (Figure 8.9a) [96]. Taken together, these observations strongly indicate that the P-cluster of VFe protein is electronically and structurally different from its counterpart in MoFe protein [78, 94].

8.4.2.3 **The FeVco**

FeVco has been characterized in both the protein-bound and solvent-extracted states by EPR, XAS/EXAFS, magnetic circular dichroism (MCD), and Mossbauer spectroscopy [97–102]. The outcome of these studies suggests that FeVco resembles FeMo-co in overall structure and function, but it differs somewhat from the latter in electronic properties and cofactor–protein interactions. In the isolated state, FeVco displays an $S = 3/2$ EPR signal that is similar to that displayed by the isolated FeMo-co, although the lineshape of this signal is significantly broader than that of the latter [102]. Interestingly, this FeVco-associated signal is somewhat "tightened" in its native VFe protein environment, rendering its g values ($g = 4.32$, 3.77) closer to those of its counterpart in the MoFe protein ($g = 4.31, 3.67$) despite a much decreased signal intensity of the former relative to that of the latter [94, 102]. Moreover, the protein-bound FeVco displays an additional $g = 5.50$ signal that is absent from the spectrum of protein-bound FeMo-co, isolated FeMo-co, or isolated FeVco [102]. This $g = 5.50$ signal could represent an $S = 3/2$ species that originates from the interactions between FeVco and P-cluster or from the interactions between FeVco and its surrounding protein environment. Fe K-edge XAS analysis of the isolated FeVco indicates the presence of a well-ordered sphere of solvent molecules around the cofactor [102]. Such a sphere is absent in the case of isolated FeMo-co, suggesting electronic differences between FeMo-co and FeVco in the extracted states (Figure 8.9b). Furthermore, EXAFS analysis reveals that the isolated FeVco adopts an octahedral geometry, whereas the isolated FeMo-co assumes a tetrahedral/trigonal pyramidal structure. This observation indicates that the core structure of FeVco is slightly elongated than that of FeMo-co, which is further supported by the presence of a longer Fe – heterometal backscattering path in FeVco (Fe–V path: 2.88 Å) than that in FeMo-co (Fe–Mo path: 2.68 Å).

8.4.3
Catalytic Features of V Nitrogenase

V nitrogenase resembles its Mo counterpart in catalytic properties. Both V- and Mo nitrogenases likely use a similar "Fe protein cycle" during catalysis, as kinetic analysis reveals similar rates of complex formation, interprotein electron transfer, and ATP hydrolysis by the two nitrogenases. Moreover, both nitrogenases reduce the same substrates, such as N_2, H^+, C_2H_2, and HCN/CN^-, to identical or similar products [89]. On the other hand, V nitrogenase exhibits some catalytic features that are notably distinct from those of its Mo counterpart. V nitrogenase reduces C_2H_2 much more poorly than Mo nitrogenase, with a K_m value for C_2H_2 more than 10-fold higher than the latter [5]. Moreover, in addition to carrying out the two-electron reduction of C_2H_2 to C_2H_4, V nitrogenase produces a minor amount of C_2H_6 from C_2H_2 without forming C_2H_4 as an intermediate of this reaction [5, 103, 104]. V nitrogenase diverts a bigger portion of the total electron flux (~60%) than Mo nitrogenase (<10%) to the reduction of H^+ in the presence of C_2H_2 (Figure 8.10) [105]. Similarly, V nitrogenase is less efficient in N_2 reduction, and it favors H_2 formation in the presence of N_2. Contrary to Mo nitrogenase, which produces H_2 and NH_3 at a ratio of 1 : 2 in the presence of N_2, V nitrogenase generates H_2 and NH_3 at a ratio of ~1 : 1 (Figure 8.10) [94, 106]. The rationale for the lower efficiency of N_2 reduction by V nitrogenase, as well as the question of whether H_2 evolution is coupled to NH_3 formation in this case, remains unclear

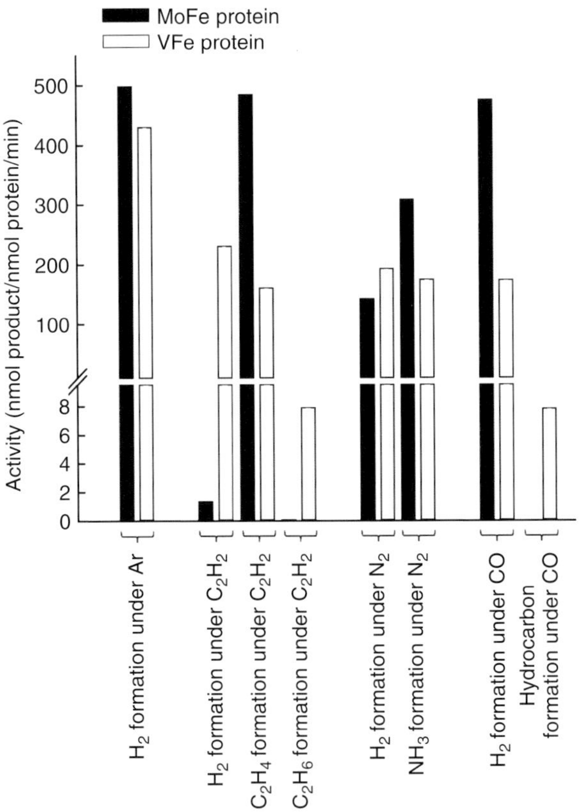

Figure 8.10 Specific activities of Mo- and V nitrogenases. Specific activities of H^+-, C_2H_2-, N_2-, and CO reduction by Mo (black) and V (white) nitrogenase. Both nitrogenases have similar H^+-reducing activities under Ar, but Mo nitrogenase is more efficient in converting C_2H_2 and N_2 to C_2H_4 and NH_3, respectively. V nitrogenase diverts a bigger portion of electrons and protons to H_2 formation in the presence of all substrates except CO. It also produces significantly more C_2H_6 and hydrocarbons (i.e., C2–C4 alkanes and alkenes) from C_2H_2 and CO.

at the moment. Nevertheless, the *in vivo* measurements of N_2 uptake and H_2 evolution reveal a similar H_2:NH_3 ratio to that in the *in vitro* assay, suggesting the physiological relevance of the observed increase of H_2 production in the presence of N_2 [105].

Despite the fact that Mo nitrogenase is far better understood than its V counterpart, the discovery of CO as a substrate of nitrogenase was accomplished only through the recent characterizations of V nitrogenase. The first indication of CO as a nitrogenase substrate came from the observation of a ~75% inhibition of H^+ reduction by V nitrogenase – contrary to a nearly negligible inhibition of the same reaction by Mo nitrogenase – in the presence of 100% CO (Figure 8.10) [94]. Given that the total electron flux and the rate of ATP hydrolysis are comparable in reactions catalyzed by both V- and Mo nitrogenases,

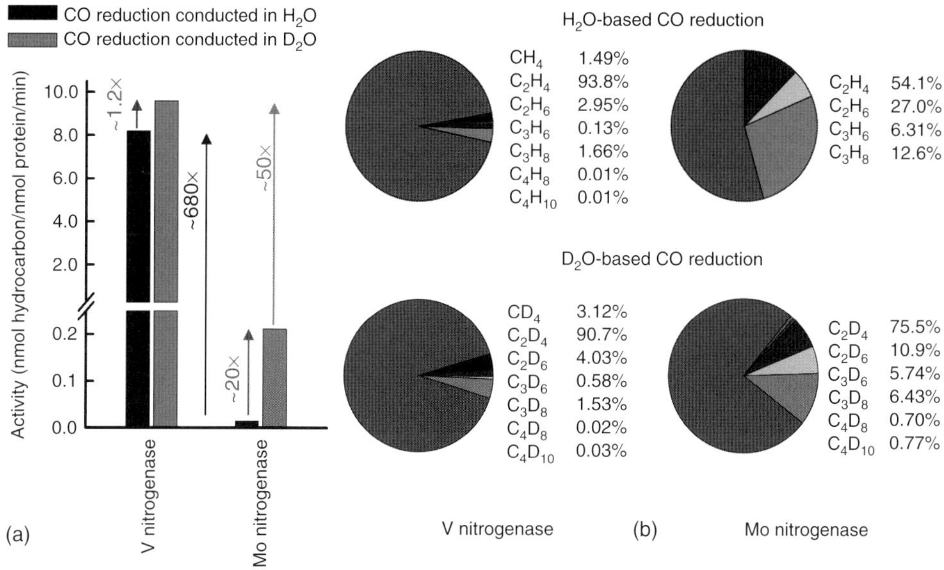

Figure 8.11 The total specific activities of CO reduction (a) and the product distribution of hydrocarbons (b) generated by Mo- and V nitrogenases in the presence of H_2O and D_2O. (a) The total activities of hydrocarbon formation from CO by Mo- and V nitrogenases in H_2O (black)- and D_2O (gray)-based reactions. In H_2O, V nitrogenase is 680-fold more active than Mo nitrogenase (indicated by black arrow), whereas in D_2O, the activity difference is decreased to 50-fold (indicated by light gray arrow). The activity of V nitrogenase is minimally impacted upon D_2O substitution (indicated by gray arrow), whereas the activity of Mo nitrogenase is significantly increased in D_2O (indicated by gray arrow). (b) Distributions of hydrocarbons in reactions catalyzed by V- and Mo nitrogenases in the presence of H_2O (top) or D_2O (bottom). The total amounts of hydrocarbons formed in V- and Mo-nitrogenase-based reactions were set as 100%, and the percentages of individual products were determined accordingly.

it was proposed that V nitrogenase redirected a significant portion of electrons and protons toward the reduction of CO. Indeed, isotopic labeling and GCMS analyses confirmed that V nitrogenase could convert CO to C1–C4 hydrocarbons (i.e., CH_4, C_2H_4, C_2H_6, C_3H_6, C_3H_8, 1-C_4H_8, and n-C_4H_{10}) in a reaction that parallels the industrial Fischer–Tropsch process [107]. Consistent with the disparate degree of proton/electron rerouting, the total amount of hydrocarbons formed by V nitrogenase was 680-fold higher than that by its Mo counterpart (Figure 8.11a) [108]. Moreover, while the same C2 and C3 products were detected in Mo-nitrogenase-catalyzed reaction, CH_4, α-C_4H_8, and n-C_4H_{10} were only detected in the V-nitrogenase-catalyzed reaction.

A closer examination has revealed more discrepancies between the reactions of CO reduction by Mo- and V nitrogenases. First, CO reduction by V nitrogenase differs from that by Mo nitrogenase in product distribution. V nitrogenase produces almost exclusively C_2H_4 from CO (~94% of the total hydrocarbons); whereas Mo nitrogenase has a more evenly distributed product profile despite having C_2H_4 as the major product (~54% of the total hydrocarbons) (Figure 8.11b) [108]. Second, activities of CO reduction by V- and Mo nitrogenases respond differently upon substitution of D_2O for H_2O. In the presence of D_2O, the product yield was increased only slightly in the case of V nitrogenase (~1.2%), but significantly in the case of Mo nitrogenase (>20-fold) (Figure 8.11a) [108, 109]. Formation of C_2D_4 by Mo nitrogenase was increased to ~75% of the total amount of hydrocarbon (Figure 8.11b). Moreover, 1-C_4D_8 and *n*-C_4D_{10} were also detected in the Mo-nitrogenase-catalyzed reaction upon D_2O substitution [108, 109]. In contrast, there was a clear, albeit slight shift from the formation of C_2D_4 to the formation of C_2D_6 and >C2 products [108]. Together, these changes render the product profiles of Mo- and V nitrogenases more similar to each other, which points to a common mechanism of CO reduction by both nitrogenases.

8.5
Catalytic Properties of Isolated FeMo-co and FeVco

Very recently, it was demonstrated that the solvent-extracted FeMo-co and FeVco could reduce CO and CN^- to alkanes and alkenes in the presence of a strong reductant, europium(II) diethylenetriaminepentaacetate [Eu^{II}-DTPA] [87]. The isolated cofactors remained sufficiently stable in this reaction medium, showing ~90% activity in reconstituting apo-MoFe protein after the first hour [87]. In the presence of [Eu^{II}-DTPA] ($E^{0'} = -1140$ mV at pH 8), both FeMo-co and FeVco were capable of reducing CO to CH_4, C_2H_4, C_2H_6, C_3H_6, C_3H_8, 1-C_4H_8, *n*-C_4H_{10}, and 1-C_5H_{10} under ambient conditions. When CN^- was used as a substrate, the same set of products was generated in reactions catalyzed by both FeMo-co and FeVco; however, additional products, including *n*-C_5H_{12}, 1-C_6H_{12}, *n*-C_6H_{14}, and *n*-C_7H_{16}, were also detected in the reaction catalyzed by FeMo-co [87].

As reported for MoFe protein- and VFe protein-based reactions, NH_3 was also formed from the reduction of CN^- at molar ratios of ~1 : 1 and ~0.9 : 1, respectively, to hydrocarbons in FeMo-co- and FeVco-based reactions [87]. However, contrary to the reactions catalyzed by the protein-bound cofactors, CH_4 was no longer the major product of CN^- reduction catalyzed by the isolated cofactors. Such a shift from the formation of CH_4 toward the production of C2–C7 alkenes/alkanes suggests that the cofactors favor the formation of C–C bonds over the reduction and protonation of substrates in the absence of their protein environment. The product profiles of CN^-- and CO reduction by isolated cofactors were similar to each other, pointing to a common C–C coupling pathway of these reactions. However, the rates of product formation in the reaction with CN^- were significantly higher than those in the reaction with CO, which could be attributed to a stabilizing effect of CN^- binding on certain redox states of the isolated cofactors.

Compared to their respective protein-bound counterparts, the isolated cofactors exhibit some unique catalytic features, particularly with regard to the reduction of CO to hydrocarbons. Both isolated cofactors are less active than protein-bound cofactors in this reaction, which is likely due to an inefficient delivery of protons and electrons to cofactor in solutions [87]. On the other hand, the tendency of the isolated cofactors toward generating larger products suggests their ability to further extend the carbon chain in the absence of steric constraints of the protein scaffold. It is also interesting to note that, contrary to the protein-bound cofactors, both isolated cofactors form CH_4 (rather than C_2H_4) as the major product of CO reduction, a behavior well documented for the industrial Fischer–Tropsch catalysts [110]. These unique features of isolated cofactors could be exploited for future biotechnological adaptations for industrial hydrocarbon production.

Acknowledgments

Work in our laboratory is supported by National Institutes of Health Grant GM-67626 (MWR) and Herman Frasch Foundation Grant 617-HF07 (MWR).

References

1. Allen, R.M., Chatterjee, R., Madden, M.S., Ludden, P.W., and Shah, V.K. (1994) *Crit. Rev. Biotechnol.*, **14**, 225.
2. Peters, J.W., Fisher, K., and Dean, D.R. (1995) *Annu. Rev. Microbiol.*, **49**, 335.
3. Howard, J.B. and Rees, D.C. (1996) *Chem. Rev.*, **96**, 2965.
4. Burgess, B.K. and Lowe, D.J. (1996) *Chem. Rev.*, **96**, 2983.
5. Eady, R.R. (1996) *Chem. Rev.*, **96**, 3013.
6. Seefeldt, L.C. and Dean, D.R. (1997) *Acc. Chem. Res.*, **30**, 260.
7. Rees, D.C. and Howard, J.B. (2000) *Curr. Opin. Chem. Biol.*, **4**, 559.
8. Smith, B.E. (1999) *Adv. Inorg. Chem.*, **47**, 159.
9. Christiansen, J., Dean, D.R., and Seefeldt, L.C. (2001) *Annu. Rev. Plant. Physiol. Plant. Mol. Biol.*, **52**, 269.
10. Lawson, D.M. and Smith, B.E. (2002) *Met. Ions Biol. Syst.*, **39**, 75.
11. Frazzon, J. and Dean, D.R. (2002) *Met. Ions Biol. Syst.*, **39**, 163.
12. Eady, R.R. (2003) *Coord. Chem. Rev.*, **237**, 23.
13. Rees, D.C., Tezcan, F.A., Haynes, C.A., Walton, M.Y., Andrade, S., Einsle, O., and Howard, J.B. (2005) *Philos. Trans. R. Soc. London, Ser. A*, **363**, 971.
14. Barney, B.M., Lee, H.I., Dos Santos, P.C., Hoffman, B.M., Dean, D.R., and Seefeldt, L.C. (2006) *Dalton Trans.*, **19**, 2277.
15. Howard, J.B. and Rees, D.C. (2006) *Proc. Natl. Acad. Sci. U.S.A.*, **103**, 17088.
16. Seefeldt, L.C., Hoffman, B.M., and Dean, D.R. (2009) *Annu. Rev. Biochem.*, **78**, 701.
17. Gruber, N. and Galloway, J.N. (2008) *Nature*, **45**, 293.
18. Galloway, J.N. and Cowling, E.B. (2002) *Ambio*, **31**, 64.
19. Smith, B.E. (2002) *Science*, **297**, 1654.
20. Oxtoby, D.W., Gillis, H.P., and Nachtrieb, N.H. (1999) *Principles of Modern Chemistry*, 4th edn, Saunders College Publishing, San Diego, CA, p. 243.
21. Bazhenova, T.A. and Shilov, A.E. (1995) *Coord. Chem. Rev.*, **144**, 69.
22. Gadkari, D., Mörsdorf, G., and Meyer, O. (1992) *J. Bacteriol.*, **174**, 6840.
23. Ribbe, M., Gadkari, D., and Meyer, O. (1997) *J. Biol. Chem.*, **272**, 26627.
24. Eady, R.R. (1995) *Met. Ions Biol. Syst.*, **31**, 363.
25. Rees, D.C. (2002) *Annu. Rev. Biochem.*, **71**, 221.
26. Burgess, B.K. (1990) *Chem. Rev.*, **90**, 1377.

27. Howard, J.B. and Rees, D.C. (1994) *Annu. Rev. Biochem.*, **63**, 235.
28. Thorneley, R.N.F. and Lowe, D.J. (1996) *J. Biol. Inorg. Chem.*, **1**, 576.
29. Georgiadis, M.M., Komiya, H., Chakrabarti, P., Woo, D., Kornuc, J.J., and Rees, D.C. (1992) *Science*, **257**, 1653.
30. Jang, S.B., Seefeldt, L.C., and Peters, J.W. (2000) *Biochemistry*, **48**, 14745.
31. Ryle, M.J. and Seefeldt, L.C. (1996) *Biochemistry*, **35**, 4766.
32. Seefeldt, L.C., Morgan, T.V., Dean, D.R., and Mortenson, L.E. (1992) *J. Biol. Chem.*, **267**, 6680.
33. Wolle, D., Dean, D.R., and Howard, J.B. (1992) *Science*, **258**, 992.
34. Gavini, N. and Burgess, B.K. (1992) *J. Biol. Chem.*, **267**, 21179.
35. Lindahl, P.A., Day, E.P., Kent, T.A., Orme-Johnson, W.H., and Münck, E. (1985) *J. Biol. Chem.*, **260**, 11160.
36. Hagen, W.R., Dunham, W.R., Braaksma, A., and Haaker, H. (1985) *FEBS J.*, **187**, 146.
37. Hagen, W.R., Eady, R.R., Dunham, W.R., and Haaker, H. (1985) *FEBS J.*, **189**, 250.
38. Onate, Y.A., Finnegan, M.G., Hales, B.J., and Johnson, M.K. (1993) *Biochim. Biophys. Acta*, **1164**, 113.
39. Lowe, D.J. and Smith, B.E. (1985) *Biochem. Soc. Trans.*, **13**, 579.
40. Lowe, D.J. and Thorneley, R.N.F. (1984) *Biochem. J.*, **224**, 895.
41. Watt, G.D. and McDonald, J.W. (1985) *Biochemistry*, **24**, 7226.
42. Lanzilotta, W.N., Ryle, M.J., and Seefeldt, L.C. (1995) *Biochemistry*, **34**, 10713.
43. Guo, M., Sulc, F., Ribbe, M.W., Farmer, P.J., and Burgess, B.K. (2002) *J. Am. Chem. Soc.*, **124**, 12100.
44. Nyborg, A.C., Johnson, J.L., Gunn, A., and Watt, G.D. (2000) *J. Biol. Chem.*, **275**, 39307.
45. Schindelin, H., Kisker, C., Schlessman, J.L., Howard, J.B., and Rees, D.C. (1997) *Nature*, **387**, 370.
46. Kim, J. and Rees, D.C. (1992) *Science*, **257**, 1677.
47. Kim, J. and Rees, D.C. (1992) *Nature*, **360**, 553.
48. Chan, M.K., Kim, J., and Rees, D.C. (1993) *Science*, **260**, 792.
49. Peters, J.W., Stowell, M.H., Soltis, S.M., Finnegan, M.G., Johnson, M.K., and Rees, D.C. (1997) *Biochemistry*, **36**, 1181.
50. Sorlie, M., Christiansen, J., Lemon, B.J., Peters, J.W., Dean, D.R., and Hales, B.J. (2001) *Biochemistry*, **40**, 1540.
51. Einsle, O., Tezcan, F.A., Andrade, S.L.A., Schmid, B., Yoshida, M., Howard, J.B., and Rees, D.C. (2002) *Science*, **297**, 1696.
52. Schmid, B., Ribbe, M.W., Einsle, O., Yoshida, M., Thomas, L.M., Dean, D.R., Rees, D.C., and Burgess, B.K. (2002) *Science*, **296**, 352.
53. Surerus, K.K., Hendrich, M.P., Christie, P.D., Rottgardt, D., Orme-Johnson, W.H., and Münck, E. (1992) *J. Am. Chem. Soc.*, **114**, 8579.
54. Zimmermann, R., Orme-Johnson, W.H., Münck, E., Shah, V.K., Brill, W.I., Henzl, M.T., and Rawlings, J. (1978) *Biochim. Biophys. Acta*, **537**, 185.
55. Watt, G.D., Burns, A., Lough, S., and Tennent, D.L. (1980) *Biochemistry*, **19**, 4926.
56. Marcus, R.A. and Sutin, N. (1985) *Biochim. Biophys. Acta*, **811**, 265.
57. Pierik, A.J., Wassink, H., Haaker, H., and Hagen, W.R. (1993) *Eur. J. Biochem.*, **212**, 51.
58. Tittsworth, C. and Hales, B.J. (1993) *J. Am. Chem. Soc.*, **115**, 9763.
59. Lancaster, K.M., Roemelt, M., Ettenhuber, P., Hu, Y., Ribbe, M.W., Neese, F., Bergmann, U., and DeBeer, S. (2011) *Science*, **334**, 974.
60. Spatzal, T., Aksoyoglu, M., Zhang, L., Andrade, S.L., Schleicher, E., Weber, S., Rees, D.C., and Einsle, O. (2011) *Science*, **334**, 940.
61. Thorneley, R.N.F. and Lowe, D.J. (1985) in *Molybdenum Enzymes* (ed. T.G. Spiro), John Wiley & Sons, Inc., New York, p. 221.
62. Rees, D.C. and Howard, J.B. (1983) *J. Biol. Chem.*, **258**, 12733.
63. Walker, G.A. and Mortenson, L.E. (1974) *Biochemistry*, **13**, 2382.
64. Ljones, T. and Burris, R.H. (1978) *Biochemistry*, **17**, 1866.

65. Deits, T.L. and Howard, J.B. (1989) *J. Biol. Chem.*, **264**, 6619.
66. Yates, M.G. (1992) in *Biological Nitrogen Fixation* (eds G. Stacey, R.H. Burris, and H.J. Evan), Chapman & Hall, New York, p. 685.
67. Schrock, R.R. (2005) *Chem. Res.*, **38**, 955.
68. Thorneley, R.N.F., Eady, R.R., and Lowe, D.J. (1978) *Nature*, **272**, 557.
69. Barney, B.M., Laryukhin, M., Igarashi, R.Y., Lee, H.I., Dos-Santos, P.C., Yang, T.C., Hoffman, B.M., Dean, S.R., and Seefeldt, L.C. (2005) *Biochemistry*, **44**, 8030.
70. Lukoyanov, D., Dikanov, S.A., Yang, Z.Y., Barney, B.M., Samoilova, R.I., Narasimhulu, K.V., Dean, D.R., Seefeldt, L.C., and Hoffman, B.M. (2011) *J. Am. Chem. Soc.*, **133**, 11655.
71. Barney, B.M., Lukoyanov, D., Igarashi, R.Y., Laryukhin, M., Yang, T.C., Dean, D.R., Hoffman, B.M., and Seefeldt, L.C. (2009) *Biochemistry*, **48**, 9094.
72. Lukoyanov, D., Yang, Z.Y., Barney, B.M., Dean, D.R., Seefeldt, L.C., and Hoffman, B.M. (2012) *Proc. Natl. Acad. Sci. U.S.A.*, **109**, 5583.
73. Hoffman, B.M., Dean, D.R., and Seefeldt, L.C. (2009) *Acc. Chem. Res.*, **42**, 609.
74. Lukoyanov, D., Barney, B.M., Dean, D.R., Seefeldt, L.C., and Hoffman, B.M. (2007) *Proc. Natl. Acad. Sci. U.S.A.*, **104**, 1455.
75. Barney, B.M., Lukoyanov, D., Yang, T.C., Dean, D.R., Hoffman, B.M., and Seefeldt, L.C. (2006) *Proc. Natl. Acad. Sci. U.S.A.*, **103**, 17113.
76. Rivera-Ortiz, J.M. and Burris, R.H. (1975) *J. Bacteriol.*, **123**, 537.
77. Burgess, B.K. (1993) in *Molybdenum Enzymes, Cofactors and Model Systems* (eds E.I. Stiefel, D. Coucouvanis, and W.E. Newton), American Chemical Society, Washington, DC, p. 144.
78. Hu, Y., Lee, C.C., and Ribbe, M.W. (2012) *Dalton Trans.*, **41**, 1118.
79. Cameron, L.M. and Hales, B.J. (1995) in *Nitrogen Fixation: Fundamental and Applications* (eds I.A. Tikhonovich, N.A. Provorov, V.I. Romanov, and W.E. Newton), Kluwer Academic Publishers, p. 109.
80. Smith, B.E., Bishop, P.E., Dixon, R.A., Eady, R.R., Filler, W.A., Lowe, D.J., Richards, A.J.M., Thomson, A.J., Thorneley, R.N.F., and Postgate, J.R. (1985) in *Nitrogen Fixation Research Process* (eds H.J. Evans, P.J. Bottomley, and W.E. Newton), Martinus Nijhoff, Dordrecht, p. 597.
81. Conradson, S.D., Burgess, B.K., Vaughn, S.A., Roe, A.L., Hedman, B., Hodgson, K.O., and Holm, R.H. (1989) *J. Biol. Chem.*, **264**, 15967.
82. Lee, H.I., Igarashi, R.Y., Laryukhin, M., Doan, P.E., Dos Santos, P.C., Dean, D.R., Seefeldt, L.C., and Hoffman, B.M. (2004) *J. Am. Chem. Soc.*, **126**, 9563.
83. Dance, I. (2004) *J. Am. Chem. Soc.*, **126**, 11852.
84. Christie, P.D., Lee, H.I., Cameron, L.M., Hales, B.J., Orme-Johnson, W.H., and Hoffman, B.M. (1996) *J. Am. Chem. Soc.*, **118**, 8707.
85. Lee, H.I., Cameron, L.M., Hales, B.J., and Hoffman, B.M. (1997) *J. Am. Chem. Soc.*, **119**, 10121.
86. George, S.J., Ashby, G.A., Wharton, C.W., and Thorneley, R.N.F. (1997) *J. Am. Chem. Soc.*, **119**, 6450.
87. Lee, C.C., Hu, Y., and Ribbe, M.W. (2012) *Angew. Chem. Int. Ed.*, **51**, 1947.
88. Hales, B.J. (1990) *Adv. Inorg. Biochem.*, **8**, 165.
89. Eady, R.R., Richardson, T.H., Miller, R.W., Hawkins, M., and Lowe, D.J. (1988) *Biochem. J.*, **256**, 189.
90. Hales, B.J., Langosch, D.J., and Case, E.E. (1986) *J. Biol. Chem.*, **261**, 15301.
91. Bishop, P.E. and Premakumar, R. (1992) in *Biological Nitrogen Fixation* (eds G. Stacey, R.H. Burris, and H.J. Evans), Chapman & Hall, New York, p. 736.
92. Blank, M.A., Lee, C.C., Hu, Y., Hodgson, K.O., Hedman, B., and Ribbe, M.W. (2011) *Inorg. Chem.*, **50**, 7123.
93. Bergström, J., Eady, R.R., and Thorneley, R.N.F. (1988) *Biochem J.*, **251**, 165.
94. Lee, C.C., Hu, Y., and Ribbe, M.W. (2009) *Proc. Natl. Acad. Sci. U.S.A.*, **106**, 9209.
95. Blanchard, C.Z. and Hales, B.J. (1996) *Biochemistry*, **35**, 472.
96. Hu, Y., Corbett, M.C., Fay, A.W., Webber, J.A., Hedman, B., Hodgson,

K.O., and Ribbe, M.W. (2005) *Proc. Natl. Acad. Sci. U.S.A.*, **102**, 13825.

97. Ravi, N., Moore, V., Lloyd, S.G., Hales, B.J., and Huynh, B.H. (1994) *J. Biol. Chem.*, **269**, 20920.
98. Chen, J., Christiansen, J., Tittsworth, R.C., Hales, B.J., George, G.N., Coucouvanis, D., and Cramer, S.P. (1993) *J. Am. Chem. Soc.*, **115**, 5509.
99. Harvey, I., Arber, J.M., Eady, R.R., Smith, B.E., Garner, C.D., and Hasnain, S.S. (1990) *Biochem. J.*, **266**, 929.
100. Arber, J.M., Dobson, B.R., Eady, R.R., Stevens, P., Hasnain, S.S., Garner, C.D., and Smith, B.E. (1987) *Nature*, **325**, 372.
101. George, G.N., Coyle, C.L., Hales, B.J., and Cramer, S.P. (1988) *J. Am. Chem. Soc.*, **110**, 4057.
102. Fay, A.W., Blank, M.A., Lee, C.C., Hu, Y., Hodgson, K.O., Hedman, B., and Ribbe, M.W. (2010) *J. Am. Chem. Soc.*, **132**, 12612.
103. Hales, B.J., Case, E.E., Morningstar, J.E., Dzeda, M.F., and Mauterer, L.A. (1986) *Biochemistry*, **25**, 7251.
104. Eady, R.R., Robson, R.L., Richardson, T.H., Miller, R.W., and Hawkins, M. (1987) *Biochem. J.*, **244**, 197.
105. Bishop, P.E., Hawkins, M.E., and Eady, R.R. (1986) *Biochem. J.*, **238**, 437.
106. Dilworth, M.J., Eady, R.R., and Eldridge, M.E. (1988) *Biochem. J.*, **249**, 745.
107. Lee, C.C., Hu, Y., and Ribbe, M.W. (2010) *Science*, **329**, 642.
108. Hu, Y., Lee, C.C., and Ribbe, M.W. (2011) *Science*, **333**, 753.
109. Lee, C.C., Hu, Y., and Ribbe, M.W. (2011) *Angew. Chem. Int. Ed.*, **50**, 5545.
110. Khodadadi, A.A., Hudgins, R.R., and Silveston, P.L. (1996) *Can. J. Chem. Eng.*, **74**, 695.

9
Model Complexes of the Active Site of Nitrogenases: Recent Advances

Frédéric Barrière

9.1
Introduction

The activation of dinitrogen stands arguably among the most challenging endeavors facing chemists for its scientific and societal implications. Nitrogen is indeed both highly stable in its molecular form and essential for life as an element in biological molecules, such as amino- or nucleic acids, for example. The availability of nitrogen for sustaining life is made possible naturally by some prokaryotic organisms and their nitrogenases enzymes that "fix" dinitrogen from the air into the usable form of ammonium (NH_4^+). Well before the bases of biological nitrogen fixation where known, ancient empirical agricultural practices such as crops rotation developed that maintained the soil fixable nitrogen content. It is however only recently, with the birth of chemistry as a science, that it was gradually discovered that nitrogen from the air is fixed (i.e., reduced), notably in symbiosis with leguminous plants, and that this is catalyzed by microbes and their nitrogenase enzyme [1]. In recent years, protein crystallographers have successively refined at the molecular level the snapshot of the most studied of known nitrogenases and that of its active site, namely, the iron–molybdenum cofactor (FeMo-co) [2]. The active site of this enzyme is an iron–molybdenum–sulfur cluster centered on a carbide core (Fe_7MoS_9C) and linked to the protein through an apical iron-cysteinate tether and a histidine-imidazole donating bond to molybdenum at the other end. The coordination sphere of the molybdenum center is completed in a pseudooctahedral environment by the bidendate and chiral (*R*)-homocitrate *exo*-ligand (Figure 9.1).

The availability of detailed crystallographic models of the molybdenum nitrogenase has, however, not led so far to obvious clues regarding the detailed mechanism of dinitrogen activation at FeMo-co. In fact, we are still pretty much in the dark in this respect. In this context, model synthetic chemists have claimed that they could help solve this issue. First, before the advent of nitrogenase protein crystallography, comparing the spectroscopic properties of synthetic metal–sulfur clusters with those of the enzyme aimed at defining the structure of the active site. This is now apparently obsolete because we do have

Bioinspired Catalysis: Metal-Sulfur Complexes, First Edition.
Edited by Wolfgang Weigand and Philippe Schollhammer.

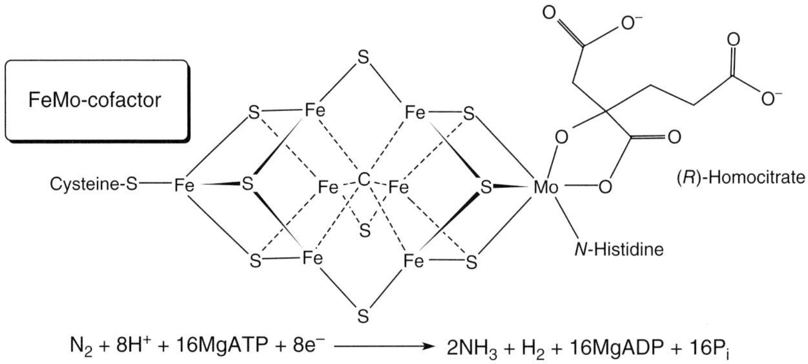

Figure 9.1 Structure and function of the FeMo-cofactor of nitrogenase [2].

a high-resolution crystallographic model of the enzyme. Besides, the artificial synthesis of a resembling structural model has not yet succeeded so far. The other truths held as self-evident by model chemistry were that establishing artificial nitrogen fixation pathways by synthetic complexes would provide precedents for mechanisms of possible biological relevance and would also provide candidates for competitive artificial catalysts. Indeed, our century-old Haber–Bosch industrial process for turning dinitrogen and dihydrogen into ammonia requires a lot of energy (high temperature, high pressure, pure dihydrogen generated from fossil fuel and an iron-based heterogeneous catalyst) [3]. In contrast, the biological conversion of dinitrogen to ammonia (with the release of 1 equiv of dihydrogen in the optimal conditions) has the advantage of operating at ambient conditions. For 1 mol of dinitrogen biologically fixed at optimal conditions, however, 16 mol of ATP have to be hydrolyzed [4], which makes the contemporary and refined Haber–Bosch process more favorable energetically [5]. Hence, modeling nitrogenase is likely to put off the most willing chemist for the following reasons: the active site of the enzyme has not been synthesized so far; extracting the FeMo-cofactor from the enzyme is possible, but it does not activate N_2 outside of the protein; activating dinitrogen is not that common and is difficult; it is rarely catalytic, and if it is, it does not last for long. Since the first report of a metal complex with a dinitrogen ligand [6] (actually derived from hydrazine), a lot of progress has been made and has been regularly reviewed [7–17], including by this author [15–17]. Hence, in this short account, we focus on a selection of the most recent and/or most relevant advances in nitrogenase model chemistry and show that there are reasons for hope and that some of its proclaimed goals might well be met in the future. This chapter covers first the progress in the structural modeling of the metal–sulfur clusters of nitrogenases. Then, the catalytic activation of dinitrogen at a single molybdenum (and tungsten) site is discussed, followed by a related section involving iron complexes. The issue of biological obligatory hydrogen evolution and the specificity of the homocitrate ligand is then addressed. This leads to the question and role of sulfur at the active site of nitrogenase and in model complexes, together with functional metal–sulfur cluster models with direct metal–metal bonds or interactions. Finally, the role of the protein scaffold is addressed and the analogy with heterogeneous catalysis raised.

9.2 Structural Models of Metal–Sulfur Clusters in the Nitrogenases

Of the three metal–sulfur clusters of the dinitrogenase complex, only the Fe_4S_4 cluster of dinitrogenase reductase (the iron protein) has been modeled for some time now [13, 17–19]. Many Fe_4S_4 clusters with usually thiolate terminal ligands at the iron centers have indeed been synthesized and studied in four redox states, including the all-ferrous state [18, 19]. In addition to the more common $[Fe_4S_4]^{1+/2+}$ states, the all-ferrous state $[Fe_4S_4]^0$ may be stabilized by the iron protein scaffold [20]. The relevance of this all-ferrous state to the biological mechanism is still being investigated as the involvement of the $[Fe_4S_4]^{0/2+}$ couple would be a mean for delivering two electrons to the molybdenum–iron protein [21]. There are several examples of synthetic iron–sulfur clusters with topology akin to the P-cluster Fe_8S_7 core structures. Recent synthetic methodologies involve either self-assembly of iron(II)-bis(amide), tetramethylthiourea, elemental sulfur, and a sterically encumbered benzenethiol [22], or the reductive coupling and monodesulfurization of two all-ferric Fe_4S_4 clusters in the presence of trialkyl phosphine [23], the latter mechanism being proposed to be of relevance to the biological maturation process of the P-cluster. Apart from the mere synthetic challenge, there is still a need for an artificial model of the active site of nitrogenase that would then be available in sufficiently large amounts and pure form as to permit reproducible studies in various media and with different spectroscopic methods. The fairly recent discovery (2002) of the presence of a light atom at the center of the FeMo-co [24] somehow reset the clock for the quest for a synthetic analog of the biological catalytic active site. The likelihood of an interstitial atom was however hinted upon (μ^6-oxide was taken as an example) in the first theoretical account on this topic [25] published soon after the first crystallographic model of FeMo-co was available (1992) [26]. There has been some debate about the identity of this light atom, initially proposed to be C, N, or O, and it now seems that there is a consensus for it being a carbon atom, or, more precisely, a carbide [2, 27, 28]. The μ^6-carbide was also detected by X-ray emission spectroscopy at the center of the Fe_8S_9 "L cluster," the isostructural precursor in the biosynthesis of FeMo-co, before apical iron substitution for molybdenum and coordination of (*R*)-homocitrate [29]. To date, there are no FeMo-co synthetic models available with a central carbide. Other light atoms (such as sulfur and oxygen) have nevertheless been introduced by self-assembly of lower nuclearity Fe–S clusters, iron-thiolate, iron-amide, or organo-iron complexes in the presence of elemental sulfur (and water for the oxide example) (Figure 9.2) [30, 31]. These clusters, however, display a distorted structure around the pseudo μ^6-sulfide or μ^4-oxide core, with longer and different bond lengths

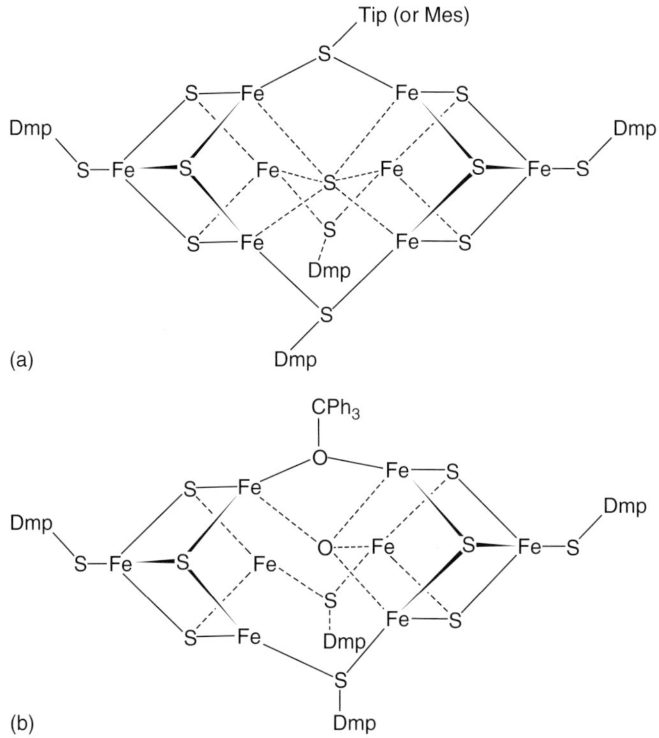

Figure 9.2 Structure of recent structural models of the FeMo-cofactor cluster core displaying either an interstitial μ^6-S atom (a) Ref. [30] within a Fe_8S_9 scaffold (note that three of the sulfur atoms in the scaffold are μ^2-thiols bridging the subcubes as opposed to the μ^2-sulfides found in FeMo-co, cf. Figure 9.1) or a μ^4-O center (b) Ref. [31]. Tip: 2,4,6-$^i Pr_3C_6H_2$; Mes: mesityl (2,4,6-Me_3Ph); Dmp: 2,6-dimesitylphenyl; and Ph: phenyl.

between the light atom core and the surrounding six iron atoms prism. This translates into longer inter-subcubes Fe–Fe distances (about 3.6 and 2.9 Å for the sulfide examples) compared with those found in FeMo-co (about 2.6 Å). Other obvious differences are the presence in the models of thiolate rather than sulfide μ^2-bridges between the two subcubes and the absence of an apical pseudooctahedral heterometal (i.e., all iron clusters and no molybdenum or vanadium). Together with the introduction of an interstitial carbide, these are the main challenges ahead for producing a closer structural model of FeMo-co. The fact that the oxide core (Figure 9.2b) is only linked to four of the six central iron atoms while the two others interact with mesityl rings from the ligands may hint at a possible flexibility of the light atom during biological catalysis and substrate binding. Despite the progress toward the faithful synthetic structural model of FeMo-co or of its subunits [32–34], no interaction with dinitrogen has ever been found with these artificial metal–sulfur clusters (this is nevertheless consistent with the fact that extracted FeMo-co has not been shown to interact with N_2

either). Dinitrogen coordination and activation are however well documented with the nitrogenase metals (molybdenum, iron, and, to a lesser extent, vanadium) as discussed in the following sections.

9.3 Functional Modeling at a Single Molybdenum Center

Functional modeling of nitrogenase can take various forms as the reactions catalyzed by the enzyme are many and their mechanisms complicated. Strictly speaking, however, we can restrict our demands to some of the following criteria: interaction with N_2 followed by stoichiometric or catalytic activation of the substrate up to ammonia release. Historically, molybdenum has been studied a lot in this context partly because it was recognized very early on that this metal was essential for nitrogen fixation [35]. The other reason is that models based on molybdenum (and tungsten) proved over the years and until recently to be significantly more efficient than those based on iron [15]. Vanadium-based functional models have also been developed owing to the existence of an alternative, but less efficient, nitrogenase where Mo is substituted by V [36]. As far as molybdenum is concerned, a large body of work based on Mo or W tetraphosphine complexes has lent support to the hypothesis that sequential reduction and protonation of dinitrogen at these metals could lead to ammonia synthesis in a catalytic manner with an artificial catalyst or in the biological system [37]. Most of the likely intermediates of the electronation/protonation sequence (namely, dinitrogen, diazenido, hydrazido, hydrazinium, nitride, imido, amido, and ammine) have been isolated or synthesized at tetraphosphino pseudooctahedral metal centers and this is now known as the *Chatt cycle*. There are examples where ammonia can be synthesized stoichiometrically as, for example, by protolysis of *cis*-$[W(N_2)_2(PMe_2Ph)_4]$ [38]. Protonation of *trans*-$[W(N_2)_2(dppe)_2]$ (dppe, bis-diphenylphosphinoethane) in tetrahydrofuran and the subsequent cathodic reduction of the hydrazido product at −2.7 V versus ferrocene permit the electrosynthesis of ammonia with part retention of the $W(dppe)_2$ moiety that hence allows the recovery of some of the bis-dinitrogen complex and the repetition of the catalytic cycle upon addition of *para*-toluenesulfonic acid ($dppe = Ph_2PC_2H_4PPh_2$) [39]. In these metal complexes, the phosphine coligands are obviously abiological and the metal oxidation states (zero in the dinitrogen complexes) too low to be compatible with the biological system. This precedent, however, initially strengthened the view that biocatalysis in the natural system could directly involve molybdenum, or, in the alternative and less efficient nitrogenases, vanadium or iron at this apical site. Later, related systems operating at higher oxidation state were developed on the basis of the piano-stool pentamethylcyclopentadienyl Mo or W trimethyl complexes. Stoichiometric ammonia synthesis was shown with a dimer bridged by a dinitrogen ligand but the isolation of mononuclear compounds with hydrazido, hydrazine, imido, amido, and ammonia was achieved [40]. Hence, a related Chatt-type dinitrogen reduction cycle can be envisioned at such a center with

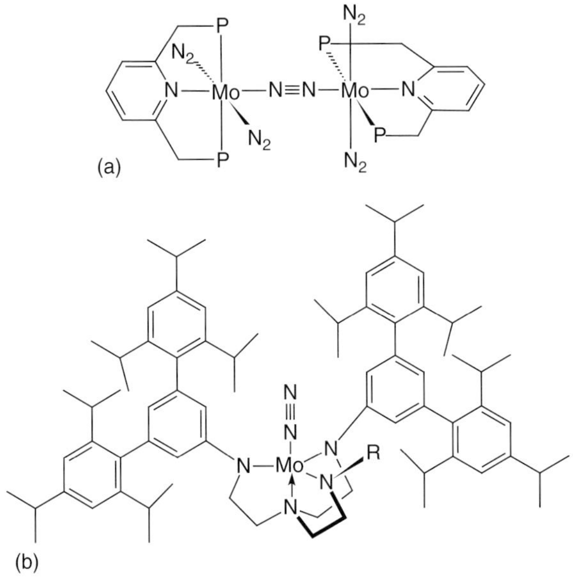

Figure 9.3 The two examples of molybdenum homogeneous catalysts for the reduction of dinitrogen to ammonia in the presence of a proton source and a chemical reductant. (a) The dimer pre-catalyst with PNP pincer ligands (P=P-$^{t}Bu_2$, [41]). (b) The triamido amine dinitrogen molybdenum complex [42]. The front R group is omitted for clarity and is the same as the other bulky substituent borne by the two other amido groups.

the following nuance: reduced dinitrogen intermediates may rearrange from an η^1 to an η^2 coordination mode that would short cut the thermodynamic cul-de-sac nitride level. Recently, the first fully homogeneous catalytic system at a molybdenum center has been reported with a bulky triamido amine ligand (Figure 9.3b) [42]. The molybdenum dinitrogen complex is formally at the +III oxidation state but nevertheless very electron rich. The bulkiness of the ligand avoids the formation of a less reactive linear dimer with dinitrogen as a bridging ligand. Protonation of the amido nitrogen of the triamido amine ligand is proposed to play a role in catalysis as proton relay. Catalysis was carried out in heptane in the absence of oxygen, using a proton source of little solubility in this solvent ([2,6-dimethylpyridinium][$B(3,5\text{-}(CF_3)_2C_6H_3)_4$]) and decamethylchromocene as the electron source ($E_{1/2} = -1.35$ V vs Ag/Ag^+ [$AgNO_3$] = 0.1 M in acetonitrile). The low proton concentration and stepwise addition of the reductant were critical to avoid significant dihydrogen evolution, a competitive reaction possibly also harmful to the catalyst itself [43–45]. Similar reaction conditions were utilized in the more recent catalytic system, also based on molybdenum but relying on a different ligand set [41, 46]. This new catalytic reaction was also carried out in a solvent of low polarity (toluene) with cobaltocene as a slightly milder reductant ($E_{1/2} = -1.15$ V vs Ag/Ag^+ [$AgNO_3$] = 0.1 M in acetonitrile) and [2,6-dimethylpyridinium][CF_3O_2SO] as the proton source. Here, the triflate counter anion ($CF_3O_2SO^-$) of the proton source (2,6-lutidinium) is critical for efficient catalysis compared to other salts of either strongly coordinating anion such as chloride or poorly coordinating anion like $[B(3,5\text{-}(CF_3)_2C_6H_3)_4]^-$. Triflate seems to play a well-balanced role in binding and protecting the molybdenum catalytic coordination site while remaining labile as not to inhibit catalysis. In this second catalytic example, dihydrogen evolution was also observed and seems unavoidable at a reduced metal site in the presence of protons. Likewise, all nitrogenases also produce a lot of dihydrogen in nonoptimal turnover conditions. The pre-catalyst for this recent system is shown in Figure 9.3a. The ligand is a PNP pincer ligand chelating a molybdenum center, whose coordination sphere is completed by two end-on dinitrogen ligands and a third dinitrogen bridging a second identical metal site. The actual catalyst is thought to be mononuclear, arising from a dissymmetric dissociation of the dimer pre-catalyst. Hence, molybdenum dinitrogen coordination and activation chemistry has culminated in catalytic ammonia synthesis processes with different ligand sets. For both of these catalytic systems, some likely intermediates in the sequential reduction and protonation of N_2 coordinated end-on at a single metal center have been characterized and related, nevertheless not identical, mechanisms proposed. The value of these catalysts is of a fundamental nature rather than for competing with the current industrial ammonia synthesis, not to mention nitrogenase itself. The reaction conditions mentioned are not very practical and, most of all, the well-crafted molecular catalysts die out after a few cycles. Regarding nitrogenase modeling and despite the biologically irrelevant ligands, the absence of both sulfur– and metal–metal interactions, these recent ammonia synthesis catalysts, together with the large and more ancient molybdenum dinitrogen chemistry, reinforce the hypothesis that Mo could play a direct role in dinitrogen activation by the most efficient of all nitrogenases. In the alternative nitrogenases, initial activation could involve the same apical site occupied by vanadium [36] or iron. Iron atoms, however, are also present at the active site in all nitrogenases, and recent works have shown that the so far less developed iron–dinitrogen chemistry might be closing in, as discussed next.

9.4
Functional Modeling at a Single Iron Center

As discussed in the preceding section, molybdenum dinitrogen chemistry developed and strengthened to the point that demonstration of catalytic ammonia synthesis was provided in organic solution with defined homogeneous molecular metal catalysts, protons, and a cathode or chemical reductants as the electron source. The view, however, that molybdenum could be the crucial site for dinitrogen activation at FeMo-co gradually lost momentum. This is because simultaneously, an increasing amount of biochemical data has been published that points to the involvement of a specific Fe_4 face of FeMo-co for substrate binding and activation. A draft mechanism based on these experiments was recently refined [4]. Particularly, it was put forward that the identified Fe_4 face and its specific surrounding amino acid residues (α70-valine acting as a substrate gatekeeper and α195-histidine as proton delivery relay) could be the site for N_2 coordination and activation. In addition, and as this hypothesis was continuously supported by enzymology studies, the Fe_4 site has concentrated the attention and efforts of computational chemists [47]. Concurrently, however, and until recently, little progress was made on the iron-based model chemistry [48–50]. Many iron phosphine dinitrogen complexes have been synthesized and are related to the molybdenum and tungsten ones discussed. Few examples of ammonia synthesis have been documented with these species. The iron (II) *trans*-$[Fe(dmpe)_2(N_2)H]^+$ (dmpe, bis-dimethylphosphinoethane) complex may be prepared through dihydrogen substitution by dinitrogen from the parent *trans*-$[Fe(dmpe)_2(H_2)H]^+$ complex (dmpe = $Me_2PC_2H_4PMe_2$). This reaction, in its own right, is interesting in the context of nitrogenase functional modeling as an obligatory release of one molecule of dihydrogen is necessary before 1 equiv of dinitrogen binds and gets reduced at the active site. Dihydrogen production by nitrogenase implies the reduction of protons and metal hydride as intermediates. The role of metal-hydrides in the obligatory H_2 release and N_2 binding is discussed later in a following section. Treatment of *trans*-$[Fe(dmpe)_2(N_2)H]^+$ with sulfuric acid allows the production of small amounts of ammonia [51]. The yield of ammonia can be increased (up to 20%) by *in situ* generation of the iron (0) dinitrogen complex $[Fe(dmpe)_2(N_2)]$ upon treatment of *trans*-$[Fe(dmpe)_2(N_2)H]^+$ with a strong base (*t*-BuOK), followed by reaction with hydrochloric acid [52]. Some *trans*-$[Fe(dmpe)_2Cl_2]$ complex may be recovered from which a putative catalytic cycle could be envisioned. There remains some doubt, however, about the identity of the catalytic species. A striking feature of this system is that the nature of the base, *t*-BuOK, and that of the acid, HCl, are critical for ammonia synthesis. This resonates with recent works [53] where ammonia was synthesized from dinitrogen with the cooperativity of low-valent iron centers and potassium ions, and immediately suggests that potassium and multimetallic activation could also be at work in the earlier system. In addition, a related dinitrogen complex has been structurally characterized, namely, $[Fe(depe)_2(N_2)]$ (depe (bis-diethylphosphinoethane) = $Et_2PC_2H_4PEt_2$), differing only in the nature of the diphosphine substituents (ethyl vs methyl) [54]. The ethyl derivative did not produce ammonia upon acid treatment but was prepared in a different manner that did not involve a base (nor potassium ion) but a reduction of the parent $[Fe(depe)_2Cl_2]$ complex with sodium-naphthalene under dinitrogen. In the more recent and better characterized work, a bis-μ-Cl β-diketiminate iron (II) was reduced with 1 equiv of potassium graphite per iron atom under an atmosphere of dinitrogen, yielding the bridging bis-nitride structure shown in Figure 9.4a [53]. This species yields 42% NH_3 (with respect to the nitrogen atoms) when exposed to hydrogen in toluene at room temperature and close to stoichiometric amounts of ammonia (82% with respect to the nitrogen atoms) upon treatment with HCl in tetrahydrofuran. This work has been discussed in the context of the Haber–Bosch industrial process for ammonia synthesis from N_2 and H_2 [1, 3],

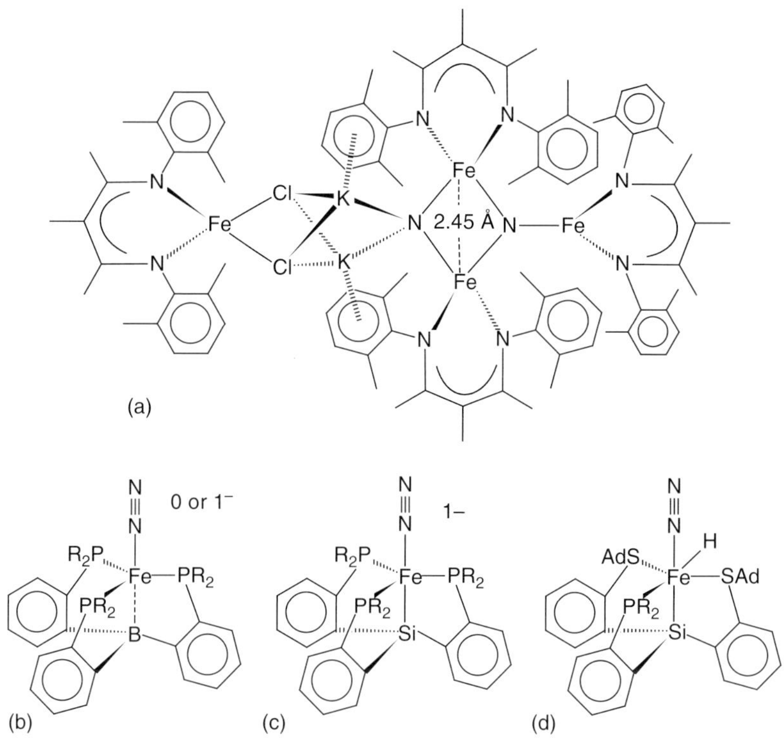

Figure 9.4 (a) The structure of the bis-nitride product obtained by reduction of a parent bis-μ-Cl β-diketiminate iron (II) with 1 equiv of potassium graphite. Treatment of this complex with dihydrogen or acid yields ammonia [53]. (b) The formally iron (0) triphosphino borane dinitrogen complex [55, 56], the first homogeneous ammonia synthesis catalyst. (c) The structurally related iron (0) triphosphino silyl dinitrogen complex [57]. (d) The iron hydride dinitrogen complex with two thiolates (SAd is *S*-adamantyl) instead of phosphines in the co-ligand [58].

and more specifically the iron-based Mittasch catalyst that also contains potassium ions as promoters [59]. With respect to nitrogenase modeling, this work represents a nice example of the involvement of multiiron centers for the activation of dinitrogen. In addition to the generation of ammonia, the isolated complex at the bis-nitride level shows a short metal–metal distance between two iron centers (2.45 Å, cf. Figure 9.4a), which is interesting cognizant of the Fe–Fe bond lengths in FeMo-co (about 2.59 Å inter-subcube and 2.69 Å intra-sub-cube). Compared to the previous related work, also based on iron β-diketiminate for dinitrogen activation [60], N_2-splitting to nitrides and subsequent ammonia synthesis was made possible here by the slight release of the steric strain borne out by the ancillary ligands. For example, when the phenyl substituent of the nitrogen atom in the β-diketiminate moiety bears isopropyl groups instead of methyl (cf. Figure 9.4a), the bridging dinitrogen ligand is only elongated in a manner akin to that previously reported [60]. Another class of ligands that has recently boosted low-valent iron dinitrogen chemistry [61] is topologically related to that of the molybdenum triamido amine ligand [42] (cf. Figure 9.3b) in that they usually yield a bipyramid trigonal environment around the metal. The chemistry at these single iron centers also aims at developing a parallel Chatt-type cycle previously established at molybdenum (cf. Section 9.3) [61]. The isolation of the key end-on dinitrogen species was possible with tris(phosphino)borane [55], tris(phophino) silyl [57], and even mixed phosphino-thiolate silyl ligands [58], as shown in Figure 9.4. In addition, the end-on dinitrogen ligand is activated and can be functionalized as demonstrated with the tris(phosphino)borane [55, 62, 63] and tris(phophino) silyl [64–66] complexes. Moreover, these trigonal bipyramid iron platforms can accommodate reactions of relevance to a putative N_2 reduction catalytic cycle, for example, the redox-induced release of NH_3 from an ammine complex with generation of the N_2 species [64, 65], or the coordination of N_2H_4, NH_3, or NH_2 at iron and the release of ammonia from protonation/reduction of the amido derivative, also yielding the dinitrogen complex as a product [63]. All of these complexes are reminiscent of some of the intermediates of the Chatt cycle. The demonstration that the coordinated N_2 is activated and can be functionalized, and the reduction/protonation reactions releasing ammonia with the recovery of the end-on dinitrogen ligand suggest that a catalytic ammonia synthesis cycle might be possible at iron with this class of ligands. This has indeed been recently demonstrated with the iron tris(phosphino)borane complex (Figure 9.4b) [56]. The reading of the article reporting this breakthrough is interesting and thought provoking. It is indeed the first demonstration of catalysis at iron, and, as such, from the model chemistry perspective, it balances the debate with examples now available both at molybdenum [42, 41] and iron mononuclear complexes [56]. It is striking that all of these catalysts rely on coligands that are strongly electron donating and therefore yield electron-rich metal centers. One other common trait of these catalysts is unfortunately their inefficiency, as they decompose only after a few cycles. Finally, the catalytic reaction conditions always require a low-polarity solvent, a rather strong chemical reductant, and a source of protons of the lutidinium type. Returning to the first homogeneous iron catalyst for ammonia synthesis [56], other original aspects are the low temperature (−78 °C) and the possible role attributed to the boron atom that could be of relevance to the biological system. In addition to the iron tris(phosphino)borane complex, other species have been considered in the same reaction conditions including the silyl analog. No other iron complex showed any catalytic activity. This suggests that the flexibility of the *trans*-boron linkage is critical and might favor the coordination and activation of the dinitrogeneous and nitrogeneous species with different electronic requirements at this conserved metal center. Moreover, the trigonal geometry of this model complex is akin to that found at the central iron atoms in FeMo-co. That the central carbon atom in FeMo-co atom could also play a role similar to that proposed for the boron atom in the model complex is an interesting and original suggestion that should fuel further studies and debates. Compared with the two related catalyses at the single molybdenum centers that use electron donors such as decamethyl chromocene (∼−0.9 V vs normal hydrogen electrode (NHE)), the iron system works only with the much stronger reductant KC_8. Reducing dinitrogen is difficult for many reasons and in particular because it requires the delivery of a sequence of protons and electrons to the substrate and its partially reduced intermediates bound to a metal center. This potentially gives rise to a number of denaturation pathways for the catalyst, including protonation at the ligand or at the metal, generation of dihydrogen (waste of electron and protons) by direct reaction of the protons and the reductant or catalyzed by the metal center, and possible subsequent reaction of dihydrogen with the catalyst. It is known that reaction of H_2 with the iron tris(phosphino)borane dinitrogen complex yields the catalytically inefficient dihydride derivative, one of which hydride bridges the iron center and the boron atom [67]. In this iron-based system, one may also speculate on the possible involvement in catalysis of the potassium ion, as discussed, with other iron complexes [51, 52]. It has to be kept in mind, however, that many other reductants including metallic potassium were tried and found inefficient in the present catalytic conditions. Hence, it could be that the apparent specificity of KC_8 as the reductant may not stem from its very low potential (very strong reductant) but rather from the possible H_2 adsorption properties of alkali metal graphite intercalation compounds that may delay for a while the inevitable catalyst denaturation by dihydrogen [68].

In 2014, we therefore have nearly equivalent homogeneous catalysts for ammonia synthesis from dinitrogen relying on the most common metals found in the nitrogenases, molybdenum and iron. The originality of the iron system is that it implies the flexibility of the ligand *trans* to dinitrogen (namely, boron) keeping in mind that a related flexibility might also occur to some extent in the molybdenum-based catalyses. From the catalysis point of view and despite the significant breakthrough they represent, these systems can be characterized as equally inefficient. From the nitrogenase modeling point of view, they also suffer from obvious drawbacks that are discussed next and can be summarized with the following items: the dihydrogen and (*R*)-homocitrate issues, the sulfur environment, and the metal–metal interactions.

9.5
The Hydrogen and Homocitrate Issues in Nitrogenase Model Chemistry

It is well known that nitrogenase evolves dihydrogen in the absence of dinitrogen (or low electron flux) and resumes ammonia synthesis when the substrate is present. In addition, and in the best-case scenario, the release of 1 mol of dihydrogen is coupled to the reduction of 1 mol of dinitrogen to ammonia. This is called the *obligatory hydrogen evolution* and is thought to be coupled to the binding of dinitrogen at FeMo-co. In addition, the chiral alkoxy-polycarboxylate (*R*)-homocitrate *exo*-ligand is also required for efficient catalysis. As a matter of fact, a mutant with the closely related citrate ligand at Mo (one less carbon in a side chain) yields a poorly performing enzyme [69]. In this context, there is apparently only one convincing model chemistry study involving a nitrogenase metal and addressing both of these issues. The study is based on a molybdenum Chatt-type complex. [15, 16, 37, 70] The dihydride $[MH_2(\eta^2\text{-}O_2CMe)(dppe)_2]^+$ (M = Mo or W) complex can be reduced quasi reversibly, consistent with the carboxylate ligand switching between the η^2 and η^1 coordination modes, respectively. The tungsten η^1 complex can be further reduced to the anion and trapped by protonation of the metal, yielding a trihydride derivative with a retained η^1-carboxylate ligand as established crystallographically. The analogous trihydride molybdenum complex is thermally unstable and undergoes dihydrogen loss, yielding the monohydride complex with an η^2-coordinated carboxylate ligand. In the presence of excess acid and under reduction, both dihydride complexes $[MH_2(\eta^2\text{-}O_2CMe)(dppe)_2]^+$ (M = Mo or W) partake in a dihydrogen evolution cycle. At low concentration of protons, the η^1-O_2CMe ligand dissociates from $[MoH_2(\eta^1\text{-}O_2CMe)(dppe)_2]^+$ and exposes a site at which dinitrogen (or dihydrogen) can bind. The dinitrogen complex $[MoH_2(N_2)(dppe)_2]$ is thermally unstable and loses H_2 to yield the known *trans*-$[Mo(N_2)_2(dppe)_2]$ species. This reactivity is not observed for the tungsten analog, where the metal–oxygen bond is stronger. The behavior of these complexes, and in particular that of the Mo complex summarized in Figure 9.5, has been integrated in the first steps of the Lowe and Thorneley cycle for nitrogenase function. It suggests that a carboxylate ligand (homocitrate in nitrogenase) can act as a leaving group at Mo, thereby providing a coordination site for N_2. This model also implies that hydrogen release from a metallic hydride is not necessary for N_2 binding and that obligatory H_2 evolution could occur after substrate binding. Finally, in the absence of the substrate and in the presence of protons, the molybdenum center would also sustain a hydrogen evolution cycle without carboxylate decoordination. Of relevance to this study, it is worth mentioning that efforts to produce a biological FeW-cofactor have not demonstrated that it could effectively reduce dinitrogen, although hydrogen evolution was still observed [71]. Further refinement of the model should involve synthetic metal–sulfur clusters and the homocitrate ligand itself. While the former is difficult and is discussed in a different section, the latter is not straightforward either as polycarboxylates can easily bridge metal centers. Recently, an example of a bis-homocitrate dioxo molybdenum complex has been synthesized and characterized [72]. The coordination mode through the α-alkoxy and α-carboxy group is similar to that found in nitrogenase. In addition, the homocitrate ligand was found to be labile as deduced from ^1H and ^{13}C NMR studies in D_2O. As far as metal–sulfur clusters are concerned, an alternative to synthetic models is the study of biological cofactors extracted from the protein. Kinetic studies of different substrate substitution by thiophenolate at the wild-type FeMo-co or at the mutant strain containing citrate instead of homocitrate (NifV$^-$ from *Klebsiella pneumoniae*) showed significant difference when imidazole was present as a ligand at molybdenum [73]. On the basis of molecular mechanics modeling of the protein structure, this has been interpreted as hydrogen bonding between the homocitrate ligand and imidazole. This weak interaction would not be possible with the shorter citrate. Decoordination of the carboxylate ligation from Mo could then open a substrate coordination site while

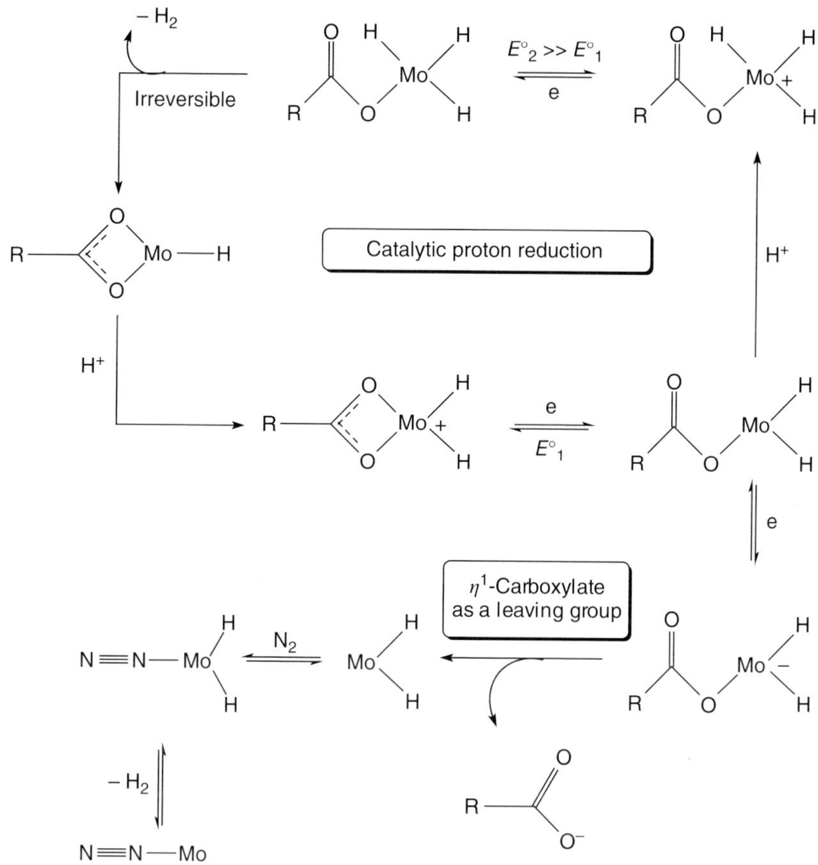

Figure 9.5 Reactivity of the dihydride $[MoH_2(\eta^2\text{-}O_2CMe)(dppe)_2]^+$ suggesting that η^1-carboxylate could serve as a leaving group at the molybdenum center of FeMo-co opening a coordination site at which N_2 could bind before the release of H_2. (Adapted from Refs [37, 70].)

retaining key hydrogen bonds with other amino acid side chain residues such as α-Gln191 and α-Lys426 (as examined with *Azotobacter vinelandii*) [74]. Of course, the involvement of protons and hydrides could intervene differently in the biological catalysis, in particular, at the preferred central Fe_4 face of FeMo-co [4]. Interestingly, the relevance of metal hydride and dihydrogen in the activation of dinitrogen to ammonia in homogeneous, heterogeneous, and biological processes has recently been reviewed [75]. It has been argued that the homocitrate ligand would simply be crucially important to control a proton delivery chain to the cofactor [76]. These alternative suggestions should also fuel a more complex and relevant model chemistry at synthetic metal–sulfur clusters. Finally, the comparative study of the kinetics of proton transfer to cuboidal metal–sulfur clusters containing either the $\{Fe_4S_4\}^{2+}$ or the $\{MoFe_3S_4\}^{3+}$ core in the presence of a nucleophile (thiophenolate or halogenide) has been carried out [77]. The proton presumably binds at sulfur, and the nucleophiles are known to substitute other thiols or halogenides at the iron sites. The rate of proton transfer after the binding of the nucleophile is similar for both type of clusters and assigned to an equal increased basicity of the clusters by thiolate binding. However, this rate is slower before nucleophile binding and it is also two orders of magnitude slower with the $\{MoFe_3S_4\}^{3+}$ core compared with the $\{Fe_4S_4\}^{2+}$ core. On the basis of these findings, it is proposed that irrespective of the binding site of dinitrogen at FeMo-co or of the intimate hydrogen evolution mechanism, the presence of the molybdenum heterometal at the active site could make the enzyme a poorer hydrogenase and a better nitrogenase. This would also be consistent with the known limiting stoichiometries of the Mo- and all Fe nitrogenases (respectively 1 and 7.5 mol of H_2 produced per mole of N_2 reduced).

9.6 Sulfur- and Metal–Metal Interaction in Functional Models of Nitrogenase

Common shortcomings of most of the synthetic functional models of nitrogenase is the absence of a sulfur environment; another one being the lack of metal–metal bonds. If iron–sulfur clusters have proved to be ubiquitous in nature for catalysis, electron transfer, transport, and delivery [78–82], from the nitrogenase model molecular chemistry point of view, the combination of these elements has not brought much success so far. There have been indeed rather few sulfur-containing metal–dinitrogen complexes reported in the literature and even fewer showing ammonia synthesis or N_2 activation [83], a fact that can be explained by the poorer π-basic properties of sulfur. One notable exception in this context is the bis-dinitrogen complex of molybdenum bearing a tetrathio[crown]ether ligand [84], which is structurally related to the Chatt-type complexes. The yield of ammonia is however low in contrast to the reactivity of some of its Mo or W(0) tetraphosphine congeners. In addition, and regarding nitrogenase modeling, the use of thioethers is not rigorously relevant for taking into account the sulfide environment of metals in FeMo-co. The introduction of thiolate ligands represents an improvement for modeling especially if one accepts that the inorganic sulfide at FeMo-co may be protonated during catalysis. This approach, however, led mostly to diazo chemistry involving already partially reduced ligands such as hydrazine and others [15, 85], including at vanadium complexes [86]. Other examples are available where a given metal complex activates dinitrogen while its sulfur analog will not. Indeed, the three-coordinate Mo(III) anilide complex $Mo(NRAr)_3$ (where $R{=}C(CD_3)_2CH_3$ and $Ar{=}C_6H_3Me_2$) has been shown to split the N≡N bond and produce two $N{\equiv}Mo(NRAr)_3$ nitrides by reductive cleavage [87, 88]. By contrast, the thiolate liganded tricoordinate analog $Mo(SAd)_3$ (Ad = 1-adamantyl) generated transiently does not interact with N_2 [89]. An example of a thiolate-containing dinitrogen complex has been discussed in Section 9.4 and Figure 9.4d [58] but so far has not been shown to significantly activate N_2 to the extent found in the related triphosphine borane iron catalyst (Figure 9.4b) [55, 56]. It can be argued that key features lacking in these latter mononuclear models are metal–metal bond(s). To address this issue, we consider first two related body of work that involve thiolate-bridged dinuclear metal complexes based either on molybdenum [90–92] or on iron [93–95]. The trithiolato-bridged dicyclopentadienyl-molybdenum type complexes [90], such as $[Mo_2Cp_2(\mu\text{-}SMe)_3(\mu\text{-}Cl)]$ pictured in Figure 9.6a, have been shown to bind and interconvert substrates of relevance to nitrogenase either at mononuclear or dinuclear metal sites. Of particular interest is the cyclic reduction of hydrazine derivatives into an amine derivative and ammonia at a conserved dinuclear Mo_2 site. The peculiarity of this model chemistry is that the bimetallic cluster contains a short metal–metal bond (Mo–Mo = 2.6–2.8 Å) that is retained in the $\{Mo_2Cp_2(\mu\text{-}SR)_3\}$ core throughout the proposed cycle (Figure 9.6a). This is in contrast to the related approaches developed on thiolato-bridged di-iron complexes that also bind and activate hydrazine or other more reduced azo derivatives such as amide in some cases up to ammonia formation [93–95]. With $[Cp^*{}_2Fe_2(\mu\text{-}SMe)_3]$, for example, reaction of phenylhydrazine (1.5 equiv) yields aniline and ammonia together with the bridging η^2-phenyldiazene di-iron thiolate complex. This binding induces a significant elongation of the Fe–Fe distance from 2.7 Å in the parent complex to 3.2 Å in the diazene bridged complex $[Fe_2Cp^*{}_2(\mu\text{-}SMe)_2(\mu\text{-}\eta^2\text{-}RN{=}NH)]$, Figure 9.6b [93]. Hence, this system works most likely akin to a di-iron metallic pseudochelator rather than the reactive di-molybdenum metallic edge discussed earlier. Both examples may be of relevance to nitrogenase catalysis depending on the extent of FeMo-co flexibility and structural change during biocatalysis and to the identity of the metal(s) involved in activation at the active site. A potentially interesting extension of this modeling chemistry would be the development of related hetero-bimetallic Fe–Mo complexes to consider the relevant Fe–Mo edges in FeMo-co. Interestingly, the reactivity of such a Mo–Fe edge has been assessed theoretically at FeMo-co [96]. Theoretical modeling has also been carried out for the di-iron system with inclusion of dinitrogen binding and activation [94]. Putative N_2 complexes can always be studied theoretically, but the chemical reality so far remains that these metal–sulfur models involve only reduced diazo intermediates and follow "the rocket-fuel approach" rather than fixing dinitrogen. There are, however, examples of dinitrogen coordination at metals sharing a direct interaction but not with relevant nitrogenase metals and coligands [10, 75]; see, for example, the side-on/end-on bridging N_2 at a dinuclear tantalum complex (Ta–Ta = 2.8 Å) [97].

Higher nuclearity FeS or MoFeS clusters that better model the network of metal–metal bonds in FeMo-co do not interact with N_2 either [12–15]. Nevertheless, interesting alternative substrate binding and catalysis has been established at the molybdenum center, whose first and second coordination is well reproduced in some of the synthetic models, as pictured in Figure 9.7a,b. It is striking to note that the first polymetallic-sulfur cluster that binds dinitrogen and based on a $RuIr_3S_4$ core shows large metal–metal distances between the Ru–N≡N center and the iridium atoms (Ru–Ir = 3.5–3.6 Å) [98]. It seems

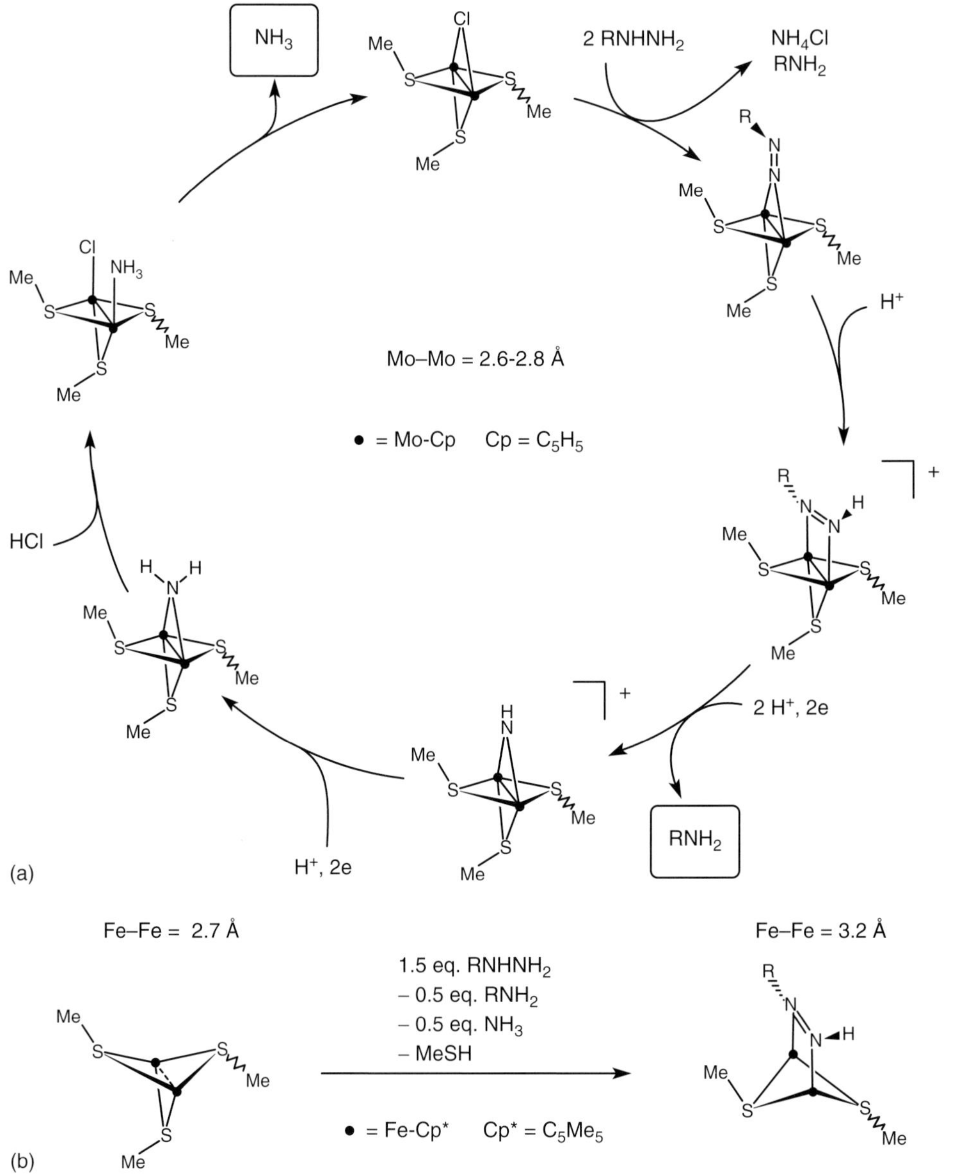

Figure 9.6 (a) A cycle proposed for the reduction of phenylhydrazine to aniline based on the reactivity observed at a retained thiolate-bridged di-molybdenum center. Note that the metal–metal distance remains short in the different isolated complexes [91, 92]. (b) Initial reaction of a substituted hydrazine with a trithiolato-bridged di-iron center. Note the significant elongation of the Fe–Fe distance upon coordination of the bridging diazene [93].

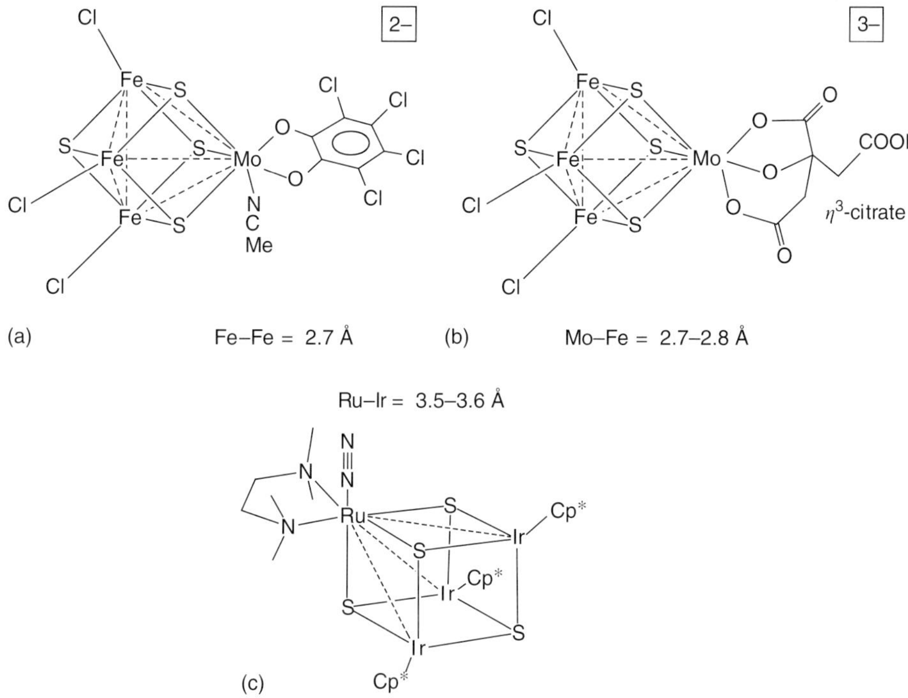

Figure 9.7 (a,b) Cubane-type clusters with the $\{MoFe_3S_4\}^{3+}$ core efficient for hydrazine reduction to ammonia at the molybdenum center. (a) Note the labile acetonitrile solvent ligand at Mo [32]. (b) Note that catalysis at Mo suggests partial decoordination of the η^3-citrate ligand [33]. (c) The first metal–sulfur cluster with an end-on N_2 ligand [98]. Note the long Ru–Ir distance, here Ru–Ir = 3.612 Å; 3.621 Å and 3.497 Å compared with about 2.75 Å for a Ru–Ir bond. Cp* is permethylated cyclopentadienyl (η^5-C_5Me_5).

that classical end-on N_2 binding is possible at a polymetallic-sulfur cluster but requires a quasi mononuclear site isolated from strong interaction with the other metals [15] (Figure 9.7c).

The recent and refined structural model of FeMo-co will certainly stimulate the quest for more faithful mimics of the central iron centers, thought by some to be at the heart of biocatalysis. Nevertheless, progress has been made recently toward the synthesis of metal–sulfur models possessing some structural features that begin to take into account the peculiar carbide-centered iron prism of FeMo-co, as discussed in Section 9.2 [14, 17]. From a functional point of view, the iron centers of these models have not been shown to be reactive toward dinitrogen. Therefore, in an ironic narrative epanalepsis, we are back in Section 9.2, calling for more accurate structural models with the knowledge that these will most probably show no catalysis of dinitrogen fixation, in line with the properties of extracted FeMo-co. The missing link, which we had in mind all along, is obviously the supramolecular protein environment of the active site. We attempt to address this point and its relation to heterogeneous catalysis in the following concluding section.

9.7 Surface Chemistry and the Supramolecular Protein Environment

Nitrogen fixation through biological, homogeneous, and heterogeneous catalysis proceeds through different mechanisms and catalysts but nevertheless must overcome in all cases the initial high stability of the nitrogen–nitrogen triple bond, and, as such, at least at the first steps of catalysis, share a common crucial challenge [75]. The elementary steps of some heterogeneous catalyses at surfaces, including the reaction of dinitrogen with dihydrogen, have been established in great detail [99]. At the active sites of solid catalysts, a high input of thermal energy is usually needed to effect adsorption, activation, and combination of gases. In some cases, studies of metal surface reactivity at low temperature with scanning tunneling microscopy allow the identification of the reaction steps. This approach also makes the reaction happen step-by-step through the controlled delivery of energy input through electron injection with the scanning tunneling microscope tip [100]. This has led to an inspiring parallel with enzymatic catalysis and particularly with nitrogenase [101]. The mild temperature at which the biological reaction occurs must be balanced by the way the enzyme controls and delivers the high energy released from ATP hydrolysis in the form of electron injection to the active site, protons, and dinitrogen. Hence, the "*surface science of enzyme*" concept should help to understand how nitrogenase works at the molecular level [102]. It is in addition striking to note that the inorganic metal sulfides of iron and molybdenum have known reactivity of relevance to nitrogenase. Indeed, fresh iron sulfide can react at near-ambient conditions with hydrogen sulfide forming iron disulfide, dihydrogen, and ammonia, a reaction which has been proposed to be of prebiotic relevance [103]. On the other hand, molybdenum disulfide, the catalyst *par excellence* for hydrodesulfurization is being studied as an alternative to platinum for the electrocatalytic production of dihydrogen [104]. A catalytically active mononuclear complex with a side MoS_2 structure has been proposed to model the reactive edge of inorganic molybdenum disulfide [104]. Recently, a common porous oxide supporting a ruthenium metal catalyst for heterogeneous ammonia synthesis has been modified (reduced) to contain electrides in some the cages (vacancies) of the unit cell [105, 106]. The modified material is one order of magnitude more efficient for ammonia synthesis. This has been attributed to the promoter effect of the electrides that exacerbate the electron donation from ruthenium to the π-antibonding orbitals of dinitrogen. The other important property of the modified porous oxide material is its high affinity for dihydrogen, which is reduced as hydrides in the oxide vacancies. This consequently prevents or limits inhibition of the ruthenium catalyst from a too high dihydrogen coverage. We hence again hit upon two important and universal aspects of dinitrogen activation that are critical regardless of the type of catalysis. First, we may wonder if there is an equivalent "promoter" in the structure of FeMo-co (and the central carbide would be the obvious candidate) as in the latter electride example, or as discussed for the role of potassium ion and oxide in the Haber–Bosch process and in the reactivity of some iron model complexes. Second, the possible importance of potential hydrogen scavenger (e.g., possibly KC_8) in the reactivity of iron model complexes [53, 64] was also pointed out earlier and one kinetic study on model metal–sulfur clusters was discussed [77]. The way FeMo-co deals with the delivery of protons and electrons, to form ammonia from dinitrogen but also dihydrogen in a continuous or obligatory manner and the possible involvement of hydrides is still an open question. The FeMo-co cluster might well behave as an elementary heterogeneous catalyst but it is also wrapped by a polypeptide scaffold. We know that many amino acid residues around and further away from FeMo-co are essential for catalysis. Site-directed mutagenesis has indeed produced mutants with altered catalytic properties. From the biochemistry and enzymology studies of these mutants, somehow, divergent suggestions can be proposed regarding the detailed biological mechanism of nitrogen fixation. One thing that seems established is that the catalytic properties of the unique FeMo-co metal–sulfur cluster is revealed or allowed by the specific molybdenum–iron protein surrounding. This should lead to the integration in the model complexes of ligands with a second sphere with specific properties such as electroactivity, acid–base character, or switchable weak bonding properties, for example. This approach has already been developed to model some of the active sites of hydrogenases as the synthetic and catalytic challenges are much more at reach than for nitrogenases [107]. Another related approach that has been implemented with simpler di-iron active sites such as those of hydrogenases is the testing of the activity of apo-proteins after binding of different synthetic sub-site analogs [108, 109]. This could culminate in the design of an artificial protein sequence to modulate and improve catalysis at the active sites [110, 111]. One challenge and limitation of these approaches applied to nitrogenases is that we do need synthetic models with a structure that is close enough, if not identical, to FeMo-co.

9.8 Conclusion and Outlook

This short survey of current nitrogenase modeling chemistry has focused on the need for accurate structural models taking into account the most recent data on the biological cluster, which is the presence of a central carbide at the heart of FeMo-co [2]. Apart from the synthetic challenge that is interesting for metal–sulfur chemistry in its own right, this would allow the investigation of the reactivity of FeMo-co in a large array of experimental conditions and with repeatable and reproducible experiments. As pointed out at the end of the last section, a structural model could also be of interest for studies involving mutated apo-proteins. Regarding functional modeling, the reactivity of metal–sulfur clusters is at best only indirectly relevant as none of these clusters have been shown to activate dinitrogen so far. There are however two exceptions in this context. First, the proposal based on detailed kinetic studies that the heterometal (Mo) in iron–sulfur cluster could make hydrogen evolution less efficient [77]. Second, the publication of the first metal–sulfur cluster containing an end-on dinitrogen ligand showed that the ruthenium-N_2 fragment is elongated from the rest of the strongly interacting iridium–sulfur cluster [98]. When metal–sulfur clusters catalyze the reduction of alternative nitrogenase substrates or reduced diazo intermediates, the heterogeneous apical metal site is involved. This is partly because we do not have a model of the carbide-centered hexa-iron prism whose reactivity could be tested. Lower nuclearity thiolate-bridged bimetallic clusters based either on molybdenum [92] or iron [93] have yielded interesting chemistry involving diazo intermediates and ammonia synthesis. There are obvious similarities between these models but also striking differences that should fuel further studies such as the development of heterogeneous Fe–Mo dinuclear models. The nature of the metals is obviously different in both systems but most importantly the retaining or not of strong metal–metal interaction during catalysis is intriguing. More generally, the combination of both sulfur and metal–metal interaction seems detrimental to artificial homogeneous N_2 coordination and activation. The specific role of homocitrate has been discussed within the framework of one model study that suggests its partial decoordination from molybdenum followed by N_2 binding and the release of H_2 [70]. More generally, the subtle involvement of protons, hydrides, and dihydrogen is a common feature of any nitrogen fixation process [75] and has been discussed along different sections of this account. Regarding mononuclear models with abiological models, the most recent achievements have been the publication of two molybdenum [42, 41]- and one iron [56]-based catalysts for dinitrogen reduction to ammonia. We have discussed the merits and limits of these catalysts and their potential relevance to nitrogenase. Regarding the nature of the metal, one must keep in mind that different nitrogenases exist with an apical molybdenum, vanadium, or iron atom, and a gradual loss of efficiency for N_2 fixation so that this apical site could well be intimately involved in initial dinitrogen coordination and activation. The first iron-based catalyst nevertheless balances the debate. It provides a precedent that can be put forward to advocate one of central Fe_4 face of FeMo-co being the active site as seems to be the consensus of the most recent enzymology studies [4]. In the last section, some common issues shared with biology and surface chemistry have been put forward and these could foster approaches to better understand the enzyme [102]. Finally, the other critical component of metalloenzyme is obviously the polypeptide. Notwithstanding the difficulty of producing mutated protein in large amounts, the synthesis of a structural model of FeMo-co could lead to new studies involving reconstructed enzymes or even artificial enzymes. In the past decades, knowledge of the biological system has progressed immensely and so has model chemistry. It is however blatant how little is established regarding biological catalysis at the molecular level. The combination of metals and sulfur is ubiquitous in nature and also allows ammonia synthesis with the appropriate and complex protein environment.

This has not been the right match for homogeneous catalysis so far. A common picture is emerging for the challenges common to biological, heterogeneous, and homogeneous nitrogen fixation catalysis. It is hoped that this short survey has demonstrated that model chemistry can contribute positively to the debate and research on nitrogenase catalysis. Indeed, many challenges lie ahead for the chemists not only in understanding the biological systems but also in developing a simple and efficient, homogeneous artificial catalyst for nitrogen fixation. On the basis of the results discussed here, one can foresee that the design of a simple metal complex, sulfur-free or not, with or without nitrogenase relevant metals [97, 112, 113], adapted to a designed artificial protein or recognition polymer could be an approach to yield a more efficient homogeneous artificial catalyst. The story of nitrogen fixation is not over and continues to be fascinating [1, 48, 114–116]. In biology, a long-term prospect would be the engineering and growing of cereals and other nonleguminous crops associated with nitrogen-fixing microorganism [115]. On its side, and in addition to the broader challenge of artificial nitrogen fixation, model chemistry will continue its quest to explain how the biological system proceeds (may be) [117].

References

1. Leigh, G.J. (2004) *The World's Greatest Fix: A History of Nitrogen in Agriculture*, Oxford University Press.
2. Spatzal, T., Aksoyoglu, M., Zhang, L., Andrade, S.L.A., Schleicher, E., Weber, S., Rees, D.C., and Einsle, O. (2011) *Science*, **334**, 940.
3. Smil, V. (2004) *Enriching the Earth: Fritz Haber, Carl Bosch, and the Transformation of World Food Production*, The MIT Press.
4. Hoffman, B.M., Lukoyanov, D., Dean, D.R., and Seefeldt, L.C. (2013) *Acc. Chem. Res.*, **46**, 587–595.
5. Adams, C. (1999) *Chem. Ind. (London)*, 740–743.
6. Allen, A.D. and Senoff, C.V. (1965) *J. Chem. Soc., Chem. Commun.*, 621–622.
7. Chatt, J., Dilworth, J.R., and Richards, R.L. (1978) *Chem. Rev.*, **78**, 589–625.
8. Bazhenova, T.A. and Shilov, A.E. (1995) *Coord. Chem. Rev.*, **144**, 69–145.
9. Fryzuk, M.D. and Johnson, S.A. (2000) *Coord. Chem. Rev.*, **200-202**, 379–409.
10. MacKay, B.A. and Fryzuk, M.D. (2004) *Chem. Rev.*, **104**, 385–401.
11. Hinrichsen, S., Broda, H., Gradert, C., Söncksen, L., and Tuczek, F. (2012) *Annu. Rep. Prog. Chem. Sect. A: Inorg. Chem.*, **108**, 17–47.
12. Malinak, S.M. and Coucouvanis, D. (2001) *Prog. Inorg. Chem.*, **49**, 599–662.
13. Venkateswara Rao, P. and Holm, R.H. (2004) *Chem. Rev.*, **104**, 527–559.
14. Lee, S.C. and Holm, R.H. (2004) *Chem. Rev.*, **104**, 1135–1157.
15. Barrière, F. (2003) *Coord. Chem. Rev.*, **236**, 71–89.
16. Barrière, F., Durrant, M.C., and Pickett, C.J. (2004) in *Nitrogenase and Chemical Models*, Chapter 7 (eds B.E. Smith, R.L. Richards, and W.E. Newton), Kluwer Academic Publishers, Amsterdam, pp. 161–199.
17. Barrière, F. (2005) in *Encyclopedia of Inorganic Chemistry*, 2nd edn (ed. R.B. King), John Wiley & Sons, Inc., pp. 3637–3650.
18. Scott, T.A., Berlinguette, C.P., Holm, R.H., and Zhou, H.-C. (2005) *Proc. Natl. Acad. Sci. U.S.A.*, **102**, 9741–9744.
19. Deng, L. and Holm, R.H. (2008) *J. Am. Chem. Soc.*, **130**, 9878–9886.
20. Strop, P., Takahara, P.M., Chiu, H.-J., Angove, H.C., Burgess, B.K., and Rees, D.C. (2001) *Biochemistry*, **40**, 651–656.

21. Jacobs, D. and Watt, G.D. (2013) *Biochemistry*, **52**, 4791–4799.
22. Ohki, Y., Imada, M., Murata, A., Sunada, Y., Ohta, S., Honda, M., Sasamori, T., Tokitoh, N., Katada, M., and Tatsumi, K. (2009) *J. Am. Chem. Soc.*, **131**, 13168–13178.
23. Ohki, Y., Tanifuji, K., Yamada, N., Cramer, R.E., and Tatsumi, K. (2012) *Chem. Asian J.*, **7**, 2222–2224.
24. Einsle, O., Tezcan, F.A., Andrade, S.L.A., Schmid, B., Yoshida, M., Howard, J.B., and Rees, D.C. (2002) *Science*, **297**, 1696–1700.
25. Deng, H. and Hoffmann, R. (1993) *Angew. Chem. Int. Ed. Engl.*, **32**, 1062–1065.
26. Kim, J. and Rees, D.C. (1992) *Science*, **257**, 1677–1682.
27. Ramaswamy, S. (2011) *Science*, **334**, 914–915.
28. Lancaster, K.M., Roemelt, M., Ettenhuber, P., Hu, Y., Ribbe, M.W., Neese, F., Bergmann, U., and DeBeer, S. (2011) *Science*, **334**, 974–977.
29. Lancaster, K.M., Hu, Y., Bergmann, U., Ribbe, M.W., and DeBeer, S. (2013) *J. Am. Chem. Soc.*, **135**, 610–612.
30. Ohki, Y., Hashimoto, T., and Tatsumi, K. (2010) *Inorg. Chem.*, **49**, 6102–6109.
31. Ohta, S., Ohki, Y., Hashimoto, T., Cramer, R.E., and Tatsumi, K. (2012) *Inorg. Chem.*, **51**, 11217–11219.
32. Palermo, R.E., Singh, R., Bashkin, J.K., and Holm, R.H. (1984) *J. Am. Chem. Soc.*, **106**, 2600–2612.
33. Demadis, K.D. and Coucouvanis, D. (1995) *Inorg. Chem.*, **34**, 436–448.
34. Chen, X.-D., Duncan, J.S., Verma, A.K., and Lee, S.C. (2010) *J. Am. Chem. Soc.*, **134**, 15884–15886.
35. Bortels, H. (1930) *Arch. Mikrobiol.*, **1**, 333–342.
36. Rehder, D. (2000) *J. Inorg. Biochem.*, **80**, 133–136.
37. Pickett, C.J. (1996) *J. Biol. Inorg. Chem.*, **1**, 601–606.
38. Chatt, J., Pearman, A.J., and Richards, R.L. (1975) *Nature*, **253**, 39–40.
39. Pickett, C.J. and Talarmin, J. (1985) *Nature*, **317**, 652–653.
40. Schrock, R.R., Kolodziej, R.M., Liu, A.H., Davis, W.H., and Vale, M.G. (1990) *J. Am. Chem. Soc.*, **112**, 4338–4345.
41. Arashiba, K., Miyake, Y., and Nishibayashi, Y. (2011) *Nat. Chem.*, **3**, 120–125.
42. Yandulov, D.V. and Schrock, R.R. (2003) *Science*, **301**, 76–78.
43. Schrock, R.R. (2003) *Chem. Commun.*, 2389–2391.
44. Schrock, R.R. (2008) *Angew. Chem. Int. Ed.*, **47**, 5512–5522.
45. Weare, W.W., Dai, X., Byrnes, M.J., Chin, J.M., Schrock, R.R., and Müller, P. (2006) *Proc. Natl. Acad. Sci. U.S.A.*, **103**, 17099–17106.
46. Schrock, R.R. (2011) *Nat. Chem.*, **3**, 95–96.
47. Dance, I. (2012) *Dalton Trans.*, **41**, 4859–4865.
48. McLeod, K.C. and Holland, P.L. (2013) *Nat. Chem.*, **5**, 559–565.
49. Crossland, J.L. and Tyler, D.R. (2010) *Coord. Chem. Rev.*, **254**, 1883–1894.
50. Hazari, N. (2010) *Chem. Soc. Rev.*, **39**, 4044–4056.
51. Leigh, G.J. and Jimenez-Tenorio, M. (1991) *J. Am. Chem. Soc.*, **113**, 5862–5863.
52. Hall, D.A. and Leigh, G.J. (1996) *J. Chem. Soc., Dalton Trans.*, 3539–3541.
53. Rodriguez, M.M., Bill, E., Brennessel, W.W., and Holland, P.L. (2011) *Science*, **334**, 780–783.
54. Hirano, M., Akita, M., Morikita, T., Kubo, H., Fukuoka, A., and Komiya, S. (1997) *J. Chem. Soc., Dalton Trans.*, 3453–3458.
55. Moret, M.-E. and Peters, J.C. (2011) *Angew. Chem. Int. Ed.*, **50**, 2063–2067.
56. Anderson, J.S., Rittle, J., and Peters, J.C. (2013) *Nature*, **501**, 84–88.
57. Mankad, N.P., Whited, M.T., and Peters, J.C. (2007) *Angew. Chem. Int. Ed.*, **46**, 5768–5771.
58. Takaoka, A., Mankad, N.P., and Peters, J.C. (2011) *J. Am. Chem. Soc.*, **133**, 8440–8443.
59. Mittasch, A. (1951) *Geschichte der Ammoniaksynthese*, Verlag Chemie, Weinheim.
60. Smith, J.M., Sadique, A.R., Cundari, T.R., Rodgers, K.R., Lukat-Rodgers, G., Lachicotte, R.J., Flaschenriem, C.J.,

Vela, J., and Holland, P.L. (2006) *J. Am. Chem. Soc.*, **128**, 756–769.

61. Peters, J.C. and Mehn, M.P. (2006) in *Activation of Small Molecules*, Chapter 3 (ed. W.B. Tollman), Wiley-VCH Verlag GmbH, pp. 81–119.
62. Moret, M.-E. and Peters, J.C. (2011) *J. Am. Chem. Soc.*, **133**, 18118–18121.
63. Anderson, J.S., Moret, M.-E., and Peters, J.C. (2013) *J. Am. Chem. Soc.*, **135**, 534–537.
64. Lee, Y., Mankad, N.P., and Peters, J.C. (2010) *Nat. Chem.*, **2**, 558–565.
65. Field, L.D. (2010) *Nat. Chem.*, **2**, 520–521.
66. Suess, D.L.M. and Peters, J.C. (2013) *J. Am. Chem. Soc.*, **135**, 4938–4941.
67. Fong, H., Moret, M.-E., Lee, Y.L., and Peters, J.C. (2013) *Organometallics*, **32**, 3053–3062.
68. Herold, A. and Saehr, D. (1960) *Compt. Rend.*, **250**, 545–546.
69. Mayer, S.M., Gormal, C.A., Smith, B.E., and Lawson, D.M. (2002) *J. Biol. Chem.*, **277**, 35263–35266.
70. Hughes, D.L., Ibrahim, S.K., Pickett, C.J., Querné, G., Laouénan, A., Talarmin, J., Queiros, A., and Fonseca, A. (1994) *Polyhedron*, **13**, 3341–3348.
71. Siemann, S., Schneider, K., Oley, M., and Müller, A. (2003) *Biochemistry*, **42**, 3846–3857.
72. Zhou, Z.-H., Wang, H., Yu, P., Olmstead, M.M., and Cramer, S.P. (2013) *J. Inorg. Biochem.*, **118**, 100–106.
73. Grönberg, K.L.C., Gormal, C.A., Durrant, M.C., Smith, B.E., and Henderson, R.A. (1998) *J. Am. Chem. Soc.*, **120**, 10613–10621.
74. Durrant, M.C., Francis, A., Lowe, D.J., and Newton, W.E. (2006) *Biochem. J.*, **397**, 261–270.
75. Jia, H.-P. and Quadrelli, E.A. (2014) *Chem. Soc. Rev.*, **43**, 547–564.
76. Dance, I. (2012) *Dalton Trans.*, **41**, 7647–7659.
77. Bell, J., Dunford, A.J., Hollis, E., and Henderson, R.A. (2003) *Angew. Chem. Int. Ed.*, **42**, 1149–1152.
78. Beinert, H., Holm, R.H., and Munck, E. (1997) *Science*, **277**, 653–659.
79. Stephens, P.J. and Jollie, D.R. (1996) *Chem. Rev.*, **96**, 2491–2513.
80. Dos Santos, P.C. and Dean, D.R. (2008) *Proc. Natl. Acad. Sci. U.S.A.*, **105**, 11589–11590.
81. Hernandez, J.A., Curatti, L., Aznar, C.P., Perova, Z., Britt, R.D., and Rubio, L.M. (2008) *Proc. Natl. Acad. Sci. U.S.A.*, **105**, 11679–11684.
82. Hu, Y. and Ribbe, M.W. (2013) *J. Biol. Chem.*, **288**, 13173–13177.
83. Sellmann, D., Hautsch, B., Rösler, A., and Heinemann, F.W. (2001) *Angew. Chem. Int. Ed.*, **40**, 1505–1507.
84. Yoshida, T., Adachi, T., Ueda, T., Kaminaka, M., Sasaki, N., Higuchi, T., Aoshima, T., Mega, I., Mizobe, Y., and Hidai, M. (1989) *Angew. Chem., Int. Ed. Engl.*, **28**, 1040–1042.
85. Richards, R.L. (1996) *Coord. Chem. Rev.*, **154**, 83–97.
86. Hsu, H.-F., Chu, W.-C., Hung, C.-H., and Liao, J.-H. (2003) *Inorg. Chem.*, **42**, 7369–7371.
87. Cummins, C.C. (1998) *Chem. Commun.*, 1777–1786.
88. Peters, J.C., Cherry, J.-P.F., Thomas, J.C., Baraldo, L., Mindiola, D.J., Davis, W.M., and Cummins, C.C. (1999) *J. Am. Chem. Soc.*, **121**, 10053–10067.
89. Agapie, T., Odom, A.L., and Cummins, C.C. (2000) *Inorg. Chem.*, **39**, 174–179.
90. Barrière, F., Le Mest, Y., Pétillon, F.Y., Poder-Guillou, S., Schollhammer, P., and Talarmin, J. (1996) *J. Chem. Soc., Dalton Trans.*, 3967–3976.
91. Pétillon, F.Y., Schollhammer, P., and Talarmin, J. (1999) *Inorg. Chem.*, **38**, 1954–1955.
92. Le Grand, N., Muir, K.W., Pétillon, F.Y., Pickett, C.J., Schollhammer, P., and Talarmin, J. (2002) *Chem. Eur. J.*, **8**, 3115–3127.
93. Chen, Y., Zhou, Y., Chen, P., Tao, Y., Li, Y., and Qu, J. (2008) *J. Am. Chem. Soc.*, **130**, 15250–15251.
94. Luo, Y., Li, Y., Yu, H., Zhao, J., Chen, Y., Hou, Z., and Qu, J. (2012) *Organometallics*, **31**, 335–344.
95. Li, Y., Li, Y., Wang, B., Luo, Y., Yang, D., Tong, P., Zhao, J., Luo, L., Zhou, Y., Chen, S., Cheng, F., and Qu, J. (2013) *Nat. Chem.*, **5**, 320–326.
96. Durrant, M.C. (2002) *Biochemistry*, **41**, 13934–13945.

97. Fryzuk, M.D., Johnson, S.A., and Rettig, S.J. (1998) *J. Am. Chem. Soc.*, **120**, 11024–11025.
98. Mori, H., Seino, H., Hidai, M., and Mizobe, Y. (2007) *Angew. Chem. Int. Ed.*, **46**, 5431–5434.
99. Ertl, G. (2008) *Angew. Chem. Int. Ed.*, **47**, 3524–3535.
100. Hahn, J.R. and Ho, W. (2001) *Phys. Rev. Lett.*, **87**, 166102.
101. Nørskov, J.K. (2001) *Nature*, **414**, 405–406.
102. Rod, T.H. and Nørskov, J.K. (2002) *Surf. Sci.*, **500**, 678–698.
103. Dörr, M., Käßbohrer, J., Grunert, R., Kreisel, G., Brand, W.A., Werner, R.A., Geilman, H., Apfel, C., Robl, C., and Wiegand, W. (2003) *Angew. Chem. Int. Ed.*, **42**, 1540–1543.
104. Karunadasa, H.I., Montalvo, E., Sun, Y., Majda, M., Long, J.R., and Chang, C.J. (2012) *Science*, **335**, 698–702.
105. Kitano, M., Inoue, Y., Yamazaki, Y., Hayashi, F., Kanbara, S., Matsuishi, S., Yokoyama, T., Kim, S.-W., Hara, M., and Hosono, H. (2012) *Nat. Chem.*, **4**, 934–940.
106. Giamello, E. (2012) *Nat. Chem.*, **4**, 869–870.
107. Rakowski DuBois, M. and DuBois, D.L. (2009) *Chem. Soc. Rev.*, **38**, 62–72.
108. Berggren, G., Adamska, A., Lambertz, C., Simmons, T.R., Esselborn, J., Atta, M., Gambarelli, S., Mouesca, J.-M., Reijerse, E., Lubitz, W., Happe, T., Artero, V., and Fontecave, M. (2013) *Nature*, **499**, 66–69.
109. Bethel, R.D. and Darensbourg, M.Y. (2013) *Nature*, **499**, 40–41.
110. Reig, A.J., Pires, M.M., Snyder, R.A., Wu, Y., Jo, H., Kulp, D.W., Butch, S.E., Calhoun, J.R., Szyperski, T.G., Solomon, E.I., and DeGrado, W.F. (2012) *Nat. Chem.*, **4**, 900–906.
111. Berry, S.M. (2012) *Nat. Chem.*, **4**, 868–869.
112. Shima, T., Hu, S., Luo, G., Kang, X., Luo, Y., and Hou, Z. (2013) *Science*, **340**, 1549–1552.
113. Semproni, S.P., Knobloch, D.J., Milsmann, C., and Chirik, P.J. (2013) *Angew. Chem. Int. Ed.*, **52**, 5372–5376.
114. Tanabe, Y. and Nishibayashi, Y. (2013) *Coord. Chem. Rev.*, **257**, 2551–2564.
115. Beatty, P.H. and Good, A.G. (2011) *Science*, **333**, 416–417.
116. Peplow, M. (2013) A fixation with nitrogen, RSC Chemistry World (Apr. 18 2013).
117. Leigh, G.J. (2003) *Science*, **301**, 55–56.

10
A Unified Chemical Mechanism for Hydrogenation Reactions Catalyzed by Nitrogenase

Ian Dance

10.1
Introduction

10.1.1
Nitrogenase: the Enzyme

The microbial enzyme nitrogenase naturally catalyzes the reaction $N_2 + 6H^+ + 6e^- \rightarrow 2NH_3$ [1–6]. It is remarkable that this is effected under mild ambient conditions that contrast the high pressures and temperatures required in the best industrial (Haber–Bosch) practice [7, 8]. Nitrogenase also co-catalyzes the reduction of ubiquitous protons to H_2, and is able to catalyze the hydrogenation of many nonphysiological substrates.

The nitrogenase enzyme comprises two proteins: a larger protein denoted the MoFe protein, and a smaller protein, the Fe protein. The MoFe protein is almost a symmetrical dimer, containing four protein subunits and two copies each of two metal sulfide clusters that are crucial for the mechanism. The catalysis site is the iron–molybdenum cofactor (FeMo-co), which is about 13 Å distant from the P-cluster that is involved in electron transfer to FeMo-co. The Fe protein contains two protein subunits bridged by a $Fe_4S_4(cys)_4$ cluster on the protein surface, and two nucleotide (MgATP/ADP) binding sites. When the two proteins are docked, the Fe_4S_4 cluster is about 14 Å from the P-cluster.

The overall biochemical mechanism involves a sequence (called the *Fe protein cycle*) in which the Fe protein with two bound MgATP associates with the MoFe protein. Conformational changes in the Fe protein initiate hydrolysis of the two MgATP, associated with transfer of one electron from the Fe_4S_4 cluster to the P-cluster and on to FeMo-co [4, 9, 10]. The Fe protein then dissociates and is regenerated with fresh MgATP, and the Fe_4S_4 cluster is reduced [4, 6, 11, 12].

Bioinspired Catalysis: Metal-Sulfur Complexes, First Edition.
Edited by Wolfgang Weigand and Philippe Schollhammer.

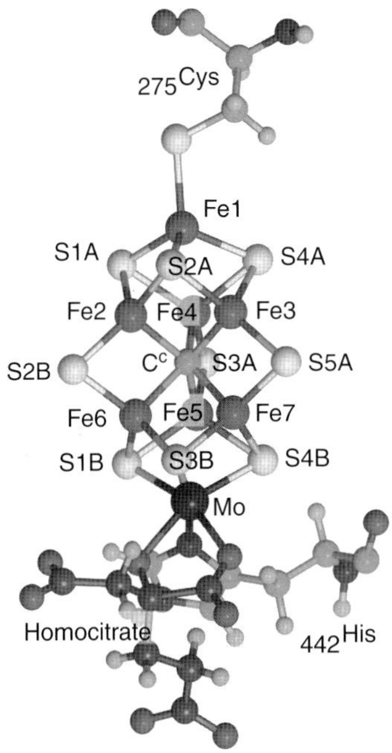

Figure 10.1 The structure of FeMo-co (PDB 3U7Q) [13]. (*R*)-homocitrate carbon atoms are shaded darker grey. This view direction, normal to the Fe2, Fe3, Fe6, Fe7 face, is the standard view direction used for most other figures.

10.1.2
FeMo-co

With this supply of electrons, the chemical catalytic reactions occur within the MoFe protein, at the FeMo-co site. The structure [13] of resting FeMo-co is shown in Figure 10.1. A central carbon atom (C^c) is surrounded by six core Fe atoms (Fe2–Fe7) in trigonal prismatic geometry, with three doubly bridging μ-S atoms on the axial edges of the prism. Each trigonal face of the prism is connected by three triply bridging μ_3-S atoms to another metal, Fe at the top and Mo at the bottom. Variants of nitrogenase contain V or Fe in place of Mo [14–16]. The CFe_7MoS_9 cluster can also be conceived as two cubanoid M_4S_3C clusters fused at the shared C, and additionally bridged with $(\mu\text{-S})_3$. The structure of FeMo-co is completed by (*R*)-homocitrate (2-hydroxybutane 1,2,4-tricarboxylate) that functions as an obligatory bidentate ligand to Mo. Two protein side chains ligate FeMo-co: at Fe1, tetrahedral four-coordination is completed with thiolate sulfur of α-275^{Cys}, while six-coordination of Mo is completed with the imidazole side chain of α-442^{His}.[1)] While the FeMo-co cluster is unique, the coordination stereochemistries of the metals within it are conventional.

1) Amino acids are numbered according to species *Azotobacter vinelandii*, crystal structure PDB 3U7Q.

The surprising feature of FeMo-co is the six-coordinate carbon atom at its center. This central atom was first thought to be nitrogen, but there is now strong evidence that it is carbon [13, 17], together with evidence for the biosynthetic pathway for its incorporation [18, 19].

10.1.3
Where Does the Catalysis Occur on FeMo-co?

Before the revelation of the structure of FeMo-co, it was widely believed, on the basis of model chemistry, that Mo would be involved in the activation and hydrogenation of N_2. However, Mo in FeMo-co is coordinatively saturated, and the central six Fe atoms are coordinatively unsaturated [20]. It was recognized early [21] that the front face of FeMo-co – Fe2-S2A-Fe3-S5A-Fe7-S3B-Fe6-S2B – is likely to be the locus of its chemical reactivity, because the cofactor is anchored to surrounding protein through a set of hydrogen bonds to the rear sulfur atom S3A, restricting access to Fe4 and Fe5. These hypotheses have been confirmed and refined experimentally in seminal papers by Seefeldt *et al.* [22–24]. The identity of residue α-70, located over Fe2 and Fe6, has the most profound influence on the reactivity of nitrogenase and on the sizes of substrates that can be reduced [25, 26], and it is now well established that Fe2 and Fe6 are the metal atoms involved in the chemical steps that enable the catalysis effected by nitrogenase [27].

10.2
Investigations of Mechanism

A principal objective of experimental investigation of the mechanism of action of nitrogenase is the trapping and characterization of intermediates. This has proved to be a somewhat elusive goal, for a number of reasons. The enzyme reduces protons, which are impossible to avoid, thereby thwarting attempts to interrupt turnover under normal conditions using modified substrates and surrogate intermediates. The even-electron count of approximately half of the intermediates confounds spin-dependent spectroscopies. Despite this, freeze quenching of EPR-active intermediate states of the enzyme, coupled with electron-nuclear double resonance (ENDOR) spectroscopy, has yielded data for intermediates containing allyl alcohol ($H_2C{=}CHCH_2OH$), N_2, diazene (HN=NH), methyl diazene, hydrazine, and protons, bound to modified proteins [27–38]. Crystallographic characterization of any intermediate in the chemical catalytic cycle of nitrogenase has not yet been achieved, and there is still no definitive structural data about any intermediate bound to FeMo-co. The structure of FeMo-co is unprecedented, and close synthetic analogs are not yet available, so mechanistic insight from model chemistry is limited.

In 2000, Rees and Howard [39] opined: "Rationality notwithstanding, anyone studying nitrogenase should be excused for occasionally wondering whether a Faustian bargain might be required to establish the mechanism of dinitrogen reduction by this enzyme." However, theoretical investigations, specifically density functional (DF) simulations of structure, potential energy surfaces, and reaction trajectories, can bypass these impediments and have yielded a large amount of information about the structure and coordination chemistry of FeMo-co [40–61], and have led to proposed mechanisms of catalysis [62–73].

10.2.1 Density Functional Simulations

The DF methodology used for the investigations described here is based on Bernard Delley's program DMol3 [74–77], which has the advantage (evident in speed and accuracy) of using numerical rather than analytical basis sets. Details and descriptions of my validations of this methodology against experimental data have been published [53, 60, 78, 79]. The basic model of FeMo-co used in simulations is that in Figure 10.2, which retains the essential coordination of all metal atoms.

The electronic structure of FeMo-co is complex, as expected for a cluster with eight metal atoms and >40 d electrons, and in DF calculations the electronic and spin states of all species are specified and managed via the distribution of signs and magnitudes of spin densities on the metal atoms [59]: see also Section 10.2.3. For the determination of transition states (TS) and reaction profiles, conventional methods often fail because of the complexity of the vibrational structure of FeMo-co and its derivatives, and the multiple geometrical changes (e.g., variations of Fe–C^c distances, S3B–Fe distances, and Fe coordination stereochemistry) during a reaction trajectory. Therefore, I have developed a pragmatic procedure that works reliably for the determination of transition states in reaction steps [80, 81].

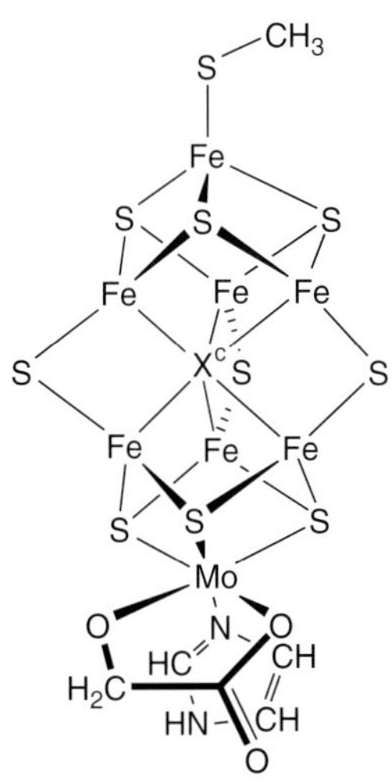

Figure 10.2 The model of FeMo-co used in DF simulations: X = C or N.

10.2.2
The Coordination Chemistry of FeMo-co

Numerous calculations have revealed that the central Fe atoms of FeMo-co can be ligated, with relevant ligands such as N_2, H, H_2, C_2H_2, CO, yielding the various coordination stereochemistries illustrated in Figure 10.3. The ligand can be approximately trans to the Fe–C^c bond in a coordination position labeled exo (Figure 10.3a) or approximately cis to the Fe–C^c in a coordination position labeled endo (Figure 10.3b). These result in trigonal bipyramidal (a) or square pyramidal (b) coordination of Fe. Exo-ligation of Fe can weaken the Fe–C^c interaction to the extent that it is no longer bonding (about 3 Å in length), resulting in approximately tetrahedral coordination of Fe (Figure 10.3c). The distance between ligated Fe and C^c has been calculated to vary in the range 2.1–3 Å. The first experimental (EXAFS, extended X-ray absorption fine structure) evidence for C^c–Fe elongation (Fe coordinated by allyl alcohol) has recently been reported [27]. Movement of ligated Fe away from C^c is not surprising, because a six-coordinate central C atom is over-coordinated, and can shed a bond to Fe. Intermediates with calculated structures in which *two* C^c–Fe distances are elongated beyond 2.5 Å are rare.

Another variation in the coordination of Fe arises because a ligand can be positioned on the L–Fe–C^c arc between the exo and endo positions, that is between Figure 10.3a,b, usually resulting in trigonal bipyramidal coordination of Fe, as illustrated in Figure 10.3d. Coordination at both the exo and endo positions of

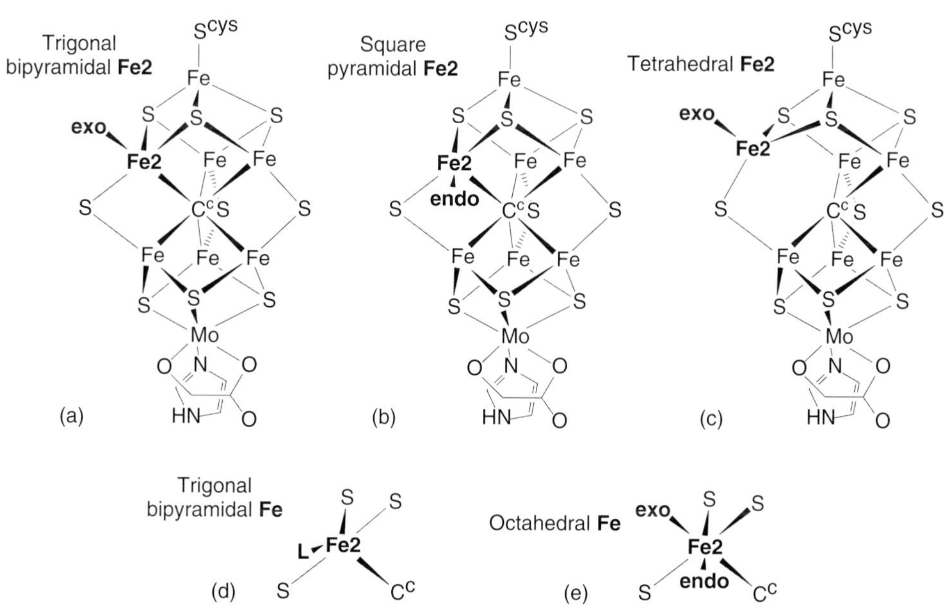

Figure 10.3 (a–e) The coordination positions and coordination stereochemistries that can occur at the active Fe atoms of FeMo-co, illustrated for Fe2.

Fe generates octahedral coordination, Figure 10.3e. Ligation at the endo positions of both Fe2 and Fe6 can occur only if the ligand is bridging.

10.2.3 Electronic Structure of FeMo-co

It is notable that even though FeMo-co is an unusual cluster structure, the various local coordination stereochemistries of the metal atoms are normal. The electronic structure of FeMo-co and its derivatives is more complex. The appropriate conceptual base for interpretation of this electronic structure is electron delocalization over all 18 $C^cFe_7MoS_9$ atoms, and polar covalent bonding between them, because the atomic partial charges are relatively small [55]. To illustrate this point, calculated (Mulliken) charges for an intermediate of FeMo-co with bound H atoms are Fe +0.41 to +0.67, Mo +0.99, S −0.47 to −0.77, H −0.07 to −0.17. Therefore, concepts of charge and oxidation state for individual atoms are less relevant. The redox level of FeMo-co and its intermediates is best described in terms of the total charge, and a theoretical method for assessing this from electrochemical data has been described [61, 82]. The calculations of intermediate structures and of reaction trajectories do not specify or conceive of oxidation states of individual metal atoms (or of C^c, or S), but specify only the total charge of the FeMo-co model.

The numerous spin and electronic states of FeMo-co and its derivatives have been described in detail [40, 59, 61, 70]; it is sufficient here to point out that the calculated model of FeMo-co (with C centering and retaining full coordination of Fe1 and Mo) has 10 electronic states (9 $S = 3/2$, 1 $S = 1/2$) within 10 kcal mol^{-1} of the ground state, which is $S = 3/2$ as observed.

10.3 Hydrogen Supply for the Reactions of Nitrogenase

10.3.1 Multiple Protons are Needed for Catalytic Reaction Cycles

Nitrogenase is also capable of reducing a large number of nonphysiological substrates, in addition to N_2 and H^+ [1, 6, 83, 84]. Table 10.1 lists the products and reaction equations for some of these reactions. The recent discoveries that nitrogenase can reduce CO to small hydrocarbons are a surprising addition to this collection of reactions [85–88].

One characteristic of these reactions of nitrogenase is that multiple protons are needed. The hydrogenation of N_2 requires 6 protons, and, for instance, the conversion of CO to propane requires 14 protons.

What can be the source of these multiple protons that are transferred to the substrate during the course of each catalytic cycle? Inspection [89] of the protein surrounding FeMo-co in the very high-resolution (1.0 Å) crystal structure PDB

Table 10.1 Some substrates and reactions of nitrogenase.

Substrate	Products	Equations
N_2	NH_3	$N_2 + 6H^+ + 6e^- \rightarrow 2NH_3$
N_2H_4	NH_3	$N_2H_4 + 2H^+ + 2e^- \rightarrow 2NH_3$
HCN	$CH_4 + NH_3$	$HCN + 6H^+ + 6e^- \rightarrow CH_4 + NH_3$
C_2H_2	C_2H_4, C_2H_6	$C_2H_2 + 2H^+ + 2e^- \rightarrow C_2H_4$
		$C_2H_2 + 4H^+ + 4e^- \rightarrow C_2H_6$
$CH_2{=}C{=}CH_2$	$CH_2{=}CHCH_3$	$CH_2 = C = CH_2 + 2H^+ + 2e^- \rightarrow CH_2 = CHCH_3$
CO	CH_4, C_2H_6, C_3H_8	$3CO + 14H^+ + 14e^- \rightarrow C_3H_8 + 3H_2O$

3U7Q (PDB, protein data bank) (and other high-resolution structures), reveals the surrounds to be essentially hydrophobic and anhydrous, and incapable of provision of the multiple protons required. There are no water molecules that could be involved in multiple proton transfers, and the residues flanking the active site, α-71^{Val}, α-70^{Val}, α-69^{Gly}, α-66^{Gly}, α-65^{Ala}, and α-61^{Gly}, are all hydrophobic. In addition, the main helix embracing the active site has its acidic or basic residues and associated water molecules on the distal side of the helix, insulated from the active site. Detailed and pictorial justification of the conclusion that the immediately surrounding protein is incapable of providing multiple protons to intermediates at the active site is provided in the full paper [89].

10.3.2
The Proton Supply Chain

However, analysis of the highest quality crystal structures of the MoFe protein reveals a chain of water molecules that leads from the hydrated surface of the protein to a key sulfur atom, S3B, of FeMo-co [53, 89, 90]. Figure 10.4 shows the characteristics of this chain of hydrogen-bonded water molecules. The first half of the water chain from the protein surface inward is branched and slightly variable, and demonstrably able to accommodate small exogenous molecules such as proton donors: this domain is called the *proton bay*. The second half, from the proton bay to S3B, is a single chain of eight water molecules, intimately hydrogen bound with three oxygen atoms (O3, O6, O5) of homocitrate (reflecting the strict requirement for homocitrate). This is the proton wire, and it is strictly conserved in all protein structures, reflecting its significance. An atomic-level Grotthuss mechanism for serial translocation of protons through this wire has been developed, with a cycle of six stages that move a proton from W8 to S3B [89]. All steps in the cycle involve either linear movement of a proton along a hydrogen bond, or angular movement of H around the O atom of the water molecule to which it belongs. This is a standard Grotthuss mechanism involving two phases, proton hopping along hydrogen bonds and water molecule reorientation [91–93]. The geometrical feasibility of this proton translocation along the proton wire has been demonstrated [89].

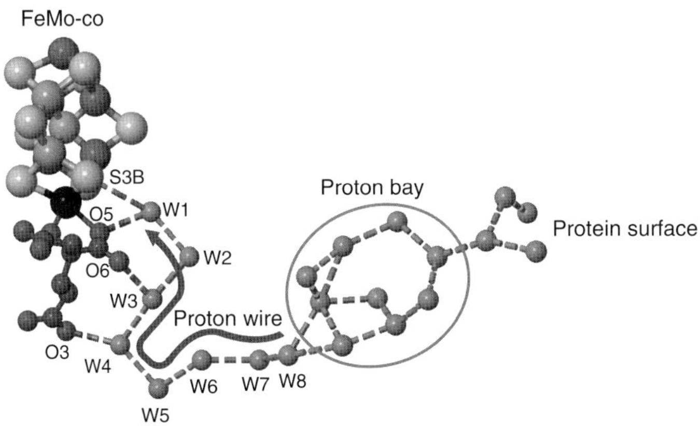

Figure 10.4 The water chain from the protein surface to atom S3B of FeMo-co, involving atoms O3, O6, and O5 of homocitrate. Water molecules are light grey, broken lines are hydrogen bonds. The proton bay (PDB 3U7Q) and the strictly conserved proton wire from W8 to W1 are marked.

This proton supply mechanism can easily provide – from the medium outside the protein – the multiple protons required for each complete catalytic cycle of substrate reduction. Figure 10.5 shows the location of the proton supply path, relative to the electron supply path via the P-cluster. The pathways are geometrically separated, as required. This figure also shows the putative location of the pathway by which N_2 (and other nonpolar substrates) access the FeMo-co active site. This suggestion is based on the locations of two Xe atoms when the MoFe protein is crystallized in the presence of Xe (D.C. Rees, personal communication, 2006) [5]. Igarashi and Seefeldt [4] also proposed this as the ingress path for gas substrates, after independent exploration of the protein structure for a hydrophobic channel.

10.3.3
Hydrogenation of FeMo-co

The coupling of proton transfer and electron transfer steps in chemistry [94, 95] and biochemistry [96, 97] is widely discussed. It is postulated that this coupling occurs in nitrogenase in the following manner. Upon electron transfer to FeMo-co, S3B becomes more negatively charged and basic [53], triggering the final step of proton translocation along the supply path, to S3B. This generates what is effectively a hydrogen atom bound to S3B: the calculated partial charge of this H is about +0.1. This H atom can invert around S3B and then migrate to Fe6, and further to S2B and Fe2. This migration of H away from S3B subsequently allows S3B to attract the next proton from the supply chain, in conjunction with the next electron transfer to FeMo-co. This cycle for serial generation of hydrogen atoms on FeMo-co is illustrated in Scheme 10.1. It is not evident whether proton transfer to S3B triggers electron transfer to FeMo-co, or the reverse, but there is a clear

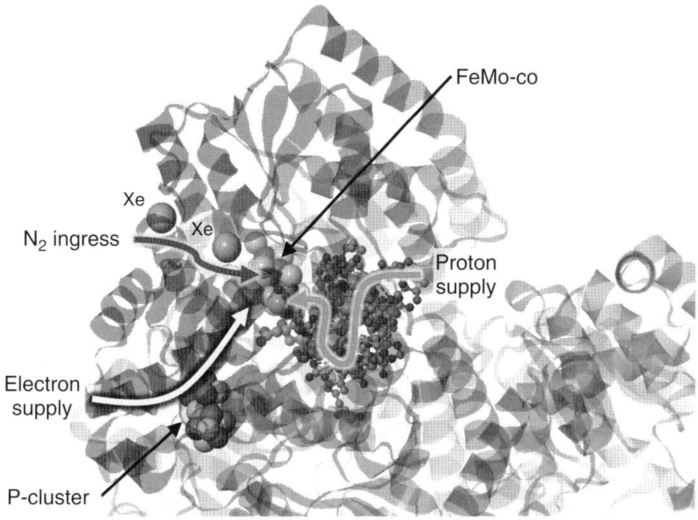

Figure 10.5 Subunit α and parts of subunits β and δ of the MoFe protein, with one FeMo-co and one P-cluster. The proton supply pathway from the surface of the protein to FeMo-co is marked, as is the electron supply path from Fe_4S_4 to the P-cluster to FeMo-co. The locations of two trapped Xe atoms (D.C. Rees, personal communication, 2006) are marked, indicating a putative channel for ingress of substrate N_2.

basis for coupled proton and electron transfer to FeMo-co. The potential energy barriers for the steps involving conformational change at S3B are calculated to be $<10\,\text{kcal mol}^{-1}$ [53, 98].

10.3.4 Hydrogen Atom Migration over FeMo-co

After the first H atom generated on FeMo-co at S3B has inverted around S3B toward Fe6, it is in a position not unlike an S3B–H–Fe6 bridge. Transfer of this H atom to Fe6 in either the endo or exo coordination positions is a straightforward step, with a potential energy barrier of $2.5-6\,\text{kcal mol}^{-1}$ (depending on the electronic/spin state). This is the beginning of a possible cascade of further H-atom migration steps, in which the H atom can move to S2B and to the endo or exo positions on Fe2 [53]. As soon as this H atom has moved to Fe6 or beyond, S3B is available for generation of a second H atom in the same manner as the first. This ability to replenish H atoms on S3B once prior H atoms have migrated away is a repetitive property, with the capacity to load FeMo-co with multiple H atoms. There are several possibilities for the H-atom migration steps, and depending on the sequences of these steps there are multiple possibilities for collections of H atoms bound to atoms S3B, Fe6, S2B, and Fe2 and including an H-atom bridge between Fe6 and Fe2. For the first H atom transferred from S3B, there are six

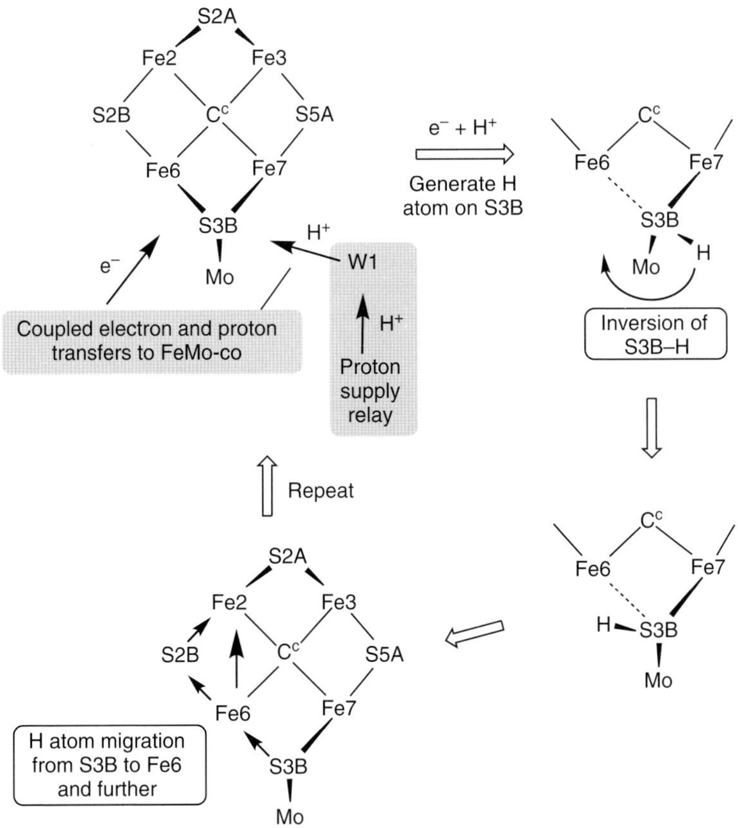

Scheme 10.1 Cycle for serial H-atom generation on FeMo-co by proton-electron coupling at S3B.

stable product structures – *exo*-Fe6-H, *endo*-Fe6-H, S2B–H, Fe6–H–Fe2, *endo*-Fe2-H, and *exo*-Fe2-H. For two, three, or four H atoms on FeMo-co, there are more possibilities. A full account of all stable structures, and their electronic and spin states, and of the energy profiles of the migration steps, is provided elsewhere [53, 98].

The important conclusion from these calculations is that the active face of FeMo-co can function as a store of H atoms. Various H-atom storage configurations on FeMo-co are possible. These states correspond to the E_nH_n stages ($n = 1-4$) of progressive reduction of FeMo-co that were first proposed by Thorneley and Lowe [99, 100] to interpret their extensive kinetic data on the reactivity of nitrogenase. The early part of this Thorneley–Lowe mechanistic scheme is shown in Scheme 10.2, emphasizing the E_1H_1, E_2H_2, E_3H_3, and E_4H_4 stages, which are states of the MoFe protein reaction cycle, each of which is generated by one complete Fe protein cycle introducing one electron. The various multiply hydrogenated forms of FeMo-co resulting from the calculations [53, 98] represent the structural possibilities for the E_nH_n kinetic states, and further emphasize that there are numerous structural possibilities for each of the E_nH_n states. Thorneley and Lowe measured the evolution of H_2, from kinetic states E_2H_2, E_3H_3, and E_4H_4 (Scheme 10.2): the structural interpretation of this is straightforward in general concept, because many of the hydrogenated forms of FeMo-co with two or more H atoms have those H atoms in contiguous locations, from which H_2 can be formed (see Section 10.7.1) with low activation barriers [53, 98].

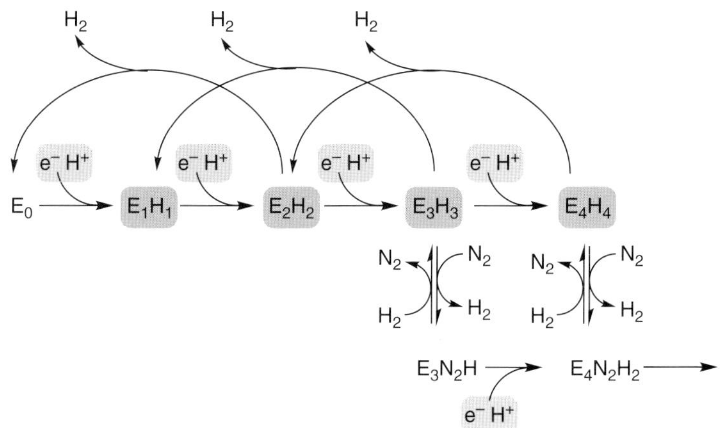

Scheme 10.2 Early sections of the Thorneley–Lowe mechanism.

10.4
FeMo-co in Nitrogenase as a General Hydrogenating Machine

10.4.1
Modes of Substrate Binding to FeMo-co

The binding of substrates to the Fe atoms of FeMo-co follows the normal principles of ligand coordination. For instance, N_2 can bond to Fe6 or Fe2 in η^1 (end-on) or η^2 (side-on) modes, at the endo and exo coordination sites. Similarly, calculated structures for η^1-CO, η^2-C_2H_2, and η^2-H_2 and for hydrogenated intermediates such as N_2H_2, N_2H_4, and η^2-C_2H_4 coordinated to Fe of FeMo-co are analogous to those established in standard coordination complexes. If an H atom and substrate are bound to the same Fe atom of FeMo-co, then they occupy clear endo and exo positions (the substrate can be in either), while a substrate bound alone to Fe can also occupy a position between endo and exo. One Fe atom of FeMo-co can bear both η^2-H_2 and N_2 (η^1 or η^2).

Ligand H can bridge Fe2 and Fe6, but bridging by N_2 or CO through a single donor atom (N or C) between two Fe atoms of FeMo-co is generally unstable, with reversion to η^1 coordination in the endo position of one of those Fe atoms. Intermediates in the hydrogenation of N_2 can form stable single-donor-atom bridges between Fe2 and Fe6, as described in Section 10.5.2.3.

As well as the structural aspect of substrate binding, in many cases the TS for association was determined, and the potential energy barriers calculated for association and dissociation of the substrate (or product) [54, 71, 73]. The identity of the central atom (C^c or N^c) has a marginal influence on the coordination of reaction substrates and intermediates at the surrounding Fe atoms. Early simulations of the coordination chemistry at Fe were made with N^c, and subsequently repeated with C^c once this identity was established experimentally: the differences in structure and association dynamics were relatively small [61]. In addition, calculations show that the influence of the identity of the six-coordinate metal in FeMo-co (Mo or V) on substrate binding at Fe is secondary.

Ligation at one or two central Fe atoms of FeMo-co can substantially weaken the bonding between that Fe atom and the central C atom, lengthening Fe–C^c (up to about 3 Å), with concomitant minor distortions of FeMo-co. The central C atom has an important role in stabilizing the structure of FeMo-co, because calculations of models with the central atom absent reveal serious major distortions of FeMo-co when coordinated by N_2 or CO. A more subtle role of the central atom is in modulating details of ligation at both Fe2 and Fe6. This is the phenomenon of coordinative allosterism [53, 55, 61], in which additional ligation at one Fe atom influences and connects with the coordinative properties of a second Fe atom, via C^c.

10.4.2
Vectorial Hydrogenation of FeMo-co in Relation to Substrate Binding

FeMo-co can accumulate and store multiple H atoms, through multiple series of H-atom migration steps, as illustrated in Figure 10.6a. FeMo-co can also bind substrates at Fe6 or Fe2, and such bound substrate can be surrounded by H atoms. The reduction of the substrate is then readily conceived as transfers of H atoms to the substrate. This concept is illustrated in Figure 10.6b. Multiple H atoms can be located around the bound substrate, and positioned so that steps forming substrate-H bonds are geometrically favorable. In this way, the enzyme possesses general hydrogenating machinery: (i) FeMo-co provides sites that can bind substrates using normal coordination chemistry, (ii) FeMo-co also provides surrounding H atoms able to hydrogenate the substrate and its intermediates, and (iii) there is no limit to the number of H atoms provided to a substrate for its complete hydrogenation.

This general mechanism for addition of H atoms to substrate and intermediates has several important attributes. First, the H-atom supply and migration is vectorial, being sourced at S3B and moving through Fe6 to S2B and Fe2: H-atom replenishment can move only in this direction. Therefore, once the substrate is bound at Fe6, H-atom migration past Fe6 is effectively blocked, and the H atoms

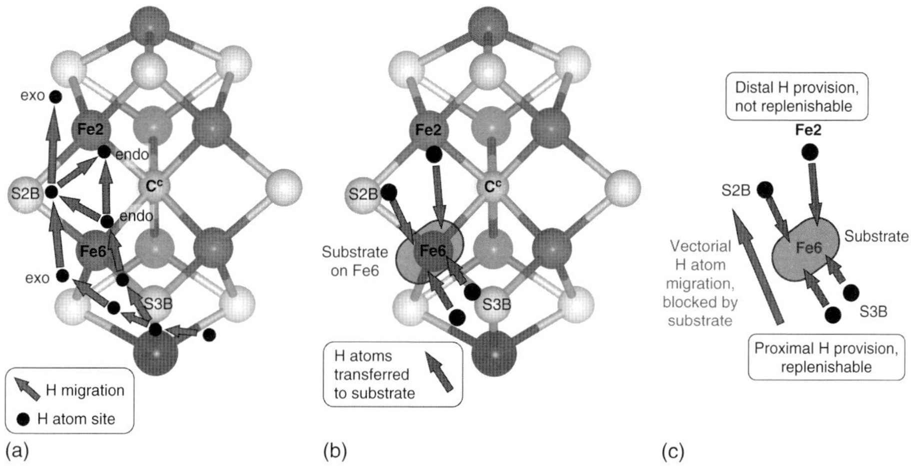

Figure 10.6 Fundamental concepts for the hydrogenation of FeMo-co and for the hydrogenation of substrates. (a) The H-atom binding sites and migration pathways for multiple hydrogenation of FeMo-co. (b) Pathways for H-atom transfers to substrate bound to Fe6. (c) Substrate or intermediate bound at Fe6 blocks replenishment of H atoms on the distal side but not on the proximal side.

surrounding the substrate are differentiated as proximal and distal relative to the migration vector (see Figure 10.6c). Proximal H atoms can be replenished via S3B once they have transferred to the substrate, while distal H atoms are not replenishable. Second, the number and distribution of H atoms on FeMo-co influences the activation energies for association and dissociation of the substrate. This influence occurs because H atoms on FeMo-co change its geometry and electronic structure, as well as the availability of coordination positions. Thus, there are important mutual influences between H-atom accumulation and substrate binding to FeMo-co. This provides a mechanism for the selection of the hydrogenation level, E_0, E_1H_1, E_2H_2, E_3H_3, at which various substrates are first bound, and for the levels that can lead to the product. These properties add a structural dimension to the Thorneley–Lowe kinetic scheme (Scheme 10.2). They are also a key part of the interpretation [68] of other kinetic data, in which N_2 is a weak competitive inhibitor of C_2H_2 reduction, but C_2H_2 is an effective noncompetitive inhibitor of N_2 reduction [101–103].

10.4.3
The Intramolecular Hydrogenation Paradigm for the Catalytic Reactivity of FeMo-co

The capacity of FeMo-co within nitrogenase to hydrogenate substrates involves relatively simple reaction steps in which an H atom from an S–H group or an Fe–H group is transferred to an atom of the contiguously bound substrate. Repetition of these H-atom transfers from S–H or Fe–H to the partially hydrogenated substrate can continue, until the overall reaction is complete.

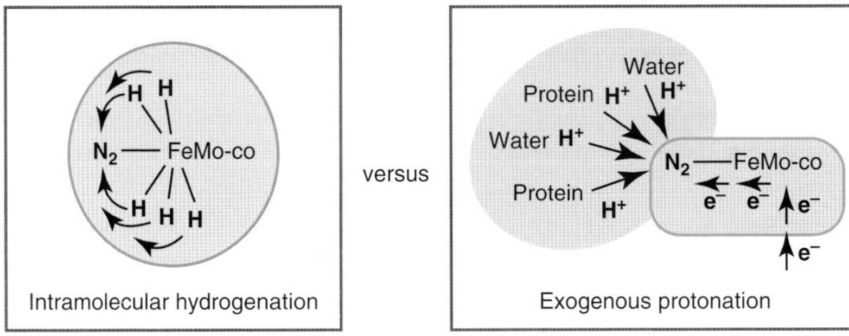

Figure 10.7 The contrast between the intramolecular hydrogenation paradigm and the exogenous protonation paradigm, using N_2 as substrate.

Additional bond-breaking steps for the partially hydrogenated substrate (such as N–N scission in the reduction of N_2) are also part of the cycle. The formation of H_2 involves the same types of reactions, that is, S–H + Fe–H, or Fe–H + Fe–H. Examples of these steps are detailed below.

The important characteristic of these steps through which substrate is converted to product is that they are all intramolecular. This general mechanistic framework, applicable to all reactions of nitrogenase, is one of intramolecular hydrogenation. The *intramolecular hydrogenation* paradigm is distinctly different from an alternative mechanistic concept, which involves electronation of the substrate and *exogenous protonation* of the substrate. This alternative concept assumes that electrons are transferred to FeMo-co, then to the bound substrate, and that the more electron-rich (basic) substrate is protonated from the protein surrounds and/or water within the protein surrounds. The difference between the two mechanistic paradigms, intramolecular hydrogenation and exogenous protonation, is emphasized in Figure 10.7. As already mentioned, the surroundings of FeMo-co in the nitrogenase protein are hydrophobic and anhydrous, and there is no evident way in which multiple exogenous protonation steps could occur. One advantageous attribute of intramolecular hydrogenation in nitrogenase is the capacity for multiple, sequential, intramolecular hydrogenation steps, as required for many substrates. Exogenous protonation is unable to meet this requirement. A second advantageous attribute of intramolecular hydrogenation is postulated hydrogen atom quantum tunneling in each of the fundamental steps that comprise the mechanism: this is discussed further in Section 10.6.

The intramolecular hydrogenation paradigm and the exogenous protonation paradigm as discussed here for nitrogenase and related systems are both subsumed under the widely applicable concept of proton-coupled electron transfer [95, 104]. The two paradigms differ in the location and timing of the coupled transfers of electrons and protons. In the intramolecular hydrogenation mechanism, the coupling is proposed to occur at S3B, generating a hydrogen atom for subsequent transfer as such during the hydrogenation of substrates. In the exogenous protonation mechanism, the coupling occurs later, on the substrate.

10.5 Chemical Mechanisms for the Catalysis of Substrate Hydrogenation at FeMo-co

It is now appropriate to provide more substance to the hypothesis that the catalytic hydrogenation effected at FeMo-co in nitrogenase occurs through repeated intramolecular H-atom transfer steps. The most developed mechanism is that for the physiological reaction, conversion of N_2 to $2NH_3$ [71]. Before describing the choreography of this reaction, which must have at least 16 steps, it is necessary to understand how the substrate, N_2, can be coordinated to FeMo-co.

10.5.1 How Does N_2 Bond to FeMo-co?

An early investigation dealt with the binding of N_2, and its dependence on the nature of the prior hydrogenation of FeMo-co [54]. A search for energy-minimized structures of FeMo-co coordinated by N_2, together with varying degrees and types of H/H_2 coordination at Fe and/or hydrogenation of S atoms, and restricting coordination to Fe2 and Fe6, revealed 94 different local energy wells, encompassing η^2- and η^1-bound N_2. No energy minima were found for N_2 bridging two, three, or four Fe atoms of the front face of FeMo-co. The significance and stability of 57 representative modes of binding N_2 was assessed by calculations of the TS along the association/dissociation pathway. The majority of these calculations were made with N-centered FeMo-co, before the availability of evidence for C centering, but exploratory calculations with C-centered FeMo-co indicated that the identity of the central atom was not a major factor in N_2 binding dynamics [54, 61].

The principal observation from these results is the impressive diversity of geometrical structure, and of association dynamics. FeMo-co is a remarkably capable agent for the coordination of small molecules, as would be expected: it is more capable than a single-metal coordination site. In general, the barriers to association of N_2 depend mainly on the location of the hydrogenation and the location of N_2 coordination, while dissociation barriers depend primarily on whether N_2 is η^2- or η^1-coordinated, and secondarily on the location of the hydrogenation. Detailed analyses of correlations between structure and dynamics of the binding of N_2 to FeMo-co are published [54, 55].

These structures for bound N_2 were calculated first without the surrounding protein, and then were further evaluated for their specific interactions with the surrounding protein, and in particular residue α-70. Experimentally, mutation of α-70 from valine to isoleucine (i.e., expanding the side chain from $CH(CH_3)_2$ to $CH(CH_3)(CH_2CH_3)$) severely diminishes the ability to reduce N_2 (and of all other substrates except H^+) [26, 105]. This is a valuable datum for determination of the more probable modes of binding of N_2 in the wild-type enzyme. The examination of specific interactions between the N_2-ligated FeMo-co models and the protein in which it would be embedded asked (i) whether the various models for N_2 plus

H/H_2 coordination of FeMo-co would fit into the protein, (ii) what movements of surrounding protein might be required for N_2 binding, and (iii) what energy penalties might these incur. The procedure involved substituting the ligated FeMo-co for unligated FeMo-co in the protein (using a relevant section of PDB 1M1N containing 1332 residues plus 1032 water molecules and with all hydrogen atoms included), and examining the consequent structural changes and strain energies. There is insufficient space here to describe all of the insights resulting from these *in silico* transplantation experiments: the favored position for binding of N_2 was the endo position of Fe6 [54].

Hinnemann and Nørskov [51] calculated that exo-η^1 coordination of N_2, with three H atoms on the three equatorial μ-S atoms was most stable. The calculations of Kastner, Blochl, and coworkers [65, 70] on the binding of N_2 led them instead to disrupted structures with severed Fe–S bonds and nonbridging Fe–SH groups.

10.5.2 Proposed Intimate Chemical Mechanism for the Catalysis of Hydrogenation of N_2 to NH_3 at FeMo-co

10.5.2.1 Possibilities

The overall catalysis must have at least 16 processes, which are (without regard for temporal sequence) the binding of N_2, the collection and migration of six H atoms, six transfers of H atoms to N atoms, scission of the N–N bond, and the dissociations of two NH_3 molecules. There are numerous possibilities for the sequence of mechanistic steps, arising because (i) N_2 can be bound in η^1 or η^2 mode, (ii) N_2 binding can be at Fe2 or Fe6, in endo or exo coordination positions (but the most likely is endo coordination at Fe6), (iii) the number of H atoms assembled on FeMo-co before N_2 binding is variable, but probably 3 or 4 [99, 100, 106, 107], and there are several possible placements for these H atoms, (iv) there are multiple possibilities for the crucial first H to N transfer and for the structure of the intermediate containing N_2H, (v) then there are multiple possibilities for the second H to N transfer, with HNNH or NNH_2 as possible formulations for the second intermediate, and several coordination structures involving Fe2 or Fe6 or both, and (vi) at each stage there is the possibility of introduction of a new H atom via S3B.

10.5.2.2 Preliminary Simulations

Given this plethora of mechanistic possibilities, it is fortunate that DF simulations can evaluate them with some efficiency and accuracy. Preliminary surveys of diverse possibilities for the first intramolecular transfer of H to bound N_2, a difficult step in view of the electronic structure of N_2, and the thermochemistry of its hydrogenation, allowed rejection of many options. In particular, the first transfer is difficult if N_2 is η^1-bound, and it is believed that this mode of binding N_2 does not lead to NH_3 but instead accounts for the inhibitory effects of N_2 on reactions with other substrates. The best framework for the first H-transfer step involves η^2-bound N_2 at the endo position of Fe6, which, as shown in

Scheme 10.3a, is stereochemically well positioned for multiple H-transfer steps (arrows). The preliminary investigations also revealed intermediates that are favorably formed and suitable for subsequent H-transfers, as well as a favorable N–N bond-breaking step. Thus, the overall mechanism (with alternative paths at some stages) was developed around these auspicious intermediates, tracing the sequence of steps with the lowest potential energy barriers. The key to the mechanism is the progression of structures shown in Scheme 10.3b, in which the N–N entity is first located on Fe6, then (in increasingly hydrogenated intermediates) bridges to Fe2, then spans between Fe2 and Fe6, before breaking the N–N bond. Note the continuing diagonal motif for the N–N vector, and that there is minimal movement of N atoms throughout this sequence.

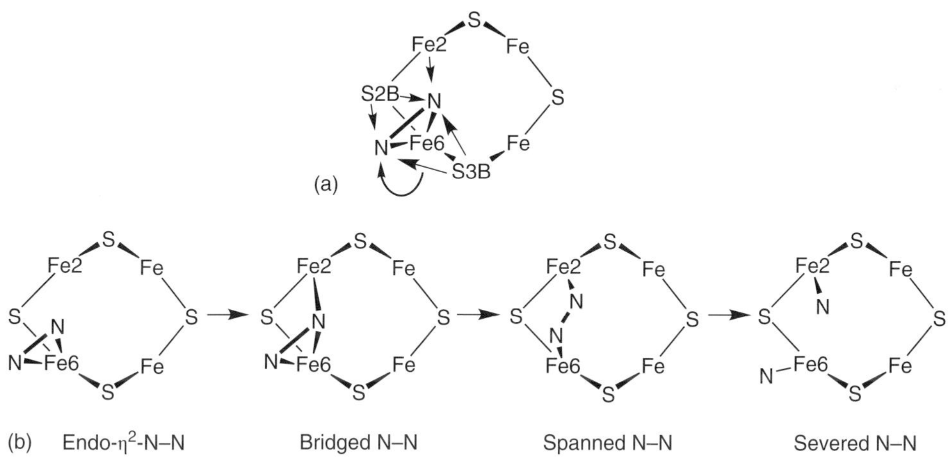

Scheme 10.3 (a) The possible H-atom transfers from Fe2, S2B, Fe6 and S3B to diagonally bound N_2. (b) The diagonal orientation of the N-N entity maintained through the mechanism.

10.5.2.3 Complete 21-Step Choreography

The full mechanism with 21 steps has been described in detail [71]. Scheme 10.4 outlines the main intermediates in the cycle, omitting some of the steps in which new H atoms are introduced. The addition of the first H atom to the inner N atom of N_2 causes it to bridge (without barrier) between Fe6 and Fe2, which then favors addition of the second H to form *cis*-diazene HNNH, also partially bridging Fe2 and Fe6 (both of which possess normal octahedral stereochemistry). At this point, the distal H atoms on S2B and Fe2 are added sequentially, to form the hydrazine intermediate that spans Fe2 and Fe6. In this intermediate, the stereochemistries of both NH_2 moieties of coordinated hydrazine are normal tetrahedral, and the coordination stereochemistries of Fe2 (square pyramidal) and Fe6 (octahedral) are normal. At this point, the N–N bond can lengthen and sever, with a potential energy barrier of 14 kcal mol^{-1}, and at the same time the NH_2 group forming in the endo position of Fe6 undergoes inversion, which favors its acceptance of the *exo*-Fe6-H atom to form NH_3 that spontaneously dissociates from Fe6 during this step. Finally, a new H atom is introduced from S3B onto the NH_2 bonded to Fe2, forming NH_3 that dissociates with a small barrier of 6 kcal mol^{-1}.

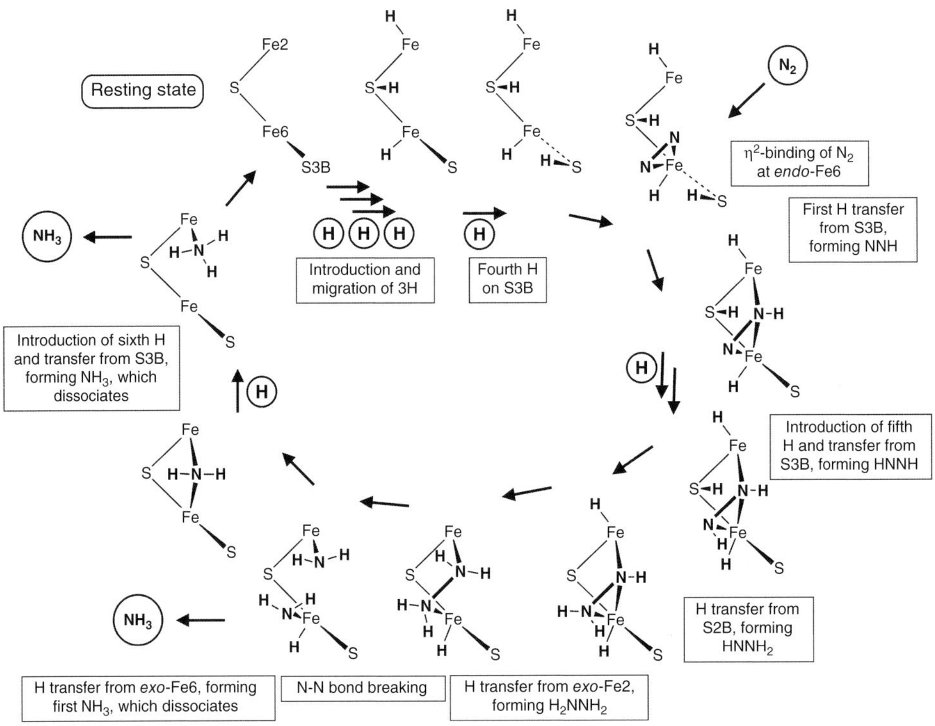

Scheme 10.4 Outline of the main mechanism proposed for hydrogenation of N_2 to NH_3.

This is a chemically satisfying mechanism, in that the structures are all consistent with standard coordination chemistry (although these intermediates have not been replicated in synthetic models). If it is reasonable geometrically, then it must be asked whether the mechanism is feasible kinetically. The three largest calculated potential energy barriers in the cycle have values of 24, 21, and 17 kcal mol^{-1} [71], which appear to imply reaction rates slower that the experimental turnover of nitrogenase, which is 1–5 s^{-1} [99, 108, 109]. The potential energy barriers as calculated here are incomplete as indicators of kinetics because zero-point energies and entropies are not included. Because the steps are intramolecular, the entropy changes are expected to be small. However, there is another factor that needs to be considered carefully, and that is the possibility that quantum tunneling of hydrogen atoms accelerates the H-transfer steps.

10.6 Hydrogen Tunneling in the Nitrogenase Mechanism

10.6.1 Characteristics of H Atom Tunneling in Enzyme Reactions

Current research indicates that the majority of enzymatic reactions involving H-atom transfers occur, in part or in full, by quantum mechanical tunneling [110–119]. Hydrogen tunneling, a quantum mechanical phenomenon, is possible when there is overlap of the probability functions of the reactant and product states. Tunneling is often evaluated in terms of the de Broglie wavelength, which for a hydrogen atom with ambient (300 K) thermal energy of 2.5 kcal mol^{-1} is about 0.5 Å: the concept is that as the H atom ascends a barrier it can tunnel through it when the barrier width is of similar or smaller dimension [118, 120].

The occurrence of tunneling in many enzymatic reactions involving transfer of H atoms, generalized as D–H···A → D···H–A (donor atom D is often carbon), has been investigated and established largely through kinetic isotope effects and their temperature dependence [110, 113, 120, 121]. Two important aspects of this tunneling are the width of the barrier and the coupling of local vibrational dynamics with the D–H–A reaction coordinate, affecting the symmetry and width of the barrier. During a hydrogen atom transfer reaction, tunneling can occur when the D–A separation is reduced to about 2.7 Å from the normal van der Waals separation of >3.2 Å. There is considerable discussion about the roles of protein motions that affect the D–A distance [119, 122].

Detailed experimental [123] and theoretical [124] investigations of a rate-determining C–H to O transfer in the oxidative deamination of tryptamine by aromatic amine dehydrogenase concludes that >99% of the proton transfer goes via a combination of classical activation and proton tunneling instead of via a purely classical route. The activation free energy barrier is lowered by ~3 kcal mol^{-1}. The proton tunnels ~0.5 Å through the barrier, in which the donor–acceptor (C–O) distance is 2.7 Å. For a different reaction, the hydrogen transfer from nicotinamide adenine dinucleotide (NADH) to enzyme-bound flavin mononucleotide (FMN) in flavoprotein morphinone reductase, theoretical calculations suggest that 99% of the transfer proceeds by tunneling, and that the reactants approach by classical activation to a D–A distance of 2.73 ± 0.01 Å, at which point the hydride tunnels a distance of ~0.4 Å [125].

Computational investigations of these hydrogen tunneling reactions show that there is usually a preparatory stage during which the D–H and A functionalities move toward each other until the D–A distance is close to that (about 2.7 Å) at which tunneling occurs. This preparation to a "tunneling-ready" state could be a local vibration or protein conformational motion [114, 118, 126–129].

10.6.2
Characteristics of H-Atom Transfer in Nitrogenase

In this context, the H-transfer steps in the mechanism proposed in Section 10.5.2.3 appear to be excellent candidates for acceleration by hydrogen quantum tunneling. The step involving the first transfer of H from S3B to N of bound N_2 has been examined for its conformance with the geometrical characteristics of hydrogen tunneling. To my knowledge, there has been no prior research on tunneling in reactions involving S–H to N or Fe–H to N transfer steps. In contrast to the C–H-transfers studied for enzymes, H/D isotope exchange for S–H and Fe–H reactants would complicate experimental measurements of kinetic isotope effects in the H-transfer steps proposed for nitrogenase.

Figure 10.8 provides details of the trajectory for the first transfer of H, from S3B to the inner N atom of N_2 η^2-bound at the endo position of Fe6 [72]. There is not a single bond-breaking/bond-making reaction coordinate (as is evident from the structures of the reactant, TS, and the product), and four significant geometrical changes are therefore plotted with the change in energy: they are S3B–H breaking, the N–H making, the S3B–N transfer distance, and the Fe6–S3B distance. Three phases of transformation are recognized. In the preparatory phase, the S3B–H bond is intact, but the S3B–H entity flaps toward Fe6, shortening the S3B–N transfer distance from 3.4 to 2.9 Å, and shortening N–H but without any bond formation. In the H-transfer phase through the TS, the N–H bond is formed fully while the S3B–H bond is broken. This involves movement of the H atom by 0.6 Å, while the separation of the atoms S3B and N that bear the transfer is relatively constant at 2.8 Å. The completion phase involves conformational change of coordinated N–N–H to its equilibrium geometry, and re-formation of the bond between S3B and Fe6.

These characteristics are similar to those of established hydrogen tunneling reactions as outlined earlier: (i) the H-transfer barrier is narrow, involving about 0.6 Å movement of H to complete the bond breaking and bond making; (ii) the preparatory phase, to the tunneling-ready state, involves a low energy "flapping" vibrational motion of S3B–H, which decreases the separation between the donor (S3B) and acceptor (N) atoms; (iii) this separation between the donor and acceptor atoms is constant during the transfer; and (iv) the D–A separation of 2.8 Å for S3B and N is equivalent to the 2.7 Å separation of C and O in the enzymes described in Section 10.6.1, due to the larger size of sulfur. Therefore, it is postulated that this step would be accelerated by quantum tunneling of hydrogen [72].

Similar analysis of the third H-transfer, converting the HNNH intermediate to the H_2NNH intermediate (see Scheme 10.4) involving S2B–H to N transfer, reveals a preparatory flapping vibration of S2B–H, and a narrow H-transfer barrier of about 0.6 Å. Most of the steps of the complete mechanism (Scheme 10.4) have similar relatively small movements of the transferring H atom through the barrier, and therefore appear to be excellent prospects for extensive hydrogen tunneling as part of the catalytic rate enhancement. This is a consequence of

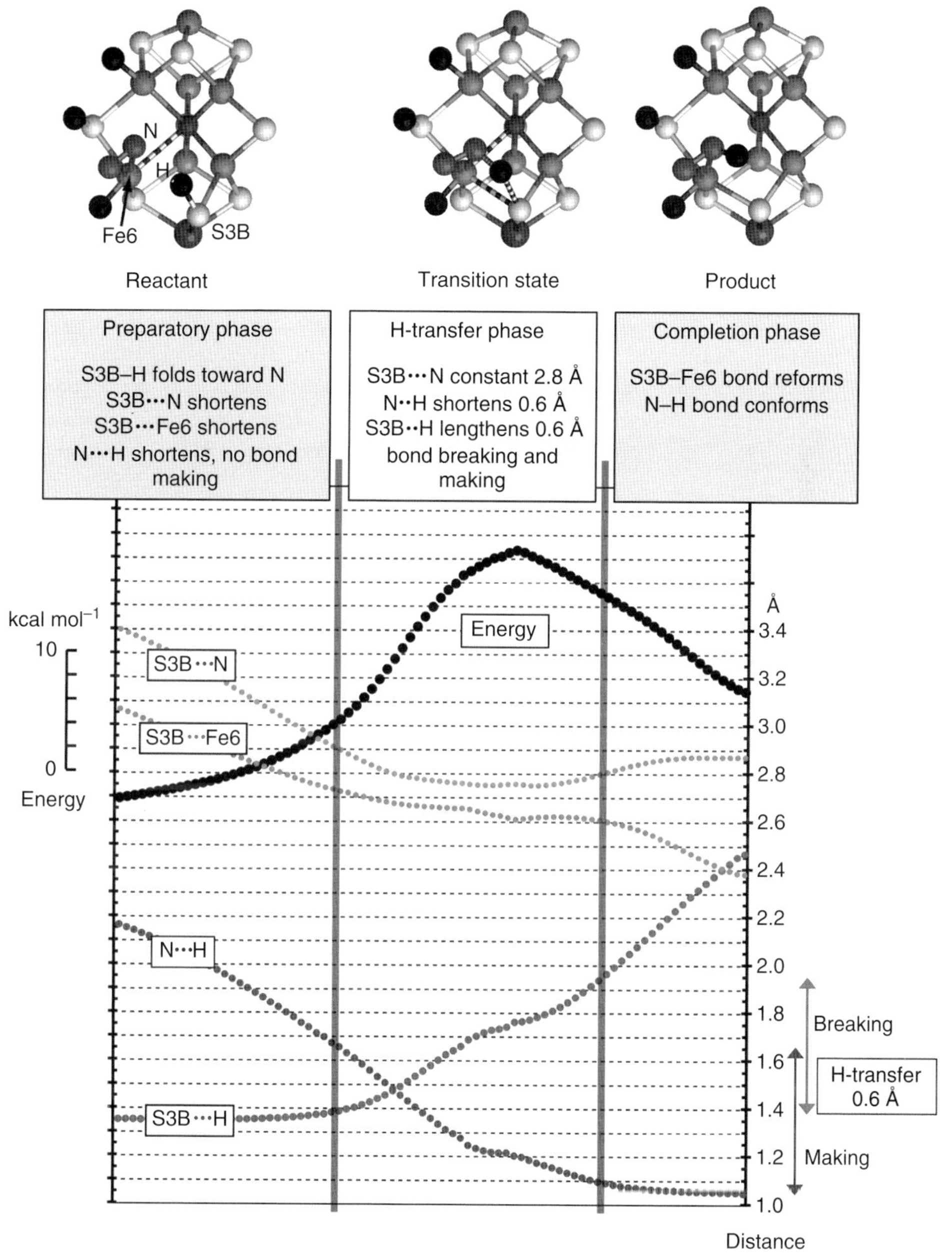

Figure 10.8 Details of the energy (black, left ordinate, kilocalories per mole) and relevant geometry (right ordinate, angstrom) changes during the first H-transfer from S3B to N. The bond-breaking S3B-H and bond-forming N-H distances are grey; non-bonding S3B–N and S3B–Fe6 distances are light grey. The preparatory, H-transfer, and completion phases are discussed in the text.

the intramolecular character of the mechanism, and is it is probable that similar hydrogen tunneling will be part of the mechanism for hydrogenation of other substrates on the surface of FeMo-co. Experimental assessment of this tunneling will be difficult, but further calculations can estimate the likely magnitude of the tunneling rate enhancement.

10.7 Intramolecular Hydrogenation of Other Substrates

The essence of the mechanism proposed for the nitrogenase-catalyzed hydrogenation of N_2 to NH_3 is applicable to all other substrates. The general concept is that hydrogen atoms assembled on FeMo-co can transfer readily, and probably with tunneling, to a substrate bound at Fe6 or Fe2. For some substrates containing an oxygen atom, water is a co-product, and these reactions require an additional type of step (outlined below) in which water is eliminated.

10.7.1 Formation of Dihydrogen

Dihydrogen is an inevitable product of nitrogenase turnover, in variable proportions depending on turnover conditions, but with at least one H_2 generated for each $N_2 \rightarrow 2NH_3$ cycle [130]. The reaction is formally written as $2H^+ + 2e^- \rightarrow H_2$. Within the intramolecular hydrogenation paradigm, the formation of H_2 is readily explicable simply as the combination of two H atoms on FeMo-co. These two H atoms may be present as Fe–H and S–H, or can be both Fe–H. Scheme 10.5 shows three examples of H-atom combination steps which form H_2 coordinated to Fe, and the dissociation of such H_2 from FeMo-co has a low potential energy barrier of $\leq 5\,kcal\,mol^{-1}$ [53]. In this mechanistic context of contiguous H atoms forming H_2, which is then readily dissociable, and with reference to the variety of forms of hydrogenated FeMo-co possessing contiguous H atoms, it is not surprising that H_2 is always a product of turnover. Indeed, this inevitable formation of H_2 could be regarded as strong evidence for the proposed hydrogenation of FeMo-co at multiple sites. The effect of N_2 in diminishing the production of H_2 is easily accommodated in this mechanistic paradigm because bound N_2 occupies sites involved in some of the H_2 formation steps.

10.7.2 Hydrogenation of Alkynes

As an alternative substrate, acetylene is significant for a number of reasons. First, it is rapidly reduced by nitrogenase to ethylene, and has long been used for practical activity assay of nitrogenase enzymes. Second, it has the potential to monitor the cis/trans stereochemistry of addition, and the hydrogenation of C_2D_2 to form *cis*- or *trans*-$C_2H_2D_2$ has been reported for wild-type and mutant enzymes

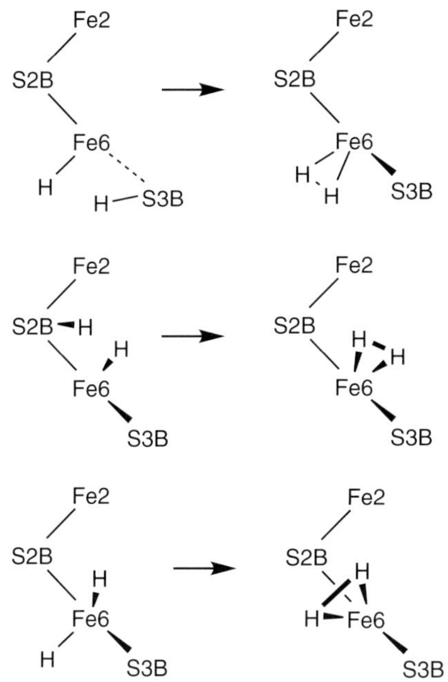

Scheme 10.5 Possibilities for H_2 formation.

[1, 131, 132]: wild-type nitrogenase effects 96% cis-hydrogenation. This stereochemical data about the mechanism of reaction is not provided by N_2 and all other substrates except other alkynes and cyclopropene [133, 134]. Third, there is substantial accumulated biochemical data on the mutual interactions of acetylene and other substrates and inhibitors [101–103, 135, 136]. Thorneley and Lowe [137] measured the kinetics of the reduction of acetylene by *Klebsiella pneumoniae* nitrogenase and concluded that productive involvement of C_2H_2 begins at the E_1H_1 level of prehydrogenation of FeMo-co, but can also occur from the E_2H_2 state, and that after binding of C_2H_2 a further two cycles (Fe protein cycles) of H atom introduction are required until C_2H_4 is released.

Hypotheses for the mechanism of the double hydrogenation of C_2H_2 are readily developed, by analogy with the N_2 mechanism, as illustrated in Figure 10.9. Coordination of C_2H_2 in standard η^2 mode at the endo position of Fe6 leaves it positioned and primed for two sequential H-transfer steps from S3B to cis positions, accounting for the observed stereospecificity. The presence of *exo*-Fe6-H enforces endo coordination of η^2-C_2H_2 at Fe6, and cis addition. The E_2H_2 state of FeMo-co could have an H atom on S2B or Fe2 when it binds C_2H_2, and if this distal H (see Figure 10.6c) adds to η^2-bound C_2H_2 together with a proximal H from S3B then trans addition results [53]. Some nitrogenases also produce a small amount of ethane, and the transfer steps in Figure 10.9 provide the mechanistic basis for this.

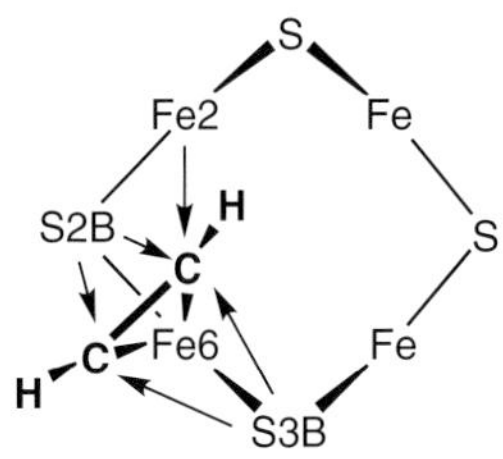

Figure 10.9 Generalized model for intramolecular hydrogenation of C_2H_2 bound to Fe6.

The elaborated alkyne HC≡C-CH_2OH (propargyl alcohol) has yielded detailed experimental information about the reduction of alkynes, because when it interacts with α-$70^{Val→Ala}$ modified nitrogenase an intermediate can be freeze-trapped and characterized spectroscopically [30]. ENDOR investigation concluded that the trapped intermediate was the doubly hydrogenated product, allyl alcohol (H_2C=C-CH_2OH), η^2-bound to an Fe atom. Further experiments using also propargyl amine (HC≡C-CH_2NH_2), and with additional mutation of residue α-195^{His}, and measurement of pH dependence [29], enabled the site of binding of the product alkenes H_2C=CH-CH_2X (X = OH or NH_2) to be localized in more detail on Fe6, with the X group hydrogen bonded to Nϵ of α-195^{His}. This work was supported by detailed exploratory DF calculations of the possibilities and *in silico* transplantations of the optimized structures into the protein. The most probable structure for H_2C=CH-$CH_2NH_3^+$ bound to Fe6 involves hydrogen bonds with S2B and Nϵ of α-195^{His} [49]. When the protein with this intermediate in place was viewed (as a Corey–Pauling–Koltun (CPK) space-filling graphic) along the substrate ingress channel described by Igarashi and Seefeldt [4], the intermediate was just visible, supporting both the proposed structure of the intermediate and the pathway for ingress of the propargyl reactant [49].

Experiments with C_2H_2 and N_2 as co-substrates have yielded mechanistically valuable data. In the wild-type enzyme, N_2 is a weak competitive inhibitor of C_2H_2 reduction, but C_2H_2 is an effective noncompetitive inhibitor of N_2 reduction [101–103]. This behavior changes in the α-$69^{Gly→Ser}$ mutant, which has diminished C_2H_2 activity (but essentially unchanged N_2 and H^+ activity), and the inhibition of N_2 activity by C_2H_2 is competitive in contrast to the noncompetitive inhibition of wild-type protein [135, 136]. The α-$195^{His→Gln}$ mutant is unable to reduce N_2, but N_2 is a competitive inhibitor of C_2H_2 reduction [138], indicating that N_2 is still binding, but unproductively. In addition, many experimental data have led to postulation of high-affinity and low-affinity binding sites and pathways for reaction with C_2H_2 [132, 135, 139–141]. A unified explanation of these observations has been developed [68]. The essence of the model arises from the prehydrogenation of FeMo-co, as in Figure 10.6. The positions of H atoms influence the binding of substrates N_2 and C_2H_2, and the bound substrate subsequently blocks further migration of H atoms past the binding site. It is proposed that Fe6 is the high-affinity site for C_2H_2, and Fe2 is the low-affinity site, and that the different experimental kinetic characteristics of these two sites are not due to fundamental differences in their binding of C_2H_2 but to differences in the preparatory location of H atoms on FeMo-co. See the original paper [68] for details of the models and the full explanations.

10.7.3 Hydrogenation of D_2: the HD Reaction

A mechanistically important substrate for nitrogenase is D_2, which is reduced to 2HD [142], according to the equation: $D_2 + 2H^+ + 2e^- \rightarrow 2HD$. This is traditionally called the *HD reaction* of nitrogenase. Note that even though protons are involved, this reaction is different from the reduction of protons. The HD reaction is the hydrogenation of D_2 with D–D breaking, and in this sense is analogous to the hydrogenation of N_2. Also, the HD reaction of nitrogenase is *not* isotope exchange (i.e., $D_2 + H_2 \rightarrow 2HD$), as is effected by hydrogenase enzymes [143]. The HD reaction of nitrogenase is *not* catalyzed isotope exchange between D_2 and water [144].

The HD reaction of nitrogenase is not just a sideline reaction, but is significant for investigation of the mechanism of the $N_2 \rightarrow NH_3$ functionality of nitrogenase, because the HD reaction requires N_2 (and only N_2, not any of the other nonphysiological substrates of nitrogenase), and because H_2 (and D_2) inhibit $N_2 \rightarrow NH_3$ activity [99, 144–148].

The core of the mechanism proposed for the HD reaction follows directly from the intramolecular hydrogenation paradigm, as illustrated in Figure 10.10a, with two H to D transfers from S3B [98]. The D atom(s) are bonded only to Fe; and as the H to D transfer steps are effectively irreversible, there is no opportunity for D to exchange into the environment. It is unlikely that the H donor to D_2 is located on Fe rather than S3B, because this would set up the steps shown in Figure 10.10b, steps that are likely to be reversible and would enable the exchange $H_2 + D_2 \leftrightarrow 2DH$, which is contrary to experiment. It is postulated that the role of N_2 in the HD reaction is coordination at Fe6, thereby blocking the operation of the steps in Figure 10.10b. A further postulate is that the blocking N_2 is η^1 coordinated to Fe6, and therefore is not on the path to NH_3. The formation of NH_3 by nitrogenase in the presence of both N_2 and D_2 is postulated to occur by the mechanism already described (Section 10.5.2.3).

10.7.4 Hydrogenation of CO and CO_2

Carbon monoxide has long been known as an inhibitor of nitrogenase, but in recent years it has been recognized as a substrate, being hydrogenated to small hydrocarbons. Vanadium nitrogenase converts CO to CH_4, C_2H_4, C_2H_6, C_3H_6, C_3H_8, C_4H_8, and C_4H_{10} [85, 86], while molybdenum nitrogenase converts CO to all of these hydrocarbons except CH_4, but in very much reduced (about 1%) yields and rate [86]. C_2H_4 is the dominant product for both enzymes. Substitution of the α-70^{Val} residue of molybdenum nitrogenase with alanine or glycine substantially increases the yields of products, and double substitution

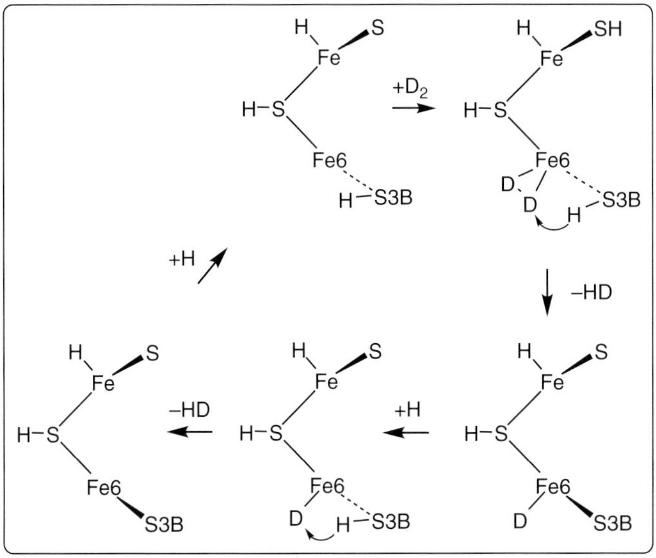

(a)

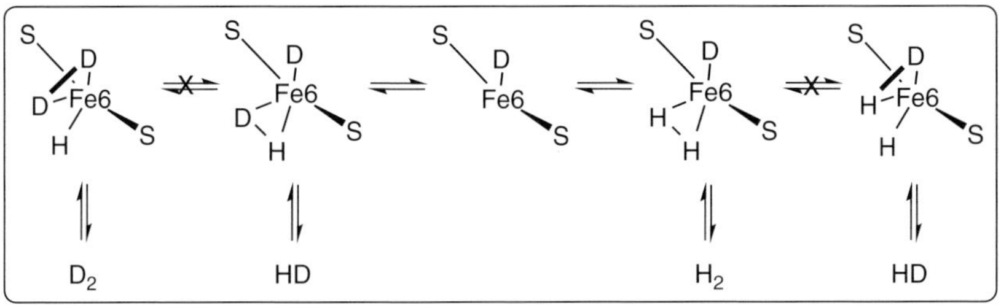

(b)

Figure 10.10 (a) The essence of the intramolecular hydrogenation steps of the HD reaction, in which S3B–H is the donor to D_2. This is not the complete mechanism – see text. (b) A discounted alternative mechanism with Fe–H as donor to D_2, which would enable $H_2 + D_2 \leftrightarrow 2DH$ exchange, contrary to experiment.

α-70$^{Val\rightarrow Ala}$ + α-191$^{Gln\rightarrow Ala}$ allows traces of CH_4 to be detected [88]. Relationships between V- and Mo nitrogenase and their reduction of CO have been discussed [16]. It is established that CO is the carbon source in these reactions, and that the hydrogenation of CO does not use H_2, an inevitable co-product, but uses protons and electrons in the same way that nitrogenase does with other substrates [86, 87]. A new facet of this reactivity of nitrogenase has developed with the recent report that Mo nitrogenase, doubly substituted as α-70$^{Val\rightarrow Ala}$ + α-195$^{His\rightarrow Gln}$ is able to reduce carbon *dioxide* to CH_4, and to reductively couple CO_2 and C_2H_2 to produce H_2C=CH-CH_3 [149]. The extent of reduction of CO_2 to CH_4 is very small: under optimal conditions, only 2% of the electrons generate CH_4, and 98% generate H_2.

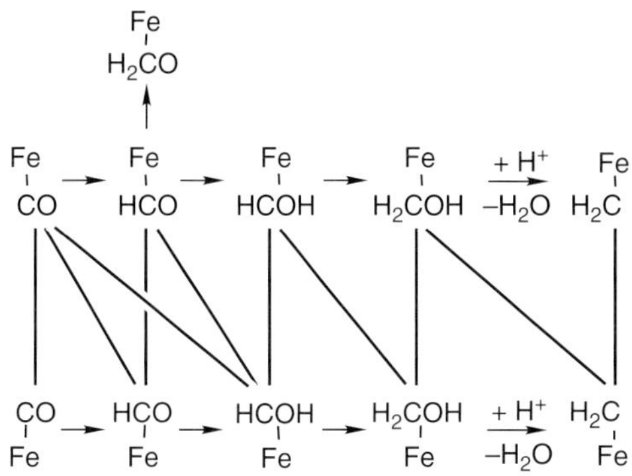

Figure 10.11 The conceptual foundation for hydrogenation and coupling of CO by nitrogenase.

Despite being minor components of the reactivity of nitrogenase, these reactions in which CO is hydrogenated raise many questions about mechanism that need to be answered. The standard intramolecular hydrogenation paradigm is applicable for the generation of C–H bonds. In addition to this hydrogenation, a new type of reaction step is required: C–C *bond formation* contrasts the N–N *bond breaking* of physiological nitrogenase chemistry. A reaction stoichiometry such as $3CO + 14H^+ + 14e^- \rightarrow C_3H_8 + 3H_2O$ will contain more steps than does $N_2 + 6H^+ + 6e^- \rightarrow 2NH_3$. In addition, removal of oxygen (presumably as water) is involved.

General consideration of the mechanism is based on Figure 10.11 and some preliminary simulations of possibilities [73]. Two Fe sites (Fe2, Fe6) can coordinate CO. At each there can be H-transfers to CO and its progressively hydrogenated forms, presented as the horizontal steps in Figure 10.11. It is likely that hydrogenations occur at both C and O atoms of CO, because if H_2CO is generated it dissociates and interrupts the progress to products. It is likely that protonation of the bound COH function will be required for the elimination of water. At some point, C–C bond formation occurs: the connecting lines between the two halves in Figure 10.11 illustrate the variety of possibilities for C–C bond formation.

The first endeavors in finding a feasible pathway through this maze have been published [73]. For the initial hydrogenation of bound CO, a difficult step, H-atom transfer from *endo* Fe_6–H to *endo* Fe_2–CO is the most promising, with a potential energy barrier of 15 kcal mol^{-1}. The conformation of HCO bound to Fe2 is determined by an intramolecular hydrogen bond (O···H–S2B) with S2B–H. Then, this S2B–H can transfer across the hydrogen bond with a barrier of 10 kcal mol^{-1} to form bound HCOH. Different types of C–C bond formation steps have been simulated, with calculated potential energy barriers of 8, 3, 9, or 12 kcal mol^{-1} [73]. Barriers calculated for some postulated steps are quite large, and much further work is needed. However, the simulations have shown that there are feasible steps for all of the requisite components of the overall reactions, namely, (i) the binding of CO, (ii) the initial hydrogenation of CO, (iii) continued hydrogenation of CO at both C and O, (iv) elimination of O as H_2O, and (v) C–C bond formation. The simulations have also shown (vi) that intermediate organic fragments can migrate around the active face of FeV-co, and (vii) that hydrogen bonding between COH functions and S or SH components of FeV-co can occur and contribute to the stabilization and orientation of intermediates [73]. The evolving mechanistic picture involves intramolecular hydrogenation as a key component, as in all other reactions of nitrogenase, and requires a serial hydrogenating ability for FeV-co plus protein analogous to that described for Mo nitrogenase. The structural homology between the V- and Mo nitrogenases and the functional similarities and differences have been reviewed recently [16].

The additional dimension for the reactions of CO and CO_2 to form hydrocarbons is protonation of bound HCOH or H_2COH moieties in order to remove O as water (see Figure 10.11). The simulations so far [73] have involved an exogenous proton, conveyed to COH as H_3O^+, via the "water pool" in the vicinity of homocitrate bound to V, as in Scheme 10.6. Two, three, or four such steps would be required for the production of C_2, C_3, or C_4 hydrocarbons, with the more numerous H atoms in the products being supplied by serial intramolecular hydrogenation. For the N_2 reaction of nitrogenase, the water pool is probably the NH_3 egress path (Section 10.3.2), but in the CO and CO_2 reactions the water pool may be a H_3O^+ supply path, because it is unlikely to be the egress path for hydrophobic hydrocarbon products. It is suggested that the hydrocarbon products could leave via the N_2 ingress path, and that both CO ingress and small hydrocarbon egress occur through the same hydrophobic channel, consistent with the comparative inefficiency of the CO reactions.

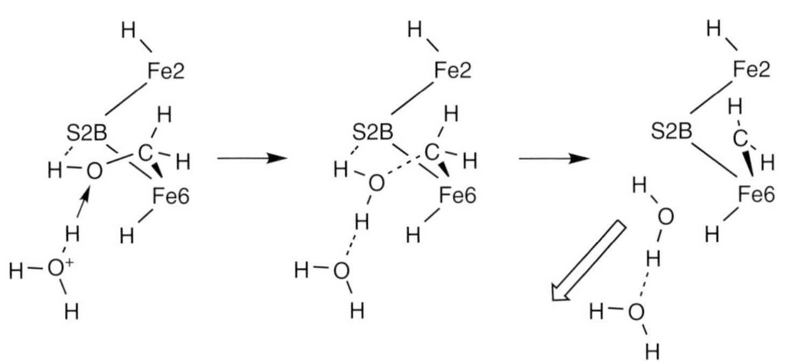

Scheme 10.6 Model for protonation of bound CH_2OH and elimination of water.

In this context, results on the reactivity of mutants of Mo nitrogenase are informative [88, 149]. Diminution of the size of the α-70 side chain is important for the reactions of CO and CO_2, as expected for mechanisms involving the binding of more than one of these reactant molecules in the Fe2/Fe6 domain. For the CO reaction, α-$195^{\text{His}\rightarrow\text{Gln}}$ substitution (together with α-$70^{\text{Val}\rightarrow\text{Ala}}$) greatly *decreases* the rates of product formation [88], but for the CO_2 reaction, α-$195^{\text{His}\rightarrow\text{Gln}}$ + α-$70^{\text{Val}\rightarrow\text{Ala}}$ was *required* for CH_4 production. Residue α-195^{His} is within hydrogen bonding distance of S2B, and the details of the involvement of residue α-195 in hydrogen bonding with S2B–H and/or with incoming H_3O^+ (as in Scheme 10.6) are expected to be crucial in the dehydration step(s).

10.8
Interpretation of the Structure of FeMo-co and Its Surrounds

Here, I review my current interpretation of the structure of FeMo-co and its surrounds, in relation to function [55]. Figure 10.12 depicts key points.

1) The full structure of FeMo-co, as two cubanoid M_4S_3 entities linked by four atoms ($(\mu\text{-S})_3 + C^c$), is required to provide geometrical integrity for the reactive $Fe_4(\mu\text{-S})_2(\mu_3\text{-S})_2$ face, which would otherwise be mechanically unstable.
2) FeMo-co is anchored to protein at three places, α-275^{Cys} at the top Fe, α-442^{His} at the bottom Mo, and at the back via hydrogen bonds to equatorial S atoms.
3) The central atom C prevents extreme distortion of FeMo-co and connects the Fe atoms where coordinative reaction occurs, thus transmitting coordinative information between them.
4) The central Fe atoms, with trigonal-pyramidal coordination in the resting state, each provide one or two adaptable coordination sites for the reactants

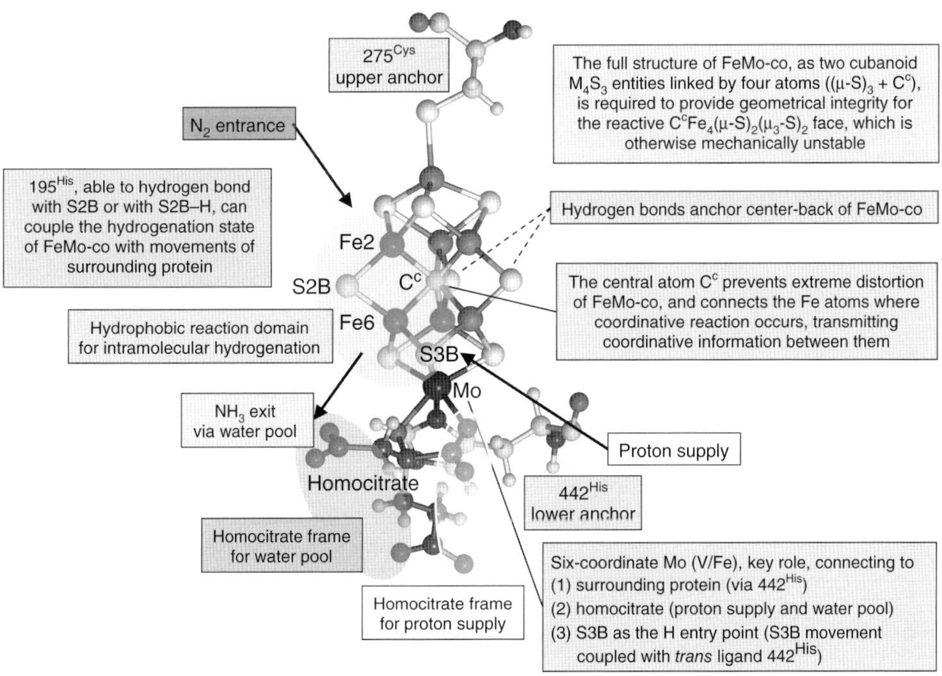

Figure 10.12 Interpretations of the structural attributes of FeMo-co, in relation to its functions.

and intermediates. Two of these central Fe atoms, Fe2 and Fe6, lie below the most influential amino acid α-70^{Val} and are believed to be intimately involved in catalysis. Other central Fe atoms are less accessible for catalytic functions.

5) The Mo atom has multiple roles, in connecting with the surrounding protein, linking homocitrate (which maintains the proton supply chain) with S3B as the H entry point, and orientating these features in relation to Fe6 as a principal coordination site. The significance of a six-coordinate metal atom at this site in FeMo-co is clear.
6) Homocitrate has two roles that explain its essentiality: one side, involving three O atoms, frames the crucial last stage of the proton supply chain; the other side, with four O atoms, frames the water pool.
7) The protein surrounding the Fe6 and Fe2 catalytic sites is hydrophobic, favoring the incoming nonpolar physiological substrate N_2, and allowing the H-atom transfers to substrate (steps that are also relatively nonpolar) to be controlled intramolecularly by FeMo-co.
8) Amino acid α-195^{His}, which is able to hydrogen bond with S2B or S2B–H, has a role in connecting and regulating the hydrogenation state of FeMo-co with movements of surrounding protein.

10.9 Mimicking Nitrogenase

Among the recent results for hydrogenation of CO to hydrocarbons is a report [150] that the FeMo-co and FeV-co cofactors, separated from the protein and stabilized in *N*-methylformamide solution [151, 152], will also convert CO to hydrocarbons when a strong homogeneous reductant is present, at pH 8. This draws attention, again, to the question of whether the catalytic power of nitrogenase might be replicated in a synthetic chemical system.

Structurally pure FeMo-co has not been synthesized as such, but this is not necessarily an obstacle to attempts to mimic the functions of FeMo-co. The previous sections identified the key components of the catalytic site, and their roles in effecting the catalysis. Therefore, approaches that replicate these functional characteristics could be sufficient. In a recent paper [153], the attributes required for a catalyst system mimicking FeMo-co in nitrogenase were considered in two parts, first for the cluster core which effects the intramolecular H-atom transfer steps, and then for the ligand surrounds which must control the proton transfer to the cluster and inhibit alternative interfering reactions. A number of well-characterized existing clusters [154–158] with the minimal attributes for the catalytic steps of substrate binding and H-atom transfer were identified, and considered for further development. This included theoretical simulations of the catalytic abilities of the models, by evaluation of structures in which the intermediates from FeMo-co mechanism were transposed onto the synthetic model clusters. Modifications of the ancillary ligands were proposed and tested, *in silico*.

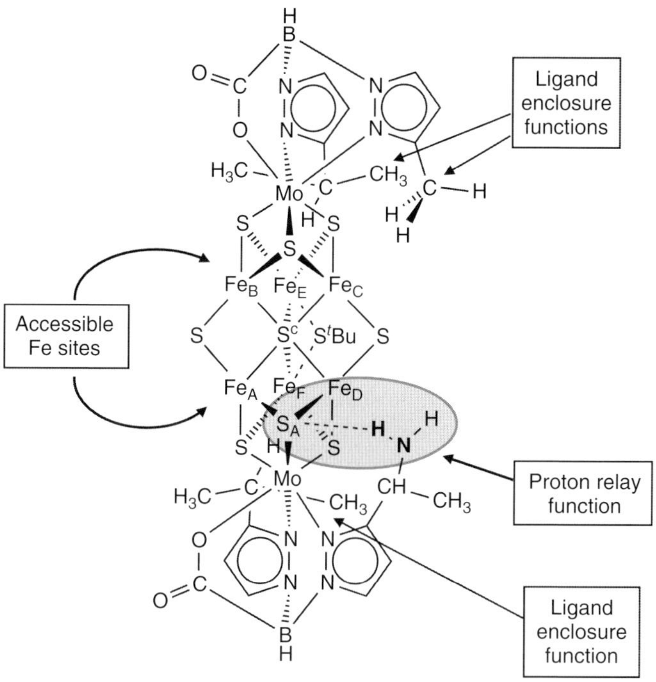

Figure 10.13 The optimized structure for mimic model **M1C**.

The main outcome from this investigation was the molecule **M1C** shown in Figure 10.13. It is predicted that this molecule, or one similar to it, is well equipped to catalyze the hydrogenation of N_2 to NH_3, given an appropriate proton source and an appropriate source of electrons (a homogeneous reductant or an electrochemical cell). Suggestions for the synthetic approach to **M1C**, and for preparatory experimental and theoretical investigations of the suitability of synthetic mimic models, have been advanced [153].

10.10 Summary and Epilog

In this chapter, I have overviewed computational investigations of the structure and reactivity of the active site of nitrogenase. Evolving from these is a comprehensive theory for the mechanisms of the diverse hydrogenation reactions effected by this enzyme. Because nitrogenase must be able to add six hydrogen atoms to the physiological substrate N_2, and more than six H atoms to some other substrates, a key component of the theory provides for the accumulation of multiple H atoms on FeMo-co, and their replenishment as required during a complete catalytic cycle. The intramolecular hydrogenation paradigm is generally applicable to all reactions of nitrogenase. It is proposed that quantum hydrogen tunneling is intrinsic to the intramolecular hydrogenation processes, and contributes to the remarkable ability of nitrogenase to reduce recalcitrant substrates such as N_2 and CO. The theoretical simulations provide a general structural elaboration of the scheme proposed by Thorneley and Lowe to account for detailed kinetic data on the main reactions of nitrogenase.

The reaction chemistry of nitrogenase is a suite of complex molecular dances, each with many steps and much changing of partners. The main $N_2 \rightarrow NH_3$ dance has been choreographed, using DF simulations of many possibilities, and rejection of awkward steps. How will the critics – the experimentalists – react? Nitrogenase is coy, a shy dancer, and inherently hides some of its performance from critical experimentalists.

Measurable attributes of proposed mechanisms, such as the electron spin densities and vibrational properties of intermediates, can be calculated as guidance for further experiment design and interpretation. While this article focuses on the conceptual frameworks, the papers cited contain numerical and geometrical details, and suggest experimental tests.

Acknowledgments

My research described here was made possible by awards from the National Computational Infrastructure Facility at the Australian National University, and by the University of New South Wales. I thank Lance Seefeldt, Chris Pickett, Bill Newton, Marcus Ribbe, and Steve Cramer for sharing information and ideas.

Figures were generated with Jmol, an open-source Java viewer for chemical structures in 3D (*http://www.jmol.org*), and with CrystalMaker (*http://www.crystalmaker.com*).

References

1. Burgess, B.K. and Lowe, D.J. (1996) Mechanism of molybdenum nitrogenase. *Chem. Rev.*, **96**, 2983–3011.
2. Howard, J.B. and Rees, D.C. (1996) Structural basis of biological nitrogen fixation. *Chem. Rev.*, **96**, 2965–2982.
3. Christiansen, J., Dean, D.R., and Seefeldt, L.C. (2001) Mechanistic features of the Mo-containing nitrogenase. *Annu. Rev. Plant. Physiol. Plant. Mol. Biol.*, **52**, 269–295.
4. Igarashi, R.Y. and Seefeldt, L.C. (2003) Nitrogen fixation: the mechanism of the Mo-dependent nitrogenase. *Crit. Rev. Biochem. Mol. Biol.*, **38** (4), 351–384.
5. Rees, D.C., Tezcan, F.A., Haynes, C.A., Walton, M.Y., Andrade, S., Einsle, O., and Howard, J.A. (2005) Structural basis of biological nitrogen fixation. *Philos. Trans. R. Soc. London, Ser. A*, **363**, 971–984.
6. Yang, Z.-Y., Danyal, K., and Seefeldt, L.C. (2011) Mechanism of Mo-dependent nitrogenase. *Methods Mol. Biol.*, **766**, 9–29.
7. Smil, V. (2001) *Enriching the Earth: Fritz Haber, Carl Bosch, and the Transformation of World Food Production*, MIT Press, Cambridge, MA.
8. Schlogl, R. (2003) Catalytic synthesis of ammonia – a "never-ending story"? *Angew. Chem. Int. Ed.*, **42** (18), 2004–2008.

9. Seefeldt, L.C. and Dean, D.R. (1997) Role of nucleotides in nitrogenase catalysis. *Acc. Chem. Res.*, **30**, 260–266.
10. Sen, S., Igarashi, R., Smith, A., Johnson, M.K., Seefeldt, L.C., and Peters, J.W. (2004) A conformational mimic of the MgATP-bound "on state" of the nitrogenase iron protein. *Biochemistry*, **43**, 1787–1797.
11. Seefeldt, L.C., Hoffman, B.M., and Dean, D.R. (2012) Electron transfer in nitrogenase catalysis. *Curr. Opin. Chem. Biol.*, **16** (1–2), 19–25.
12. Danyal, K., Mayweather, D., Dean, D.R., Seefeldt, L.C., and Hoffman, B.M. (2010) Conformational gating of electron transfer from the nitrogenase Fe protein to MoFe protein. *J. Am. Chem. Soc.*, **132** (20), 6894–6895.
13. Spatzal, T., Aksoyoglu, M., Zhang, L., Andrade, S.L.A., Schleicher, E., Weber, S., Rees, D.C., and Einsle, O. (2011) Evidence for interstitial carbon in nitrogenase FeMo cofactor. *Science*, **334**, 940.
14. Eady, R.R. (1996) Structure–function relationships of alternative nitrogenases. *Chem. Rev.*, **96**, 3013–3030.
15. Eady, R.R. (2003) Current status of structure function relationships of vanadium nitrogenase. *Coord. Chem. Rev.*, **237**, 23–30.
16. Hu, Y., Lee, C.C., and Ribbe, M.W. (2012) Vanadium nitrogenase: a two-hit wonder? *Dalton Trans.*, **41** (4), 1118–1127.
17. Lancaster, K.M., Roemelt, M., Ettenhuber, P., Hu, Y., Ribbe, M.W., Neese, F., Bergmann, U., and DeBeer, S. (2011) X-ray emission spectroscopy evidences a central carbon in the nitrogenase iron-molybdenum cofactor. *Science*, **334**, 974–977.
18. Wiig, J.A., Hu, Y., Lee, C.C., and Ribbe, M.W. (2012) Radical SAM-dependent carbon insertion into the nitrogenase M-cluster. *Science*, **337** (6102), 1672–1675.
19. Boal, A.K. and Rosenzweig, A.C. (2012) A radical route for nitrogenase carbide insertion. *Science*, **337** (6102), 1617–1618.
20. Seefeldt, L.C., Dance, I.G., and Dean, D.R. (2004) Substrate interactions with nitrogenase: Fe versus Mo. *Biochemistry*, **43** (6), 1401–1409.
21. Dance, I.G. (1994) The binding and reduction of dinitrogen at an Fe_4 face of the FeMo cluster of nitrogenase. *Aust. J. Chem.*, **47**, 979–990.
22. Dos Santos, P.C., Igarashi, R., Lee, H.-I., Hoffman, B.M., Seefeldt, L.C., and Dean, D.R. (2005) Substrate interactions with the nitrogenase active site. *Acc. Chem. Res.*, **38** (3), 208–214.
23. Barney, B.M., Lee, H.-I., Dos Santos, P.C., Hoffman, B.M., Dean, D.R., and Seefeldt, L.C. (2006) Breaking the N_2 triple bond: insights into the nitrogenase mechanism. *Dalton Trans.*, 2277–2284.
24. Dos Santos, P.C., Mayer, S.M., Barney, B.M., Seefeldt, L.C., and Dean, D.R. (2007) Alkyne substrate interaction within the nitrogenase MoFe protein. *J. Inorg. Biochem.*, **101** (11–12), 1642–1648.
25. Seefeldt, L.C., Hoffman, B.M., and Dean, D.R. (2009) Mechanism of Mo-dependent nitrogenase. *Annu. Rev. Biochem.*, **78**, 701–722.
26. Sarma, R., Barney, B.M., Keable, S., Dean, D.R., Seefeldt, L.C., and Peters, J.W. (2010) Insights into substrate binding at FeMo-cofactor in nitrogenase from the structure of an α-70^{Ile} MoFe protein variant. *J. Inorg. Biochem.*, **104**, 385–389.
27. George, S.J., Barney, B.M., Mitra, D., Igarashi, R.Y., Guo, Y., Dean, D.R., Cramer, S.P., and Seefeldt, L.C. (2012) EXAFS and NRVS reveal a conformational distortion of the FeMo-cofactor in the MoFe nitrogenase propargyl alcohol complex. *J. Inorg. Biochem.*, **112**, 85–92.
28. Benton, P.M.C., Laryukhin, M., Mayer, S.M., Hoffman, B.M., Dean, D.R., and Seefeldt, L.C. (2003) Localization of a substrate binding site on the FeMo-cofactor in nitrogenase: trapping propargyl alcohol with an α-70-substituted MoFe protein. *Biochemistry*, **42**, 9102–9109.
29. Igarashi, R., Dos Santos, P.C., Niehaus, W.G., Dance, I.G., Dean, D.R., and Seefeldt, L.C. (2004) Localization of a

catalytic intermediate bound to FeMo-cofactor of nitrogenase. *J. Biol. Chem.*, **279** (33), 34770–34775.

30. Lee, H.-I., Igarashi, R., Laryukhin, M., Doan, P.E., Dos Santos, P.C., Dean, D.R., Seefeldt, L.C., and Hoffman, B.M. (2004) An organometallic intermediate during alkyne reduction by nitrogenase. *J. Am. Chem. Soc.*, **126**, 9563–9569.
31. Barney, B.M., Laryukhin, M., Igarashi, R.Y., Lee, H.-I., Dos Santos, P.C., Yang, T.-C., Hoffman, B.M., Dean, D.R., and Seefeldt, L.C. (2005) Trapping a hydrazine reduction intermediate on the nitrogenase active site. *Biochemistry*, **44**, 8030–8037.
32. Barney, B.M., Yang, T.-C., Igarashi, R., Dos Santos, P.C., Laryukhin, M., Lee, H.-I., Hoffman, B.M., Dean, D.R., and Seefeldt, L.C. (2005) Intermediates trapped during nitrogenase reduction of $N\equiv N$, CH_3-N=NH, and H_2N-NH_2. *J. Am. Chem. Soc.*, **127**, 14960–14961.
33. Igarashi, R.Y., Laryukhin, M., Dos Santos, P.C., Lee, H.-I., Dean, D.R., Seefeldt, L.C., and Hoffman, B.M. (2005) Trapping H-bound to the nitrogenase FeMo-cofactor active site during H_2 evolution: characterization by ENDOR spectroscopy. *J. Am. Chem. Soc.*, **127** (17), 6231–6241.
34. Barney, B.M., Lukoyanov, D., Yang, T.-C., Dean, D.R., Hoffman, B.M., and Seefeldt, L.C. (2006) A methyldiazene (HN=N-CH_3)-derived species bound to the nitrogenase active-site FeMo cofactor: implications for mechanism. *Proc. Natl. Acad. Sci. U.S.A.*, **103** (46), 17113–17118.
35. Barney, B.M., McClead, J., Lukoyanov, D., Laryukhin, M., Yang, T.C., Dean, D.R., Hoffman, B.M., and Seefeldt, L.C. (2007) Diazene (HN=NH) is a substrate for nitrogenase: insights into the pathway of N_2 reduction. *Biochemistry*, **46** (23), 6784–6794.
36. Barney, B.M., Lukoyanov, D., Igarashi, R.Y., Laryukhin, M., Yang, T.-C., Dean, D.R., Hoffman, B.M., and Seefeldt, L.C. (2009) Trapping an intermediate of dinitrogen (N_2) reduction on nitrogenase. *Biochemistry*, **48** (38), 9094–9102.
37. Lukoyanov, D., Barney, B.M., Dean, D.R., Seefeldt, L.C., and Hoffman, B.M. (2007) Connecting nitrogenase intermediates with the kinetic scheme for N_2 reduction by a relaxation protocol and identification of the N_2 binding state. *Proc. Natl. Acad. Sci. U.S.A.*, **104** (5), 1451–1455.
38. Doan, P.E., Telser, J., Barney, B.M., Igarashi, R.Y., Dean, D.R., Seefeldt, L.C., and Hoffman, B.M. (2011) ^{57}Fe ENDOR spectroscopy and "electron inventory" analysis of the nitrogenase E_4 intermediate suggest the metal-ion core of FeMo-cofactor cycles through only one redox couple. *J. Am. Chem. Soc.*, **133** (43), 17329–17340.
39. Rees, D.C. and Howard, J.B. (2000) Nitrogenase: standing at the crossroad. *Curr. Opin. Chem. Biol.*, **4**, 559–566.
40. Lovell, T., Li, J., Liu, T., Case, D.A., and Noodleman, L. (2001) FeMo cofactor of nitrogenase: a density functional study of states M^N, M^{OX}, M^R, and M^I. *J. Am. Chem. Soc.*, **123**, 12392–12410.
41. Szilagyi, R.K., Musaev, D.G., and Morokuma, K. (2001) Theoretical studies of biological nitrogen fixation. I. Density functional modeling of the Mo-site of the FeMo-cofactor. *Inorg. Chem.*, **40**, 766–775.
42. Lovell, T., Li, J., Case, D.A., and Noodleman, L. (2002) FeMo cofactor of nitrogenase: energetics and local interactions in the protein environment. *J. Biol. Inorg. Chem.*, **7**, 735–749.
43. Lovell, T., Li, J., Case, D.A., and Noodleman, L. (2002) Binding modes for the first coupled electron and proton addition to FeMoco of nitrogenase. *J. Am. Chem. Soc.*, **124**, 4546–4547.
44. Lovell, T., Torres, R.A., Han, W.-G., Liu, T., Case, D.A., and Noodleman, L. (2002) Metal substitution in the active site of nitrogenase MFe_7S_9 ($M = Mo^{4+}$, V^{3+}, Fe^{3+}). *Inorg. Chem.*, **41**, 5744–5753.
45. Noodleman, L., Lovell, T., Liu, T., Himo, F., and Torres, R.A. (2002) Insights into properties and energetics of iron–sulfur proteins from simple clusters to nitrogenase. *Curr. Opin. Chem. Biol.*, **6**, 259–273.

46. Schimpl, J., Petrilli, H.M., and Blochl, P.E. (2003) Nitrogen binding to the FeMo-cofactor of nitrogenase. *J. Am. Chem. Soc.*, **125**, 15772–15778.
47. Vrajmasu, V., Munck, E., and Bominaar, E.L. (2003) Density functional study of the electric hyperfine interactions and the redox-structural correlations in the cofactor of nitrogenase. Analysis of general trends in ^{57}Fe isomer shifts. *Inorg. Chem.*, **42**, 5974–5988.
48. Lovell, T., Liu, T., Case, D.A., and Noodleman, L. (2003) Structural, spectroscopic, and redox consequences of a central ligand in the FeMoco of nitrogenase: a density functional theoretical study. *J. Am. Chem. Soc.*, **125**, 8377–8383.
49. Dance, I. (2004) The mechanism of nitrogenase. Computed details of the site and geometry of binding of alkyne and alkene substrates and intermediates. *J. Am. Chem. Soc.*, **126**, 11852–11863.
50. Hinnemann, B. and Norskov, J.K. (2003) Modeling a central ligand in the nitrogenase FeMo cofactor. *J. Am. Chem. Soc.*, **125**, 1466–1467.
51. Hinnemann, B. and Nørskov, J.K. (2004) Chemical activity of the nitrogenase FeMo cofactor with a central nitrogen ligand: density functional study. *J. Am. Chem. Soc.*, **126**, 3920–3927.
52. Hinnemann, B. and Norskov, J.K. (2004) Structure of the FeFe-cofactor of the iron-only nitrogenase and possible mechanism for dinitrogen reduction. *Phys. Chem. Chem. Phys.*, **6**, 843–853.
53. Dance, I. (2005) The hydrogen chemistry of the FeMo-co active site of nitrogenase. *J. Am. Chem. Soc.*, **127**, 10925–10942.
54. Dance, I. (2007) The mechanistically significant coordination chemistry of dinitrogen at FeMo-co, the catalytic site of nitrogenase. *J. Am. Chem. Soc.*, **129** (5), 1076–1088.
55. Dance, I. (2007) Elucidating the coordination chemistry and mechanism of biological nitrogen fixation. *Chem. Asian J.*, **2**, 936–946.
56. Pelmenschikov, V., Case, D.A., and Noodleman, L. (2008) Ligand-bound S = 1/2 FeMo-cofactor of nitrogenase: hyperfine interaction analysis and implication for the central ligand X identity. *Inorg. Chem.*, **47** (14), 6162–6172.
57. Xie, H., Wu, R., Zhou, Z., and Cao, Z. (2008) Exploring the interstitial atom in the FeMo cofactor of nitrogenase: insights from QM and QM/MM calculations. *J. Phys. Chem. B*, **112** (36), 11435–11439.
58. Harris, T.V. and Szilagyi, R.K. (2011) Nitrogenase structure and function relationships by density functional theory. *Methods Mol. Biol.*, **766**, 267–291.
59. Dance, I. (2011) Electronic dimensions of FeMo-co, the active site of nitrogenase, and its catalytic intermediates. *Inorg. Chem.*, **50** (1), 178–192.
60. Dance, I. (2011) Calculated vibrational frequencies for FeMo-co, the active site of nitrogenase, bearing hydrogen atoms and carbon monoxide. *Dalton Trans.*, **40** (24), 6480–6489.
61. Dance, I. (2012) Ramifications of C-centering rather than N-centering of the active site FeMo-co of the enzyme nitrogenase. *Dalton Trans.*, **41** (16), 4859–4865.
62. Dance, I.G. (1997) Calculated details of a mechanism for conversion of N_2 to NH_3 at the FeMo cluster of nitrogenase. *J. Chem. Soc., Chem. Commun.*, 165–166.
63. Rod, T.H., Hammer, B., and Norskov, J.K. (1999) Nitrogen adsorption and hydrogenation on a $MoFe_6S_9$ complex. *Phys. Rev. Lett.*, **82** (20), 4054–4057.
64. Rod, T.H. and Norskov, J.K. (2000) Modelling the nitrogenase FeMo cofactor. *J. Am. Chem. Soc.*, **122**, 12751–12763.
65. Kastner, J., Hemmen, S., and Blochl, P.E. (2005) Activation and protonation of dinitrogen at the FeMo cofactor of nitrogenase. *J. Chem. Phys.*, **123** (7), 074306.
66. Kastner, J. and Blochl, P.E. (2005) Model for acetylene reduction by nitrogenase derived from density functional theory. *Inorg. Chem.*, **44**, 4568–4575.
67. Cao, Z., Zhou, Z., Wan, H., and Zhang, Q. (2005) Enzymatic and catalytic reduction of dinitrogen to ammonia:

density functional theory characterization of alternative molybdenum active sites. *Int. J. Quantum Chem.*, **103**, 344–353.

68. Dance, I. (2006) Mechanistic significance of the preparatory migration of hydrogen atoms around the FeMo-co active site of nitrogenase. *Biochemistry*, **45** (20), 6328–6340.
69. McKee, M.L. (2007) Modeling the nitrogenase FeMo cofactor with high-spin $Fe_8S_9X^+$ (XN, C) clusters. Is the first step for N_2 reduction to NH_3 a concerted dihydrogen transfer? *J. Comput. Chem.*, **28** (8), 1342–1356.
70. Kastner, J. and Blochl, P.E. (2007) Ammonia production at the FeMo cofactor of nitrogenase: results from density functional theory. *J. Am. Chem. Soc.*, **129** (10), 2998–3006.
71. Dance, I. (2008) The chemical mechanism of nitrogenase: calculated details of the intramolecular mechanism for hydrogenation of η^2-N_2 on FeMo-co to NH_3. *Dalton Trans.*, 5977–5991.
72. Dance, I. (2008) The chemical mechanism of nitrogenase: hydrogen tunneling and further aspects of the intramolecular mechanism for hydrogenation of η^2-N_2 on FeMo-co to NH_3. *Dalton Trans.*, 5992–5998.
73. Dance, I. (2011) How does vanadium nitrogenase reduce CO to hydrocarbons? *Dalton Trans.*, **40** (20), 5516–5527.
74. Delley, B. (1990) An all-electron numerical method for solving the local density functional for polyatomic molecules. *J. Chem. Phys.*, **92** (1), 508–517.
75. Delley, B. (1995) in *Modern Density Functional Theory: A Tool for Chemistry* (eds J.M. Seminario and P. Politzer), Elsevier, Amsterdam, pp. 221–254.
76. Delley, B. (2000) From molecules to solids with the DMol3 approach. *J. Chem. Phys.*, **113** (18), 7756–7764.
77. Accelrys Software Inc. (2010) DMol3, *accelrys.com/products/datasheets/dmol3.pdf* (accessed 15 March 2014).
78. Dance, I.G. (1996) in *Transition Metal Sulfur Chemistry: Biological and Industrial Significance* (eds E.I. Stiefel and K. Matsumoto), American Chemical Society, Washington, DC, pp. 135–152.
79. Dance, I. (1998) Understanding structure and reactivity of new fundamental inorganic molecules: metal sulfides, metallocarbohedrenes, and nitrogenase. *J. Chem. Soc., Chem. Commun.*, 523–530.
80. Dance, I. (2008) A pragmatic method for location of transition states and calculation of reaction paths. *Mol. Simul.*, **34** (10), 923–929.
81. Dance, I. (2011) A pragmatic method for location of transition states and calculation of reaction paths: erratum. *Mol. Simul.*, **37** (3), 257.
82. Dance, I. (2006) The correlation of redox potential, HOMO energy, and oxidation state in metal sulfide clusters and its application to determine the redox level of the FeMo-co active-site cluster of nitrogenase. *Inorg. Chem.*, **45** (13), 5084–5091.
83. Lowe, D.J., Fisher, K., Thorneley, R.N.F., Vaughn, S.A., and Burgess, B.K. (1989) Kinetics and mechanism of the reaction of cyanide with molybdenum nitrogenase from Azotobacter vinelandii. *Biochemistry*, **28** (21), 8460–8466.
84. Newton, W.E. and Dilworth, M.J. (2011) Assays of nitrogenase reaction products. *Methods Mol. Biol.*, **766**, 105–127.
85. Lee, C.H., Hu, Y., and Ribbe, M.W. (2010) Vanadium nitrogenase reduces CO. *Science*, **329** (5992), 642.
86. Hu, Y., Lee, C.C., and Ribbe, M.W. (2011) Extending the carbon chain: hydrocarbon formation catalyzed by vanadium/molybdenum nitrogenases. *Science*, **333** (6043), 753–755.
87. Lee, C.C., Hu, Y., and Ribbe, M.W. (2011) Tracing the hydrogen source of hydrocarbons formed by vanadium nitrogenase. *Angew. Chem. Int. Ed.*, **50** (24), 5545–5547.
88. Yang, Z.-Y., Dean, D.R., and Seefeldt, L.C. (2011) Molybdenum nitrogenase catalyzes the reduction and coupling of CO to form hydrocarbons. *J. Biol. Chem.*, **286** (22), 19417–19421.
89. Dance, I. (2012) The controlled relay of multiple protons required at the active

site of nitrogenase. *Dalton Trans.*, **41** (25), 7647–7659.

90. Durrant, M.C. (2001) Controlled protonation of iron–molybdenum cofactor by nitrogenase: a structural and theoretical analysis. *Biochem. J.*, **355**, 569–576.
91. Agmon, N. (1995) The Grotthuss mechanism. *Chem. Phys. Lett.*, **244** (5–6), 456–462.
92. Pomes, R. and Roux, B. (1998) Free energy profiles for H^+ conduction along hydrogen-bonded chains of water molecules. *Biophys. J.*, **75** (1), 33–40.
93. Ball, P. (2008) Water as an active constituent in cell biology. *Chem. Rev.*, **108** (1), 74–108.
94. Hammes-Schiffer, S. (2001) Theoretical perspectives on proton-coupled electron transfer reactions. *Acc. Chem. Res.*, **34** (4), 273–281.
95. Huynh, M.H.V. and Meyer, T.J. (2007) Proton-coupled electron transfer. *Chem. Rev.*, **107**, 5004–5064.
96. Dempsey, J.L., Winkler, J.R., and Gray, H.B. (2010) Proton-coupled electron flow in protein redox machines. *Chem. Rev.*, **110**, 7024–7039.
97. Siegbahn, P.E.M. and Blomberg, M.R.A. (2010) Quantum chemical studies of proton-coupled electron transfer in metalloenzymes. *Chem. Rev.*, **110**, 7040–7061.
98. Dance, I. (2014) In preparation for publication.
99. Thorneley, R.N.F. and Lowe, D.J. (1985) in *Molybdenum Enzymes* (ed. T.G. Spiro), Wiley- Interscience, New York, pp. 221–284.
100. Thorneley, R.N.F. and Lowe, D.J. (1996) Nitrogenase – substrate binding and activation. *J. Biol. Inorg. Chem.*, **1** (6), 576–580.
101. Dilworth, M.J. (1966) Acetylene reduction by nitrogen fixing preparations from *Clostridium pasteurianum*. *Biochim. Biophys. Acta*, **127**, 285–294.
102. Rivera-Ortiz, J.H. and Burris, R.H. (1975) Interactions among substrates and inhibitors of nitrogenase. *J. Bacteriol.*, **123** (2), 537–545.
103. Hwang, J.C., Chen, C.H., and Burris, R.H. (1973) Inhibition of nitrogenase-catalyzed reactions. *Biochim. Biophys. Acta*, **292** (1), 256–270.
104. Hammes-Schiffer, S. (2010) Introduction: proton-coupled electron transfer. *Chem. Rev.*, **110** (12), 6937–6938.
105. Barney, B.M., Igarashi, R.Y., Dos Santos, P.C., Dean, D.R., and Seefeldt, L.C. (2004) Substrate interaction at an iron-sulfur face of the FeMo-cofactor during nitrogenase catalysis. *J. Biol. Chem.*, **279** (51), 53621–53624.
106. Thorneley, R.N.F. and Lowe, D.J. (1984) The mechanism of *Klebsiella pneumoniae* nitrogenase action. Pre-steady-state kinetics of an enzyme-bound intermediate in N_2 reduction and of NH_3 formation. *Biochem. J.*, **224**, 887–894.
107. Lowe, D.J. and Thorneley, R.N.F. (1984) The mechanism of *Klebsiella pneumoniae* nitrogenase action. The determination of rate constants required for the simulation of the kinetics of N_2 reduction and H_2 evolution. *Biochem. J.*, **224**, 895–901.
108. Duyvis, M.G., Wassink, H., and Haaker, H. (1998) Nitrogenase of *Azotobacter vinelandii*: kinetic analysis of the Fe protein redox cycle. *Biochemistry*, **37**, 17345–17354.
109. Fisher, K., Newton, W.E., and Lowe, D.J. (2001) Electron paramagnetic resonance analysis of different *Azotobacter vinelandii* nitrogenase MoFe-protein conformations generated during enzyme turnover: evidence for S = 3/2 spin states from reduced MoFe-protein intermediates. *Biochemistry*, **40**, 3333–3339.
110. Sutcliffe, M.J. and Scrutton, N.S. (2002) A new conceptual framework for enzyme catalysis. Hydrogen tunnelling coupled to enzyme dynamics in flavoprotein and quinoprotein enzymes. *Eur. J. Biochem.*, **269** (13), 3096–3102.
111. Kohen, A. (2003) Kinetic isotope effects as probes for hydrogen tunneling, coupled motion and dynamics contributions to enzyme catalysis. *Prog. React. Kinet. Mech.*, **28** (2), 119–156.

112. Liang, Z.-X. and Klinman, J.P. (2004) Structural bases of hydrogen tunneling in enzymes: progress and puzzles. *Curr. Opin. Struct. Biol.*, **14** (6), 648–655.

113. Masgrau, L., Basran, J., Hothi, P., Sutcliffe, M.J., and Scrutton, N.S. (2004) Hydrogen tunneling in quinoproteins. *Arch. Biochem. Biophys.*, **428**, 41–51.

114. Hammes-Schiffer, S. (2006) Hydrogen tunneling and protein motion in enzyme reactions. *Acc. Chem. Res.*, **39** (2), 93–100.

115. Murakawa, T., Okajima, T., Kuroda, S., Nakamoto, T., Taki, M., Yamamoto, Y., Hayashi, H., and Tanizawa, K. (2006) Quantum mechanical hydrogen tunneling in bacterial copper amine oxidase reaction. *Biochem. Biophys. Res. Commun.*, **342** (2), 414–423.

116. Pu, J., Gao, J., and Truhlar, D.G. (2006) Multidimensional tunneling, recrossing, and the transmission coefficient for enzymatic reactions. *Chem. Rev.*, **106**, 3140–3169.

117. Dybala-Defratyka, A., Paneth, P., Banerjee, R., and Truhlar, D.G. (2007) Coupling of hydrogenic tunneling to active-site motion in the hydrogen radical transfer catalyzed by a coenzyme B_{12}-dependent mutase. *Proc. Natl. Acad. Sci. U.S.A.*, **104** (26), 10774–10779.

118. Hay, S. and Scrutton, N.S. (2008) H-transfers in Photosystem II: what can we learn from recent lessons in the enzyme community? *Photosynth. Res.*, **98** (1–3), 169–177.

119. Nagel, Z.D. and Klinman, J.P. (2010) Update 1 of: tunneling and dynamics in enzymatic hydride transfer. *Chem. Rev.*, **110** (12), PR41–PR67.

120. Nagel, Z.D. and Klinman, J.P. (2009) A 21st century revisionist's view at a turning point in enzymology. *Nat. Chem. Biol.*, **5**, 543–550.

121. Cha, Y., Murray, C.J., and Klinman, J.P. (1989) Hydrogen tunneling in enzyme reactions. *Science*, **243** (4896), 1325–1330.

122. Nashine, V.C., Hammes-Schiffer, S., and Benkovic, S.J. (2010) Coupled motions in enzyme catalysis. *Curr. Opin. Chem. Biol.*, **14** (5), 644–651.

123. Masgrau, L., Roujeinikova, A., Johannissen, L.O., Hothi, P., Basran, J., Ranaghan, K.E., Mulholland, A.J., Sutcliffe, M.J., Scrutton, N.S., and Leys, D. (2006) Atomic description of an enzyme reaction dominated by proton tunneling. *Science*, **312**, 237–241.

124. Masgrau, L., Ranaghan, K.E., Scrutton, N.S., Mulholland, A.J., and Sutcliffe, M.J. (2007) Tunneling and classical paths for proton transfer in an enzyme reaction dominated by tunneling: oxidation of tryptamine by aromatic amine dehydrogenase. *J. Phys. Chem. B*, **111** (11), 3032–3047.

125. Pang, J., Hay, S., Scrutton, N.S., and Sutcliffe, M.J. (2008) Deep tunneling dominates the biologically important hydride transfer reaction from NADH to FMN in morphinone reductase. *J. Am. Chem. Soc.*, **130** (22), 7092–7097.

126. Faulder, P.F., Tresadern, G., Chohan, K.K., Scrutton, N.S., Sutcliffe, M.J., Hillier, I.H., and Burton, N.A. (2001) QM/MM studies show substantial tunneling for the hydrogen-transfer reaction in methylamine dehydrogenase. *J. Am. Chem. Soc.*, **123** (35), 8604–8605.

127. Alhambra, C., Luz Sanchez, M., Corchado, J., Gao, J., and Truhlar, D.G. (2002) Quantum mechanical tunneling in methylamine dehydrogenase. *Chem. Phys. Lett.*, **355** (4–6), 388–394.

128. Pudney, C.R., Hay, S., Sutcliffe, M.J., and Scrutton, N.S. (2006) α-secondary isotope effects as probes of "tunneling-ready" configurations in enzymatic H-tunneling: insight from environmentally coupled tunneling models. *J. Am. Chem. Soc.*, **128** (43), 14053–14058.

129. Johannissen, L.O., Scrutton, N.S., and Sutcliffe, M.J. (2008) The enzyme aromatic amine dehydrogenase induces a substrate conformation crucial for promoting vibration that significantly reduces the effective potential energy barrier to proton transfer. *J. R. Soc. Interface*, **5** (Suppl. 3), 225–232.

130. Simpson, F.B. and Burris, R.H. (1984) A nitrogen pressure of 50 atmospheres does not prevent evolution of hydrogen by nitrogenase. *Science*, **224**, 1095–1096.

131. Benton, P.M.C., Christiansen, J., Dean, D.R., and Seefeldt, L.C. (2001) Stereospecificity of acetylene reduction catalyzed by nitrogenase. *J. Am. Chem. Soc.*, **123**, 1822–1827.
132. Han, J. and Newton, W.E. (2004) Differentiation of acetylene-reduction sites by stereoselective proton addition during *Azotobacter vinelandii* nitrogenase-catalyzed C_2D_2 reduction. *Biochemistry*, **43**, 2947–2956.
133. Burgess, B.K. (1993) Nitrogenase structure, function, and genetics. *ACS Symp. Ser.*, **535**, 145–169.
134. Smith, B.E. (1999) Structure, function, and biosynthesis of the metallosulfur clusters in nitrogenases. *Adv. Inorg. Chem.*, **47**, 159–218.
135. Christiansen, J., Cash, V.L., Seefeldt, L.C., and Dean, D.R. (2000) Isolation and characterisation of an acetylene-resistant nitrogenase. *J. Biol. Chem.*, **275** (15), 11459–11464.
136. Christiansen, J., Seefeldt, L.C., and Dean, D.R. (2000) Competitive substrate and inhibitor interactions at the physiologically relevant active site of nitrogenase. *J. Biol. Chem.*, **275** (46), 36104–36107.
137. Lowe, D.J., Fisher, K., and Thorneley, R.N.F. (1990) *Klebsiella pneumoniae* nitrogenase. Mechanism of acetylene reduction and its inhibition by carbon monoxide. *Biochem. J.*, **272**, 621–625.
138. Kim, C.H., Newton, W.E., and Dean, D.R. (1995) Role of the MoFe protein alpha subunit histidine-195 residue in FeMo-cofactor binding and nitrogenase catalysis. *Biochemistry*, **34**, 2798–2808.
139. Lowe, D.J., Eady, R.R., and Thorneley, R.N.F. (1978) Electron-paramagnetic resonance studies on nitrogenase of *Klebsiella pneumoniae*. *Biochem. J.*, **173**, 277–290.
140. Davis, L.C., Henzl, M.T., Burris, R.H., and Orme-Johnson, W.H. (1979) Iron-sulfur clusters in the molybdenum-iron protein component of nitrogenase. Electron paramagnetic resonance of the carbon monoxide inhibited state. *Biochemistry*, **18**, 4860–4869.
141. Shen, J., Dean, D.R., and Newton, W.E. (1997) Evidence for multiple substrate-reduction sites and distinct inhibitor-binding sites from an altered *Azotobacter vinelandii* nitrogenase MoFe protein. *Biochemistry*, **36** (16), 4884–4894.
142. Hoch, G.E., Schneider, K.C., and Burris, R.H. (1960) Hydrogen evolution and exchange, and conversion of N_2O to N_2 by soybean root nodules. *Biochim. Biophys. Acta*, **37** (2), 273–279.
143. Leroux, F., Dementin, S., Burlat, B., Cournac, L., Volbeda, A., Champ, S., Martin, L., Guigliarelli, B., Bertrand, P., Fontecilla-Camps, J.C., Rousset, M., and Leger, C. (2008) Experimental approaches to kinetics of gas diffusion in hydrogenase. *Proc. Natl. Acad. Sci. U.S.A.*, **105** (32), 11188–11193.
144. Burgess, B.K., Wherland, S., Newton, W.E., and Stiefel, E.I. (1981) Nitrogenase reactivity: insight into the nitrogen-fixing process through hydrogen-inhibition and HD-forming reactions? *Biochemistry*, **20**, 5140–5146.
145. Jackson, E.K., Parshall, G.W., and Hardy, R.W.F. (1968) Hydrogen reactions of nitrogenase. Formation of the molecule HD by nitrogenase and by an inorganic model. *J. Biol. Chem.*, **243** (19), 4952–4958.
146. Newton, W.E., Bulen, W.A., Hadfield, K.L., Stiefel, E.I., and Watt, G.D. (1977) in *Recent Developments in Nitrogen Fixation* (eds W.E. Newton, J.R. Postgate, and C. Rodriguez-Barrueco), Academic Press, London, pp. 119–130.
147. Guth, J.H. and Burris, R.H. (1983) Inhibition of nitrogenase-catalyzed NH_3 formation by H_2. *Biochemistry*, **22**, 5111–5122.
148. Li, J.-L. and Burris, R.H. (1983) Influence of pN_2 and pD_2 on HD formation by various nitrogenases. *Biochemistry*, **22**, 4472–4480.
149. Yang, Z.-Y., Moure, V.R., Dean, D.R., and Seefeldt, L.C. (2012) Carbon dioxide reduction to methane and coupling with acetylene to form propylene catalyzed by remodeled nitrogenase. *Proc. Natl. Acad. Sci.*, **109** (48), 19644–19648.
150. Lee, C.C., Hu, Y., and Ribbe, M.W. (2012) ATP-independent formation of hydrocarbons catalyzed by isolated

nitrogenase cofactors. *Angew. Chem. Int. Ed.*, **51** (8), 1947–1949.

151. Smith, B.E., Durrant, M.C., Fairhurst, S.A., Gormal, C.A., Gronberg, K.L.C., Henderson, R.A., Ibrahim, S.K., Gall, T.L., and Pickett, C.J. (1999) Exploring the reactivity of the isolated iron-molybdenum cofactor of nitrogenase. *Coord. Chem. Rev.*, **185–186**, 669–687.
152. Fay, A.W., Blank, M.A., Lee, C.C., Hu, Y., Hodgson, K.O., Hedman, B., and Ribbe, M. (2010) Characterization of isolated nitrogenase FeVco. *J. Am. Chem. Soc.*, **132** (38), 12612–12618.
153. Dance, I. (2010) Mimicking nitrogenase. *Dalton Trans.*, **39**, 2972–2983.
154. Berlinguette, C.P., Miyaji, T., Zhang, Y., and Holm, R.H. (2006) Precursors to clusters with the topology of the P^N cluster of nitrogenase: edge-bridged double cubane clusters $[(Tp)_2Mo_2Fe_6S_8L_4]^z$: synthesis, structures, and electron transfer series. *Inorg. Chem.*, **45** (5), 1997–2007.
155. Hlavinka, M.L., Miyaji, T., Staples, R.J., and Holm, R.H. (2007) Hydroxide-promoted core conversions of molybdenum – iron – sulfur edge-bridged double cubanes: oxygen-ligated topological P^N clusters. *Inorg. Chem.*, **46** (22), 9192–9200.
156. Zuo, J.-L., Zhou, H.-C., and Holm, R.H. (2003) Vanadium-iron-sulfur clusters containing the cubane-type $[VFe_3S_4]$ core unit: synthesis of a cluster with the topology of the P^N cluster of nitrogenase. *Inorg. Chem.*, **42** (15), 4624–4631.
157. Zhang, Y. and Holm, R.H. (2003) Synthesis of a molecular $Mo_2Fe_6S_9$ cluster with the topology of the P^N cluster of nitrogenase by rearrangement of an edge-bridged $Mo_2Fe_6S_8$ double cubane. *J. Am. Chem. Soc.*, **125**, 3910–3920.
158. Zhang, Y. and Holm, R.H. (2004) Structural conversions of molybdenum-iron-sulfur edge-bridged double cubanes and P^N-type clusters topologically related to the nitrogenase P-cluster. *Inorg. Chem.*, **43** (2), 674–682.

11 Binding Substrates to Synthetic Fe–S-Based Clusters and the Possible Relevance to Nitrogenases

Richard A. Henderson

11.1 Introduction

Fe–S-based clusters are the catalytic sites in certain metalloenzymes including the nitrogenases [1]. Studies on the nitrogenases have led to a detailed understanding of the structural composition and characteristics of the component Fe–S clusters. Nonetheless, it is, arguably, the most fundamental aspects of the action of nitrogenases that are still a mystery. Where and how do substrates bind to the Fe–S-based clusters that comprise the active site and what is the mechanism by which these bound substrates are converted into products? Answering these questions is crucial to understanding how nitrogenases operate. This is a long-standing, scientific problem. Because of the structural complexity and multistep nature of the reactions of the nitrogenases, it is often difficult to answer these questions by studies on the enzymes alone. Consequently, the investigation of synthetic Fe–S-based clusters, whose structures mimic those of the enzymes' active sites, plays an important role in addressing these questions. The much lower molecular weight and structural simplicity of the model systems allow us, at least in principle, to answer these questions in greater detail. However, despite several decades of work, involving studies on both the nitrogenases and synthetic Fe–S-based clusters, our understanding of the binding and activation of substrates at Fe–S clusters is rudimentary and imprecise. Kinetic and mechanistic studies on both the enzyme and model compounds have been performed since the 1960s, but more than 50 years later, despite a significant international effort, we still do not know how dinitrogen (or the so-called alternative substrates, such as H^+, N_3^-, CN^-, N_2O, MeNC, and C_2H_2) [2], bind to the active site. Nor do we know the pathways by which bound substrates are converted into products.

This chapter summarizes various studies involving, and related to, the binding of small molecules and ions to synthetic Fe–S-based clusters, in particular focusing on those investigations that are pertinent to understanding the action of the nitrogenases. Because the role of nitrogenases is to convert dinitrogen into ammonia, this chapter would ideally concentrate on dinitrogen binding to synthetic clusters. Unfortunately, at the moment no such clusters exist and so this chapter is

Bioinspired Catalysis: Metal-Sulfur Complexes, First Edition.
Edited by Wolfgang Weigand and Philippe Schollhammer.

concerned principally with some of the so-called alternate substrates of nitrogenase. The chapter starts by outlining how experiments and theoretical calculations on the enzyme have resulted in the current understanding of how the enzyme operates. This is followed by detailing the synthetic and mechanistic studies on how and where substrates bind to synthetic clusters.

Studies on the enzyme reveal that binding of dinitrogen is complicated by the formation of dihydrogen, which presumably involves protonation of the cofactor. The involvement of protons with the nitrogenases is particularly intriguing, and problematic, as protons are both substrates (being reduced to dihydrogen) and reactants (protonating dintrogen and the alternate substrates to convert them into products). Thus, the enzyme has to balance the binding of dinitrogen with the binding and reduction of protons. Protonation of Fe–S-based clusters is commonplace and has been studied in some detail. However, in this chapter, we only discuss those aspects of protonation pertinent to substrate binding as protonation of synthetic Fe–S-based clusters is the subject of a recent review dedicated exclusively to this single reaction [3]. Other reviews on the mechanisms of the reactions of synthetic Fe–S-based clusters have been published over the past few years. The reader is referred to these for more wide-ranging presentations and discussion of the chemistry of this class of compounds [4].

11.2
Mechanism of Nitrogenases

The nitrogenases are a group of metalloenzymes that convert dinitrogen into ammonia under ambient conditions. There are three nitrogenases: Mo nitrogenase (the most studied and the only one characterized by X-ray crystallography); V nitrogenase; and Fe-only nitrogenase. All three nitrogenases comprise two essential proteins: the Fe protein and the MFe protein (M = Mo, V, or Fe). The Fe protein (~65 000) is the smaller of the two proteins and contains a single $\{Fe_4S_4\}$ cuboidal cluster that mediates electron transfer from the external reductant (flavodoxin or ferredoxin) to the larger MFe protein (mol. wt. ~ 220 000 for Mo nitrogenase) [5]. The MFe protein contains two unique Fe–S-based clusters: P-clusters that have a $\{Fe_8S_7\}$ core and FeM-cofactors which have a $\{MFe_7S_9\}$ (M = Mo, V, or Fe) core, as shown in Figure 11.1. The P-cluster is believed to be effectively a capacitor that subsequently transfers electrons to the active site (FeM-cofactor), where the dinitrogen is bound and transformed into ammonia [6].

Figure 11.1 shows the electron transfer pathway associated with the nitrogenases, from flavodoxin through to substrate. Thorneley and Lowe [7] first described the complicated mechanistic sequence of protein–protein interactions and redox state changes of the proteins involved in the catalytic cycle for Mo nitrogenase. The main features of their mechanism are as follows. The Fe protein is first reduced by the external reductant (ferredoxin or flavodoxin) and then binds to the oxidized MoFe protein. Electron transfer from the Fe protein to the MoFe

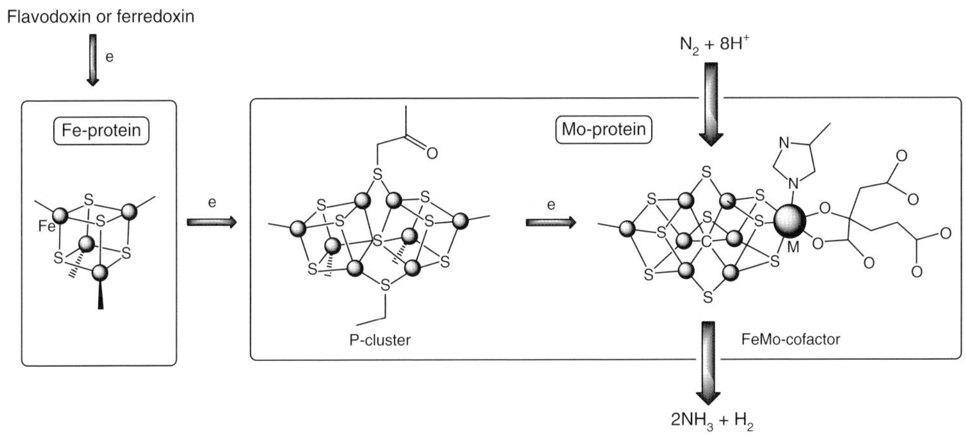

Limiting stoichiometries of nitrogenases	
Mo nitrogenase:	$N_2 + 8H^+ + 8e \longrightarrow 2NH_3 + H_2$
V nitrogenase:	$N_2 + 12H^+ + 12e \longrightarrow 2NH_3 + 3H_2$
Fe nitrogenase:	$N_2 + 21H^+ + 21e \longrightarrow 2NH_3 + 7.5H_2$

Figure 11.1 Composition of nitrogenases and structures of Fe–S-based clusters associated with each protein.

protein ensues and then the proteins dissociate, allowing the oxidized Fe protein to be reduced again by the external reductant and the MoFe protein to bind and transform the substrate. This sequence of reactions between the two proteins is repeated at least six times in order to supply sufficient electrons to transform dinitrogen into ammonia. The rate-limiting step of nitrogenase is dissociation of the Fe- and MoFe proteins ($k \sim 6.4\,s^{-1}$). An important feature of the mechanism of the Mo nitrogenase is that dinitrogen cannot bind to the resting (isolated) state of the MoFe protein. Three electrons have to be transferred to the MoFe protein before dinitrogen can bind. The dinitrogen is bound and transformed at the FeMo-cofactor, which has a composition of $\{MoFe_7S_9X((R)\text{-homocitrate})\}$ ($X = C$, N, or O; recent studies indicate $X = C$, and this is adopted throughout this chapter) [8].

It is when we come to describe the binding and transformation of dinitrogen, including the intermediates involved, that the mechanism of nitrogenase becomes less clear. The vagueness is because of the structural complexity of the FeMo-cofactor and the lack of directly analogous clusters in chemistry. Even superficial consideration of the mechanism of the nitrogenases raises fundamental questions. (i) The structure of the FeMo-cofactor determined from X-ray crystallography is the resting state: a state of the MoFe protein, which is incapable of binding dinitrogen. Three electrons have to be added to the MoFe protein before dinitrogen will bind. Does the structure of the FeMo-cofactor change when these three electrons are added? (ii) The MoFe protein binds dinitrogen and converts it rapidly into ammonia. Thus, FeMo-cofactor can exist in a variety of redox states with dinitrogen and various intermediates bound. Furthermore, the multimetal nature of FeMo-cofactor allows, in principle, the intermediates in dinitrogen transformation to adopt a variety of coordination modes. What are the structures of the cofactor and the mode of coordination of the substrate and intermediates at each stage in the transformation?

11.2.1 Detecting Substrates and Intermediates Bound to the Enzyme

In the 1970s, the first intermediate in the transformation of dinitrogen into ammonia was detected in the Mo nitrogenase [9]. The intermediate was detected by quenching the enzyme during the pre-steady state period of the catalysis using an acid or alkali quench. Addition of either acid or alkali to the reacting enzyme results in the formation of hydrazine. This hydrazine is not the result of the enzyme-producing free hydrazine, but rather is produced from an enzyme-bound intermediate. The maximum concentration of the intermediate was reached after about 200 s and corresponded to 36% of the dinitrogen-reducing MoFe protein (after allowing for the dihydrogen production). At longer times, the concentration of the intermediate decreases such that at about 600 s no intermediate is detectable. This is consistent with the accumulation at long times of the product-inhibiting MgADP. By comparison with the known chemistry of nitrogenous ligands bound to mononuclear Mo or W phosphine complexes, it was proposed that the hydrazine-producing intermediate was most likely a bound hydrazide (i.e., metal-NNH_2) [10].

More recently, using a combined biochemical, genetic, and spectroscopic approach [11], other intermediates have been detected and give further insight into the mechanism of the enzyme. These studies indicate the following. (i) The binding site on FeMo-cofactor is associated with a specific face in the center of the cluster, as shown in Figure 11.2. (ii) That a valine side chain (α70V) controls access of the substrate to this face. (iii) That a histidine (α195H) close to the cofactor is involved in proton delivery. Using altered MoFe proteins, in which either or both α70V and α195H have been substituted, it has been possible to show the following new characteristic reactivities. Changing the valine to the smaller alanine allows substrates to bind to the cofactor which are excluded by the wild-type enzyme (i.e., propyne, 1-butyne, and propargyl alcohol) and changing the valine to the larger isoleucine stops all substrate transformation except proton reduction.

One of the most notable features about these studies on altered nitrogenases is that it has made possible the detection of intermediates in the transformation of alkynes and dinitrogen, and the reduction of protons. Using a combination of ^{13}C and $^{1,2}H$ Mims 35 GHz electron-nuclear double resonance (ENDOR) spectroscopy, an intermediate produced during the reduction of propargyl alcohol ($HOCH_2C{\equiv}CH$) has been shown to have an intact three-carbon backbone with no two carbon atoms equivalent. Furthermore, one of the terminal carbons contains two hydrogen atoms, one derived from the substrate and the other derived from

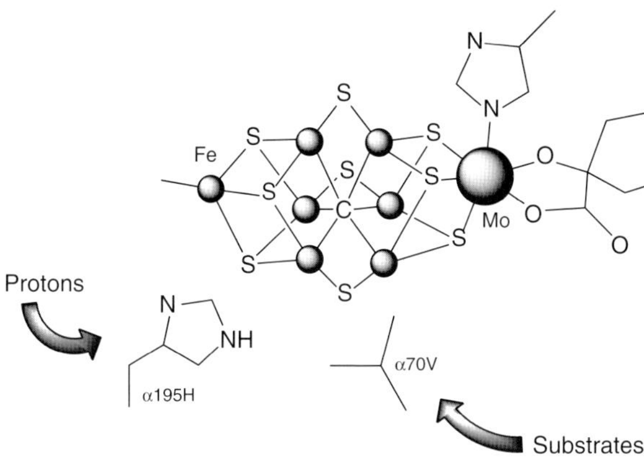

Figure 11.2 Indication of the pathways for the access of substrates and protons to FeMo-cofactor.

the solvent. In addition, another solvent-derived hydrogen is bound to another carbon. These features are consistent with the detected intermediate being the bound product, $HOCH_2CH{=}CH_2$. A major advantage in studying such carbon-based substrates is that the very slow rate of exchange of C–H bonds with H^+ in solution permits distinction between the hydrogens that were originally bound to carbon and those derived from the solvent.

Using a combination of freeze-quench trapping experiments [12] together with ^{15}N and ^{1}H ENDOR spectroscopies, intermediates in the transformation of dinitrogen have been detected. In wild-type nitrogenase, an unprotonated nitrogen intermediate has been detected and in certain altered nitrogenases intermediates containing NH_x residues from reactions with MeNNH, N_2H_2, and N_2H_4 have been detected. Although originally believed to all be different intermediates, more recent studies using X/Q band electron paramagnetic resonance (EPR) and $^{1,2}H$, $^{14,15}N$ ENDOR/HYSCORE (hyperfine sublevel correlation)/ESEEM (electron spin echo envelope modulation) spectroscopies have shown that the intermediates detected with MeNNH, N_2H_2, and N_2H_4 are all the same. Furthermore, in this intermediate, the N–N bond has been broken. It has been suggested that this common intermediate is either =NH, $-NH_2$, or $-NH_3$.

Substituting the α70V for isoleucine results in an enzyme that can only reduce protons to dihydrogen. This allows the detection of intermediates in the proton reduction pathway and $^{1,2}H$ ENDOR spectroscopy has detected two chemically equivalent hydrides (hydrogens attached to metal sites). The involvement of metal hydrides in nitrogen fixation was first proposed in the 1980s in the context of dinitrogen binding (displacement of dihydrogen from a metal-dihydride or metal-dihydrogen site) [13]. We return to the possible involvement of metal hydrides in dinitrogen binding and the competition between dinitrogen transformation and dihydrogen production in Section 11.3.6.3.

Finally, stopped-flow Fourier transform infrared (FTIR) spectroscopy has been used to detect carbon monoxide bound to the enzyme [14]. These studies indicate that there is more than one carbon monoxide binding site. At low concentrations of carbon monoxide (stoichiometric equivalence of carbon monoxide and FeMo-cofactor centers), a transient is observed ($\upsilon_{CO} = 1904\,cm^{-1}$). The concentration of this species peaks at about 7 s. Because inhibition of nitrogen fixation by CO is rapid, it appears that it is the species giving rise to this IR peak which is responsible for inhibiting the Mo nitrogenase. At high concentrations of carbon monoxide, longer lived species are formed ($\upsilon_{CO} = 1936\,cm^{-1}$, which peaks at 55 s, and $\upsilon_{CO} = 1958\,cm^{-1}$, which peaks at 100 s). These species persist for more than 10 min.

Recently, it has been shown that carbon monoxide is a substrate for the V nitrogenase producing a mixture of the light hydrocarbons, ethylene, ethane, propylene, and propane (i.e., Fischer–Tropsch chemistry) [15]. The observation that carbon monoxide is a substrate for the V nitrogenase is particularly intriguing as carbon monoxide inhibits the Mo nitrogenase.

In summary, the following species have been detected bound to Mo nitrogenases: (i) a "hydrazine-level" intermediate (hydrazide); (ii) an intermediate formed after N–N cleavage (imide, amide, or ammonia); (iii) bound $CH_2{=}CHCH_2OH$ product formed from propargyl alcohol; (iv) metal hydrides; and (v) multiple carbon monoxide ligands.

11.2.2 Exploring Intermediates in the Enzyme Mechanism Using Calculations

Calculations have helped guide our understanding of intermediates in nitrogen fixation for several years. Early theoretical studies focused on the coordination geometries of the various nitrogenous intermediates, notably diazenide (NNH), hydrazide(2-) (NNH_2), and hydrazide(1-) ($NHNH_2$) [16]. After the determination of the X-ray structure of FeMo-cofactor, theoreticians' interest changed to the coordination mode for dinitrogen binding to the active site [17]. In hindsight, these studies are of limited use because the presence of the X = C atom inside the central cavity of the FeMo-cofactor had not been appreciated at that stage.

Currently, the most sophisticated calculations on the binding and transformation of dinitrogen have been developed by Dance [18] using density functional theory (DFT) calculations. These calculations describe (i) the behavior of FeMo-cofactor as a substrate-binding site; (ii) the sequence by which dinitrogen and hydrogen bind to FeMo-cofactor; and (iii) the sequence of intermediates leading from dinitrogen binding through to ammonia production. An elaborate mechanism has resulted from the calculations, in which the large numbers of potential bond-making and bond-breaking steps have each been investigated by simulation. By assessing the activation energies for all possible steps, the energetically most viable sequence of steps have been identified, the key steps of which are shown in Figure 11.3. The original calculations were performed assuming that the central

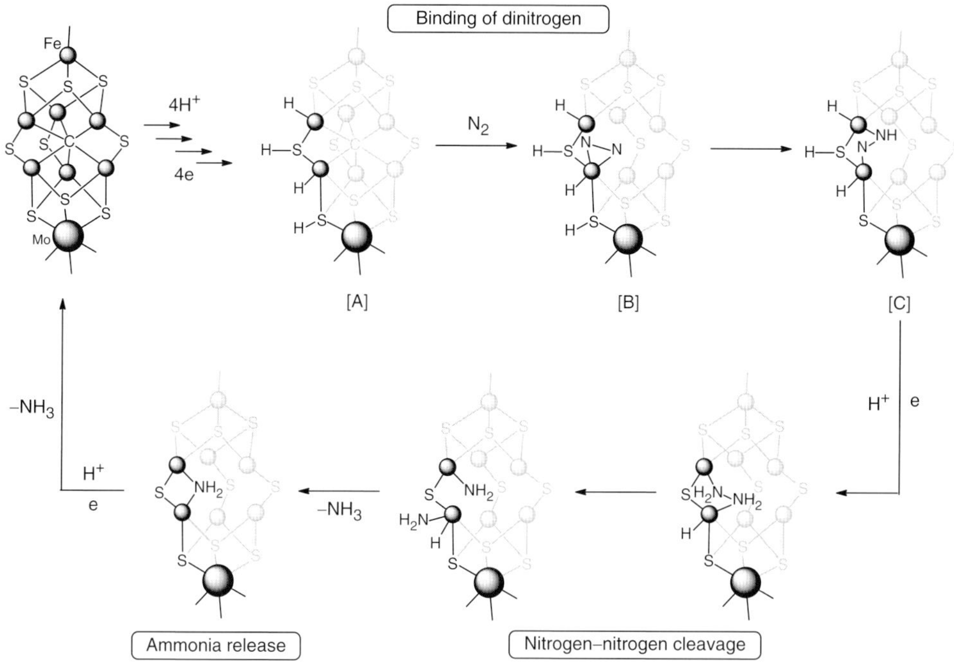

Figure 11.3 Mechanism for the binding and transformation of dinitrogen at FeMo-cofactor as suggested by DFT calculations. For clarity, nonparticipating cluster atoms are shown in shadow and from structure [**B**] onwards the central carbon is not included.

X atom was nitrogen; however, recent studies have indicated that this X atom is carbon, and Dance [19] reported new calculations where X = C indicate that the same mechanistic picture is valid.

The mechanism shown in Figure 11.3 introduces a quite new concept of how dinitrogen activation occurs on FeMo-cofactor. For a long time, it was assumed that the enzyme action involves initial coordination of the substrate followed by sequential delivery of coupled electrons and protons. In contrast, the Dance mechanism suggests that initially there is a sequential relay of multiple hydrogen atoms to FeMo-cofactor through coupled electron and proton transfer steps. This results in the state [A] shown in Figure 11.3, containing two Fe–H bonds and two protonated sulfurs. This species is capable of evolving dihydrogen. Provided dihydrogen is not produced, dinitrogen binds to a specific Fe atom within the FeMo-cofactor (as shown in Figure 11.3[B]), followed by intramolecular hydrogen atom transfer to bound dinitrogen and the sequence of steps resulting in ammonia formation. This mechanism identifies two key Fe atoms on FeMo-cofactor, which are involved in the transformation of dinitrogen. These are the same Fe atoms as proposed from the studies on the enzyme [11]. At first sight, it might appear that there are three equivalent pairs of such Fe atoms around the "girdle" of FeMo-cofactor. However, this is not the case. Each of the central faces of FeMo-cofactor is unique because of the Mo-coordinated histidine and (*R*)-homocitrate and the dispositions of the noncoordinated amino acid side chains in the active site cavity.

It is intriguing that the calculations indicate that the preferred binding mode of dintrogen is side-on. This is a rare coordination mode for dinitrogen; end-on coordination is much more common [20]. Furthermore, activation of this side-on coordinated dinitrogen is by intramolecular hydrogen transfer to form (initially) the diazenido species [C] bridged between two Fe sites. Unusual coordination modes of dinitrogen between two metal sites have been observed in binuclear complexes [21] such as $[\{Ta(P_2N_2)\}_2(\mu\text{-}H)_2(\mu,\eta^2\text{-}N_2)]$ ($P_2N_2 = PhP\{CH_2SiMe_2NPh\}_2$).

In concluding this section, it is interesting to note that both the DFT calculations and experiments on Mo nitrogenase favor dinitrogen binding at Fe sites of FeMo-cofactor. The obvious question that springs to mind is what is the role of the Mo? We return to this issue in Section 11.3.6.3.

11.3
Studies on Synthetic Clusters

Although FeMo-cofactor has yet to be synthesized in the laboratory, there are many other synthetic Fe–S-based clusters, which contain at least some of the structural features of FeMo-cofactor. Studies on this large family of synthetic Fe–S-based clusters have revealed reactivity characteristics that are so general they might reasonably be expected to be evident in the reactions of natural Fe–S-based clusters such as FeMo-cofactor. In this section, the structures and mechanisms of substrates binding to synthetic Fe–S-based clusters are presented.

11.3.1
Evidence for Substrates Bound to Synthetic Clusters

That small molecules and ions bind to Fe–S-based clusters originates from the discovery that a variety of different systems involving synthetic Fe–S-based clusters reduce dinitrogen, hydrazine, protons, or acetylene, either stoichiometrically or catalytically. Catalytic reduction of both dinitrogen and hydrazine to ammonia has been reported by electrochemically reduced cuboidal clusters in protic solvents, at a Hg electrode [22]. The reduction of hydrazine can be performed in water or MeOH/thf (thf, tetrahydrofuran) mixtures containing the cuboidal $[Fe_4S_4(SR)_4]^{2-}$ or dicubane $[\{MoFe_3S_4(SR)_3\}_2(\mu\text{-}SPh)_3]^{3-}$ (R = Ph or CH_2CH_2OH) as suspensions [23].

Holm [24] first showed that $[Fe_4S_4(SPh)_4]^{3-}$ can reduce protons to dihydrogen, or acetylene to ethylene (stereospecifically producing *cis*-$C_2H_2D_2$ in the presence of deuterons). Subsequently, the kinetics of the reduction of protons (supplied as $lutH^+$, lut = 2,6-dimethylpyridine) and transformation of acetylene by $[Fe_4S_4(SPh)_4]^{3-}$ have been investigated [25]. The kinetics indicate that $[Fe_4S_4(SPh)_4]^{3-}$ must be triprotonated before it is capable of evolving dihydrogen or converting acetylene into ethylene. We return to this system in Section 11.3.6.3 when discussing the problem of substrate binding versus dihydrogen production.

Transformation of hydrazine to ammonia or acetylene to ethylene [26] occurs at single cubane clusters such as $[MoFe_3S_4Cl_3(polycarboxylate)(NCMe)]^{2-}$ and $[VFe_3S_4Cl_3(dmf)_3]^-$ (dmf, dimethylformamide), using $lutH^+$ salts as acid and $[Co(\eta^5\text{-}C_5H_5)_2]$ as reductant. Precipitation of the ammonium salts prohibited a detailed kinetic study. It has been suggested that the heterometal is the preferred binding site, but substrates can also bind and be transformed at the Fe sites with the transformation at Fe being appreciably slower than at the heterometal. Thus, clusters where the heterometal has no vacant site (or labile ligand) still transform hydrazine or acetylene, albeit slowly. For example, $[VFe_3S_4Cl_3(HBpz_3)]^{2-}$ ($HBpz_3$, hydrotris(pyrazolyl)borate) does not transform the hydrazine or acetylene because it contains the nonlabile $HBpz_3$ ligand which occupies all potential substrate-binding sites on the vanadium atom. The problem with these experiments and their interpretation is that the nonlabile ligands on the heterometal must have an effect on the reactivity of the Fe sites. The general picture that emerges is illustrated in Figure 11.4, where both the Fe and heterometal can be

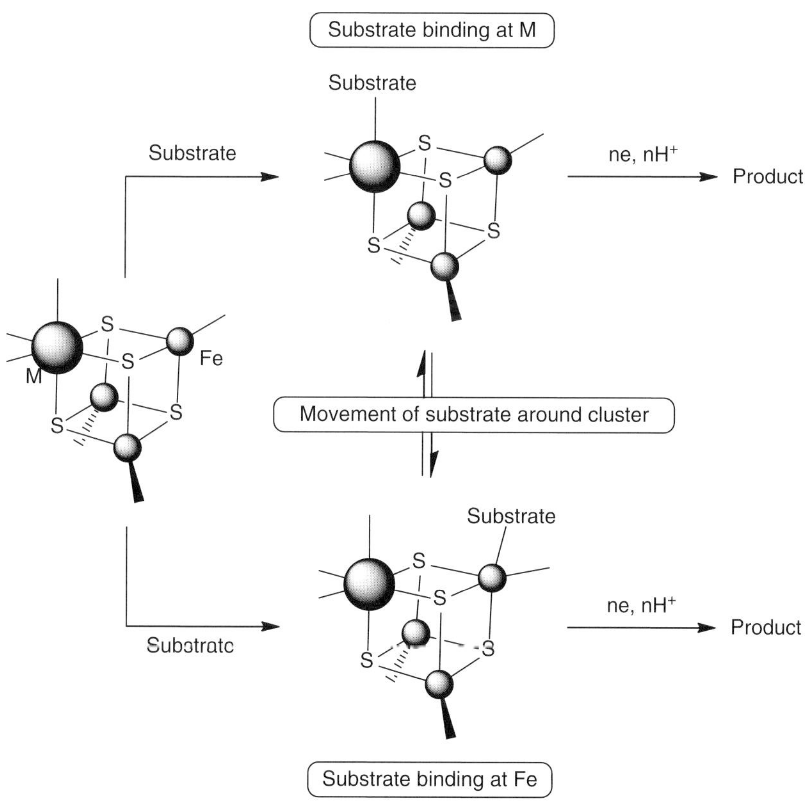

Figure 11.4 Possible substrate-binding sites and interconversions leading to transformation at Fe–S-based clusters.

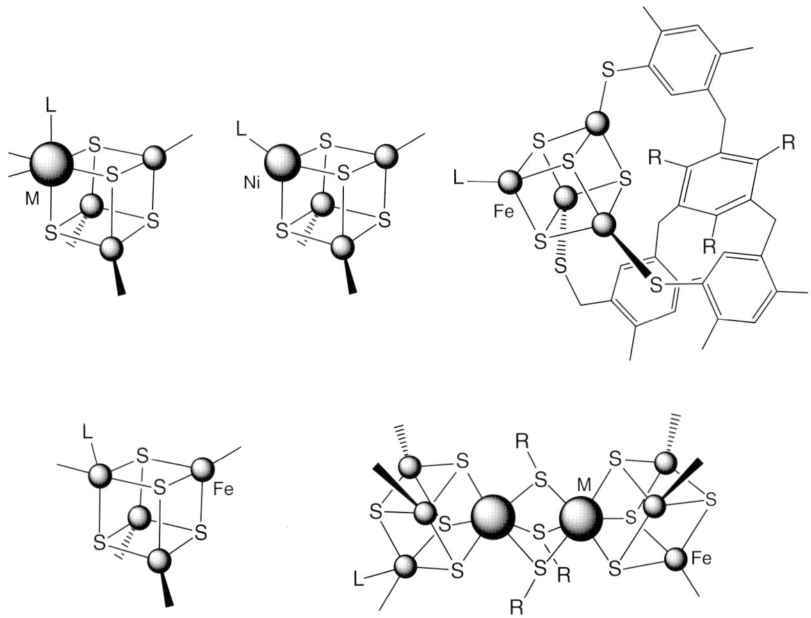

Figure 11.5 Structures of various synthetic Fe–S-based clusters indicating binding of substrate (L) to different sites. The tridentate cavitand thiolate ligand ($ArS_3{}^{3-}$) is shown bound to a $\{Fe_4S_4\}$ cluster in the top right hand corner. In the clusters on the top row, L binds by substitution, whereas for the clusters in the bottom row L binds directly to intact cluster with no substitution.

the sites of binding and transformation, and the relative efficacies of each site are affected by all the ligands on all sites. There is also the, as yet unproven, possibility that substrates can intramolecularly transfer between different metal sites.

More direct evidence that small molecules bind to synthetic Fe–S-based clusters comes from the few examples of clusters that have been isolated and contain substrates or products bound as indicated in the top row of Figure 11.5. Thus, $[VFe_3S_4Cl_3(bipy)(PhNHNH_2)]^-$ has been characterized by X-ray crystallography and contains the hydrazine bound to V in an end-on manner [26a]. In addition, $[MoFe_3S_4Cl_3(tccat)(NCMe)]^{2-}$ (tccat, tetrachlorocatecholate) reacts with *cis*-MeN=NMe to give free $MeNH_2$ and $[MoFe_3S_4Cl_3(tccat)(NH_2Me)]^{2-}$, where $MeNH_2$ is bound to Mo [27]. In analogous reactions, this same cluster binds CN^- or $N_3{}^-$ exclusively to the Mo site. In a further example [28], the nickel-containing cuboidal cluster $[NiFe_3S_4(SEt)_3(PPh_3)]^{2-}$ contains a Ni–PPh_3 bond and the PPh_3 ligand is readily replaced by other tertiary phosphines, Bu^tNC or CN^-.

Whether the structures of the clusters presented on the top row of Figure 11.5 have a mechanistic significance is questionable. Just because a cluster has been isolated with a substrate bound to Mo does not mean Mo is the catalytically active site. The isolated cluster may correspond to a "thermodynamic pit" and, in reality, the substrate is bound and transformed at a more reactive Fe site, but such species are transient and labile and so allude isolation or detection. Indeed, coordination of small molecules or ions (substrates) to Fe sites by displacing terminal thiolate or halide ligands in Fe–S-based clusters is commonplace.

Investigating substrates binding to the Fe sites in Fe–S-based clusters is complicated by the multiple equivalent Fe sites where substitution can occur resulting, ultimately, in all such sites undergoing reaction. In order to limit the reactivity of cuboidal clusters to a single Fe site, use has been made of semirigid tridentate cavitand thiolate ligands (ArS_3^{3-}), such as that shown in Figure 11.5 [29]. The ArS_3^{3-} ligand coordinates three of the Fe sites in a cuboidal $\{Fe_4S_4\}$ cluster, producing a site-differentiated cluster that contains only a single Fe to which the substrate can bind. Studies on these clusters allow investigation into the intrinsic substrate affinity of these clusters. Using $[Fe_4S_4(ArS_3)Cl]^{2-}$ reactions with a variety of substrates produce new clusters by displacement of the single chloro-ligand. To summarize a large body of work, clusters of the type $[Fe_4S_4(ArS_3)X]^{2-}$ have been identified (either by crystallization or in solution); when X = CN^-, N_3^-, thiolate, phenolate, methoxide, the ligands are monodentate, but when X = acetate, dithiocarbamate, pyridine-2-thiolate, or benzene-1,2-dithiolates the ligands are bidentate. The reaction of $[Fe_4S_4(ArS_3)Cl]^{2-}$ with RNC results in three isonitriles coordinated to the unique Fe site.

11.3.2 Mechanisms of Substrates Binding to Fe–S-Based Clusters

In all the examples of substrates binding to synthetic Fe–S-based clusters, presented in Section 11.3.1, the substrate binds to Fe or heterometal by displacing another ligand (i.e., substitution). However, as outlined in Sections 11.2.1 and 11.2.2, the mechanism proposed for nitrogenases involves the substrate binding to a Fe site of FeMo-cofactor with no displacement of another ligand. In this section, the mechanistic characteristics of substrates binding to synthetic Fe–S-based clusters are presented. The description includes both substitution reactions at Fe–S-based clusters and transient binding of substrates to the intact cluster, which involves no substitution. An important limitation of the mechanistic descriptions that follows is that they relate only to substitution at tetrahedral Fe sites in Fe–S-based clusters. To date, there have been no quantitative studies on the substitution at heterometal sites in Fe–S-based clusters.

The classification of substitution reactions was originally described for mononuclear complexes and is described as associative (A), dissociative (D), and interchange (I_a and I_d) mechanisms. These same designations are appropriate for the substitution reactions of terminal ligands in polymetallic clusters [30]. However, there is additional mechanistic behavior unique to clusters as a consequence of their polymetallic composition. We address this feature in Section 11.3.3.

The first study [31] on the kinetics and mechanism of substitution of synthetic Fe–S clusters was the thiolate exchange reactions of $[Fe_4S_4(SR)_4]^{2-}$ (R = Et or Bu^t) with 4-R′C_6H_4SH (R′ = NH_2, Me, or NO_2). The reactions exhibited second-order kinetics: first-order dependencies on the concentrations of both the cluster and

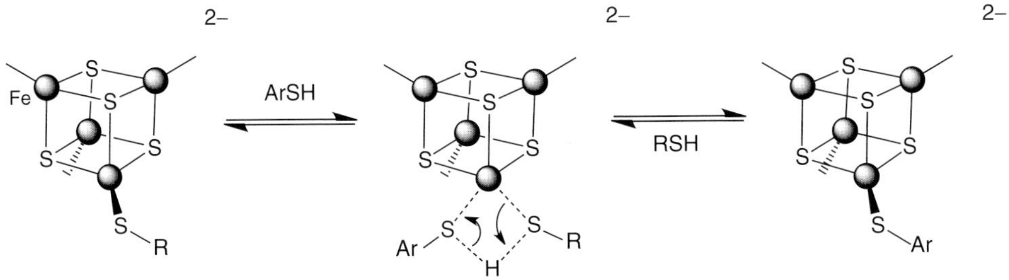

Figure 11.6 Proposed mechanism for the reaction of $[Fe_4S_4(Salkyl)_4]^{2-}$ with arylthiols. For clarity, here and in the subsequent figures, only the terminal ligand undergoing replacement is shown.

thiol. The authors noted that the rate of the reactions correlated with the acidity of the arylthiol, and proposed the mechanism shown in Figure 11.6.

In this mechanism, the initial step involves formation of an intermediate in which the sulfur of the attacking thiol interacts with the Fe site, while simultaneously the proton of this thiol interacts with the sulfur of the terminal alkyl thiolate ligand. Subsequently, the newly formed alkylthiol dissociates and the arylthiolate binds to the Fe. In order to avoid the complication of the role of the proton in this mechanism, and to exclusively focus on the substitution reactions, the substitution reactions of terminal ligands on Fe–S-based clusters have been studied with various nonprotic nucleophiles (substrates).

The kinetics of the substitution reactions of $[Fe_4S_4(SR)_4]^{2-}$ (RS = alkyl or aryl thiolate) [32] with $R'S^-$ exhibit a first-order dependence on the concentration of the cluster, but is independent of the concentration of thiolate nucleophile, as indicated in Figure 11.7. These kinetics are consistent with the dissociative mechanism shown in Figure 11.7, in which the coordinated thiolate ligand initially dissociates from one of the four equivalent Fe sites to produce a coordinatively unsaturated Fe site, which is rapidly attacked by the nucleophile. The dissociation of the thiolate ligand (k_1) is rate-limiting. The same elementary steps would operate for the binding of any substrate (L).

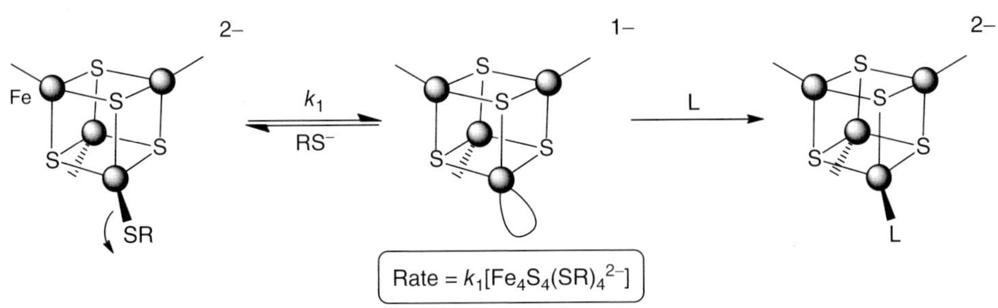

Figure 11.7 Dissociative substitution mechanism and associated rate law for binding substrates to Fe sites in Fe–S-based clusters. The lobe shown in the middle cluster represents a vacant site on the Fe. The charge on the final cluster will clearly depend on the charge of L. In this generic representation, L is a neutral molecule.

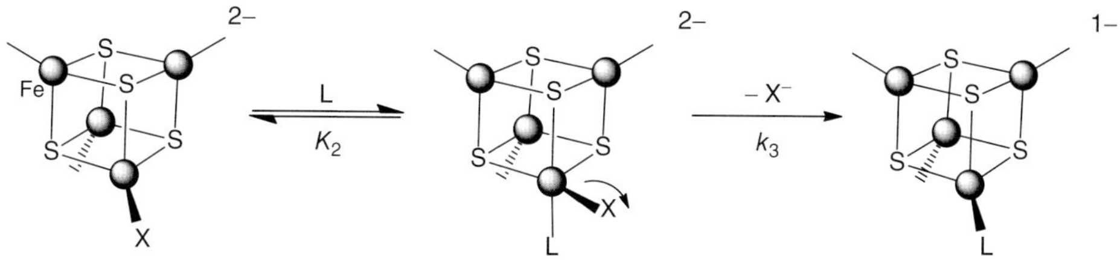

$$\text{Rate} = \frac{k_3K_2[\text{L}][\text{Fe}_4\text{S}_4\text{X}_4{}^{2-}]}{1 + K_2[\text{L}]}$$

Figure 11.8 Associative substitution mechanism and associated rate law for binding substrates to Fe sites in Fe–S-based clusters. The charges on the intermediate and final clusters will clearly depend on the charge of L. In this generic representation, L is a neutral molecule.

The kinetics of the substitution reactions of $[Fe_4S_4X_4]^{2-}$ (X = Cl, Br, or I) with PhS^- show a more complex behavior [33], exhibiting a first-order dependence on the concentration of the cluster, but a complicated dependence on the concentration of nucleophile. At low concentrations of PhS^-, the rate of substitution exhibits a first-order dependence on the concentration of thiolate. As the concentration of PhS^- is increased, the rate increases in a nonlinear manner, such that at high concentrations of thiolate the rate becomes independent of the concentration of PhS^-. There is nothing special about PhS^- as a nucleophile, and analogous kinetics could be expected with other substrates (L). The nonlinear dependence on the concentration of L is consistent with the associative mechanism shown in Figure 11.8 in which L binds at an Fe site (K_2). Subsequent dissociation of the chloro-ligand (k_3) results in the monosubstituted cluster. At high concentrations of L, the exclusive solution cluster species is $[Fe_4S_4Cl_4(L)]^{2-}$, so the rate of the reaction becomes independent of the concentration of L and the observed rate constant corresponds to the dissociation of the chloro-ligand. Mathematically, the rate law is that shown in Figure 11.8. Analysis of the kinetic data allows determination of both the equilibrium constant for the nucleophile binding to the cluster and the rate of dissociation of the chloro-ligand. Although the kinetics indicate that $[Fe_4S_4Cl_4(L)]^{2-}$ accumulates in solution at high concentrations of L, it has proved difficult to obtain unambiguous spectroscopic evidence for this intermediate. $[Fe_4S_4Cl_4(L)]^{2-}$ would not be associated with a diagnostic IR or ^1H NMR spectrum. Detection by UV–visible spectrophotometry using a stopped-flow spectrophotometer is possible. However, the electronic spectrum of Fe–S clusters is dominated by transitions associated with the $\{Fe_4S_4\}$ core, and the addition of a single nucleophile to $[Fe_4S_4Cl_4]^{2-}$ makes little difference to the electronic spectrum.

It is pertinent to mention here that studies on the reaction between the tetrahedral $[FeCl_4]^-$ and 4-$RC_6H_4S^-$ follow kinetics analogous to that of $[Fe_4S_4Cl_4]^{2-}$. However, in this case, the five-coordinate intermediate $[FeCl_4(SC_6H_4R\text{-}4)]^{2-}$, has been detected spectrophotometrically [34]. The visible absorption spectrum of the intermediate is rather featureless but contains a broad peak at $\lambda = 500$ nm.

The reason why $[Fe_4S_4(SR)_4]^{2-}$ undergoes substitution by a dissociative mechanism while the structurally similar $[Fe_4S_4Cl_4]^{2-}$ undergoes substitution by an associative mechanism is probably because the thiolate ligands in the parent cluster present a larger steric barrier to the attacking nucleophile.

The electronic factors influencing the associative substitution mechanism have been investigated in studies between $[Fe_4S_4Cl_4]^{2-}$ and $4\text{-}RC_6H_4S^-$ (R = MeO, Me, H, Cl, or CF_3) [35], where the systematic variation of the electronic effects associated with the substrate reveals the effect that the 4-R-substituent has on the binding of the nucleophile and dissociation of the leaving group from the intermediate $[Fe_4S_4(SC_6H_4R\text{-}4)Cl_4]^{3-}$. Figure 11.9 shows the Hammett plot of the rate constants for binding the nucleophile to the cluster in the associative substitution mechanism of $[Fe_4S_4Cl_4]^{2-}$ (k_2^R). It is obvious from Figure 11.9 that k_2^R increases as the 4-R-substituent becomes more electron withdrawing. This is, at first sight, unexpected. It might be anticipated that electron-releasing groups, which increase the electron density at the sulfur and hence increase its nucleophilicity, would increase the rate. That k_2^R is facilitated by electron-withdrawing groups is consistent with these substituents stabilizing the transition state for the binding step. As the anionic nucleophile approaches the dianionic $[Fe_4S_4Cl_4]^{2-}$, there must be an unfavorable buildup of negative charge in the transition state. Electron-withdrawing 4-R-substituents dissipate this unfavorable negative charge and facilitate the rate of thiolate binding.

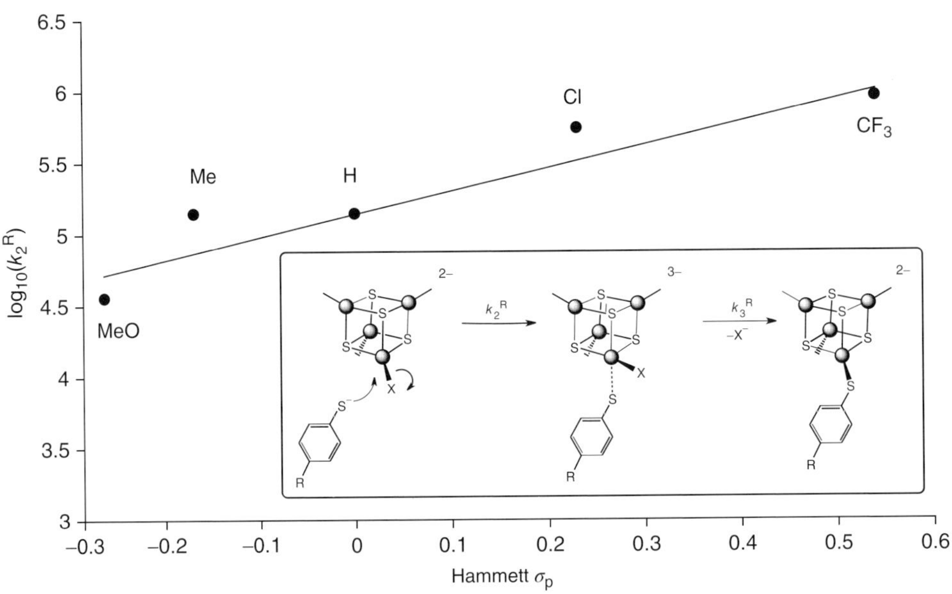

Figure 11.9 Hammett plot showing the effect that the 4-R-substituent has on the rate constant for $4\text{-}RC_6H_4S^-$ binding to $[Fe_4S_4Cl_4]^{2-}$.

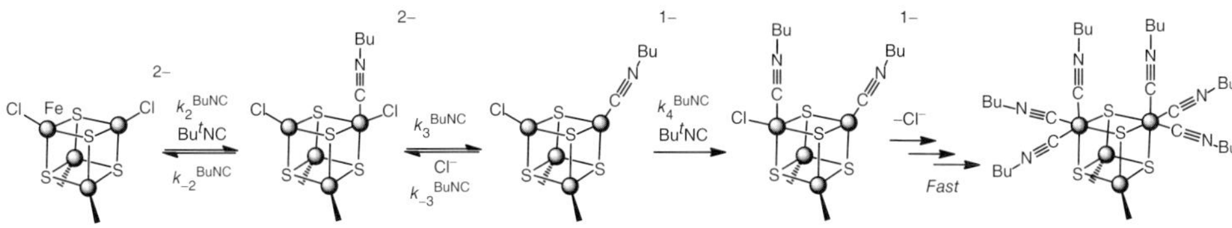

$$\text{Rate} = \frac{k_2^{\text{BuNC}} k_2^{\text{BuNC}} k_4^{\text{BuNC}} [\text{BuNC}]^2 [\text{Fe}_4\text{S}_4\text{Cl}_4{}^{2-}]}{k_4^{\text{BuNC}} [\text{BuNC}](k_3^{\text{BuNC}} + k_{-2}^{\text{BuNC}}) + k_{-2}^{\text{BuNC}} k_{-3}^{\text{BuNC}} [\text{Cl}^-]}$$

Figure 11.10 Rate law and mechanism for the reaction of Bu^{t}NC with $[Fe_4S_4Cl_4]^{2-}$ to form $[Fe_4S_4Cl_2(CNBu^t)_6]$.

Perhaps unexpectedly, the dissociation of the chloro-ligand from $[Fe_4S_4Cl_4]^{2-}$ ($k_3{}^R$) is also facilitated by electron-withdrawing groups. That dissociation of the chloro-ligand is also facilitated by electron-withdrawing groups is consistent with the transition state for chloride dissociation in which the leaving group is being pushed electronically by the bound thiolate. Electron-withdrawing substituents allow the 4-$RC_6H_4S^-$ ligand to approach the Fe more closely (shorter Fe–S bond distance) and hence "push out" the chloro-group more readily.

In the reactions of certain substrates (notably isonitriles), more than one substrate binds to a single Fe site in the cluster. We have already alluded to studies with site-differentiated clusters where three Bu^{t}NC molecules bind to the unique Fe site [29]. It has also been shown that $[Fe_4S_4Cl_4]^{2-}$ react with an excess of Bu^{t}NC to form $[Fe_4S_4Cl_2(CNBu^t)_6]$, which contains two Fe sites ligated by three Bu^{t}NC and two Fe sites ligated by chloride [36], as shown in Figure 11.10. However, the kinetics of this reaction reflect only two of the six Bu^{t}NC binding [37]. The kinetics of the reaction exhibit a first-order dependence on the concentration of $[Fe_4S_4Cl_4]^{2-}$ but the dependence on the concentration of Bu^{t}NC is complicated and the reaction is inhibited by the chloride ion, as indicated by the rate law in Figure 11.10. These kinetics are consistent with the mechanism shown in Figure 11.10. It seems likely that the displacement of a chloro-ligand occurs after the attack of the first Bu^{t}NC at each Fe site. The binding of the remaining Bu^{t}NC molecules is fast.

11.3.3 Mechanisms Peculiar to Clusters

Because clusters contain several metal sites, a mechanism is possible in which the nucleophile attacks one site but the leaving group departs from another. Thus, for associative mechanisms of clusters, in addition to A and I_a mechanisms, we should also consider a mechanism designated A_R and I_{aR} (where the subscript R indicates the attack of the nucleophile and departure of the leaving group are associated with Remote sites). As illustrated in Figure 11.8, the binding of the PhS^- to one Fe site most likely results in dissociation of the chloro-ligand bound to the same Fe.

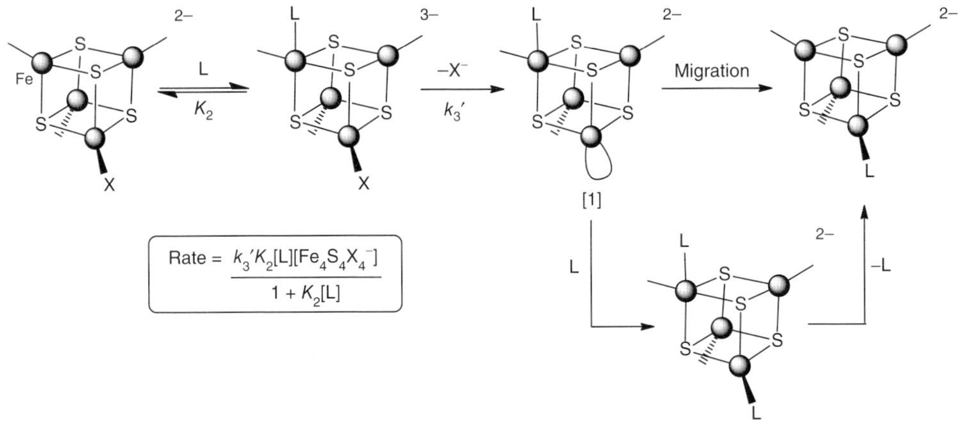

Figure 11.11 Rate law and mechanism for substitution of Fe–S-based clusters involving attack of substrate and loss of leaving group from different metal sites.

However, the lability of all the Fe sites are perturbed by the binding of the thiolate and dissociation of the leaving group can occur from any Fe site. Such a pathway is shown in Figure 11.11, and would be kinetically indistinguishable from the more direct route, provided intramolecular substrate transfer is rapid or another substrate rapidly attacks the vacant site and the initially bound substrate then rapidly dissociates from the cluster. It is not clear how the A and A_R mechanisms can be distinguished experimentally except by isolation of the intermediate [**1**].

Finally, in this section, it is pertinent to note that the binding of a nucleophile to a cluster can perturb the lability of all Fe sites in a cluster. In some cases, this results in unexpected reactivity. In several cases (e.g., $[Fe_4S_4(Salkyl)_4]^{2-}$ and $[Fe_2S_2Cl_4]^{2-}$), increasing the concentration of the nucleophile decreases the rate of the reaction! [38] We return to this point in Section 11.3.5.

11.3.4 Influence of Cluster Composition on Substrate Binding

Obviously, clusters contain more than one metal, and in some cases more than one type of metal. Consequently, one metal and its ligands can modulate the reactivity of the other metals within the cluster core. Extensive studies have developed the synthetic, structural, and electron-transfer chemistry of a series of isostructural clusters of general formulae $[\{MFe_3S_4X_3\}_2(\mu\text{-}SR)_3]^{n-}$ (M = V, Nb [39], Mo, W [40], Re [41], Fe [42], Co [43], Ni [24]; X = halide or thiolate) shown in Figure 11.5. Studies have shown that in $[\{MFe_3S_4X_3\}_2(\mu\text{-}SR)_3]^{n-}$ the bridging thiolate ligands are substitutionally inert; thus, substitution reactions are restricted to the terminal ligands on the Fe sites. Single cuboidal clusters containing heterometals are also known but, as indicated earlier, in these clusters the heterometal can contain labile ligands, making interpretation of the kinetics difficult and ambiguous. By ensuring that M is incapable of being the site of any reaction (surrounded by nonlabile ligands and coordinatively saturated), investigation of the reactivities of this series of clusters allow quantification of how M affects the reactivity of the Fe sites. Studies on the dicubane clusters $[\{MFe_3S_4(SR)_3\}_2(\mu\text{-}SR)_3]^{3-}$ (M = Mo or W; R = Et or Ph) show that while the analogous $[Fe_4S_4(SR)_4]^{2-}$ clusters undergo substitution by a dissociative mechanism (Section 11.3.2), the Mo- or W-containing clusters undergo substitution by a mixture of dissociative and associative mechanisms [44]. Thus, the formal replacement of a Fe–SR residue in the cuboidal $[Fe_4S_4(SR)_4]^{2-}$ by Mo or W (and its associated coligands) apparently makes the Fe sites in $\{MFe_3S_4\}$ clusters behave as though they are electron deficient, facilitating binding of the substrate and resulting in an associative pathway for substitution.

The reactions of the isostructural clusters $[\{MFe_3S_4Cl_3\}_2(\mu\text{-}SEt)_3]^{3-}$ (M = Mo or W) with both acids and nucleophiles have been studied [45]. The kinetics of the reactions of both clusters follow the usual acid-catalyzed associative substitution kinetics typical of cuboidal clusters [3, 4a, 4b]. Comparison of the results allows us to quantify the effect that changing M from Mo to W has on the rate of proton transfer and rate of binding of PhS^-. Both the rate of proton transfer and rate of binding of PhS^- are appreciably slower for the W complex than the corresponding Mo-containing cluster. It is interesting that, apparently, merely changing one of the core metal atoms from Mo and W in these clusters facilitates the binding of both anions (thiolate) and cations (proton). This indicates that the dominant influence of the metal cannot be electronic as electronic factors that facilitate binding of anions would disfavor binding of cations. Furthermore, comparing the reactivities of $[\{WFe_3S_4Cl_3\}_2(\mu\text{-}X)_3]^{3-}$ (X = SEt or OMe) allows us to quantify the effect that changing the spectator ligands bound to the W sites has on the reactivity of the cluster. Both the rate of thiolate binding and the rate of proton transfer are faster with the X = OMe derivative. Again, this is inconsistent with the dominant effect of the bridging ligands on the reactivity being electronic. It has been suggested that upon reaction with thiolates or protons, reorganization of the cluster bond lengths and angles is an appreciable barrier to both processes. The effect of the nonparticipating bridging X ligands is markedly smaller than the effect of changing the metal content of the cluster core. This seems intuitively reasonable because the bridging X ligands are remote from the Fe sites where the substitution reactions occur, whereas the Mo or W is part of the cluster core. In Section 11.3.6.3, we return to consider these studies further, in the context of the role of the heterometal in nitrogen fixation by the FeM-cofactor.

11.3.5 Transient Binding of Substrates to Clusters

So far the discussion has focused on substrates binding to clusters by displacing terminal ligands. In this section, we address the much simpler process in which the substrate simply adds to an intact cluster and, importantly, no substitution

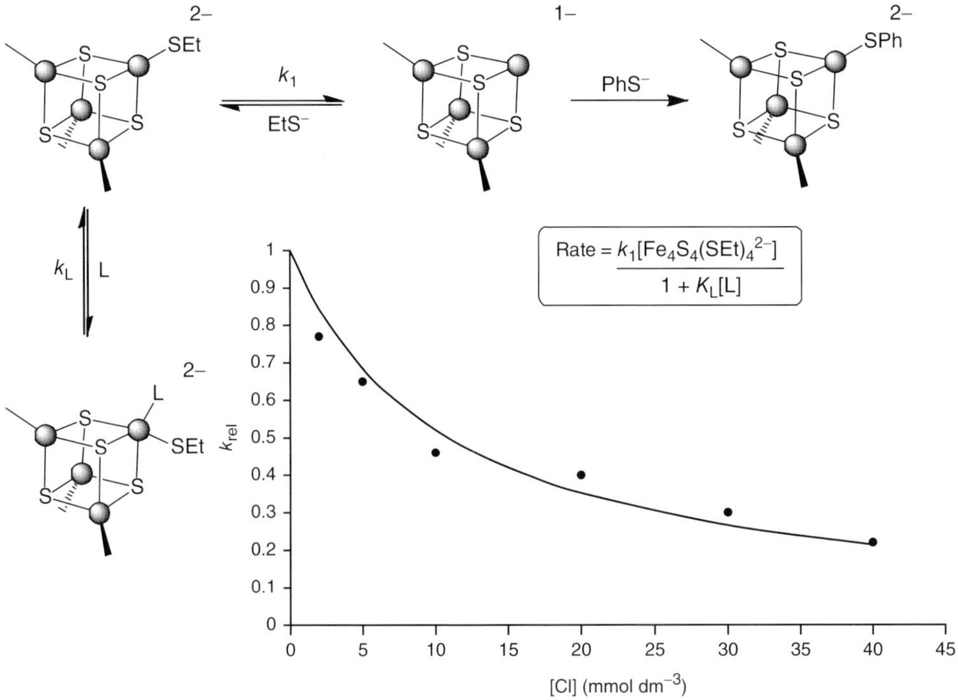

Figure 11.12 Example of reactions used to measure the binding of substrates (L) to Fe–S-based clusters. Graph shows the effect of binding Cl^- on the rate of substitution of $[Fe_4S_4(SEt)_4]^{2-}$.

occurs. An example of such a reaction is shown in Figure 11.12, for the binding of L to $[Fe_4S_4(SEt)_4]^{2-}$. The $[Fe_4S_4(SEt)_4(L)]^{2-}$ accumulates to detectable amounts in solution at high concentrations of L, but are short-lived species with L binding and dissociating rapidly.

A general problem in monitoring the binding of substrates to Fe–S-based clusters is the small or negligible change in the electronic spectrum of the cluster when the substrate binds. For this reason, an indirect method has been developed to measure the binding of substrates to the cluster [32]. The methodology is based on kinetics. In outline, changes in the rate of substitution of the terminal ligands report on the binding of substrates to the cluster. For the reactions shown in Figure 11.12, at a constant concentration of nucleophile (PhS^-), the rate of substitution of $[Fe_4S_4(SEt)_4]^{2-}$ will be a constant. Introduction of any substrate (e.g., Cl^-) that binds to the cluster will perturb the electron density within the cluster and result in a change to the rate of substitution. By varying the concentration of the substrate and analyzing the response to the rate of substitution allows us to determine: (i) the effect that binding a substrate to the cluster has on the terminal ligand's lability; (ii) the number of substrates binding to the cluster and (iii) the binding constant for substrate binding to the cluster.

The binding of substrates to Fe–S-based clusters is likely to occur at an Fe site to produce a five-coordinate Fe. So far it has not been possible to isolate such species. However, there are Fe–S-based clusters in which the Fe sites have coordination numbers higher than four [29, 36, 46].

For the binding of L to $[Fe_4S_4(SEt)_4]^{2-}$ as illustrated in Figure 11.12, the kinetic data can be fitted to the rate law shown in the figure, where k_1 is the observed rate constant for substitution at a fixed concentration of PhS^-, and K_L is the equilibrium constant for the substrate binding to the cluster. Two features indicate that the effects typified by the data in Figure 11.12 are attributable to the binding of the substrates [32] to $[Fe_4S_4(SEt)_4]^{2-}$. (i) The extent of the inhibition is a function of the concentration and nature of the substrate. (ii) The inhibition is specific for $[Fe_4S_4(SEt)_4]^{2-}$. The same substrates over the same concentration range show no effect on $[Fe_4S_4(SPh)_4]^{2-}$.

With $[Fe_4S_4(SEt)_4]^{2-}$, addition of the substrates N_2, H_2, C_2H_2, C_2H_4, or PhCCH have no effect on the rate of substitution, indicating that these substrates do not bind to the cluster. However, an important point is that substrate affinities will change with the redox state of the cluster. For example, alkynes must bind to $[Fe_4S_4(SEt)_4]^{3-}$ as they are transformed by this cluster into the corresponding alkenes.

In the studies with $[Fe_4S_4(SEt)_4]^{2-}$, binding of all substrates studied (Cl^-, Br^-, CN^-, N_3^-, N_2O, or CO) resulted in a decrease in the rate. As a consequence of the limited concentration range of L that can be used ($[L] < 40\,mmol\,dm^{-3}$), it was not possible to tell whether binding the substrate to the cluster completely switches off the substitution, or whether the binding of substrate results in a cluster with a slow, but finite, rate of substitution. However, studies with another cluster $[Fe_6S_9(SEt)_2]^{4-}$ shows more complicated kinetic behavior as shown in Figure 11.13, which gives further insight into these binding reactions.

Kinetic studies show that $[Fe_6S_9(SEt)_2]^{4-}$ exhibits the same substitution and protonation characteristics as other synthetic Fe–S-based clusters [47]. The rate of substitution of $[Fe_6S_9(SEt)_2]^{4-}$ with PhS^- in the presence of $NHEt_3^+$ is perturbed by the addition of $L = Cl^-$, CN^-, Bu^tNC, and imidazole as shown by the data in Figure 11.13. Binding of Cl^- and CN^- inhibit the substitution, while imidazole increases the rate of substitution and Bu^tNC has only a slight (inhibitory) effect. An important feature of these data is that while binding of Cl^-, CN^-, and Bu^tNC suppresses the rate of substitution of the cluster, the substitution is not completely stopped. Thus, at high concentrations of the substrate, when all clusters in solution have a substrate bound, the rate of the substitution reaction is slow, but not zero.

This kinetic behavior is consistent with the mechanism shown in Figure 11.13. After protonation of the cluster, substitution of the ethanethiol occurs, but binding of the substrate L to the cluster (K_L) competes with this substitution. Consequently, substitution can occur by two pathways: the k_1 pathway and another where the cluster is both protonated and has L bound (k_1^L pathway). Analysis of the data shows the binding affinities (K_L) follow the order

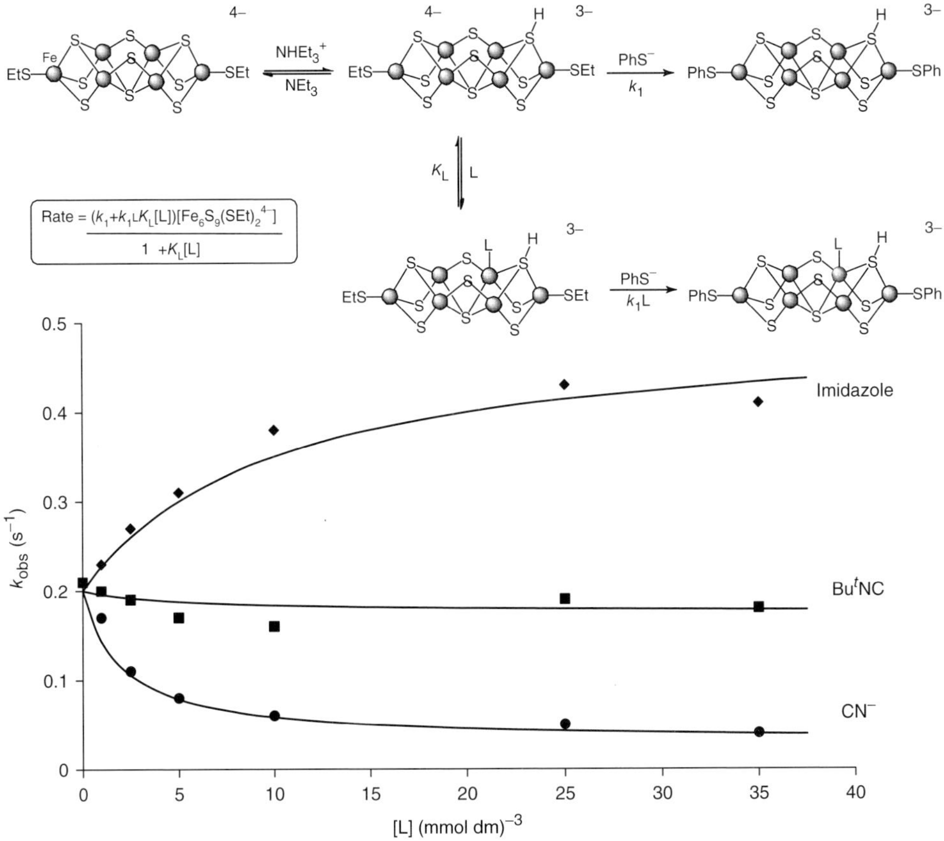

Figure 11.13 Graph showing the effect that binding various substrates (L = imidazole, Bu^{t}NC, or CN^-) to the cluster has on the rate of substitution of $[Fe_6S_9(SEt)_2]^{4-}$. Also shown is the proposed mechanism and generic rate law for these reactions.

$CN^- > Cl^- >$ Bu^{t}NC > imidazole, but all the substrates have similar binding affinities with the tightest being only five times larger than the weakest. Thus, $[Fe_6S_9(SEt)_2]^{4-}$ shows no significant discriminatory preference for binding the different types of substrates. It is notable that the anionic substrates (Cl^- and CN^-) have the most marked inhibitory effect on the rate of the reaction, while the neutral Bu^{t}NC has a very minor inhibitory effect and imidazole actually increases the rate of substitution.

Finally, in this section, we should consider the coordination modes for the substrates binding to Fe–S-based clusters. As alluded to earlier, DFT calculations [18, 19] favor the unusual side-on coordination of dinitrogen to FeMo-cofactor. Structural investigation of synthetic Fe–S-based clusters containing bound substrates shows no unusual behavior, such as the substrate interacting with more than one metal site. Rather, the substrates (which include CN^-, N_3^-, Bu^{t}NC, and MeCN) are coordinated to the clusters in the common end-on manner.

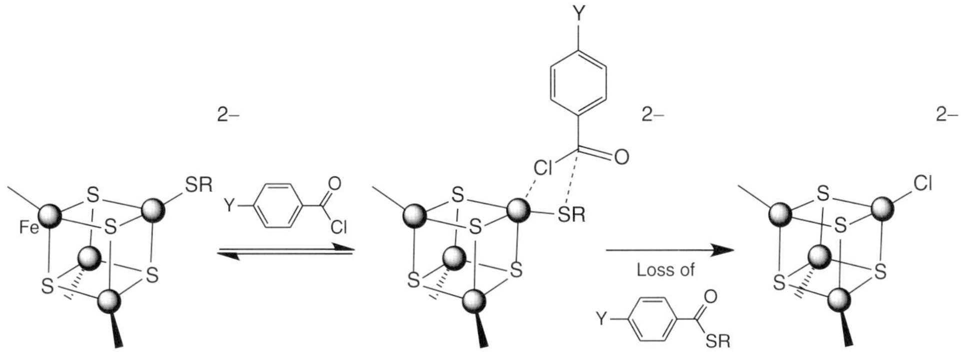

Figure 11.14 Mechanism of the reaction between acid chlorides and $[Fe_4S_4(SR)_4]^{2-}$ to produce $[Fe_4S_4Cl_4]^{2-}$, showing proposed structure of intermediate where acid chloride is bound to the cluster.

Studies on the mechanism of the reactions between $[Fe_4S_4(SR)_4]^{2-}$ (R = Et or Bu^t) and $4\text{-}YC_6H_4COCl$ (Y = MeO, H, or Cl) to form $[Fe_4S_4Cl_4]^{2-}$ suggest that binding of the substrate (acid chloride) to the cluster [48] can involve multiple interactions such as that shown in Figure 11.14. By varying the 4-Y-substituent on the acid chloride, it is found that the binding of the acid chloride to the cluster is favored by electron-withdrawing 4-Y-substituents. It has been proposed that the acid chloride is bound to one of the Fe atoms of the cluster through the oxygen or chlorine atom, while the sulfur of the coordinated thiolate interacts with the carbonyl carbon as shown in Figure 11.14. Such multiple interactions could be a feature of the binding of multifunctional substrates to Fe–S-based clusters.

11.3.5.1 **Influence of Metal Composition on Transient Binding Affinities of Substrates**

Using the kinetic methods described in Section 11.3.5, the values of K_L for both the $[Fe_4S_4(SEt)_4]^{2-}$ and $[\{MoFe_3S_4(SEt)_3\}_2(\mu\text{-}SEt)_3]^{3-}$ have been determined [32]. These data are presented in Figure 11.18, and the most notable feature is that, for the same substrate, K_L is always significantly larger for the Mo-containing cluster. The terminal ligands on Fe are the same in both clusters and the cluster geometries are essentially the same. Thus, it appears that the presence of Mo in the cluster core modulates the substrate-binding affinity of the cluster. It seems most likely that the substrate binds to Fe sites in both clusters. The data indicates that the presence of a Mo in the $\{MoFe_3S_4\}$ cluster makes the Fe sites behave as though they were more electron deficient than in the $\{Fe_4S_4\}$ cluster (see Section 11.3.4). The observation that Mo modulates the reactivity of the Fe sites in Fe–S-based clusters may be significant in considering the reactivity of the enzyme nitrogenase, and we return to this point in Section 11.3.6.3.

11.3.5.2 **Rates of Binding Substrates to Clusters**

The rates of binding substrates to clusters can be determined directly from analysis of the kinetics of certain protonation reactions [35, 45, 49] and, more generally, using a stopped-flow, sequential-mix method [37]. The sequential-mix

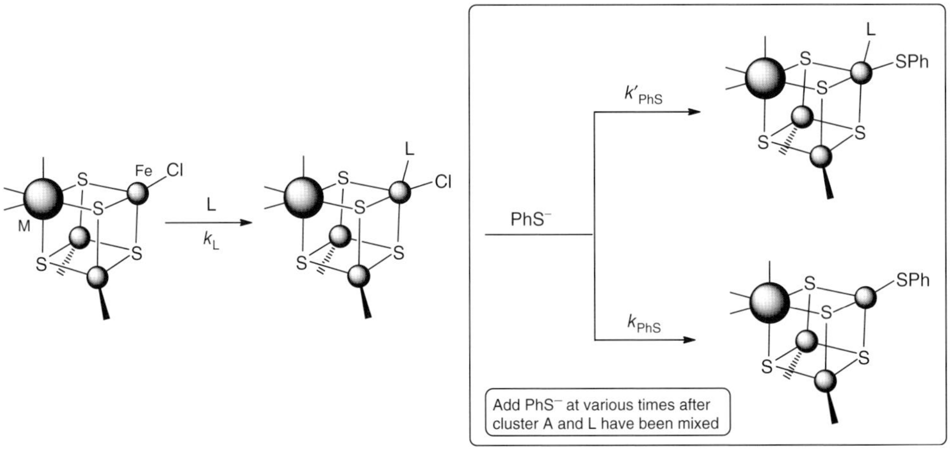

Figure 11.15 Outline of the reactions used in stopped-flow sequential-mix experiments to measure the rate of binding of substrate (L) to Fe–S-based clusters. The rate of the substitution reaction (shown in box) effectively reports on the status of the cluster as this rate depends on whether L is bound to the cluster (k'_{PhS}) or not (k_{PhS}).

method uses the observation that the rates of substitution of the terminal ligands on the cluster are sensitive to whether or not L is bound to the cluster, as shown in Figure 11.15.

There have been few studies measuring the rates of substrate binding. The rates of substrate binding [35, 37] to $[Fe_4S_4Cl_4]^{2-}$ are as follows: Bu^tNC, $k = 1.7 \times 10^3$ dm^3 mol^{-1} s^{-1}; $Et_2NCS_2^-$, $k \geq 3 \times 10^5$ dm^3 mol^{-1} s^{-1}, and PhS^-, $k = 1.4 \times 10^5$ dm^3 mol^{-1} s^{-1}. Although rapid, these rate constants are markedly slower than the diffusion-controlled limit, indicating that there is a significant barrier for binding substrates to the cluster. One barrier to substrate binding is probably the buildup of charge as the substrate approaches the anionic cluster (see Section 11.3.2). A further barrier must be the necessary geometrical and electronic reorganization of the ligands around the site to which the substrate binds. Because the binding site is part of a cluster, any reorganization at one metal will be transmitted to other atoms in the cluster core. A similar proposal has been used to explain why protonation of Fe–S-based clusters occurs at rates significantly slower than the diffusion-controlled limit [4a,b, 49].

11.3.6
Protonation of Clusters

In the absence of any other substrate, both nitrogenases and reduced synthetic clusters transform protons into dihydrogen. It seems likely that a necessary prerequisite to dihydrogen production is protonation of the clusters. Clearly, however, protons are also essential reactants in the transformation of substrates. Thus, we need to consider the dual role of the proton, as both substrate (which binds to the cluster and is a precursor to dihydrogen formation), and as a reactant (which binds to other coordinated substrates during the transformation into product). Before considering the mutual effects of substrates and protons binding to Fe–S-based clusters, the characteristics of protonation of clusters are outlined.

11.3.6.1 Kinetics and Thermodynamics of Protonation of Clusters

There have been extensive studies on the protonation of synthetic Fe–S-based clusters and the reader is referred to recent reviews in this area [3, 4a,b]. Binding of a proton to synthetic Fe–S-based clusters is associated with a negligible spectroscopic change. For this reason, protonation of the clusters has been established by studies in which the protonation is coupled to a substitution reaction. The substitution of terminal ligands in Fe–S-based clusters is invariably associated with a large spectroscopic change. Effectively, by analyzing the effect acid has on the rate of substitution gives information about the spectroscopically silent protonation step. The major limitation of this approach is that a protonation will only be detected if it affects the lability of the terminal ligands. It is possible that Fe–S-based clusters can protonate on a variety of sites, not all of which affect the lability. Such protonations would go undetected by the kinetic method.

Protonation of synthetic clusters is commonplace and the key features of this reaction are as follows: (i) protonation with $NHEt_3{}^+$ (pK_a = 18.4 in MeCN) results in single proton transfer to the cluster. When coupled to a substitution reaction of the terminal ligands, protonation by $NHEt_3{}^+$ is invariably more rapid than the act of substitution. (ii) The pK_a of the synthetic clusters all fall in the narrow range 17.9–18.9 in MeCN, irrespective of nuclearity, cluster composition, terminal ligands, and overall charge. This observation indicates that the protonation site is common to all synthetic Fe–S-based clusters and it has been suggested this is a core sulfur. (iii) Protonation of synthetic Fe–S-based clusters is inherently slower than the diffusion-controlled limit ($k \leq 1 \times 10^7$ dm^3 mol^{-1} s^{-1}). In studies with the very weak acid, pyrrH$^+$ (pyrrH$^+$ = pyrrolidinium ion; pK_a = 21.5 in MeCN), proton transfer to the clusters is sufficiently slow to be studied. Studies on a wide range of synthetic clusters show that the rates of proton transfer are markedly dependent on the cluster. It has been suggested that reorganization of the cluster dimensions (bond lengths and angles) is a significant barrier to proton transfer. (Note added in press. Recent DFT calculations indicate that, in cuboidal clusters, protonation of a core sulfur site is concomitant with Fe-SH bond cleavage [50]).

11.3.6.2 Timing of Binding Protons and Substrates to Clusters

There is a complication associated with the binding of substrates to Fe–S-based clusters in the presence of acid, which is important to consider if we are to understand how substrates bind to nitrogenases. The transformation of the nitrogenase substrates requires the presence of protons. Earlier studies showed that protons also bind to the Fe–S-based clusters. Consequently, it is necessary to address the question of whether protons or substrate bind preferentially to the clusters and what effect does protonation of the cluster have on substrate binding, and binding of substrate have on the protonation of the cluster. The binding of protons (ultimately forming H_2) and substrate (to form product) has been the source of speculative mechanisms concerning the action of nitrogenase for many years, including attracting the attention of several theoretical studies. The factors controlling the proton affinities and rates of proton transfer to a variety of synthetic Fe–S-based clusters have been established [3]. By establishing the affinities and rates of substrate binding to the same synthetic clusters, it would be possible to control (by variation of the concentrations of acid and substrate) which of those species binds preferentially to the cluster. By judicious choice of conditions, it is possible to measure (i) the affinity and rates of substrate binding to a singly protonated cluster and (ii) the rates of proton transfer to a cluster containing a bound substrate.

Little work has been done in this area. However, the substitution reaction of $[Fe_2S_2Cl_4]^{2-}$ with PhS^- in the presence of the weak acid $pyrrH^+$ involves both protonation and binding of the thiolate [49c]. The rates of each elementary step exhibit first-order dependencies on acid or PhS^-, respectively. Consequently, the mechanism is complicated because which step comes first (protonation or binding of PhS^-) depends on the relative concentrations of acid and thiolate. Thus, when $[PhS^-] > [pyrrH^+]$, the reaction follows a pathway in which thiolate binds before protonation, and when $[PhS^-] < [pyrrH^+]$, the rate of protonation is faster than the rate of thiolate binding. These two pathways are shown in Figure 11.16.

Studies with $[Fe_4S_4Cl_4]^{2-}$ have shown that initial binding of various substrates to the cluster affect the rate of proton transfer from $pyrrH^+$ to the cluster in different ways [49d]. Binding PhS^- or Bu^tNC results in an increase in the rate of proton transfer ($k^{PhS}/k = 75$, where k is the rate constant for protonation of $[Fe_4S_4Cl_4]^{2-}$ and k^{PhS} is the rate constant for protonation of $[Fe_4S_4Cl_4(SPh)]^{3-}$). In contrast, binding of halide or EtS^- produces a cluster that is protonated more slowly than

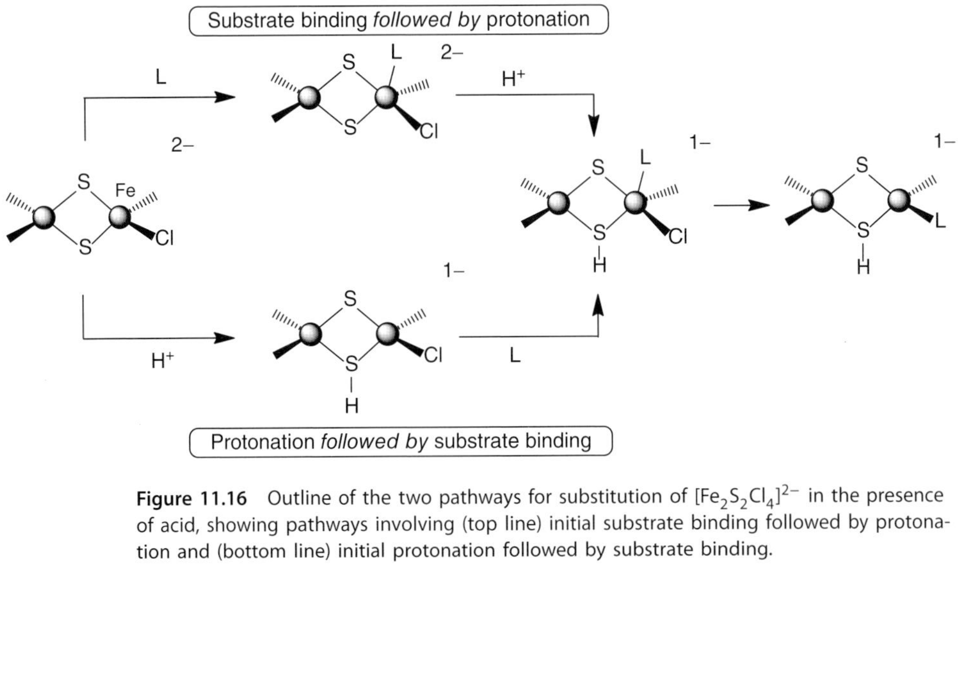

Figure 11.16 Outline of the two pathways for substitution of $[Fe_2S_2Cl_4]^{2-}$ in the presence of acid, showing pathways involving (top line) initial substrate binding followed by protonation and (bottom line) initial protonation followed by substrate binding.

the parent cluster ($k^I/k = 0.029$, $k^{Br}/k = 0.82$, $k^{EtS}/k = 0.31$). It has been suggested that the bound substrate subtly affects the dimensions or reorganizational flexibility of the cluster, thus affecting the rate of proton transfer.

11.3.6.3 The Problem of Binding Substrate Versus Reduction of Protons

Because the function of nitrogenases is to convert dinitrogen into ammonia, the system necessarily operates with a flux of electrons and protons being delivered to the active site. Consequently, production of dihydrogen by reduction of protons is a side reaction that is evident both in the absence and presence of dinitrogen. Studies on the Mo-nitrogenase showed that dihydrogen production could never be completely eliminated even at high pressures of dinitrogen, and under such conditions about one dihydrogen molecule is produced for every molecule of dinitrogen transformed. The limiting stoichiometries of the V- and Fe-only nitrogenases have now also been established, as shown in Figure 11.1. Clearly, the V- and Fe-only nitrogenases are not so efficient in routing electrons into ammonia production, and waste more electrons in forming dihydrogen, than the Mo nitrogenase.

An early interpretation of the limiting stoichiometry of the Mo nitrogenase was that when dinitrogen bound to FeMo-cofactor, it did so by displacing dihydrogen (either two hydrides or a dihydrogen ligand) [13]. With the discovery of the V- and Fe-only nitrogenases and the stoichiometrically higher amount of dihydrogen produced (Figure 11.1), this proposal lost support; after all, it was easy to reconcile that a single dinitrogen displaced a single dihydrogen but not that a single dinitrogen displaced several dihydrogens. It seems more likely that the limiting stoichiometries of the various nitrogenases reflect the relative abilities of the three cofactors to direct electron flux away from dihydrogen production and into dinitrogen reduction. The studies on synthetic Fe–S-based clusters outlined below support this proposal.

The reaction of $[Fe_4S_4(SPh)_4]^{3-}$ with an excess of the acid $lutH^+$ produces a triprotonated cluster [25], which evolves dihydrogen as described by Equation 11.1. The introduction of acetylene results in the triprotonated cluster producing ethylene according to the stoichiometry of Equation 11.2.

$$[Fe_4S_2(SH)_2(SPh)_3(SHPh)] \rightarrow [Fe_4S_3(SH)(SPh)_3(SHPh)] + 0.5H_2 \qquad (11.1)$$

$$[Fe_4S_2(SH)_2(SPh)_3(SHPh)] + 0.5C_2H_2 \rightarrow [Fe_4S_3(SH)(SPh)_3(SHPh)] + 0.5C_2H_4 \qquad (11.2)$$

As the concentration of acetylene is increased, progressively more ethylene, and less dihydrogen, is produced. However, the production of dihydrogen can never be entirely suppressed, even under conditions when all clusters have acetylene bound. This indicates that the cluster with acetylene bound can also produce dihydrogen. The limiting product distribution is about 70% ethylene and about 30% dihydrogen, as shown in Figure 11.17. This result indicates that the $[C_2H_4]$:$[H_2]$ ratio reflects the inability of the bound acetylene to route all the electrons toward its reduction. This behavior is clearly similar to that of the nitrogenases, which in the presence of dinitrogen always produce mixtures of dihydrogen and ammonia.

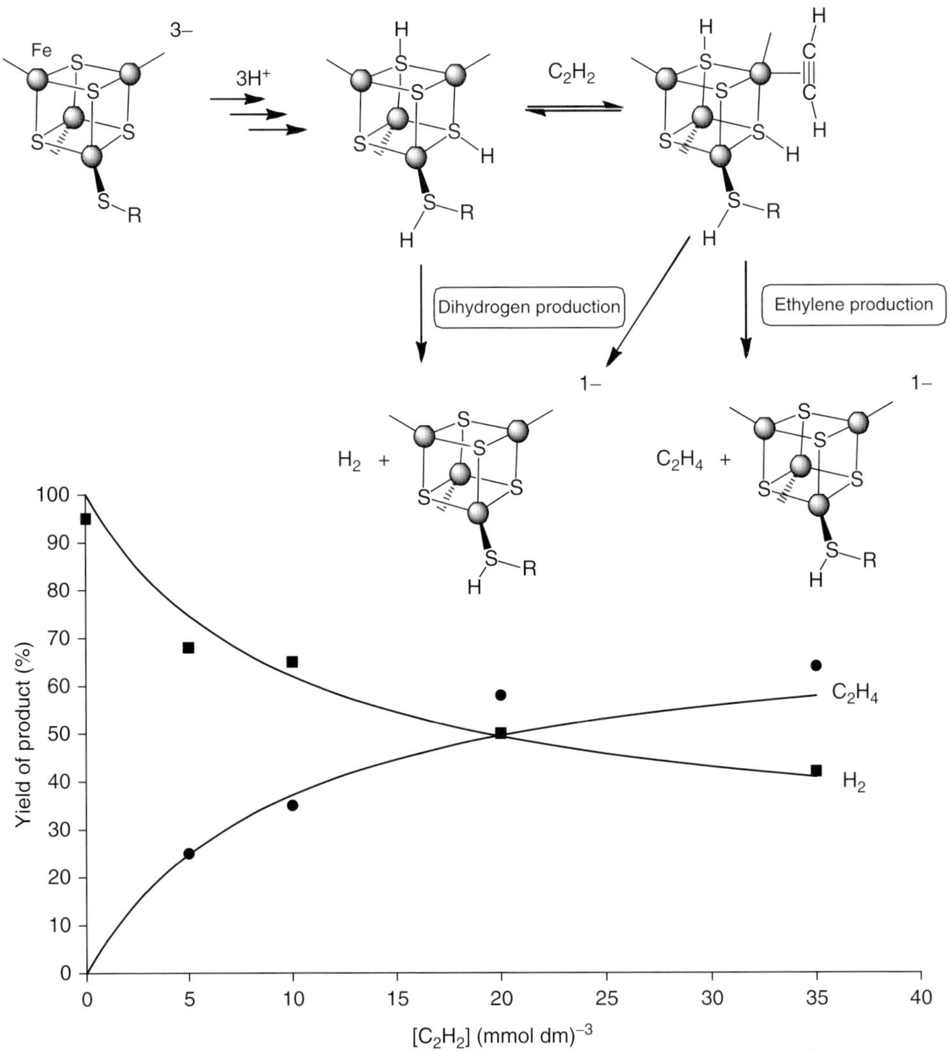

Figure 11.17 Pathways for the transformation of acetylene into ethylene and reduction of protons to dihydrogen by $[Fe_4S_4(SR)_4]^{3-}$. Graph shows that dihydrogen is produced even under conditions where each cluster has an acetylene bound.

There remains the question why the limiting stoichiometries of the three nitrogenases are so different. If, as is generally believed, the only difference between the active sites of the three nitrogenases amounts to a single metal atom in a cluster containing eight metal atoms, then this one metal must have a significant effect on the reactivity. Furthermore, both current studies on the Mo-based enzyme and DFT calculations (Sections 11.2.1 and 11.2.2) indicate that the binding site is a Fe atom of cofactor. Thus, the effect of the Mo must be to modulate

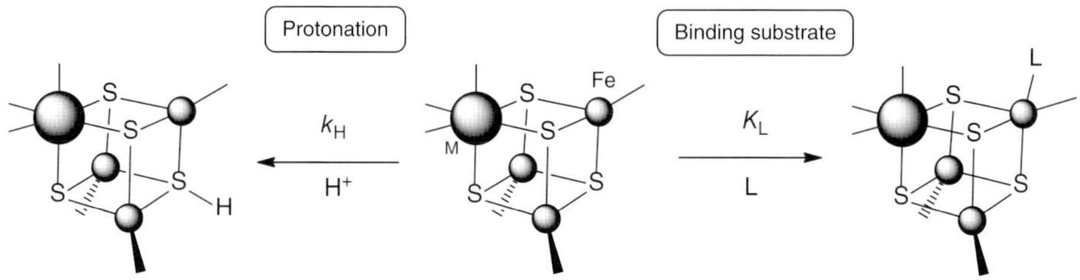

M	k_H (dm^3 mol^{-1} s^{-1})
Fe	2.4×10^4
Mo	2.5×10^2
W	8.5×10^2

	K_L (dm^3 mol^{-1})	
L	M = Fe	M = Mo
Cl^-	202	1550
Br^-	96	850
PhS^-	68	129
CO	~300	~800
N_2O	~300	~1030
Bu^tNC	>2000	

Figure 11.18 Effects of the heterometal on the binding affinities of substrates (right-hand side) and rate of protonation (left-hand side) of cuboidal $\{MFe_3S_4\}$ clusters.

the reactivity of the Fe sites. We have already seen in Sections 11.3.4 and 11.3.5.1 how the heterometal in cuboidal $\{MFe_3S_4\}$ clusters affect the reactivity of Fe sites. Now we bring these observations together to offer an explanation of the various limiting stoichiometries of the nitrogenases presented in Figure 11.1.

Systematic studies on cuboidal $\{MFe_3S_4\}^{n+}$ (M = Fe, Mo, or W) clusters have established the rate constants for proton binding and the equilibrium constants for substrate binding to the various clusters shown in Figure 11.18. Other studies on the synthetic cuboidal clusters show that the presence of the heterometal does not discriminate between the rates of proton and substrate binding to the cluster or affect the proton affinity (pK_a) of the cluster. However, there is some evidence that changing the heterometal affects the substrate binding affinity of the cluster, as shown in Figure 11.18 [32, 33, 44, 47]. Irrespective of the nature of the substrate, it is clear that the Mo-containing cluster has a higher affinity for the substrate. The rate constants for protonation of the same cuboidal clusters [49b] show that the rates of protonation of the $\{MFe_3S_4\}$ clusters are appreciably slower than protonation of $\{Fe_4S_4\}$ clusters.

Consider what characteristic reactivities are required for a system to be an efficient nitrogen fixer operating in a protic environment, delivering electrons to a metal-containing site that becomes sufficiently reduced to bind dinitrogen. The site needs to minimize the production of dihydrogen and maximize the production of ammonia. There are several ways this could be accomplished. (i) The rate of binding and transforming dinitrogen could be much faster than the rate of protonation and dihydrogen production (i.e., total kinetic control). (ii) The affinity of the cluster for dinitrogen could be much higher than its affinity for protons (i.e., total thermodynamic control). (iii) A mixture of kinetic and thermodynamic control. It is this last approach that appears to operate in the synthetic Fe–S-based clusters, where the presence of Mo (or W) in the cluster suppresses the rate of protonation but favors the affinity of the cluster to bind small molecules and ions.

11.4
Studies on Extracted FeMo-Cofactor

The investigations detailed in this chapter have focused on synthetic Fe–S-based clusters. However, a pragmatist would point out that the structure of FeMo-cofactor is unique so, perhaps, the reactivity is unique also, and different from that of synthetic clusters. To date, FeMo-cofactor has not been prepared in the laboratory. The unique structural features of FeMo-cofactor make this synthesis one of the most challenging problems in contemporary inorganic chemistry. However, FeMo-cofactor can be extracted from the MoFe protein and into *N*-methylformamide (NMF) [51]. This extraction must involve the cleavage of the Fe-cysteinate and Mo-histidine bonds, and it is reasonable to assume that the extracted cofactor has solvent bound at these two positions. It has been suggested, on the basis of electrochemical studies, that neutral NMF is bound at Mo and anionic *N*-methylformidate at Fe. Free from the polypeptide and the interactions between uncoordinated amino acid side chains of the protein and cofactor, which presumably must modulate the cluster's reactivity, we are in a position to investigate the intrinsic reactivity of FeMo-cofactor.

11.4.1
Evidence for Substrates Binding to Extracted FeMo-Cofactor

For some time there has been evidence that nitrogenase substrates including CN^-, N_3^-, H^+, and Bu^tNC can bind to the extracted FeMo-cofactor [51]. However, the spectroscopic changes associated with the cofactor when these substrates bind are small or even negligible. Using an extension of the kinetic method developed with the synthetic Fe–S-based clusters to measure substrate binding [52] (Section 11.3.5), it has been possible to map out where a variety of different substrates bind to FeMo-cofactor in the semi-reduced redox state, as shown in Figure 11.19.

Electrochemical studies have shown that extracted FeMo-cofactor (containing $C_6F_5S^-$ bound to the Fe remote from the Mo) catalyzes the reduction of protons to dihydrogen at $E° = -280$ mV (vs NHE (normal hydrogen electrode)). The mechanism of dihydrogen formation is shown in Figure 11.20 and involves the semi-reduced and reduced redox states of extracted FeMo-cofactor. The rate-limiting step in the transformation is evolution of dihydrogen ($k = 3\ s^{-1}$) [53]. As discussed in Section 3.6, protonation of synthetic Fe–S-based clusters is commonplace, with evidence indicating that the protonation site is a μ-S. Kinetic studies indicate that a single proton can bind to extracted FeMo-cofactor in the semi-reduced state. The similarity in the kinetics of protonation of extracted FeMo-cofactor and synthetic Fe–S-based clusters suggest protonation of extracted FeMo-cofactor at a

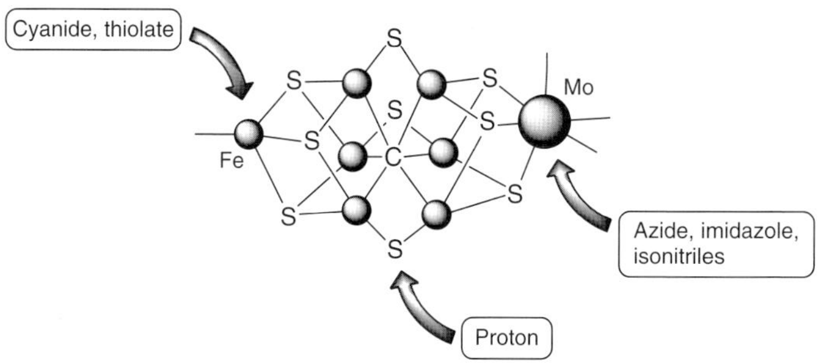

Figure 11.19 Summary of various substrate binding sites on extracted FeMo-cofactor (in semi-reduced redox state) as indicated by kinetic methods.

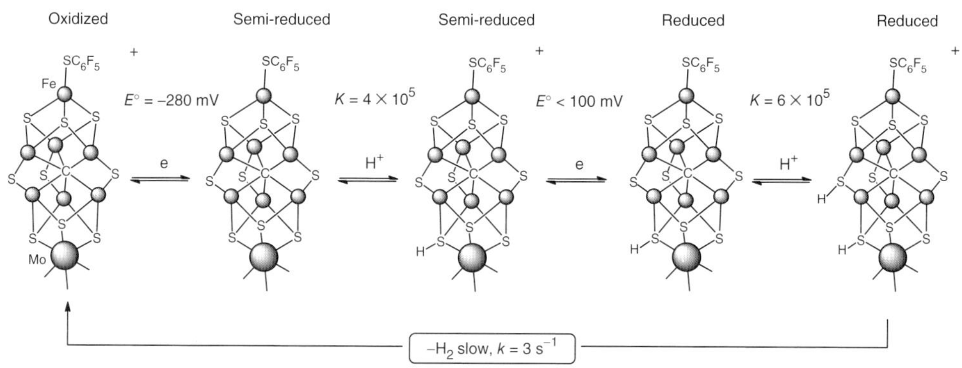

Figure 11.20 Protonation and dihydrogen production mechanism for extracted FeMo-cofactor. Positions of protons are arbitrary. Because the charge on the extracted FeMo-cofactor (semi-reduced state) is unknown, we have (arbitrarily) designated the charge of the semi-reduced state as neutral in this and subsequent figures.

core sulfur site [52, 54]. FeMo-cofactor has three distinct types of sulfur: three μ_3-S (bound to Mo), three μ_2-S (around the central girdle of the cluster), and three μ_3-S (at the end remote from Mo). The kinetic studies cannot distinguish which sulfur(s) are protonated. Indeed, it may not be protonation at a single site but at a variety of sites.

The binding of carbon monoxide to extracted FeMo-cofactor has been investigated using a spectroelectrochemical method [55]. Electrochemical studies show three reduction processes for extracted FeMo-cofactor: two successive one-electron reductions (${}^{I}E_p$ and ${}^{II}E_p$) associated with the core Fe sites and one reduction associated with the Mo (${}^{III}E_p$). Extracted FeMo-cofactor must be in the reduced state before any binding of carbon monoxide is observed. The suggested binding of carbon monoxide to extracted FeMo-cofactor is shown in Figure 11.21. In reductions encompassing only ${}^{I}E_p$ and ${}^{II}E_p$, at low concentrations of carbon monoxide, a peak at $\upsilon_{CO} = 1835\,cm^{-1}$ is observed in the IR spectrum corresponding to a single bridged carbon monoxide. At higher carbon monoxide pressures, this band disappears and a new band at $\upsilon_{CO} = 1885\,cm^{-1}$ replaces

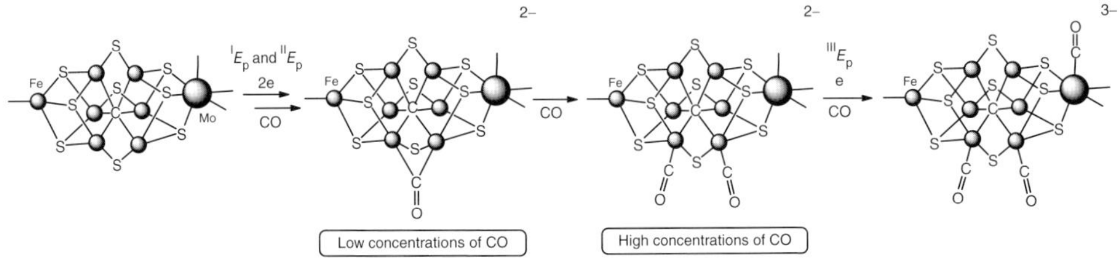

Figure 11.21 Proposed binding sites of multiple carbon monoxide molecules bound to extracted FeMo-cofactor.

it. This new band corresponds to carbon monoxide terminally bound to Fe. In reductions encompassing ${}^{III}E_p$, another band is observed at $\upsilon_{CO} = 1920\,cm^{-1}$. This band is unperturbed by the peak at $\upsilon_{CO} = 1885\,cm^{-1}$ and is proposed to correspond to terminal carbon monoxide bound to Mo.

11.4.2 Rates of Substrate Binding to Extracted FeMo-Cofactor

Using a stopped-flow, sequential-mix method, analogous to that described for synthetic Fe–S-based clusters in Section 11.3.5.2, the rates of binding of cyanide to various derivatives of extracted FeMo-cofactor (FeMoco-L; L = imidazole, NMF, or Bu^{t}NC) [54, 56] have been measured.

The reaction between cyanide and FeMoco-ImH (ImH, imidazole) is very fast (on the limit of the stopped-flow, sequential-mix technique, $k_1{}^{ImH} = 95\,s^{-1}$). The rate of the reaction is independent of the concentration of cyanide and is unaffected by the concentration of free imidazole present in solution. As the binding of FeMo-cofactor in the protein involves a histidine coordinated to Mo, it seems reasonable that the imidazole is bound to Mo in extracted FeMo-cofactor. Consequently, the kinetics observed for the reaction between FeMoco-ImH and CN$^-$ are consistent with a unimolecular reaction, most reasonably attributed to rate-limiting dissociation of the imidazole before cyanide binding, as shown on the top line of Figure 11.22.

The reaction between FeMoco-NMF and cyanide produces FeMoco-CN within the dead-time of the apparatus (2 ms), even when there is only a slight excess of cyanide ([CN$^-$]/[FeMoco-NMF] = 2.0). Consequently, the kinetics cannot be determined and the rate of cyanide binding with FeMoco-NMF can only be estimated ($k_1{}^{NMF} \geq 150\,s^{-1}$). It seems likely that the dissociation of a very labile Mo–NMF bond is the rate-limiting step, as shown on the top line of Figure 11.22.

Finally, the reaction of FeMoco-CNBu^t with an excess of cyanide shows different kinetics to that described for FeMoco-ImH or FeMoco-NMF. The rate law for the reaction of FeMoco-CNBu^t with cyanide exhibits first-order dependencies on the concentrations of cyanide and FeMoco-CNBu^t with a second-order rate constant, $k_2{}^{BuNC} = 2.5 \times 10^4\,dm^3\,mol^{-1}\,s^{-1}$. This reaction is significantly slower than the reactions with FeMoco-ImH or FeMoco-NMF. The rate of the reaction is also

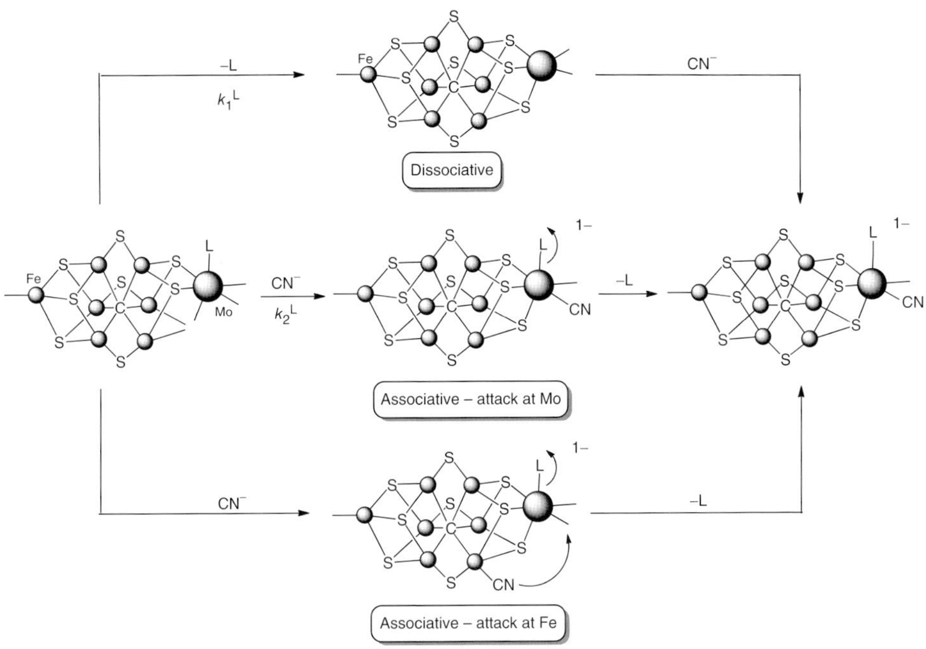

Figure 11.22 Suggested pathways for the binding of CN^- to semi-reduced extracted FeMo-cofactor derivatives.

independent of the concentration of free Bu^tNC. It has been proposed that the mechanism of this reaction is associative, involving attack of cyanide directly at the Mo and displacement of Bu^tNC (which is presumed to be coordinated to Mo), as shown in the central line of Figure 11.22. However, the kinetics are also consistent with the more complicated associative mechanism in which initial binding of cyanide occurs preferentially to a Fe site (a kinetically favored binding site). With cyanide bound to a Fe, subsequent dissociation of the Mo–$CNBu^t$ bond generates a vacant site on Mo, to which cyanide can now move. This pathway is shown on the bottom line of Figure 11.22.

A theoretical description of the energetics of cyanide binding to the different metal sites of extracted FeMo-cofactor has been calculated using DFT calculations on fragments of the FeMoco, and the metal-CN bond dissociation energies (BDEs) compared. The results of the DFT calculations indicate that cyanide prefers to bind to Mo. DFT calculations have also been used to get further insight into the pathways shown in Figure 11.22. Model structures, based on the localized geometries within FeMo-cofactor reveal marked steric congestion around Mo in FeMoco-$CNBu^t$, and mitigate against direct attack of cyanide at Mo.

The reaction of cyanide with a variety of different extracted FeMo-cofactor derivatives indicates that cyanide can bind to at least two sites on the cluster in the semi-reduced redox level: presumably, the Mo and one of the Fe atoms. In the protein, it is not clear whether a binding site on Mo is available. There is no evidence that the Mo-histidine bond is labile. However, it has been shown that in mononuclear complexes containing bidentate carboxylate groups, the carboxylate goes monodentate upon reduction [57]. Thus, there remains the possibility that in the redox state of FeMo-cofactor, which is capable of binding dinitrogen, a substrate binding site is available on the Mo because of the (*R*)-homocitrate ligand having changed from being bidentate to monodentate, only coordinating through the alkoxy-group.

11.5 The Future

Dance has stylishly summarized our current position in understanding the mechanism of nitrogenases [18b], "*The stage for the molecular dance that is biological nitrogen fixation is clearly set, the dancers are known, the rhythm is evident, and some scenes have been glimpsed, but the choreography is elusive.*" The fundamental contemporary question is how do we move forward and establish the mechanism of substrate binding and transformation of the nitrogenases? Currently, possible mechanisms for nitrogenases continue to be advanced based on chemistry established at very abiological sites. Until we have experimental information about how dinitrogen and the other substrates of nitrogenase bind (rates and binding affinities), where they bind (structure), and are then transformed (intermediates involved) at Fe–S-based clusters, the mechanism of the nitrogenases will remain incomplete.

An understanding of how FeMo-cofactor operates inside the protein has to be underpinned by studies on the structure and function correlations of synthetic Fe–S-based clusters. There are four principle objectives in this area of research, which are as follows. (i) Identification of the binding sites and coordination modes of substrates on Fe–S-based clusters and the relationship between substrate binding site and substrate transformation. (ii) To understand the dynamics of substrates binding to clusters: what factors affect how rapidly substrates bind to different sites on clusters and can substrates move around the cluster without becoming dissociated? (iii) The synthesis of a Fe–S-based cluster containing a coordinated dinitrogen. This would have important repercussions as the systematic investigation of how dinitrogen is transformed at an Fe–S cluster site could be undertaken, establishing chemical precedent for the biological process based on a biologically relevant system. (iv) The laboratory synthesis of FeMo-cofactor and study of its reactivity.

Finally, this chapter has largely not addressed the problem of how Fe–S-based clusters transform substrates. As alluded to in the preceding pages, it is known that synthetic clusters have a limited ability to transform certain substrates (notably, hydrazine, acetylene, and protons). Although there have been a limited number of kinetic studies, identification of the intermediates in these transformations are at best only educated guesses. Identification is hampered by the structural complexity of even the simplest Fe–S-based clusters, the transient nature of the intermediates, and the lack of good diagnostic spectroscopic probes for these species. The detection and structural identification of intermediates remains a major objective in this area of research.

References

1. (a) Holm, R.H., Kennepohl, P., and Solomon, E.I. (1996) *Chem. Rev.*, **96**, 2239; (b) Matsubara, H. and Saeki, K. (1992) *Adv. Inorg. Chem.*, **38**, 223; (c) Cammack, R. (1992) *Adv. Inorg. Chem.*, **38**, 281.
2. (a) Burgess, B.K. (1985) in *Molybdenum Enzymes* (ed. T.G. Spiro), John Wiley & Son, Inc., New York, p. 161; (b) Smith, B.E. (1999) *Adv. Inorg. Chem.*, **47**, 159.
3. Henderson, R.A. (2012) *BioInorg. React. Mech.*, **8**, 1.
4. (a) Henderson, R.A. (2005) *Coord. Chem. Rev.*, **249**, 1841; (b) Henderson, R.A. (2005) *Chem. Rev.*, **105**, 2365; (c) Lee, S.C. and Holm, R.H. (2004) *Chem. Rev.*, **104**, 1135; (d) Venkateswara, P. and Holm, R.H. (2004) *Chem. Rev.*, **104**, 527.
5. Howard, J.B. and Rees, D.C. (1996) *Chem. Rev.*, **96**, 2965.
6. Burgess, B. and Lowe, D.J. (1996) *Chem. Rev.*, **96**, 2983.
7. Thorneley, R.N.F. and Lowe, D.J. (1985) in *Metal Ions in Biology (Molybdenum Enzymes)*, vol. 7 (ed. T.G. Spiro), Wiley-Interscience, New York, p. 221.
8. Spatzal, T., Aksoyoglu, M., Zhang, L., Andrade, S.L.A., Schleicher, E., Weber, S., Rees, D.C., and Einsle, O. (2011) *Science*, **334**, 940.
9. Thorneley, R.N.F., Lowe, D.J., and Eady, R.R. (1978) *Nature*, **272**, 557.
10. Chatt, J., Dilworth, J.R., and Richards, R.L. (1978) *Chem. Rev.*, **78**, 589.
11. Hoffman, B.M., Dean, D.R., and Seefeldt, L.C. (2009) *Acc. Chem. Res.*, **42**, 609.
12. Lukoyanov, D., Dikanov, S.A., Yang, Z.-Y., Barney, B.M., Samoilova, R.I., Narasimhulu, K.V., Dean, D.R., Seefeldt, L.C., and Hoffman, B.M. (2011) *J. Am. Chem. Soc.*, **133**, 11655.
13. (a) Henderson, R.A. (2001) in *Recent Advances in Hydride Chemistry*, Chapter 16 (eds M. Peruzzini and R. Poli), Elsevier, Amsterdam, pp. 463–505. (b) Evans, D.J., Henderson, R.A., and Smith, B.E. (1999) in *Bioinorganic Catalysis*, 2nd edn, Chapter 7 (eds J. Reedijk and E. Bouwman), Marcel Dekker Inc., New York, pp. 153–207.
14. George, S.J., Ashby, G.A., Wharton, C.W., and Thorneley, R.N.F. (1997) *J. Am. Chem. Soc.*, **119**, 6450.
15. (a) Lee, C.C., Hu, Y., and Ribbe, M.W. (2010) *Science*, **329**, 642; (b) Lee, C.C., Hu, Y., and Ribbe, M.W. (2011) *Angew. Chem. Int. Ed.*, **50**, 5545.
16. (a) DuBois, D.L. and Hoffman, R. (1977) *Nouv. J. Chim.*, **1**, 479; (b) Yamabe, T., Hori, K., and Fukui, K. (1982) *Inorg. Chem.*, **21**, 2046; (c) Dilworth, J.R., Garcia-Rodriguez, A., Leigh, G.J., and Murrell, J.N. (1983) *J. Chem. Soc., Dalton Trans.*, 455; (d) Deeth, R.J. and Langford, S.A. (1995) *J. Chem. Soc., Dalton Trans.*, 1.
17. (a) Deng, H. and Hoffman, R. (1993) *Angew. Chem. Int. Ed.*, **32**, 1061; (b) Dance, I.G. (1994) *Aust. J. Chem.*, **47**, 979; (c) Stavrer, K.K. and Zerner, M.C. (1996) *Chem. Eur. J.*, **2**, 83; (d) Christiansen, J., Tittsworth, R.C., Hales, B.J., and Cramer, S.P. (1995) *J. Am. Chem. Soc.*, **117**, 10017; (e) Zhong, S.-J. and Liu, C.-W. (1997) *Polyhedron*, **16**, 653; (f) Dance, I. (1997) *J. Chem. Soc., Chem. Commun.*, 165.
18. (a) Dance, I. (2008) *Dalton Trans.*, 5977; (b) Dance, I. (2010) *Dalton Trans.*, **39**, 2972.
19. Dance, I. (2012) *Dalton Trans.*, **41**, 4859.
20. (a) Henderson, R.A., Leigh, G.J., and Pickett, C.J. (1983) *Adv. Inorg. Chem. Radiochem.*, **27**, 198; (b) Leigh, G.J. (2002) in *Nitrogen Fixation at the Millennium*, Chapter 11 (ed. G.J. Leigh), Elsevier, Amsterdam; (c) Schrock, R.R. (2005) *Acc. Chem. Res.*, **38**, 955; (d) Holland, P.L. (2005) *Can. J. Chem.*, **83**, 296.
21. Fryzuk, M.D. (2009) *Acc. Chem. Res.*, **42**, 127.

22. Tanaka, K., Hozumi, Y., and Tanaka, T. (1982) *Chem. Lett.*, 1203.
23. Hozumi, Y., Imasaka, Y., Tanaka, K., and Tanaka, T. (1983) *Chem. Lett.*, 897.
24. McMillan, R.S., Renauld, J., Reynolds, J.G., and Holm, R.H. (1979) *J. Inorg. Biochem.*, **11**, 213.
25. Grönberg, K.L.C., Henderson, R.A., and Oglieve, K.E. (1998) *J. Chem. Soc., Dalton Trans.*, 3093.
26. (a) Malinak, S.M., Demadis, K.D., and Coucouvanis, D. (1995) *J. Am. Chem. Soc.*, **117**, 3126; (b) Demadis, K.D., Malinak, S.M., and Coucouvanis, D. (1996) *Inorg. Chem.*, **35**, 4038.
27. (a) Malinak, S.M., Simeonov, A.M., Mosier, P.E., McKenna, C.E., and Coucouvanis, D. (1997) *J. Am. Chem. Soc.*, **119**, 1662; (b) Palermo, R.E., Singh, R., Bashkin, J.K., and Holm, R.H. (1984) *J. Am. Chem. Soc.*, **106**, 2600.
28. Ciurli, S., Ross, P.K., Scott, M.J., Yu, S.-B., and Holm, R.H. (1992) *J. Am. Chem. Soc.*, **114**, 5415.
29. (a) Scott, T.A. and Zhou, H.-C. (2004) *Angew. Chem. Int. Ed.*, **43**, 5628; (b) Weigel, J.A. and Holm, R.H. (1991) *J. Am. Chem. Soc.*, **113**, 4184.
30. Langford, C.H. and Gray, H.B. (1965) *Ligand Substitution Processes*, Benjamin, New York.
31. Dukes, G.R. and Holm, R.H. (1975) *J. Am. Chem. Soc.*, **97**, 528.
32. Henderson, R.A. and Oglieve, K.E. (1993) *J. Chem. Soc., Dalton Trans.*, 1467.
33. Henderson, R.A. and Oglieve, K.E. (1994) *J. Chem. Soc., Chem. Commun.*, 1961.
34. Garrett, B. and Henderson, R.A. (2005) *Dalton Trans.*, 2395.
35. Dunford, A.J. and Henderson, R.A. (2002) *Chem. Commun.*, 360.
36. (a) Goh, C., Weigel, J.A., and Holm, R.H. (1994) *Inorg. Chem.*, **33**, 4861; (b) Bominaar, E.L., Borshch, S.A., and Girerd, J.-J. (1994) *J. Am. Chem. Soc.*, **116**, 5362.
37. Dunford, A.J. and Henderson, R.A. (2002) *Inorg. Chem.*, **41**, 5487.
38. (a) Davies, S.C., Evans, D.J., Henderson, R.A., Hughes, D.L., and Longhurst, S. (2001) *J. Chem. Soc., Dalton Trans.*, 3470; (b) Grönberg, K.L.C. and Henderson, R.A. (1996) *J. Chem. Soc., Dalton Trans.*, 3667.
39. (a) Cen, W., Lee, S.C., Li, J., MacDonnell, F.M., and Holm, R.H. (1993) *J. Am. Chem. Soc.*, **115**, 9515; (b) Kovacs, J.A. and Holm, R.H. (1987) *Inorg. Chem.*, **26**, 702, 711; (c) Cuirli, S. and Holm, R.H. (1989) *Inorg. Chem.*, **28**, 1655; (d) Lee, S.C. and Holm, R.H. (1990) *J. Am. Chem. Soc.*, **112**, 9654; (e) Lee, S.C., Li, J., Mitchell, J.C., and Holm, R.H. (1992) *Inorg. Chem.*, **31**, 4333; (f) Hauser, C., Eckhard, B., and Holm, R.H. (2002) *Inorg. Chem.*, **41**, 1615.
40. (a) Wolff, T.E., Berg, J.M., Hodgson, K.O., Frankel, R.B., and Holm, R.H. (1979) *J. Am. Chem. Soc.*, **101**, 4140; (b) Wolff, T.E., Power, P.P., Frankel, R.B., and Holm, R.H. (1980) *J. Am. Chem. Soc.*, **102**, 4694; (c) Christou, G., Mascharak, P.K., Armstrong, W.H., Papaefthymiou, G.C., Frankel, R.B., and Holm, R.H. (1982) *J. Am. Chem. Soc.*, **104**, 2820; (d) Christou, G. and Garner, C.D. (1980) *J. Chem. Soc., Dalton Trans.*, 2354; (e) Armstrong, W.H., Mascharak, P.K., and Holm, R.H. (1982) *J. Am. Chem. Soc.*, **104**, 4373; (f) Christou, G., Garner, C.D., Miller, R.M., and King, T.J. (1979) *J. Inorg. Biochem.*, **11**, 349; (g) Palermo, R.E., Power, P.P., and Holm, R.H. (1982) *Inorg. Chem.*, **21**, 173.
41. (a) Ciurli, S. and Holm, R.H. (1991) *Inorg. Chem.*, **30**, 743; (b) Ciurli, S., Carney, M.J., Holm, R.H., and Papaefthymiou, G.C. (1989) *Inorg. Chem.*, **28**, 2696; (c) Ciurli, S., Carrie, M., and Holm, R.H. (1990) *Inorg. Chem.*, **29**, 3493.
42. (a) Holm, R.H., Phillips, W.D., Averill, B.A., Mayerle, J.J., and Herskovitz, T. (1974) *J. Am. Chem. Soc.*, **96**, 2109; (b) Reynolds, J.G., Laskowski, E.J., and Holm, R.H. (1978) *J. Am. Chem. Soc.*, **100**, 5315; (c) Cleland, W.E., Holtman, D.A., Sabat, M., Ibers, J.A., DePotis, G.C., and Averill, B.A. (1983) *J. Am. Chem. Soc.*, **105**, 6021.
43. Zhou, J., Scott, M.J., Hu, Z., Peng, G., Munck, E., and Holm, R.H. (1992) *J. Am. Chem. Soc.*, **114**, 10843.
44. Henderson, R.A. and Oglieve, K.E. (1993) *J. Chem. Soc., Dalton Trans.*, 1473.

45. Garrett, B. and Henderson, R.A. (2010) *Dalton Trans.*, 4586.
46. Liu, Q., Huang, L., Liu, H., Lei, X., Wu, D., Beisheng, K., and Lu, J. (1990) *Inorg. Chem.*, **29**, 4131.
47. Dunford, A.J. and Henderson, R.A. (2010) *J. Coord. Chem.*, **63**, 2507.
48. (a) Bates, K., Johnson, L., and Henderson, R.A. (2006) *Inorg. Chem.*, **45**, 9423; (b) Bates, K., Wouldhave, M., and Henderson, R.A. (2008) *Dalton Trans.*, 6527.
49. (a) Henderson, R.A. and Oglieve, K.E. (1999) *Dalton Trans.*, 3927; (b) Bell, J., Dunford, A.J., Hollis, E., and Henderson, R.A. (2003) *Angew. Chem. Int. Ed.*, **42**, 1149; (c) Bates, K., Garrett, B., and Henderson, R.A. (2007) *Inorg. Chem.*, **46**, 11145; (d) Bates, K. and Henderson, R.A. (2008) *Inorg. Chem.*, **47**, 5850.
50. Alwaaly, A., Dance, I. and Henderson, R. A. (2014) *Chem. Commun.*, **50**, 4799.
51. (a) Burgess, B. (1990) *Chem. Rev.*, **90**, 1377; (b) Smith, B.E., Durrant, M.C., Fairhurst, S.A., Gormal, C.A., Grönberg, K.L.C., Henderson, R.A., Ibrahim, S.K., Le Gall, T., and Pickett, C.J. (1999) *Coord. Chem. Rev.*, **185-186**, 669.
52. Grönberg, K.L.C., Henderson, R.A., Gormal, C.A., and Smith, B.E. (1997) *J. Chem. Soc., Chem. Commun.*, 713.
53. Le Gall, T., Ibrahim, S.K., Gormal, C.A., Smith, B.E., and Pickett, C.J. (1999) *Chem. Commun.*, 773.
54. Almeida, V.R., Gormal, C.A., Grönberg, K.L.C., Henderson, R.A., Oglieve, K.E., and Smith, B.E. (1999) *Inorg. Chim. Acta*, **291**, 212.
55. (a) Ibrahim, S.K., Gormal, C.A., Smith, B.E., Pickett, C.J., Vincent, K., and Best, S.P. (1999) *Chem. Commun.*, 1019; (b) Pickett, C.J., Vincent, K.A., Ibrahim, S.K., Gormal, C.A., Smith, B.E., and Best, S.P. (2003) *Chem. Eur. J.*, **9**, 76; (c) Pickett, C.J., Vincent, K.A., Ibrahim, S.K., Gormal, C.A., Smith, B.E., Fairhurst, S.A., and Best, S.P. (2004) *Chem. Eur. J.*, **10**, 4770.
56. Cui, Z., Dunford, A.J., Durrant, M.C., Henderson, R.A., and Smith, B.E. (2003) *Inorg. Chem.*, **42**, 6252.
57. (a) Pickett, C.J. (1996) *J. Biol. Inorg. Chem.*, **1**, 601; (b) Fairhurst, S.A., Henderson, R.A., Hughes, D.L., Ibrahim, S.K., and Pickett, C.J. (1995) *J. Chem. Soc., Chem Commun.*, 1569; (c) Henderson, R.A., Oglieve, K.E., Ibrahim, S.K., and Pickett, C.J. (1995) *J. Chem. Soc., Chem Commun.*, 1571.

Part IV
Miscellaneous: CO, RCN Activation, DMSO Reduction

Bioinspired Catalysis: Metal-Sulfur Complexes, First Edition.
Edited by Wolfgang Weigand and Philippe Schollhammer.

12 Sulfur-Oxygenation and Functional Models of Nitrile Hydratase

Davinder Kumar and Craig A. Grapperhaus

12.1 Introduction

The enzyme-catalyzed hydration of nitriles to amides has attracted industrial interest arising from synthetic advantages over traditional nitrile hydration routes including an aqueous environment, neutral pH, low temperatures, simple product purification, and high reaction yields. From an academic perspective, nitrile hydratase (NHase) has been the focus of intense model studies owing to the highly conserved, unusual non-corrin cobalt (III) or non-heme iron (III) active site. A series of bioinspired small-molecule complexes have been designed to elucidate the structural, spectroscopic, and functional features of NHase. Prior reviews have focused on structural and functional models of NHase [1–5]. In this chapter, we highlight selected complexes with tetra- and penta-dentate chelates that provide key insights into the oxidized sulfur environment at the enzyme active site.

12.2 Nitrile Hydratase

12.2.1 Significance

Conventionally, the hydration of nitriles to amides is promoted via acid-, or base catalysis (Scheme 12.1). Although further hydrolysis of the amide to carboxylic acid occurs in many cases, control of reagents and reaction conditions leads to selective amide formation. While simple and effective, traditional methods require harsh conditions such as high temperature, pressure, and strongly acidic/basic media. Metal-catalyzed hydrations provide a means to ease reaction conditions [6, 7]. Enzymatic nitrile hydration, which occurs at room temperature in neutral aqueous solution, can be considered a subclass of metal-based catalysis. NHase performs the initial step in nitrile assimilation in numerous soil bacteria including

Bioinspired Catalysis: Metal-Sulfur Complexes, First Edition.
Edited by Wolfgang Weigand and Philippe Schollhammer.

Agrobacterium, Rhodopseudomonas, Pseudomonas, Brevibacterium, Nocardia, and Rhodococcus [5].

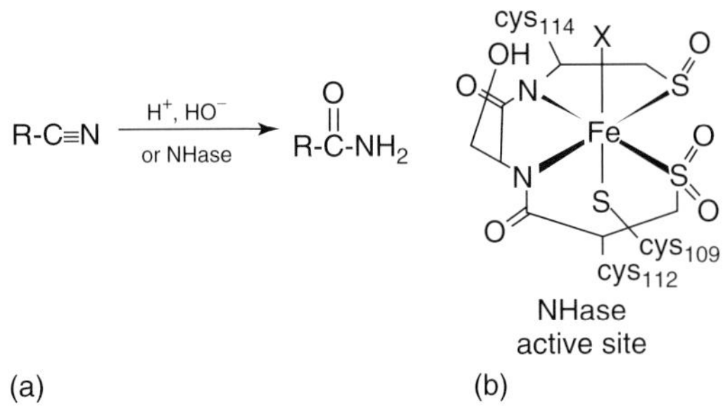

Scheme 12.1 Nitrile hydration (a) and active site of NHase (b).

The first industrial-scale organic biotransformation was the NHase-mediated hydration of acrylonitrile, which also represents the first application of biotechnology in the petroleum industry [8]. Since then, annual NHase catalyzed acrylamide production has topped 95 kton (25% of worldwide output) and at least 10 other amides have been produced on batch scales or larger [9–15]. In addition, NHase applications in the bioremediation of toxic nitriles [8, 16] have been explored with emphasis on pharmaceutical waste water [17], shale oil [18], and herbicide [19] and other contaminated soils [20]. The closely related Co-containing thiocyanate hydrolase (SCNase), which catalyzes thiocyanate degradation to carbonyl sulfide, ammonia, and hydroxide, shares a similar active site [21, 22]. Thiocyanate is a natural product of cyanide detoxification by the ubiquitous enzyme thiosulfate transferase (rhodanese) [23]. Industrially, SCNase is found in CN^--containing waste water sludge from coal gasification factories.

12.2.2 Enzyme Active Site

The active site of NHase contains either a mononuclear low-spin non-heme iron (III) or non-corrinoid cobalt (III) [1–3, 5]. Fe-NHase favors the hydration of aliphatic nitriles, while Co-NHase preferentially hydrates aromatic nitriles. With either metal, the active site contains a highly conserved -C-S-L-C-S-C- motif that serves as the metal-binding domain (Scheme 12.1). Initial X-ray crystallographic studies on Fe-NHase isolated from Brevibacterium R312 at 2.66 Å resolution finds the metal ion in a N_2S_3 primary coordination sphere consisting of deprotonated carboxamido nitrogen donors (Ser-113 and Cys-114) and cysteine sulfur donors (Cys-110, Cys-113, Cys-115) [24]. The sixth ligand-binding site is occupied by an O-atom from water or hydroxide. A higher resolution structure (1.7 Å) of NO-inactivated Fe-NHase from Rhodococcus sp. N-771 reveals posttranslational modification of two cysteine donors, resulting in three distinct oxidation states for the sulfur donors [25]. While Cys-109 is in the typical thiolate oxidation state (RS^-; oxidation state = 2−), Cys-114 has been oxidized to a sulfenate (RSO^-; oxidation state = 0) and Cys-112 is present as a sulfinate (RSO_2^-; oxidation state = 2+). The oxidized sulfur donors are co-facial with the substrate-binding site, while the thiolate sits in the trans position. The same donor arrangement, including the oxidation states of the cysteine donors, is found in Co-NHase.

Posttranslational sulfur-oxygenation is critical for hydratase activity [26]. Proposed S-oxygenation roles include modulating Lewis acidity [27], facilitating reversible NO bonding [28, 29], acting as a proximal base to activate H_2O [30–32], and altering ligand binding [33]. Density functional theory (DFT) studies by Solomon comparing the binding difference at the axial position in Fe- and Co-NHase indicate that S-oxygenation promotes tighter axial binding of nitrile for Fe-NHase because of the loss of π-bonding with the S-donors, while Co-NHase is relatively unaffected because of the lack of Co–S π-bonding [33]. Endo and coworkers [26] have shown that enzymes cultured under dark, anaerobic conditions in the presence of NO lack catalytic activity. The level of S-oxidation is also critical. Odaka [34] prepared inactive "overoxidation" mutants in which both Cys-112 and Cys-114 are oxygenated to sulfinates (RSO_2^-). NHase deactivation upon oxidation has been noted elsewhere [35]. Interestingly, a novel Fe-NHase recently characterized by high-resolution X-ray crystallography displays bis-sulfinate coordination in the putative active site [36], but the manuscript does not discuss activity of the isolated crystal.

12.2.3 Reaction Cycle

Several NHase mechanisms have been proposed including water-bound, nitrile-bound, and outer sphere pathways. While not unequivocal, most results support a nitrile-bound mechanism (Scheme 12.2) [30–32, 37]. Attack of a metal-coordinated nitrile (A) by water activated through interactions with the sulfenate yields a metal-coordinated imidate intermediate and a metal-coordinated sulfenic acid (B). Deprotonation of the acid by the imidate provides the iminol complex (C), which tautomerizes to the amide product (D). Electron paramagnetic resonance (EPR) studies on Fe–NHase confirm a low-spin Fe (III) spin state throughout the catalytic cycle, suggesting that catalytic hydration is a redox-inactive process [38].

Using time-resolved X-ray crystallography with Fe-NHase and *t*-butylisonitrile as a surrogate substrate, Odaka and coworkers [30] observed analogs of coordinated substrate (A) and product (D) complexes, strongly suggesting a nitrile-bound pathway. DFT calculations by Himo [31] confirm sulfenate as a plausible base for catalysis, but suggest the Lewis acidity of the metal does not activate nitrile; rather, it stabilizes the imidate intermediate. An alternate mechanism based on thorough kinetic investigations prefers a conserved serine residue to the sulfenate donor as the proximal base [37, 39].

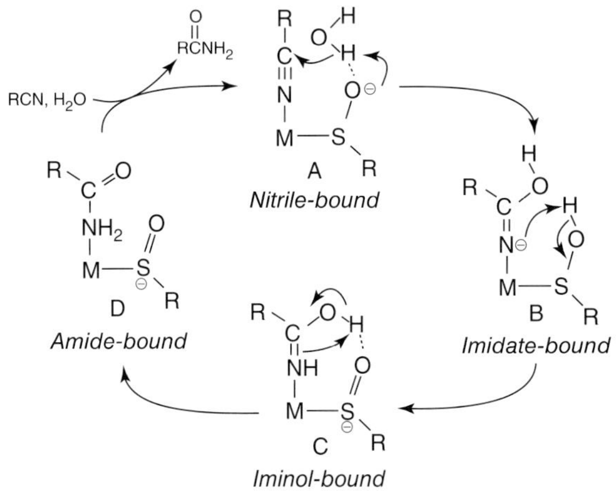

Scheme 12.2 Proposed NHase catalytic cycle.

The importance of noncovalent, second-coordination sphere effects in the regulation of substrate binding/activation is widely recognized [25, 33, 36, 37]. Conserved serine and arginine residues in close proximity to the substrate-binding site suggest their participation organizing the reactive cavity/hydration sphere affiliated with generation of an active HO^- nucleophile and expulsion of catalysis products. Serine is known to participate in proton shuttling in a broad range of biological processes and arginine often plays a "stabilization" role, but is also known to participate in H^+ shuttling [40]. The time-resolved X-ray crystal structures by Odaka [25] further support the role of sulfenate in water activation.

12.3 Small-Molecule Mimics

The synthetic analog approach utilizes small molecules to mimic key active site features. This may include structural, electronic, and/or functional aspects of the metalloenzyme [41, 42]. An advantage of the synthetic analog approach is synthetic flexibility, allowing the tuning of chemical features to match desired properties of the enzyme. This may include variation in ligand donor type, geometrical preferences, and/or choice of metal ion. In addition, small-molecule mimics are typically easy to manipulate and thoroughly characterized than metalloenzymes. A common disadvantage of small-molecule mimics is the lack of second-coordination interactions. Proposed roles for S-oxidation of cysteine in the active site of NHase include noncovalent, second-coordination sphere interactions with substrates, intermediates, and products. Small-molecule mimics with variable S-oxidation levels provide an attractive method to address these interactions.

As noted in prior reviews [1–5], NHase poses several challenges to the synthetic chemist. This review focuses on the synthesis and functional activity of S-oxygenated models employing tetra- and penta-dentate chelates. The cysteine residues of NHase include three distinct oxidation states (thiolate, sulfenate, sulfinate). Selective S-oxygenation of model complexes has proved difficult. To date, only five structurally characterized mononuclear complexes with a mixed sulfenate/sulfinate donor set are known (Scheme 12.3) [43–47]. Complexes **1** and **2** incorporate d^8-metal ions in stable four-coordinate environments. Complexes **3** and **4** are coordinatively saturated. As a result, none of these complexes are potential nitrile hydration catalysts because of the lack of a substrate-binding site. Only **5** combines the mixed sulfenate/sulfinate donor set with a variable ligand-binding site. The catalytic activity of **5** is reviewed herein.

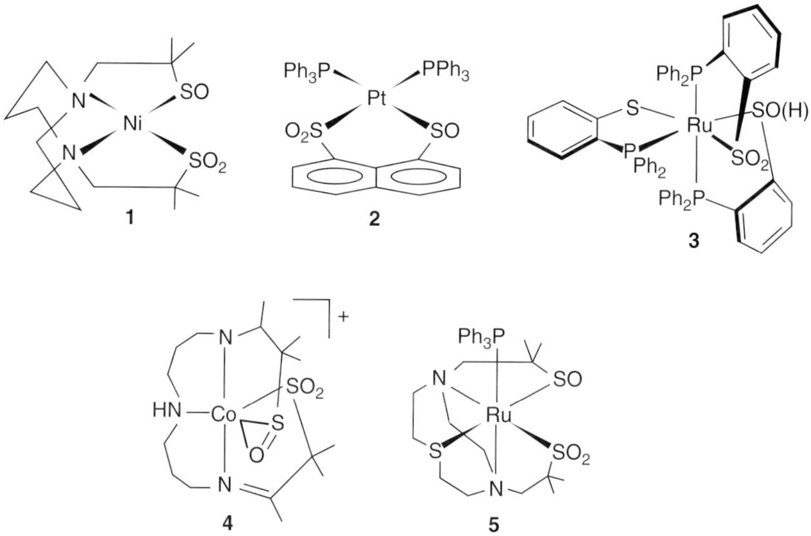

Scheme 12.3 Crystallographically characterized mononuclear sulfenate/sulfinate complexes.

In the following sections, a brief history of metal-thiolate S-oxygenation is provided, followed by selected S-oxygenation studies relevant to NHase. The NHase sections are divided by metal type and organized according to the donor atoms of the chelates. Through these studies, a total of seven functional NHase mimics with moderate activity have been discovered; four are Co(III) complexes and three are bioinspired Ru(II) complexes. Reported catalytic data is typically limited to turnover numbers (TONs) with limited substrates under a single set of reaction conditions. No detailed kinetic studies have been published, most likely due to the low TON of the catalysts. To date, no functional iron complex NHase mimic is known.

12.4 Early S-Oxygenation Studies

The oxidation of metal-thiolates to yield discrete S-oxygenates was reported approximately 65 years before the unexpected discovery of the mixed thiolate/sulfenate/sulfinate NHase donor. Since that time, several groups have contributed to the field of metal-thiolate S-oxygenation (Scheme 12.4). Schubert [48] first reported the synthesis of the sulfinate-metal complex in 1933 upon addition of aqueous hydrogen peroxide to a Co(III)-thiolate. Systematic studies of peroxide-induced S-oxygenation of Co-thiolates followed only in the late 1970s. Detailed kinetic investigations by Krueger [49] revealed a rate law that is first-order in metal-thiolate and hydrogen peroxide with a second-order rate constant (k_{SO}) ranging from 0.14 to 0.85 $M^{-1} s^{-1}$ for Co(III). Around the same time, the first structurally characterized sulfenate-(**6**) and sulfinate-metal (**7**) complexes were reported by Sargeson [50] and Deutsch and Elder [51], respectively. While initial studies employed H_2O_2 as oxidant, in 1979 Ibers [52] reported aerobic S-oxygenation of two related tris(pyrazolyl)borate Cu(II)-thiolates to yield sulfinate derivatives (**8**), noting the "*reactions may be models for the oxidation of cysteine in biological systems.*"

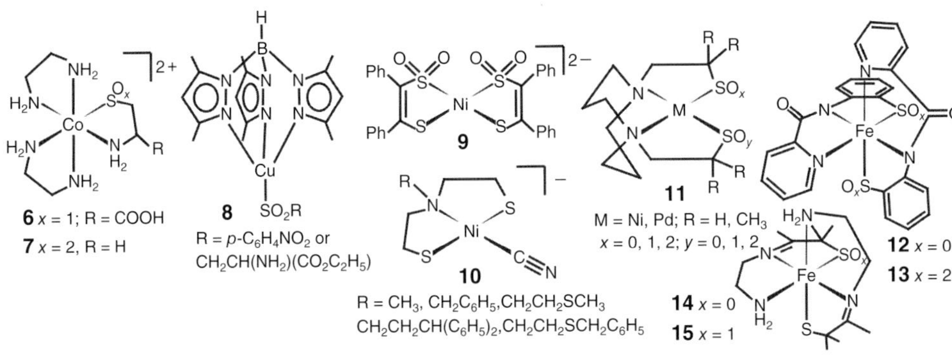

Scheme 12.4 Selected complexes from early S-oxygenation studies.

In studies that continued through the 1980s, Deutsch isolated and fully characterized a series of sulfenate/sulfinate derivatives, providing key insights into the electronic structure and spectroscopic features of metal-coordinated S-oxygenates. Notably, the sulfenate moiety is polarized toward the more basic oxygen, making the ligand an efficient H-bond acceptor [53]. In addition, Deutsch reversibly protonated the sulfenate **7** to yield a metal-coordinated sulfenic acid with a pK_a of 0.015. Such a low pK_a suggests that a metal-coordinated sulfenic acid is strongly disfavored at the active site of NHase under physiological conditions. The basicity of the coordinated sulfenate in NHase has not been quantified, but X-ray absorption spectroscopy (XAS) studies with metallopeptides [54] and NHase [55] suggest protonation of the sulfenate contrary to the expected high acidity of the coordinated sulfenic acid.

The S-oxygenation of metal-thiolates by dioxygen was further developed in the 1990s. In 1990, Schrauzer [56] reported the first bis-sulfinate complex (**9**), prepared from the dianionic nickel dithiolene precursor upon introduction of using O_2, whereas H_2O_2 addition resulted in metal-centered oxidation. Maroney [57] reported the first kinetic investigations of dioxygen-induced S-oxygenation in 1993. The enthalpy of activation for the O_2 S-oxygenation of anionic, *trans*-dithiolate-nickel(II) complexes (**10**) varies from 13.1 to 14.7 kcal mol^{-1} with an entropy of activation range of −24.2 to −18.7 e.u. Through the late 1980s and during the 1990s, Darensbourg [44, 58, 59] developed an unprecedented series of nickel and palladium S-oxygenates (**11**) with variable S-oxygenation levels including mono- and bis-sulfenates, mixed sulfenate/sulfinate, and mono- and bis-sulfinate derivatives. The different levels of S-oxygenation were achieved by a combination of synthetic strategies employing hydrogen peroxide, dioxygen, and singlet oxygen as oxidant followed by column chromatography. The synthetic utility of singlet oxygen was also exploited by Selke [60]. Notably, the $^1\Delta$ excited state of O_2 sits 22.4 kcal mol^{-1} above the $^3\Sigma$ ground state, which is just above the room temperature free energy of activation energy calculated using the kinetic parameters established by Maroney.

Early NHase models containing Fe and Co with N_2S bis-chelates successfully yielded S-oxygenates. While the bis-chelation of tridentate ligands precludes substrate binding to the metal, these complexes provided key insights. Shortly after NHase S-oxygenation was revealed in 1998, Mascharak reported the first synthetic Fe complex with carboxamide and thiolate donors. Using an N_2S tridentate ligand (L^1), the coordinatively saturated, low-spin Fe(III) complex **12** was isolated [61]. Subsequent reaction with H_2O_2 afforded the bis-sulfinate **13** as a green solid [62]. Using similar methods, Co(III) derivatives of L^1, and its aliphatic thiolate analog, were used to generate bis-sulfinate Co(III) products [63]. Much later, the bis N_2S−Fe complex **14** was used by Kovacs [64] to prepare the first sulfenate-Fe complex, **15**. Combined theoretical and spectroscopic studies with **15** reveal S-oxygenation of one thiolate to a sulfenate induces the remaining thiolate to bind more tightly, stabilizing Fe(III). Protonation of the sulfenate with strong acids further increases donation by the remaining thiolate. The results were interpreted to suggest that the unmodified *trans*-thiolate of NHase plays a role in labilizing substrate and/or tuning the pK_a of the bound water.

12.5 Sulfur-Oxygenation of Co(III) NHase Mimics

As noted, the S-oxygenation of Co-thiolates was established long before X-ray diffraction studies revealed the mixed thiolate/sulfenate/sulfinate donor set at the NHase. The synthetic tools and characterization methods developed through those early studies provided the basis for NHase model studies. Key S-oxygenation studies on Co-containing NHase mimics are described subsequently, organized according to the atom donor type of the chelate ligand.

12.5.1
N_2S_2 Co(III) Model Complexes

Several groups have employed square-planar N_2S_2 donor sets to investigate S-oxygenation reactions with Co-NHase mimics (Scheme 12.5). These ligands recreate the carboxamide nitrogen/reactive thiolate N_2S_2 plane of the NHase active site. The remaining axial positions can coordinate additional donors to yield pseudo-square-pyramidal or pseudo-octahedral complexes. The three parent ligands (L^2 to L^4) differ in the relative position of the carboxamide carbonyl and the size of the NS and NN chelate rings formed upon metal complexation. Ligand L^2 contains the carboxamide carbonyl as part of a six-member NN chelate with two five-member NS chelate rings. In contrast, L^3 and L^4 position the carboxamide carbonyl as part of NS chelates, five- and six-member rings, respectively, with a five-member (L^3) or six-member (L^4) NN chelate ring.

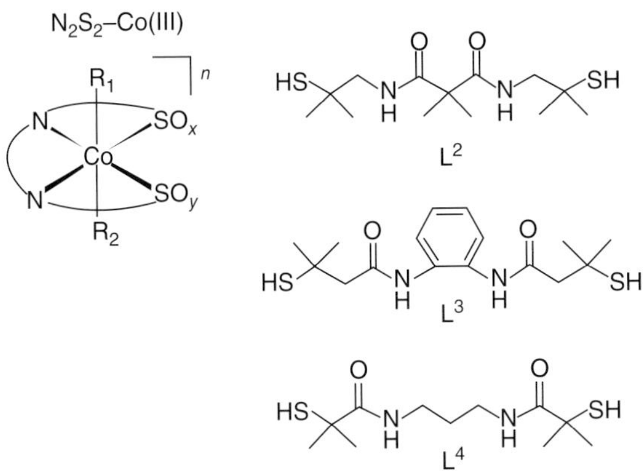

Complex	L	R_1	R_2	x	y	n
16	L^2			0	0	−1
17	L^2	^{t}BuNC	^{t}BuNC	0	0	−1
18	L^2	CN^-	CN^-	0	0	−3
19	L^2	^{t}BuNC	^{t}BuNC	2	2	−1
20	L^2	^{t}BuNC	^{t}BuNC	1	1	−1
21	L^2	CN^-	CN^-	2	2	−3
22	L^3	^{t}BuNC	^{t}BuNC	0	0	−1
23	L^3	^{t}BuNC	^{t}BuNC	2	2	−1
24	L^3	^{t}BuNC	^{t}BuNC	0	2	−1
25	L^3	CN^-	Solvent	0	0	0
26	L^3	CN^-	Cl^-	2	2	−1
27	L^3	CN^-	H_2O	2	2	0
28	L^4			0	0	−1
29	L^4	^{t}BuNC	^{t}BuNC	1	1	−1

Scheme 12.5 Selected N_2S_2–Co(III) model complexes.

The N_2S_2 ligand L^2 developed by Chottard [65] reacts with $CoCl_2$ to yield the high-spin ($S = 1$) square-planar complex Co(III) complex **16** upon aerobic oxidation of the metal center. Complex **16** is degraded by H_2O_2, but stable to air in both solution and the solid state. The high affinity of **16** for CN^- and ^{t}BuNC allows facile synthesis of the six-coordinate Co(III) complexes Na[L^2Co(^{t}BuNC)$_2$] (**17**) and Na_3[L^2Co(CN)$_2$] (**18**), which are susceptible to S-centered oxidation. Addition of 50% H_2O_2 to **17**, prepared *in situ* from **16** with excess ^{t}BuNC, affords the bis-sulfinate **19** as a yellow powder [66]. Controlled S-oxygenation employing the readily quantifiable, anhydrous H_2O_2·urea yields the orange colored bis-sulfenate **20**. Similar strategies with **17**, prepared *in situ* from **16** and NaCN, yield only the bis-sulfinate **21**.

The N_2S_2 ligand L^3 of Artaud also yields an $S = 1$ square-planar Co(III) derivative (**22**). Complex **22** reacts with O_2 in the presence of ^{t}BuNC to yield the bis-sulfinate **23**. The monosulfinate **24** was prepared via metal-free oxidation of the L^3 to disulfide by I_2, followed by S-oxidation with 2,2-dimethyl dioxirane, and S–S bond cleave with hydroxide [67]. The resulting monosulfinated ligand was then complexed with Co(III). Stoichiometric addition of CN^- to **22** yields the six-coordinate Co(III) complex $[L^3Co(CN)sol]^{2-}$ (**25**) (sol = DMF (*N*,*N*-dimethylformamide) or CH_3CN), with CN^- occupying the apical position trans to the solvent molecule. Air oxidation of **25** affords a trinuclear Co(III)-Co(II)–Co(III) complex in which the central Co(II) bridges two bis-sulfinate Co(III) derivatives. Extraction of Co(II) using 1 N HCl and chelex or KCN provides solutions of the mononuclear bis-sulfinate complexes **26** and **27** with a Cl^- or H_2O trans to an axial CN^- donor, respectively. The pK_a of the coordinated water in **27** was measured as 10.1 [68].

The L^4 ligand of Ozawa and Masuda also forms a square-planar Co(III) complex (**28**). Following a strategy similar to that used by Chottard, addition of H_2O_2·urea to **26** in the presence of ^{t}BuNC generates the bis-sulfenate **29**. Interestingly, exposure of **29** to dioxygen results in further S-oxidation to the bis-sulfinate **30**, which was not reported for **20** [69, 70].

Of the numerous N_2S_2-based Co(III) derivatives synthesized and isolated, only two have reported nitrile hydration catalytic activity. The water-bound bis-sulfinate **27** prepared by Artaud [68] hydrates acetonitrile under acidic (pH = 4.7) and neutral (pH = 7) conditions, with 7 and 14 turnovers, respectively, after 24 h at 50 °C. No further catalytic results with **27** were provided. The bis-sulfenate complex **20** of Chottard [66] also actively hydrates acetonitrile under acidic conditions (pH = 4.8, 50 TON in 6 days) despite being coordinatively saturated. Complex **20** shows no activity under basic conditions (pH = 9.0). Further, the bis-sulfinate **19** and **21** lack hydration activity under all conditions, suggesting that the sulfenate moiety is an integral part of the catalytic cycle. Hydration studies using a $H_2{}^{18}O/H_2{}^{16}O$ mixture revealed exchange of O-atoms between bulk water and the sulfinate O of **20**, but only under turnover conditions. No O-atom exchange was observed in the absence of acetonitrile substrate. This supports a proposed ligand-centered hydration cycle, with the sulfenate group acting as the nucleophilic site assisted by general acid catalysis by the acetic acid/acetate buffer. No hydration activity has been reported for complexes of L^4.

12.5.2 N_3S_2 Co(III) Model Complexes

Model complexes based on N_3S_2 chelates employ a single ligand to reproduce the protein-derived N_2S_3 donors at the NHase active site (Scheme 12.6). In these ligands, one of the cysteine thiolate donors has been substituted with a nitrogen donor. In L^5, two of the nitrogen donors are deprotonated carboxamides similar to the NHase active site donors, with the third N provided by pyridine. The N_2S_3 donor L^6 nitrogen is based on a central amine donor with pendant imine nitrogen donors and aliphatic thiolates. For both ligands, the N_3 portion of the chelate adopts a meridional arrangement. This positions the cis-thiolates in equatorial positions of a distorted trigonal bipyramidal environment for five-coordinate complexes. For six-coordinate complexes, the additional ligand occupies a position trans to one of the thiolates, leaving the other thiolate trans to the central N-donor.

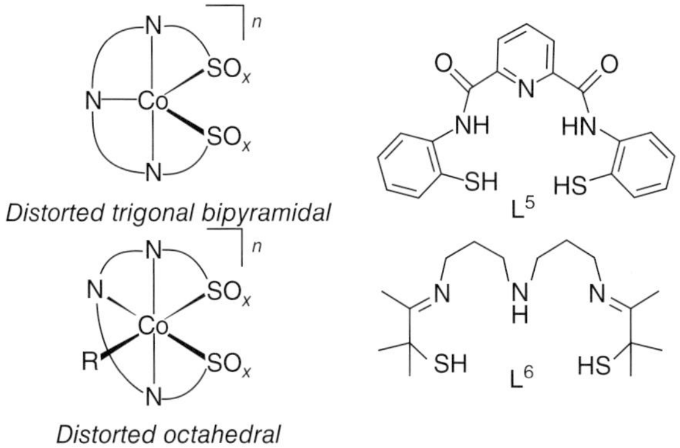

Complex	L	R	x	y	n
31	L^5	L^5Co	0	0	−2
32	L^5	CN^-	0	0	−2
33	L^5	CN^-	2	3	−2
34	L^5	CN^-	2	0	−2
35	L^5	H_2O	0	0	−1
36	L^5	H_2O	2	0	−1
37	L^6		0	0	+1
38	L^6		2	0	+1
4	L^6		2	1	+1

Scheme 12.6 Selected N_3S_2–Co(III) model complexes.

The Co(III) complex of L^5 is prepared directly from the Co(III) source $[Co(NH_3)_5Cl]Cl_2$ as a dinuclear $(Et_4N)_2[(L^5)_2Co_2]$ salt (**31**) with bridging thiolate donors [71]. The dimer is cleaved upon addition of Et_4NCN to yield the diamagnetic mononuclear complex $(Et_4N)_2[L^5Co(CN)]$ (**32**). Direct oxidation of **32** with H_2O_2 leads to the mixed sulfinate/sulfonate **33**. The sulfonate moiety is O-bound trans to the CN^- donor. In contrast, oxidation of the dinuclear complex **31** with H_2O_2 selectively oxygenates the terminal thiolate donors as the bridging thiolates are protected. The mononuclear monosulfinate **34** is isolated upon addition of NEt_4CN [27]. Complexes **32** and **34** were among the first to demonstrate the strong labilizing effect of a *trans*-thiolate, a point first emphasized by Kovacs [72]. Despite the status of low-spin Co(III) as a quintessential kinetically inert complex, the CN^- ligand in **32** and **34** is readily labile at room temperature. Complex **33**, which lacks a thiolate donor, is kinetically inert. Exchange of cyanide from **32** and **34** for water forms quasi-stable aqua derivatives that are stabilized as hydroxides at higher pH. Measured pK_a values for the aqua derivatives of **35** and **36** are 8.3 and 7.2, respectively. The lower pK_a of the sulfinate complex confirms the enhanced Lewis acidity of S-oxygenate metal complexes [27].

Complexation of L^6 to Co(III) yields a cationic, distorted trigonal bipyramidal complex (**37**) with an $S = 1$ ground state. Complex **37** exhibits a strong binding affinity toward N_3^- and SCN^-, yielding pseudo-octahedral complexes with a variable ligand trans to thiolate [45]. Oxygenation of **37** with O_2 yields the low-spin monosulfinate **38** with a concomitant change in geometry from trigonal bipyramidal to distorted square pyramidal. The oxidized sulfur sits in the apical position. Complex **38** shows no affinity toward azide or thiocyanate ligands. Further addition of H_2O_2 to **38** oxidizes the remaining thiolate to a sulfenate, which slips to a η^2-coordination in the structurally characterized **4**. Despite the side-on binding, **4** remains a rare example of a structurally characterized mixed sulfenate/sulfinate-metal complex [45].

Of the N_3S_2 Co(III) model complexes, the hydroxide derivatives of **35** and **36** reported by Mascharak [27, 71] catalytically hydrate nitriles. The hydroxide derivative of **35** displays moderate hydration activity at 50 °C in buffered acetonitrile with 15 and 18 turnovers after 2 and 4 h (pH = 9.5), respectively. The more Lewis acidic sulfinate derivative based on **36** is reported to be three times more active, although no details are provided regarding turnovers or reaction times.

The N_4S and N_4O ligands L^7 and L^8 developed by Kovacs [32] are structurally similar to N_3S_2 ligand L^6 (Scheme 12.7). Ligand L^7 is a tris(2-aminoethyl)amine derivative with one terminal amine functionalized to obtain an imine-alkoxide chelate. In L^8, the alkoxide is substituted by a thiolate. The nitrile-bound Co(III) complex $[L^7Co(NCCH_3)](PF_6)_2$ (**39**) readily reacts with ^-OH to yield an amidate-bound complex, $[L^7Co(NHC(O)CH_3)](PF_6)$ (NHC, N-heterocyclic carbene) (**40**), which is a tautomer analog of the proposed NHase imidate intermediate (B) in Scheme 12.2. Stabilization of **40** by H bonding between the amidate-O and a metal-coordinated primary amine is evident in the X-ray crystal structure. Hydration of the alkoxide complex $[L^8Co(NCCH_3)](PF_6)_2$ (**41**) proceeds to the iminol complex **42**, which is stabilized by H-bonding interactions with the alkoxide donor. Complex **42** provides a good structural analog of NHase intermediate C in Scheme 12.2 [32]. Although **39** and **41** lack the ability to completely hydrate nitriles or turnover, they model second-coordination sphere interactions proposed at the NHase active site.

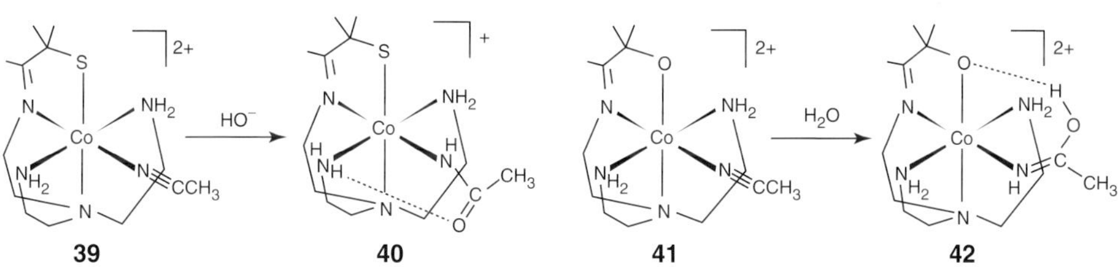

Scheme 12.7 Co(III) analogs of catalytic intermediates.

12.5.3
N_2S_3 Co(III) Model Complexes

A pentadentate N_2S_3 donor set with two aliphatic thiolates, one thioether, and two carboxamido nitrogens (L^9) reported by Ozawa and Masuda coordinates cobalt(III) yielding an air-sensitive complex that was not isolated (Scheme 12.8). Air oxidation affords the bis-sulfinate **43** with Co(III) in a distorted square-pyramidal structure. One of the sulfinate donors occupies the axial position. In aqueous solutions, UV–vis studies suggested the formation of an octahedral complex due to water coordination [73], but no hydration studies were reported. In a review by the same group, preliminary reports of the mono-sulfinate and sulfenate/sulfinate analogs of **43** were communicated. No nitrile hydration activity is reported, but these new complexes were reported to catalyze the conversion of ^{t}BuNC to ^{t}BuNH$_2$ [3].

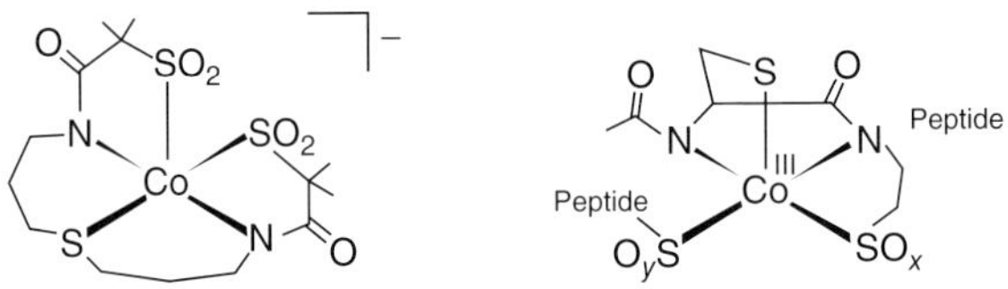

43 **44–47**

Complex	x	y
44	0	0
45	2	1
46	2	2
47	2	0

Scheme 12.8 Selected N_2S_3–Co(III) model complexes.

A series of metallopeptide maquettes reported by Shearer [54] also incorporate a Co(III) in an N_2S_3 donor environment (Scheme 12.8). The peptide NHase-m1 (AcNH-CCDLP-CGVYD-PA-COOH) was metallated using $CoCl_2$ under an inert atmosphere, yielding a green-colored solution, which was air oxidized to the brown Co(III) derivative [Co^{III}NHase-m1] (**44**). The circular dichroism (CD) and electronic spectrum of **44** are identical to Co-NHase, suggesting a similar five-coordinate square-pyramidal structure or a distorted octahedral Co (III) center. Longer air exposure yields **45**, which is proposed to include a mixed sulfenic acid/sulfinate donor set based on gel-permeation chromatography (GPC), electrospray ionization mass spectrometry (ESI-MS), and extended X-ray absorption fine structure (EXAFS) studies. The Fourier transform infrared (FT-IR) of **45** displays peaks at 1091 cm^{-1} (ν_{SO_2}) and 928 cm^{-1} (ν_{SOH}) that shift to 1052 and 889 cm^{-1} when **45** is prepared in the presence of $^{18}O_2$. Extended O_2 exposure produces the bis-sulfinate **46**. The monosulfinate derivative **47** was not experimentally observed, but is included in theoretical calculations.

The influence of S-oxygenation on substrate binding to **45–47** was investigated by DFT calculations. A comparison of water versus nitrile coordination based on the proposed structures of **45–47** finds preferential nitrile binding to **45** in the protonated, sulfenic acid form only. Deprotonation of the sulfenic acid introduces H-bonding interactions that favor the aqua derivative. Further oxidation of **45** to the bis-sulfinate **46** or computational O-atom abstraction to the hypothetical monosulfinate **47** decrease the ligand-binding affinity and the five-coordinate, substrate-free derivative is energetically favored in both cases.

Hydration studies with **45–47** support the DFT conclusion that only **45** binds nitrile. Complex **45** hydrates acrylonitrile to acrylamide, producing 58 turnovers at 25 °C under slightly basic conditions (pH = 8.0) [54]. Limited hydration activity is observed with acetonitrile as substrate and no activity is observed with 3-cyanopyridine. Both **44** and **46** lack hydration activity under all conditions reported. The results are consistent with optimal NHase activity with the mixed sulfenate/sulfinate and loss of activity upon overoxidization.

12.6
Sulfur-Oxygenation of Fe(III) NHase Mimics

Before the discovery of sulfur-oxygenated donors at the active site of NHase, it was generally established that Fe-thiolates degrade under aerobic conditions via disulfide formation coupled with metal reduction and decomplexation of the ligand [1, 41, 74]. Inspired by NHase, several S-oxygenates have been prepared by direct oxidation of iron-thiolates with dioxygen [28, 75–78]. In nearly all cases, a low-spin configuration ($S = 1/2$) Fe(III) ground state was observed in the precursor complex. Although no functional Fe-containing mimics are reported, key insights into NHase have been revealed through S-oxygenation studies. Highlights are summarized here.

12.6.1
N_2S_2–Fe(III) Model Complexes

Tetradentate N_2S_2 ligands L^{10} to L^{12} incorporate carboxamide nitrogen donors and reactive *cis*-thiolates to generate square-planar arrangements, leading to five- and six-coordinate Fe(III) models (Scheme 12.9) similar to the Co(III) systems noted with L^2 to L^4. The Fe(III) derivative of L^{10}, which is similar to L^3 but contains one carbon less in the pendant thiolate arm, was prepared by complexation of $FeCl_3$ in DMF at −15 °C, producing $(NEt_4)_2[L^{10}FeCl]$ (**48**) as a brown orange powder [79]. The five-coordinate complex has an intermediate spin ($S = 3/2$) Fe(III) in a distorted square-pyramidal geometry with a N_2S_2 basal plane and chloride in the axial position. Through a series of electrochemical studies, exchange of chloride with water, $^-$OH, and various solvents is observed. The NO derivative (**49**) is obtained by purging nitric oxide in a dichloromethane solution of **48**. Complex **49** has a $\{Fe\text{-}NO\}^6$ electronic structure according to Enemark–Feltham notation with a singlet ($S = 0$) ground state. No oxidation reactions of **48** or **49** were reported.

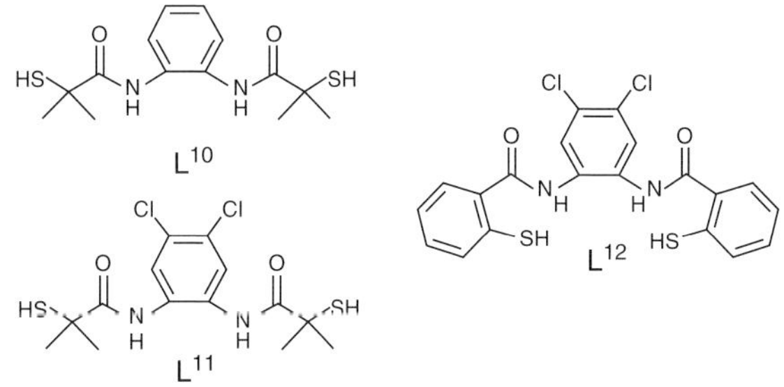

Complex	L	R_1	R_2	*x*	*y*	*n*
48	L^{10}	Cl^-		0	0	−2
49	L^{10}	NO		0	0	−1
50	L^{11}	Cl^-		0	0	−2
51	L^{11}	CN^-		0	0	−2
52	L^{11}	CN^-	CN^-	0	0	−3
53	L^{11}	CN^-	CN^-	2	2	−3
54	L^{12}	DMAP		0	0	−1
55	L^{12}	DMAP	NO	0	0	−1
56	L^{12}	DMAP	NO	2	2	−1

Scheme 12.9 Selected N_2S_2–Fe(III) model complexes.

The ligand L^{11} is similar to L^{10}, but it includes the introduction of electron-withdrawing chloro groups on the aromatic backbone. The Fe(III) complex $(Et_4N)_2[L^{11}FeCl]$ (**50**) is similar to **48** with an intermediate spin iron (III) center

($S = 3/2$) in a square-pyramidal environment. Metathesis of the coordinated chloride with NEt_4CN affords the highly hygroscopic, air sensitive $(Et_4N)_2[L^{11}Fe(CN)]$ (**51**) [78]. EPR measurements for **51** at 20 K in frozen acetonitrile solution displayed signals with g values of 4.38, 2.15, and 1.97 interpreted as a mixture of the penta-coordinate intermediate spin **51** and the low-spin six-coordinate derivative $(Et_4N)[L^{11}Fe(CN)_2]$ (**52**). While **51** decomposes under an O_2 atmosphere, the addition of 1 equiv NEt_4CN shifts equilibrium toward the low-spin **52**, yielding the green low-spin bis-sulfinate **53** [78]. The FT-IR of **53** displays intense bands at 1166 and 1050 cm^{-1}, corresponding to SO asymmetric and symmetric stretches in addition to a band at 2104 cm^{-1} for the coordinated CN^-. The results are consistent with the preferential S-oxygenation of low-spin Fe(III) complexes.

Using a ligand structure similar to L^{11} with aromatic thiolates in place of the pendant aliphatic thiolate donors, Mascharak reported the green intermediate spin, square-pyramidal complex $[L^{12}Fe(DMAP)]^-$ (DMAP, 4-dimethylaminopyridine) (**54**). Complex **54** reacts readily with nitric oxide at −40 °C to yield the red $[L^{12}Fe(NO)(DMAP)]^-$ (**55**). Oxidation of **55** with 4 (1*S*)-(+)-(10-camphorsulfonyl)oxaziridine affords the bis-sulfinate **56**. S-oxygenation results in a shift in the ν_{NO} band from 1849 to 1854 cm^{-1}. The introduction of S-oxygenation enhances NO lability such that **56** readily photodissociates NO when exposed to visible light with a quantum yield of 0.55, which is similar to the NO-inactivated NHase value of 0.48. These studies strongly support the hypothesis that S-oxygenation imparts NO lability and facilitates enzyme activation [29, 80].

12.6.2
N_3S_2–Fe(III) Model Complexes

The pentadentate N_3S_2 ligand L^5 used to generate functional Co(III) mimics was also employed by Mascharak to prepare the iron(III) derivatives $(Et_4N)[L^5Fe]$ (**57**) and $(Et_4N)_2[L^5Fe(CN)]$ (**58**) (Scheme 12.10). The five-coordinate iron(III) complex **57** finds Fe(III) in a distorted trigonal bipyramid. Complex **57** readily forms adducts with various neutral and anionic ligands including water. The pK_a

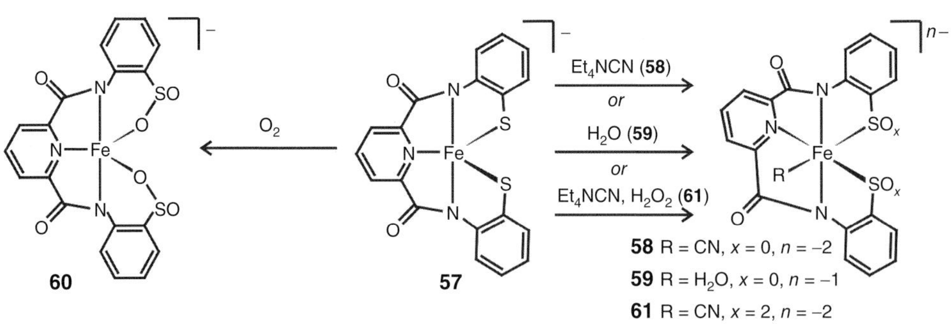

Scheme 12.10 S-oxygenation of N_3S_2–Fe(III) model complexes.

of the water-bound derivative **59**, 6.3 ± 0.4 is two orders of magnitude lower than the corresponding cobalt complex **35**, but **59** is too unstable to be used as a nitrile hydration catalyst. The six-coordinate **58** contains low-spin iron(III) in a pseudo-octahedral environment with CN^- trans to a thiolate donor. Oxygenation of **57** with various O-atom transfer reagents affords the O-bound bis-sulfinate **60**. In contrast, similar oxidations with the six-coordinate **58** yields the S-bound bis-sulfinate **61** [77]. The results suggest that S-coordination of sulfinate donors requires coordinatively saturated iron to prevent isomerization favored by the hard–hard interaction between iron(III) and oxygen.

Kovacs and coworkers prepared five- and six-coordinate iron(III) derivatives using their N_3S_2 ligand L^6 to evaluate the low-spin state of Fe(III) in NHase despite the presence of π-donating sulfur donors. Through a series of spectroscopic and computational investigations, they conclude the ligand enforced co-planar orientation of thiolate lone pairs in the model complex combined with strong σ-donating nitrogen donors enforces a low-spin environment [81].

12.6.3
N_2S_3 Fe(III) Model Complexes

Notable N_2S_3 ligand systems for Fe-NHase models include the dicarboxamide-trithiolate ligands L^{13} reported by Chottard and its structural isomer L^{14} developed by Kovacs (Scheme 12.11). Both ligands employ a symmetric structure with a central thiolate donor and a pair of identical pendant NS donors providing the most faithful reproduction of the NHase active site. Ligands L^{13} and L^{14} are quite similar, differing by the relative arrangement of the carboxamide carbons, and each coordinates Fe(III) in a square pyramidal arrangement with the central thiolate in the apical position. The N_2S_3 ligand L^{15} developed by Grapperhaus is based on the facial coordinating ligand diamine-thioether macrocycle [9] aneS_2. The pendant thiolates coordinate cis to each other leaving a co-facial site available for a sixth, variable ligand.

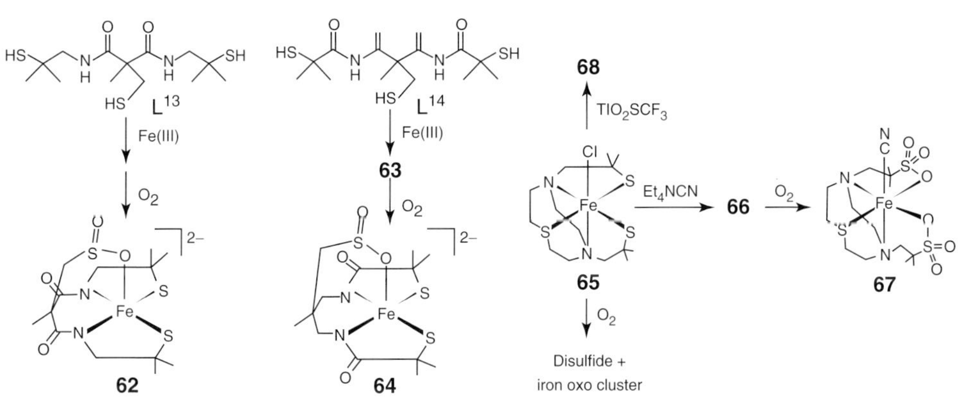

Scheme 12.11 Selected N_2S_3–Fe(III) model complexes.

The intermediate spin-state iron(III) derivative of L^{13} prepared by Chottard in 1999 [75] is extremely air sensitive precluding its isolation and characterization except as the O-bound monosulfinate derivative **62**. S-oxygenation occurs selectively at the central, apical thiolate, leaving an N_2S_2 plane defined by the remaining thiolates and the nitrogen donors. The vacant position sits trans to the O-bound sulfinate, but **62** has no affinity for additional donors and is not active for nitrile hydration. The analogous L^{14} complex of Fe(III) prepared by Kovacs is also air sensitive, but the iron(III) tri-thiolate complex **63** was isolated and thoroughly characterized. Similar to the L^{13} derivative, the metal sits in a square-pyramidal geometry with an intermediate spin ($S = 3/2$), with no affinity for additional ligands. Addition of dioxygen to acetonitrile solutions of **63** results in oxidation of the apical thiolate, yielding the O-bound sulfinate **64**. The structural and electronic properties of **64** are similar to **62** [82].

The N_2S_3 iron(III) complexes developed by Grapperhaus [4, 76, 83, 84] based on L^{15} bind a variety of ligands in the sixth coordination site, which sits trans to a nitrogen donor and co-facial with the *cis*-thiolates. The spin state of the iron complexes depends on the nature of the sixth ligand, with weak-field ligands giving high-spin complexes (**65**) and strong-field donors yielding low-spin complexes (**66**) [84]. The spin state of the iron influences the oxidation product. High-spin **46** reacts with O_2 to yield a disulfide of L^{15} and a structurally characterized iron-oxo cluster degradation product [85]. The degradation of the iron-thiolate to "rust" is consistent with early conventional wisdom regarding the air sensitivity of iron-thiolates. In contrast, low-spin **66** reacts with O_2 to yield the insoluble O-bound bis-sulfonate **67** [76]. The observed product is overoxidized with respect to NHase, but the results demonstrate the role of spin state in directing S-oxygenation. On the basis of a series of DFT calculations, this is attributed to an increased Fe–S bond covalency in low-spin iron(III) complexes relative to high-spin counterparts, which imparts an increased nucleophilic character on the coordinated thiolate.

While the chloride of **65** can be readily exchanged for other charged ligands, it cannot be displaced by neutral donors such as nitrile or water and the complex is inactive toward nitrile hydration. Weak donors can be coordinated upon metathesis of the coordinated Cl^- with thallium triflate (silver triflate was ineffective because of preferential Ag–S interactions) [86]. The weakly coordinating triflate derivative **68** exists in equilibrium between the six-coordinate $L^{15}Fe(O_3SCF_3)$ and the five-coordinate $L^{15}Fe$ complex; both are low-spin. The triflate anion is easily displaced by coordinating solvents such as DMF, water, acetonitrile, methanol, and pyridine. Complex **68** is the first NHase model to bind both nitrile and water (substrates) and amide (product). The binding preference is water > amide > acetonitrile [87]. The water-bound derivative displays two deprotonation events with measured pK_a values of 6.52 ± 0.05 and 5.4 ± 0.1 associated with the formation of μ-OH diiron and μ-O diiron complexes.

12.7
Ruthenium Complexes

Several ruthenium catalysts have been reported as nitrile hydration catalysts [6, 7]. Although not biologically relevant, they provided the basis for the "bioinspired" complexes reported by Grapperhaus employing the N_2S_3 ligand L^{15} (Scheme 12.12). Ruthenium is also a logical choice for oxidation studies as the second-row transition metal maintains a consistent low spin, which was found to promote S-oxygenation. Addition of $RuCl_2(PPh_3)_3$ to L^{15} in the presence of a base yields $L^{15}Ru(PPh_3)$ (**69**) [46]. The structure of **69** reveals distinct electronic environments for the *cis*-thiolate donors, with one sulfur trans to an amine and the trans to a thioether. The PPh_3 donor sits on the same octahedral face as the thiolate donors providing steric protection preferentially about one of the thiolates. By enforcing unique electronic and steric environments about the two thiolates, the site and extent of S-oxygenation can be regulated by the time of O_2 exposure.

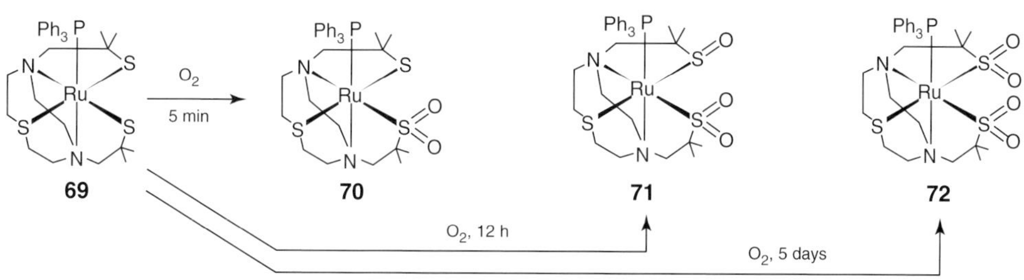

Scheme 12.12 Family of N_2S_3–Ru(II) S-oxygenates.

Exposure of **69** to O_2 for a few minutes yields the monosulfinate **70** through rapid S-oxygenation of the more sterically accessible thiolate. Oxygen exposure for several hours promotes further oxidation to the mixed sulfenate/sulfinate **71**, while reactions lasting several days are required to obtain the bis-sulfinate **72** [88]. Complexes **69–72** are unique as they represent the first family of related S-oxygenate derivatives with a variable ligand-binding site suitable for nitrile hydration. X-ray crystallographic studies reveal a systematic *increase* in the Ru–PPh_3 bond distance as a function of S-oxygenation. The trends have been reproduced by DFT calculations and XAS studies, which suggest S-oxygenation decreases the softness of the donor environment resulting in a more Lewis acidic Ru(II) center that disfavors the soft PPh_3 donor [89].

The increased Ru–P bond distance in **69–72** as a function of S-oxygenation correlates with increased PPh_3 lability in solution, thus providing an open coordination site for substrate binding. Exploiting this feature, Grapperhaus [90] reported preliminary benzonitrile hydration studies with **69–71** as catalysts. With ppm catalyst loading **69** achieves 66 ± 1 turnovers of benzonitrile to benzamide at 124 °C over 18 h with an associated turnover frequency of $3.7 \pm 0.1\ h^{-1}$. Notably, no buffer was used for carrying out these catalytic assays and all experiments were conducted under pH neutral conditions in biphasic benzonitrile: water mixtures. Varying the ratio of substrates resulted in turnovers as high as 188 ± 32 with **69**. Quantification of the benzamide product in the benzonitrile layer indicates a product inhibited with maximum turnover occurring when the concentration of benzamide reaches 31 mM in the benzonitrile layer. Metal oxidation of **69** from Ru(II) to Ru(III) reduces nitrile hydration activity, which is tentatively attributed to tight binding between the amidate/amide product and Ru(III) metal as noted by Ford [91] in similar systems.

In contrast to metal oxidation, sulfur-oxygenation enhances nitrile hydration at low nitrile:water ratios. The monosulfinate **70** and the sulfenate/sulfinate **71** display similar benzonitrile hydration results, with up to 238 ± 22 and 242 ± 23 turnovers, respectively [90]. These yields are consistent with final benzamide concentrations of 129 ± 12 and 131 ± 12 mM, respectively. The higher benzamide concentrations are interpreted as resulting from decreased product inhibition upon S-oxygenation. Detailed kinetic studies to further delineate the influence of S-oxygenation on nitrile hydration are under way.

12.8 Conclusions/Challenges

Over the past 15 years, several groups have investigated S-oxygenation reactions of cobalt- and iron-thiolate model complexes inspired by the unusual post-translational modification of the NHase active site. Although the mechanism of asymmetric S-oxygenation in NHase remains unclear, small-molecule models react with O_2 and various O-atom transfer agents providing multiple synthetic routes to yield metal S-oxygenates. Controlled oxygenation remains challenging, but partial S-oxygenation has been successfully achieved with stoichiometric O-atom transfer agents or asymmetric thiolate environments in O_2 reactions. It is noteworthy that the apical thiolate donors, preferentially oxygenated in model studies, remain unreacted in NHase. This is attributed to protection of the *trans*-thiolate in the enzyme pocket with reactivity directed toward the *cis*-thiolates, which are co-facial with the substrate-binding site. This selective S-oxygenation positions the sulfenate/suflinate donors for second-coordination sphere interactions with substrate, intermediates, and products as modeled in small-molecule complexes.

While synthetic mimics have successfully reproduced S-oxygenation, functional NHase mimics with high activity at room temperature remain elusive. As summarized in Table 12.1, to date only four functional Co(III) NHase mimics have been reported. Of these, four incorporate oxygenated sulfur donors in the first coordination sphere. To date, no functional Fe(III) NHase mimics are known, although three bioinspired Ru(II) models have recently been reported. Direct comparison of hydration activity among the seven models is problematic given the lack of consistent substrates and reactions conditions, but in all cases hydration rates are low to moderate. The lack of functional Fe(III) mimics can be attributed, in part, to

Table 12.1 Catalytic activity of functional NHase models.

Catalyst	Substrate	pH	T (°C)	Time (h)	TON	References
20	CH_3CN	4.8	4	144	50	[66]
27	CH_3CN	7	50	24	14	[68]
35[a)]	CH_3CN	9.5	50	4	18	[27]
45	CH_2CHCN	8	25	18	58	[54]
69	C_6H_5CN	—	125	18	188 ± 32	[90]
70	C_6H_5CN	—	125	18	238 ± 23	[90]
71	C_6H_5CN	—	125	18	242 ± 23	[90]

a) The sulfinate derivative **36** was reported to be three times more active.

the preferential formation of μ-oxo and μ-hydroxo complexes. Synthetic analogs with extended second-coordination spheres that stabilize substrate binding and disfavor dinuclear iron complexes remains undeveloped in the NHase model community.

Abbreviations

DFT	density functional theory
DMAP	4-dimethylaminopyridine
EPR	electron paramagnetic resonance
ESI-MS	electrospray ionization mass spectrometry
FT-IR	Fourier transform infrared
NHase	nitrile hydratase
SCNase	thiocyanate hydrolase
TON	turnover number
XAS	X-ray absorption spectroscopy
EXAFS	extended X-ray absorption fine structure

References

1. Mascharak, P.K. (2002) *Coord. Chem. Rev.*, **225**, 201–214.
2. Kovacs, J.A. (2004) *Chem. Rev.*, **104**, 825–848.
3. Yano, T., Ozawa, T., and Masuda, H. (2008) *Chem. Lett.*, **37**, 672–677.
4. O'Toole, M.G. and Grapperhaus, C.A. (2009) *ACS Symposium Series: Bioinorganic Chemistry*, vol. 1012, American Chemical Society, Washington, DC, pp. 99–113.
5. Prasad, S. and Bhalla, T.C. (2010) *Biotechnol. Adv.*, **28**, 725–741.
6. Kukushkin, V.Y. and Pombeiro, A.J.L. (2005) *Inorg. Chim. Acta*, **358**, 1–21.
7. Ahmed, T.J., Knapp, S.M.M., and Tyler, D.R. (2011) *Coord. Chem. Rev.*, **255**, 949–974.
8. Kobayashi, M. and Shimizu, S. (1998) *Nat. Biotechnol.*, **16**, 733–736.
9. Bunch, A.W. (1998) *Antonie Van Leeuwenhoek*, **74**, 89–97.

10. Hann, E.C., Eisenberg, A., Fager, S.K., Perkins, N.E., Gallagher, F.G., Cooper, S.M., Gavagan, J.E., Stieglitz, B., Hennessey, S.M., and DiCosimo, R. (1999) *Bioorg. Med. Chem.*, **7**, 2239–2245.
11. Mauger, J., Nagasawa, T., and Yamada, H. (1989) *Tetrahedron*, **45**, 1347–1354.
12. Nagasawa, T., Shimizu, H., and Yamada, H. (1993) *Appl. Microbiol. Biotechnol.*, **40**, 189–195.
13. Nagasawa, T. and Yamada, H. (1989) *Trends Biotechnol.*, **7**, 153–158.
14. Prasad, S., Raj, J., and Bhalla, T. (2007) *Indian J. Microbiol.*, **47**, 34–41.
15. Raj, J., Seth, A., Prasad, S., and Bhalla, T. (2007) *Appl. Microbiol. Biotechnol.*, **74**, 535–539.
16. Kohyama, E., Yoshimura, A., Aoshima, D., Yoshida, T., Kawamoto, H., and Nagasawa, T. (2006) *Appl. Microbiol. Biotechnol.*, **72**, 600–606.
17. Li, T., Liu, J., Bai, R., Ohandja, D.-G., and Wong, F.-S. (2007) *Water Res.*, **41**, 3465–3473.
18. Aislabie, J. and Atlas, R.M. (1988) *Appl. Environ. Microbiol.*, **54**, 2197–2202.
19. Holtze, M.S., Sørensen, S.R., Sørensen, J., and Aamand, J. (2008) *Environ. Pollut.*, **154**, 155–168.
20. Baxter, J., Garton, N., and Cummings, S. (2006) *Folia Microbiol.*, **51**, 591–597.
21. Katayama, Y., Narahara, Y., Inoue, Y., Amano, F., Kanagawa, T., and Kuraishi, H. (1992) *J. Biol. Chem.*, **267**, 9170–9175.
22. Katayama, Y., Matsushita, Y., Kaneko, M., Kondo, M., Mizuno, T., and Nyunoya, H. (1998) *J. Bacteriol.*, **180**, 2583–2589.
23. Westley, J. (1973) *Adv. Enzymol. Relat. Areas Mol. Biol.*, **39**, 327–368.
24. Huang, W.J., Jia, J., Cummings, J., Nelson, M., Schneider, G., and Lindqvist, Y. (1997) *Structure*, **5**, 691–699.
25. Shigehiro, S., Nakasako, M., Dohmae, N., Tsujimura, M., Tokoi, K., Odaka, M., Yohda, M., Kamiya, N., and Endo, I. (1998) *Nat. Struct. Biol.*, **5**, 347–351.
26. Murakami, T., Nojiri, M., Nakayama, H., Odaka, M., Yohda, M., Dohmae, N., Takio, K., Nagamune, T., and Endo, I. (2000) *Protein Sci.*, **9**, 1024–1030.
27. Tyler, L.A., Noveron, J.C., Olmstead, M.M., and Mascharak, P.K. (2003) *Inorg. Chem.*, **42**, 5751–5761.
28. Lee, C.M., Chen, C.H., Chen, H.W., Hsu, J.L., Lee, G.H., and Liaw, W.F. (2005) *Inorg. Chem.*, **44**, 6670–6679.
29. Rose, M.J., Betterley, N.M., Oliver, A.G., and Mascharak, P.K. (2010) *Inorg. Chem.*, **49**, 1854–1864.
30. Hashimoto, K., Suzuki, H., Taniguchi, K., Noguchi, T., Yohda, M., and Odaka, M. (2008) *J. Biol. Chem.*, **283**, 36617–36623.
31. Hopmann, K.H., Guo, J.D., and Himo, F. (2007) *Inorg. Chem.*, **46**, 4850–4856.
32. Swartz, R.D., Coggins, M.K., Kaminsky, W., and Kovacs, J.A. (2011) *J. Am. Chem. Soc.*, **133**, 3954–3963.
33. Dey, A., Jeffrey, S.P., Darensbourg, M.Y., Hodgson, K.O., Hedman, B., and Solomon, E.I. (2007) *Inorg. Chem.*, **46**, 4989–4996.
34. Tsujimura, M., Odaka, M., Nakayama, H., Dohmae, N., Koshino, H., Asami, T., Hoshino, M., Takio, K., Yoshida, S., Maeda, M., and Endo, I. (2003) *J. Am. Chem. Soc.*, **125**, 11532–11538.
35. Song, L., Wang, M., Yang, X., and Qian, S. (2007) *Biotechnol. J.*, **2**, 717–724.
36. Song, L., Wang, M., Shi, J., Xue, Z., Wang, M.-X., and Qian, S. (2007) *Biochem. Biophys. Res. Commun.*, **362**, 319–324.
37. Yamanaka, Y., Hashimoto, K., Ohtaki, A., Noguchi, K., Yohda, M., and Odaka, M. (2010) *J. Biol. Inorg. Chem.*, **15**, 655–665.
38. Sugiura, Y., Kuwahara, J., Nagasawa, T., and Yamada, H. (1987) *J. Am. Chem. Soc.*, **109**, 5848–5850.
39. Rao, S. and Holz, R.C. (2008) *Biochemistry*, **47**, 12057–12064.
40. Holliday, G.L., Mitchell, J.B.O., and Thornton, J.M. (2009) *J. Mol. Biol.*, **390**, 560–577.
41. Holm, R.H. and Ibers, J.A. (1977) *Iron–Sulfur Proteins*, vol. 3, Academic Press, New York.
42. Ibers, J.A. and Holm, R.H. (1980) *Science*, **209**, 223–235.
43. Dilworth, J., Zheng, Y., Lu, S., and Wu, Q. (1992) *Transition Met. Chem.*, **17**, 364–368.
44. Buonomo, R.M., Font, I., Maguire, M.J., Reibenspies, J.H., Tuntulani, T., and

Darensbourg, M.Y. (1995) *J. Am. Chem. Soc.*, **117**, 963–973.

45. Kung, I., Schweitzer, D., Shearer, J., Taylor, W.D., Jackson, H.L., Lovell, S., and Kovacs, J.A. (2000) *J. Am. Chem. Soc.*, **122**, 8299–8300.
46. Masitas, C.A., Mashuta, M.S., and Grapperhaus, C.A. (2010) *Inorg. Chem.*, **49**, 5344–5346.
47. Aucott, S.M., Milton, H.L., Robertson, S.D., Slawin, A.M.Z., Walker, G.D., and Woollins, J.D. (2004) *Chem. Eur. J.*, **10**, 1666–1676.
48. Schubert, M.P. (1933) *J. Am. Chem. Soc.*, **55**, 3336–3342.
49. Herting, D.L., Sloan, C.P., Cabral, A.W., and Krueger, J.H. (1978) *Inorg. Chem.*, **17**, 1649–1654.
50. Jackson, W.G., Sargeson, A.M., and Whimp, P.O. (1976) *J. Chem. Soc., Chem. Commun.*, 934–935.
51. Adzamli, I.K., Libson, K., Lydon, J.D., Elder, R.C., and Deutsch, E. (1979) *Inorg. Chem.*, **18**, 303–311.
52. Thompson, J.S., Marks, T.J., and Ibers, J.A. (1979) *J. Am. Chem. Soc.*, **101**, 4180–4192.
53. Lydon, J.D. and Deutsch, E. (1982) *Inorg. Chem.*, **21**, 3180–3185.
54. Shearer, J., Callan, P.E., and Amie, J. (2010) *Inorg. Chem.*, **49**, 9064–9077.
55. Dey, A., Chow, M., Taniguchi, K., Lugo-Mas, P., Davin, S., Maeda, M., Kovacs, J.A., Odaka, M., Hodgson, K.O., Hedman, B., and Solomon, E.I. (2006) *J. Am. Chem. Soc.*, **128**, 533–541.
56. Schrauzer, G.N., Zhang, C., and Chadha, R. (1990) *Inorg. Chem.*, **29**, 4104–4107.
57. Mirza, S.A., Pressler, M.A., Kumar, M., Day, R.O., and Maroney, M.J. (1993) *Inorg. Chem.*, **32**, 977–987.
58. Tuntulani, T., Musie, G., Reibenspies, J.H., and Darensbourg, M.Y. (1995) *Inorg. Chem.*, **34**, 6279–6286.
59. Grapperhaus, C.A. and Darensbourg, M.Y. (1998) *Acc. Chem. Res.*, **31**, 451–459.
60. Galvez, C., Ho, D.G., Azod, A., and Selke, M. (2001) *J. Am. Chem. Soc.*, **123**, 3381–3382.
61. Noveron, J.C., Olmstead, M.M., and Mascharak, P.K. (1998) *Inorg. Chem.*, **37**, 1138–1139.
62. Tyler, L.A., Noveron, J.C., Olmstead, M.M., and Mascharak, P.K. (1999) *Inorg. Chem.*, **38**, 616–617.
63. Tyler, L.A., Noveron, J.C., Olmstead, M.M., and Mascharak, P.K. (1999) *Inorg. Chem.*, **39**, 357–362.
64. Lugo-Mas, P., Dey, A., Xu, L., Davin, S.D., Benedict, J., Kaminsky, W., Hodgson, K.O., Hedman, B., Solomon, E.I., and Kovacs, J.A. (2006) *J. Am. Chem. Soc.*, **128**, 11211–11221.
65. Heinrich, L., Li, Y., Vaissermann, J., and Chottard, J.C. (2001) *Eur. J. Inorg. Chem.*, **2001**, 1407–1409.
66. Heinrich, L., Mary-Verla, A., Li, Y., Vaissermann, J., and Chottard, J.C. (2001) *Eur. J. Inorg. Chem.*, **2001**, 2203–2206.
67. Bourles, E., Alves de Sousa, R., Galardon, E., Giorgi, M., and Artaud, I. (2005) *Angew. Chem.*, **117**, 6318–6321.
68. Rat, M., de Sousa, R.A., Tomas, A., Frapart, Y., Tuchagues, J.-P., and Artaud, I. (2003) *Eur. J. Inorg. Chem.*, **2003**, 759–765.
69. Yano, T., Wasada-Tsutsui, Y., Arii, H., Yamaguchi, S., Funahashi, Y., Ozawa, T., and Masuda, H. (2007) *Inorg. Chem.*, **46**, 10345–10353.
70. Yano, T., Arii, H., Yamaguchi, S., Funahashi, Y., Jitsukawa, K., Ozawa, T., and Masuda, H. (2006) *Eur. J. Inorg. Chem.*, **2006**, 3753–3761.
71. Noveron, J.C., Olmstead, M.M., and Mascharak, P.K. (1999) *J. Am. Chem. Soc.*, **121**, 3553–3554.
72. Shearer, J., Kung, I.Y., Lovell, S., Kaminsky, W., and Kovacs, J.A. (2000) *J. Am. Chem. Soc.*, **123**, 463–468.
73. Ozawa, T., Ikeda, T., Yano, T., Arii, H., Yamaguchi, S., Funahashi, Y., Jitsukawa, K., and Masuda, H. (2005) *Chem. Lett.*, **34**, 18–19.
74. Ellis, K.J., Lappin, A.G., and Mcauley, A. (1975) *J. Chem. Soc., Dalton Trans.*, 1930–1934.
75. Heinrich, L., Li, Y., Vaissermann, J., Chottard, G., and Chottard, J.-C. (1999) *Angew. Chem.*, **38**, 3526–3528.
76. O'Toole, M.G., Kreso, M., Kozlowski, P.M., Mashuta, M.S., and Grapperhaus, C.A. (2008) *J. Biol. Inorg. Chem.*, **13**, 1219–1230.

77. Noveron, J.C., Olmstead, M.M., and Mascharak, P.K. (2001) *J. Am. Chem. Soc.*, **123**, 3247–3259.
78. Galardon, E., Giorgi, M., and Artaud, I. (2004) *Chem. Commun.*, 286–287.
79. Chatel, S., Chauvin, A.-S., Tuchagues, J.-P., Leduc, P., Bill, E., Chottard, J.-C., Mansuy, D., and Artaud, I. (2002) *Inorg. Chim. Acta*, **336**, 19–28.
80. Rose, M.J., Betterley, N.M., and Mascharak, P.K. (2009) *J. Am. Chem. Soc.*, **131**, 8340–8341.
81. Kennepohl, P., Neese, F., Schweitzer, D., Jackson, H.L., Kovacs, J.A., and Solomon, E.I. (2005) *Inorg. Chem.*, **44**, 1826–1836.
82. Lugo-Mas, P., Taylor, W., Schweitzer, D., Theisen, R.M., Xu, L., Shearer, J., Swartz, R.D., Gleaves, M.C., DiPasquale, A., Kaminsky, W., and Kovacs, J.A. (2008) *Inorg. Chem.*, **47**, 11228–11236.
83. Grapperhaus, C.A., Patra, A.K., and Mashuta, M.S. (2002) *Inorg. Chem.*, **41**, 1039–1041.
84. Grapperhaus, C.A., Li, M., Patra, A.K., Poturovic, S., Kozlowski, P.M., Zgierski, M.Z., and Mashuta, M.S. (2003) *Inorg. Chem.*, **42**, 4382–4388.
85. Grapperhaus, C.A., O'Toole, M.G., and Mashuta, M.S. (2006) *Inorg. Chem. Commun.*, **9**, 1204–1206.
86. Artigas, M.M., Crespo, O., Gimeno, M.C., Jones, P.G., Laguna, A., and Villacampa, M.D. (1997) *Inorg. Chem.*, **36**, 6454–6456.
87. O'Toole, M.G., Bennett, B., Mashuta, M.S., and Grapperhaus, C.A. (2009) *Inorg. Chem.*, **48**, 2300–2308.
88. Masitas, C.A., Kumar, M., Mashuta, M.S., Kozlowski, P.M., and Grapperhaus, C.A. (2010) *Inorg. Chem.*, **49**, 10875–10881.
89. Shearer, J., Callan, P.E., Masitas, C.A., and Grapperhaus, C.A. (2012) *Inorg. Chem.*, **51**, 6032–6045.
90. Kumar, D., Masitas, C.A., Nguyen, T.N., and Grapperhaus, C.A. (2013) *Chem. Commun.*, **49**, 294–296.
91. Zanella, A.W. and Ford, P.C. (1975) *Inorg. Chem.*, **14**, 42–47.

13 Molybdenum and Tungsten Oxidoreductase Models

Carola Schulzke and Ashta Chandra Ghosh

13.1 Introduction

Molybdenum- and tungsten-dependent oxidoreductases are ubiquitous, meaning they are found in almost any organism on earth from an archaic archaeon to a modern human being. For at least two reasons they form a quite unique class of enzymes: first, they are the only biological catalysts with the exception of nitrogenase, which employ molybdenum and tungsten. This is unusual because molybdenum and tungsten, in contrast to other transition metals with biological functions, which are typically 3d metals, are 4d and 5d elements. Molybdenum and tungsten are, other than, for instance, iron and copper, capable of catalyzing two electron redox reactions with only one metal center present and without a co-oxidant or co-reductant at the active site. This ability is crucial for the transformations catalyzed by the molybdenum- and tungsten-dependent oxidoreductases and quite likely the reason why these metals are used.

Second, a very peculiar bioligand is found at the active sites of these enzymes: molybdopterin (MPT). Throughout the literature, MPT (Figure 13.1) is also referred to as *pyranopterindithiolate* [1], *pterin-dithiolene* [2], *pterin-ene-dithiolate* [3], or *tungstopterin* for the tungsten-dependent enzymes [4]. The reason for the various suggested names is the potential confusion stemming from the name molybdopterin. It actually stands for only the organic ligand without the coordinated molybdenum, although the name suggests otherwise. The exact same ligand is found in both, molybdenum and tungsten enzymes. Currently, despite several attempts to change it, molybdopterin is still the most common and most widely used name and will also be employed throughout this chapter.

MPT is a hetero-tricyclic system and binds the active site metal via a dithiolene moiety, that is, with two sulfur atoms that are bound to a C=C moiety. MPT is inherently specific for molybdenum- and tungsten-dependent oxidoreductases. All enzymes of this kind carry MPT and no other enzyme uses this ligand system or anything similar.

Dithiolenes are *per se* non-innocent compounds, meaning they are redox active and their actual and formal oxidation states or charges are not necessarily the same

Bioinspired Catalysis: Metal-Sulfur Complexes, First Edition.
Edited by Wolfgang Weigand and Philippe Schollhammer.

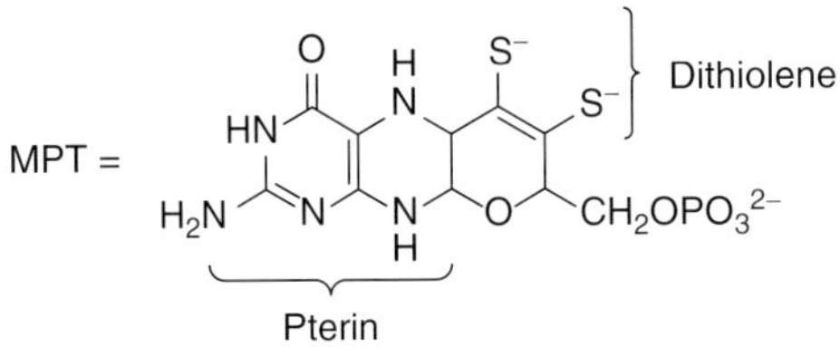

Figure 13.1 The chemical structure of molybdopterin (MPT) – the ligand in the active sites of molybdenum- and tungsten-dependent oxidoreductases.

[5–11]. Depending on the electron configuration of the coordinated metal and the overall electronic situation of the complex, the charge of a dithiolene ligand can be −2, −1, or 0, or anything in between. This allows the non-innocent ligand to fine-tune or buffer the coordinated metal's charge and facilitate redox processes of the entire complex. It has been debated, of course, that the potential ability of MPT to do so in the active site of an oxidoreductase may be a key to its efficiency. Experimental evidence, however, indicates that this is not a relevant feature, because steric hindrance of MPT movement prevents this ligand to act non-innocently [12, 13]. A prerequisite of moving electron density from the dithiolene ligand to the metal and back is the ability to modify the orbital overlap between sulfur p-orbitals with metal d-orbitals. The degree of such overlap depends on the angle at which the dithiolene is bent, the so-called folding angle [14–16]. With metals in higher formal oxidation states, and if sterically unhindered, the dithiolene ligand bends toward the metal to increase the orbital overlap between sulfur and metal; and by this, increased electron density is donated from the sulfur to the metal. In the enzymes, this is prevented by the rigidity of the active site pocket, whereas in model complexes, which mimic at least the dithiolene coordination to the metal, such active contribution of the ligand can be actually observed [17–25].

The reactions catalyzed by molybdenum- and tungsten-dependent oxidoreductases are typically referred to as *oxygen atom transfer* (OAT) reactions, that is, the formal transfer of one neutral oxygen atom from a donor to an acceptor, one of which is typically water as the source or sink of the oxygen atom. In the course of this, the oxygen acceptor atom is oxidized by two electrons and the donor atom is reduced by two electrons, respectively, as is the active site metal (see following equations for an oxidase). The regeneration of the active site is achieved by two consecutive proton-coupled electron transfer reactions (PCET).

$$\text{OAT}: \text{enz-M(VI)} = \text{O} + \text{sub} \leftrightarrows \text{enz-M(IV)} + \text{sub} = \text{O}$$

$$1 \cdot \text{PCET}: \text{enz-M(IV)} + \text{H}_2\text{O} + \text{e}^- \leftrightarrows \text{enz-M(V)} + \text{OH}^- + \text{H}^+$$

$$2 \cdot \text{PCET}: \text{enz-M(V)} + \text{OH}^- + \text{e}^- \leftrightarrows \text{enz-M(VI)} = \text{O} + \text{H}^+$$

The majority of enzymatically catalyzed reactions of the molybdenum and tungsten proteins are actually OATs. There is, however, another type of reaction catalyzed by these enzymes, the insertion of the oxygen into a C–H bond, that is, a hydroxylation rather than a typical OAT. With the exception of acetylene hydratase, all enzymatic reactions fall into one of these two categories. The formation of acetaldehyde from acetylene is the only reaction that does not involve an overall reduction or oxidation of the substrate and is a unique example for catalyses processed by the molybdenum- and tungsten-dependent oxidoreductases.

Bioinorganic chemists have developed a plethora of functional and/or structural models for these oxidoreductases, addressing distinct aspects of the enzymes. Aside from the typical OAT or hydroxylation reactions, some of these model compounds are also involved in PCET reactions, which are typically more difficult to model. An all-embracing model, however, which addresses all structural aspects of the molybdenum and tungsten cofactors including the MPT ligand still remains one of the big challenges in the bioinorganic chemistry area.

The focus of this chapter lies in an overview about the natural and artificial molybdenum and tungsten centers with dithiolene coordination or with ligands related to dithiolenes and the not always successful strategic efforts to develop accurate structural models, which are at the same time also functional models. In the following, examples for the various types of model compounds and the reactions, as observed in the natural enzymes and with artificial models, are discussed, giving an overview of the achievements of both nature and chemical scientists in employing molybdenum and tungsten in a sulfur-rich coordination sphere for the oxidation or reduction of substrates in two-electron redox processes.

13.2 Classification of Molybdenum- and Tungsten-Dependent Enzymes

Molybdenum- and tungsten-dependent enzymes catalyze fundamental metabolic reactions involving OAT to or from biological substrates in the nitrogen, sulfur, and carbon cycles. Molybdenum-containing enzymes generally need to be specified because two classes of quite distinct enzymes exist, which do employ molybdenum: the molybdenum nitrogenase enzyme constitutes the first class of molybdenum-bearing proteins. It contains a mixed metal multinuclear iron–molybdenum cofactor (MoCo) [26]. Molybdoenzymes of the second class bear the mononuclear MoCo [27] consisting of molybdenum and MPT plus coordinated small mono- to triatomic ligands (the latter being water) and/or a coordinated amino acid residue. Mononuclear tungsten enzymes are in many ways strikingly analogous to the molybdoenzymes [28] and there is only this specific class of tungsten enzymes known at present. Together they are referred to as the *molybdenum*- and *tungsten-dependent oxidoreductases*.

In terms of function, mononuclear molybdenum and tungsten enzymes can be divided into two categories: hydroxylases, which insert oxygen derived from water into a C–H bond, and oxotransferases, which transfer an oxygen atom to or from a substrate, forming or breaking a substrate oxygen double bond. The two distinct reactions are shown in Equations 13.1 and 13.2, respectively [29, 30].

$$RH + H_2O \rightleftarrows ROH + 2H^+ + 2e^- \quad (13.1)$$

$$R + H_2O \rightleftarrows R = O + 2H^+ + 2e^- \quad (13.2)$$

On the basis of the geometrical and chemical structures of the oxidized active sites and sequence homologies, Hille [27, 31] has classified the whole group of molybdenum- and tungsten-dependent oxidoreductases into certain families.

The molybdenum enzymes fall into three categories: first, the xanthine oxidase (XO) family; second, the sulfite oxidase (SO) family; and third, the dimethylsulfoxide reductase (DMSOR) family.

Despite the fact that considerably fewer tungsten enzymes are known, again three families are presently used for the tungsten-dependent enzymes: first, the aldehyde ferredoxin oxidoreductase (AOR) family; second, the formate dehydrogenase (FDH) family; and third, acetylene hydratase, which is a rather unique enzyme and simply does not fit into one of the other two tungsten enzyme families. The latter had not been discovered at the time of the original enzyme classification by Hille and was added later on.

Each family is named after the most prominent member with the prominence mainly being based on the time of discovery and crystallographic characterization. Since the first classification by Hille, despite a significant increase in the number of crystallographic studies reported for these enzymes [32], only little modification, aside from adding acetylene hydratase, has been required to the original grouping (see Figure 13.2) [33]. In the active sites of all these enzymes, the metal, molybdenum, or tungsten is coordinated to one or two oxo/hydroxo/water ligands and/or sulfide ligands and one or two MPTs [34] via its *cis*-dithiolene moiety (see Figure 13.2).

From spectroscopic studies, it is established beyond doubt that for the vast majority of these enzymes the redox reaction concomitant with OAT from water to product (or vice versa) occurs directly at the metal center in the molybdenum and tungsten oxidoreductases and the metal center cycles between the oxidation states M(VI) and M(IV) [27, 33, 35, 36]. The intermediate M(V) oxidation state is generated by a one-electron reduction of the M(VI) state in the reductases, or by a one-electron oxidation of the M(IV) state in the oxidases, respectively. It occurs via PCET during the catalytic cycle in the regeneration leg of the catalytic cycle and the catalytically active state is reestablished by a second one-electron reduction, or oxidation, respectively [37].

Knowing the exact compositions of the active sites as available from crystal structures for several members of the five original enzyme families (XO, SO, DMSOR, AOR, and FDH) increased our understanding of fundamental structure–function relationships and provided an insight into their specificities and catalytic mechanisms. In the following sections, a more detailed description of these molybdenum- and tungsten-dependent oxidoreductases is given, with an emphasis on crystallographic analyses of the enzymes and their subsequent

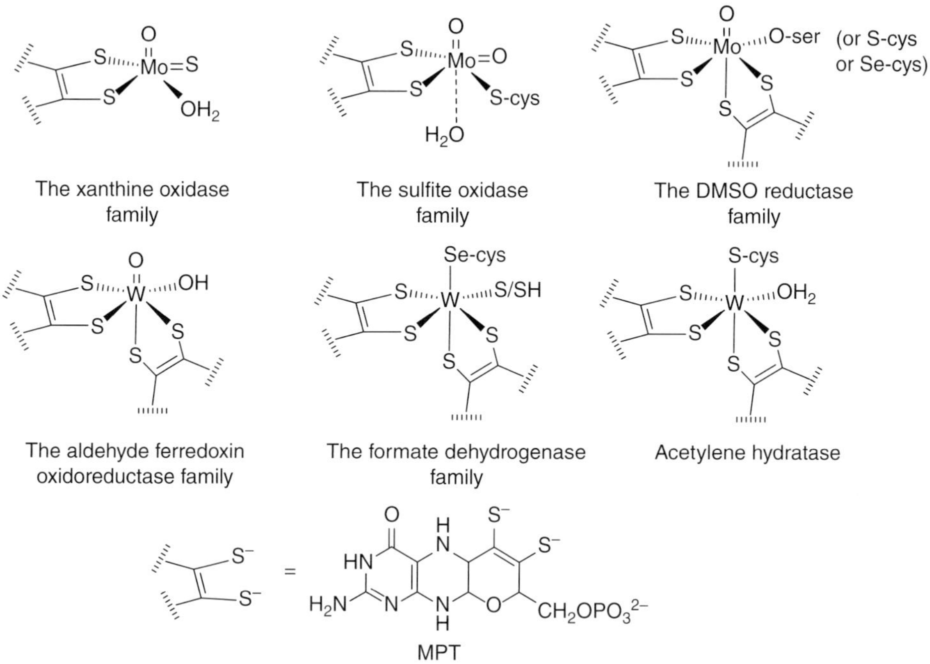

Figure 13.2 The active sites of representatives of molybdenum- and tungsten-dependent oxidoreductase families as derived from crystal structures of individual enzymes. With ser, serine residue and cys, cysteine residue less the coordinating atoms.

mechanistic implications plus selected relevant model complexes and their partly catalytic reactivities.

13.3 Ligand Systems Commonly Used in Model Studies

A number of ligand systems have been designed, synthesized, and modified to mimic at least parts of the MPT molecule in the native active sites of molybdenum and tungsten enzymes. These model ligands can be classified into two groups, namely the non-dithiolene and the dithiolene ligand systems. Among these two classes, the dithiolenes are obviously more proficient in mimicking the electronically flexible unsaturated bis-thiolate bite exhibited by the native MPT ligand. A detailed account of the early developments of general metal dithiolene complex chemistry, which was of significant interest even before the discovery of MPT in the active sites of the oxidoreductases, is available elsewhere in the literature [38]. Most of the early dithiolene work was based on exploring the many interesting aspects of dithiolene ligation such as complex geometry, the peculiar electronic structure, and redox activities. Of the most commonly used dithiolene systems [25, 39–41] in the oxidoreductase model chemistry, we can distinguish two fundamental types: the aliphatic ene-1,2-dithiolate ($R_2C_2S_2$; R = H, Me, Ph, CF_3, CN, COOMe, etc.) type and the aromatic benzene-1,2-dithiolate (bdt) type, the phenyl ring of which may carry in addition to the two thiol functions one or more substituents such as Me, Cl, $SiPh_3$, $NHCOCH_3$, and $NHCOCPh_3$ (Figure 13.3). Both types are typically represented in their fully reduced form (i.e., with a charge of −2), which is for the latter group also the most likely form because the aromatic ring system is electronically rather sturdy. On the basis of symmetry, the dithiolene systems can be further distinguished: there are symmetrically substituted dithiolenes such as $bdtCl_2$, mnt ($S_2C_2(CN)_2$), edt ($S_2C_2H_2$), $S_2C_2Ph_2$, $S_2C_2Me_2$, and so on (Figure 13.4) and unsymmetrically substituted dithiolenes such as sdt, 2-pedt, 4-pedt (Figure 13.5), with the latter being, again, more accurate models of MPT, which itself is unsymmetrically substituted.

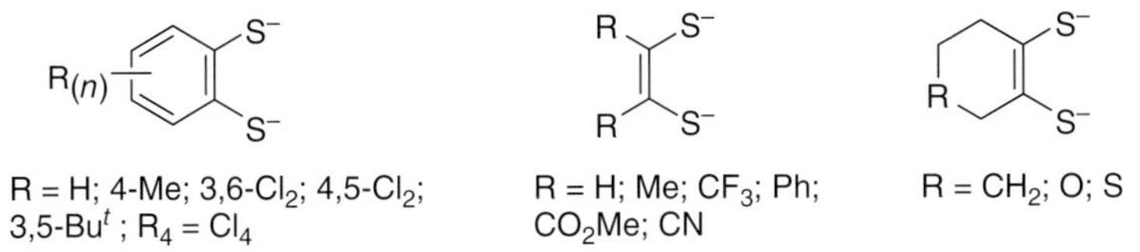

Figure 13.3 Examples of frequently used dithiolene ligand systems. (Adapted from Ref. [42] (© 2011) with permission from Elsevier [42].)

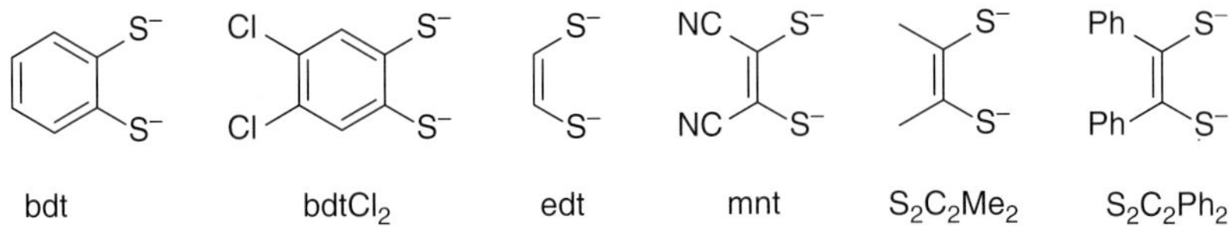

Figure 13.4 Commonly employed symmetrical dithiolene ligands constituting models for MPT with their acronyms or formulae as used throughout the literature.

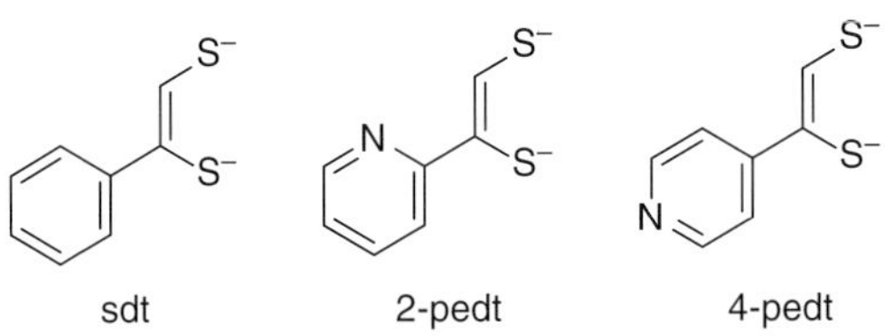

Figure 13.5 Examples for unsymmetrical dithiolene ligands constituting models for MPT with their acronyms.

13.4
Selected Molybdenum-Containing Enzymes and Relevant Modeling Chemistry

13.4.1
Enzymes of the Xanthine Oxidase (XO) Family

XO enzymes catalyze both the oxidation of hypoxanthine to xanthine and the oxidation of xanthine to uric acid (the latter is shown in Equation 13.3), which are the last steps in the purine nucleotide catabolism in human beings as well as in primates, birds, reptiles, and insects. Members of the XO family comprise several enzymes, which bear the LMo(VI)O(S) core at their active sites, where L is the MPT ligand (see Figures 13.1 and 13.2). The reactions catalyzed by XO family enzymes usually involve the cleavage of a C–H bond and the formation of a C–O–H moiety of a diverse range of aldehydes and aromatic heterocyclic compounds, as shown in Equations 13.3 and 13.4 (for aldehydes). Consequently, these enzymes are actually hydroxylases regarding their activity and not oxotransferases. In contrast to other non-molybdenum-/tungsten-dependent biological hydroxylases, the oxygen incorporated in the product is derived from water and not from dioxygen [43].

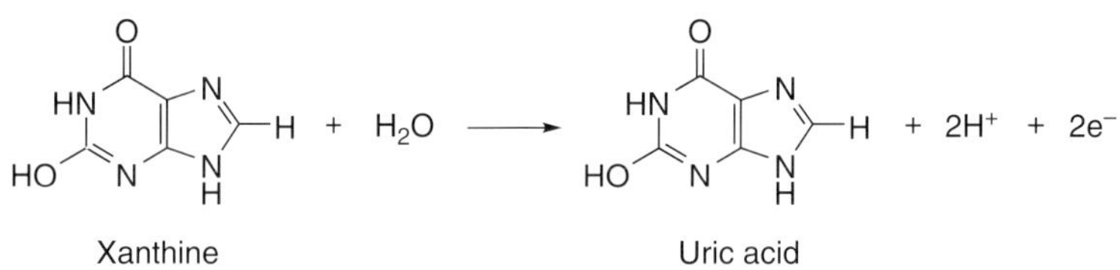

(13.3)

$$\text{R-C(O)H} + H_2O \rightarrow \text{R-C(O)OH} + 2H^+ + 2e^- \qquad (13.4)$$

13.4.1.1 Crystal Structures and Active Site Composition of XO-Related Enzymes

Before crystal structures became available, many attempts were made to explore the coordination environment of molybdenum in the XO family of enzymes, particularly making use of X-ray absorption spectroscopy techniques [44–47]. The first reported crystallographic analysis for a molybdenum hydroxylase was of the aldehyde oxidoreductase from *Desulfovibrio gigas* [48, 49]. Currently, however, XO from bovine milk [50] is the best-studied enzyme, its various structures providing in-depth metric information for this family of enzymes as well as fine geometrical details for the active sites. Bovine milk XO catalyzes, as the name says, the oxidation of xanthine to uric acid, which is the last transformation in purine degradation before excretion.

The crystallographic studies suggested an overall square-pyramidal geometry around the molybdenum center with no direct coordination to the protein (see Figure 13.6) and a $Mo(VI)O_2(OH_2)(MPT)$ composition of the active site. The dithiolene ring system is part of the equatorial S_3O plane contributing two sulfur ligands, while the catalytically essential sulfide ligand occupies another equatorial

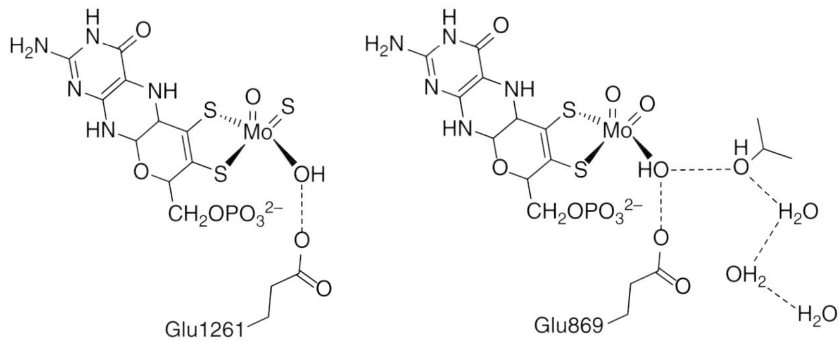

Figure 13.6 The chemical molybdenum active site structures of bovine milk xanthine oxidase [Mo, =O, =S, –OH, MPT] with the conserved glutamate 1261 [52] (a) and of the *D. gigas* aldehyde oxidoreductase [Mo, =O, =O, –OH, MPT] with the conserved glutamate 869 and the inhibitor isopropanol (b); the latter being held by hydrogen bonds to the OH ligand and to a chain of three buried water molecules [53].

position. The fivefold coordination is completed by an apical oxo group and an equatorial hydroxyl group. This OH ligand points toward the funnel that leads into the active site and is most likely the source of oxygen transferred onto the substrate [51]. The hydroxyl ligand is regenerated by water coordination and deprotonation completing the catalytic cycle. In the active site, a highly conserved glutamate (Glu1261 in bovine XO) amino acid residue establishes a hydrogen bond to the hydroxyl ligand and is most likely actively involved in the catalytic cycle being only 3.5 Å away from the metal. This general active site geometry is conserved in all members of the XO family.

13.4.1.2 **Mechanistic Implications**

On the basis of the structural studies, a mechanism has been proposed [48, 49, 54] for the reductive half-reaction of the OAT, catalyzed by XO-related enzymes (see Scheme 13.1). In the following, the oxidation of an aldehyde is used as an example. Fundamental aspects which the mechanism is based on are the following: (i) the conserved adjacent glutamate (Glu869 in case of *D. gigas* AOR) activates a coordinated water molecule that represents the labile oxygen ultimately transferred onto the carbonyl carbon of the substrate. This nucleophilic attack is facilitated by hydrogen-bonding interactions of the carbonyl oxygen with the water ligand and with a further water molecule, present in the active site cavity; (ii) the resulting carboxylic acid product, generated after hydride transfer to the sulfido group (which is unique and crucial for XO enzymes), replaces the coordinated water at the molybdenum; and (iii) the product is released from the reduced molybdenum center facilitated by the transient binding of the proximal Glu869 to the Mo(IV) center. Finally, (iv) a water molecule, present in the active site cavity, reoccupies the vacant coordination position, prompting release of the Glu869 amino acid residue.

An interesting and fascinating aspect of this mechanism is the proposal that a water ligand of molybdenum is transferred to the substrate, while a hydride (!) is transferred from the substrate to the enzyme. This mechanism differs from the

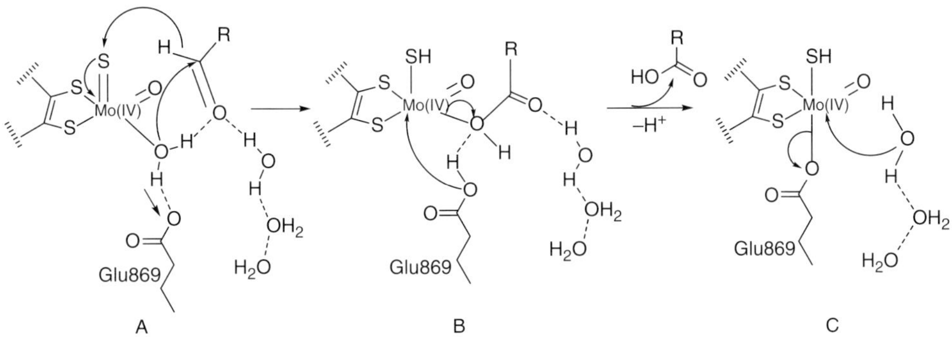

Scheme 13.1 Proposed mechanism for the reductive half-cycle of the oxygen atom transfer, catalyzed by an aldehyde oxidase as a member of the xanthine oxidase family: A, the Michaelis complex with an aldehyde substrate close to Mo(VI); B, the enzyme/carboxylic acid product complex (Mo(IV)); and C, the intermediate, after product dissociation, with Glu869 bound to the metal [55]. Curved arrows represent the movement of electron pairs toward the target atom. (Adapted from Ref. [55], (© 1997), with permission from Elsevier.)

general mechanism often proposed for MoCo-bearing enzymes, such as DMSOR, where an oxo group represents the oxygen species that is translocated between enzyme and substrate.

13.4.1.3 **Model Complexes for the Xanthine Oxidase Family**

The model chemistry for this family of enzymes is still in its infant stage and very few relevant model complexes for XO have been prepared, as the *cis*-Mo(VI)OS moiety with additional dithiolene chelation is a very uncommon structural motive in synthetic inorganic chemistry [56]. In the enzymes, the Mo(VI)OS center is believed to be converted to a Mo(VI)O(SH) species by hydride transfer from the substrate to the sulfido group and this poses considerable additional synthetic challenges [52]. Achieving a fundamental understanding of the chemical role of the terminal sulfide ligand in the catalytic turnover is still a significant issue for bioinorganic chemists working in this field. Some early examples of rather simple synthetic analogs for the XO family of enzymes include oxothiomolybdates, $[MoO_{4-n}S_n]^{2-}$ [57, 58], hydroxylamido complexes, $MoOS(ONR_2)_2$ [59, 60], and also organometallic derivatives as $Cp^*MoOS(CH_2SiMe_3)$ [61]. These pseudotetrahedral complexes are generally synthesized by sulfurization of dioxo-molybdenum(VI) analogs, most commonly using hydrogen sulfide. One of the first truly comparative spectroscopic studies of model complexes and different enzymes of the XO family was carried out by Wedd and coworkers [62]. Model complexes used in this work include $[MoOXL]^-$ and $[MoO(XH)L]$ where X = O, S and LH_2 = *N*,*N*′-dimethyl-*N*,*N*′-bis(2-mercaptophenyl)-1,2-diaminoethane (Figure 13.7). The electron paramagnetic resonance (EPR) signals of singly reduced milk XOs differ depending on the methods used for reducing the as-isolated fully oxidized species to the Mo(V) species. The different Mo(V) active sites are called *rapid type 1*, *rapid type 2*, and *slow centers of milk XO* referring

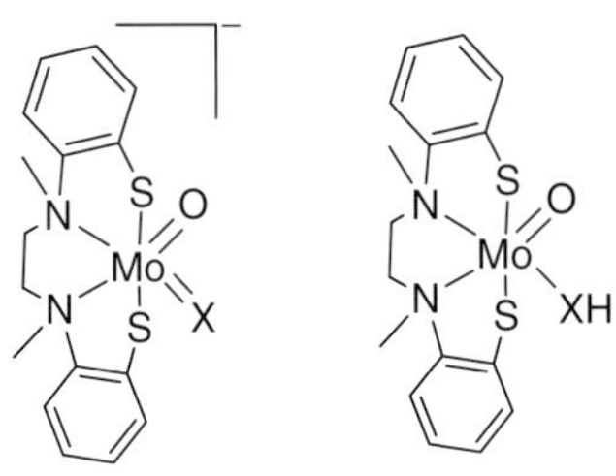

[MoOXL]$^-$ [MoO(XH)L]

Figure 13.7 Chemical structures of model complexes [MoOXL]$^-$ and [MoO(XH)L] as synthesized and spectroscopically investigated by Wedd and coworkers [62].

to the specifically observed EPR signals. These distinct enzyme EPR signals were compared to those of the synthetic species [MoOXL]$^-$ and [MoO(XH)L] (X = O, S) by definition of the ^{95}Mo hyperfine matrices from multifrequency EPR spectra. It could eventually be established that the rapid type 1 signal corresponds to a [Mo(V)O(SH)] core, the rapid type 2 signal to a [Mo(V)O(SH)(OH)] core, and the slow signal to a [Mo(V)O(OH)] core.

Holm and coworkers were also successful in developing unperturbed *cis*-Mo(VI)OS moieties, by converting simple molybdate-derived $[MoO_3S]^{2-}$ into $[MoOS(OSiPh_3)_2L]$ species (L = phen or bpy derivatives) by successive reactions with Ph_3SiCl/Et_3N and L [63]. Although spectroscopically and structurally quite important, unfortunately none of these structural models for the XO "MoOS" core mimic the chemical behavior of the enzymes, that is, rather than OAT (not to mention hydroxylation), *sulfur* atom transfer (SAT) was observed (see Scheme 13.2). This observation was attributed to the bond energy of the Mo=S bond being smaller than that of the Mo=O bond.

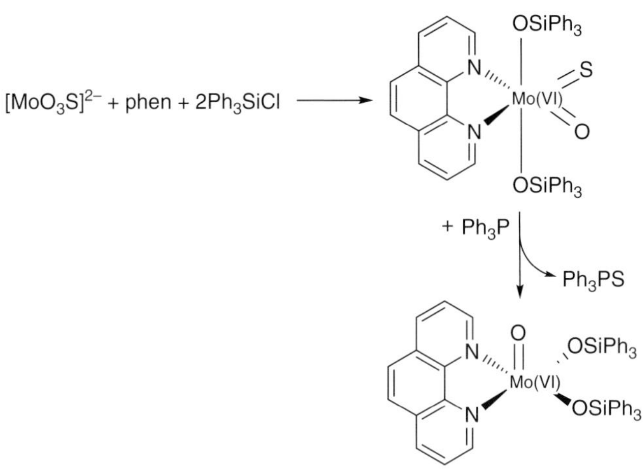

Scheme 13.2 Synthesis and reactivity of $[MoOS(OSiPh_3)_2(L)]$ model complexes with the phen complex as an example.

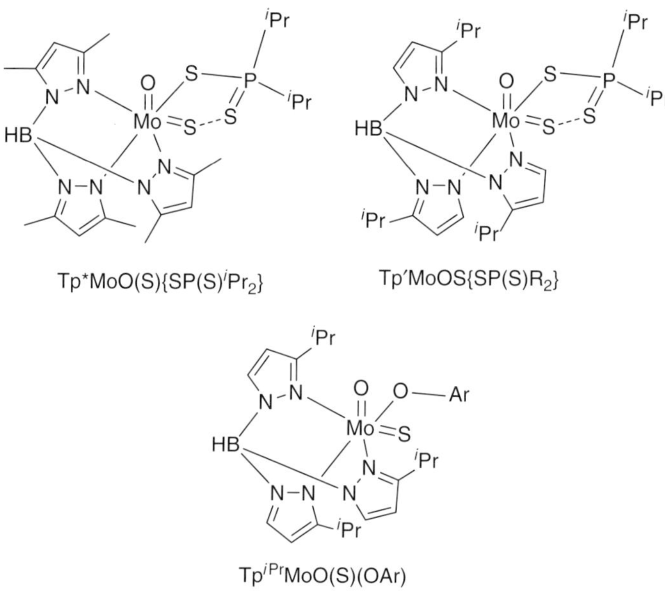

Figure 13.8 Chemical structures of model complexes [Tp*MoOS(SP(S)Pr^i_2)], [Tp′MoOS(SP(S)R_2)], and [Tp^{iPr}MoOS(OAr)].

More recently, the synthetic model chemistry for enzymes of the XO family has been enriched by the group of Young by exploiting OAT (Equation 13.5) and SAT (Equation 13.6) reactions for the preparation of mononuclear oxo-sulfido-molybdenum(VI) complexes such as [Tp*MoOS(SP(S)Pr^i_2)] (with Tp* = hydrotris-(3,5-dimethylpyrazol-1-yl)borate), [Tp′MoOS(SP(S)R_2)] (with Tp′ = hydrobis-(3-isopropylpyrazolyl-1-yl)(5-isopropylpyrazolyl-1-yl)borate; R = Pr^i, Ph), and [Tp^{iPr}MoOS(OAr)] (with Tp^{iPr} = hydrotris(3-isopropylpyrazol-1-yl)borate; OAr = phenolate) (Figure 13.8) [64, 65].

$$[Mo(IV)S]^{2+} + [O] \rightarrow [Mo(VI)OS]^{2+} \quad (13.5)$$

$$[Mo(IV)O]^{2+} + [S] \rightarrow [Mo(VI)OS]^{2+} \quad (13.6)$$

As a quite striking feature, in these sixfold coordinated complexes, the sulfido-molybdenum(VI) moiety is stabilized by an intramolecular Mo(O)S· · ·S=P interaction involving the monodentate dithiophosphinate ligand, which generates a five-membered ring. This interaction, weak as it may be, actually lengthens the Mo=S bond slightly. Interestingly, this weak S· · ·S interplay prevents the reduction of molybdenum and simultaneous oxidation of sulfur followed by a dimerization leading to a Mo(V)O(S–S)Mo(V)O species. A dimerization is consequently the fate of the third of these model complexes [Tp^{iPr}MoOS(OAr)], which does not bear a suitable stabilizing P=S moiety and is therefore rather unstable.

Such unwanted reactivity is the major obstacle in the Mo(VI)OS model chemistry and the reason why only very few structural, let alone structural and functional, models for the XO family of enzymes are available at present. Functional hydroxylation "models" known currently are rather multinuclear (polyoxometallates) or polymeric materials with molybdenum sites grafted onto them [66, 67], that is, they do lack any structural resemblance to the active sites of the enzymes. Furthermore, in contrast to the mechanism proposed for XO shown in Scheme 13.1, the oxygen being delivered to the substrate is in molybdenum oxidation catalysis typically an oxo ligand and not a hydroxide. Developing a mononuclear model compound with a structural similarity to the MoCo of the XO family enzymes and the ability to catalyze the insertion of an (hydroxide-) oxygen atom into a C–H bond remains one of the challenges for chemists working in this field and is probably not to be achieved any time soon.

13.4.2
The Sulfite Oxidase (SO) Family

The enzymes of the SO family, including the assimilatory nitrate reductases, bear a [MoO_2] unit in contrast to the [MoOS] core found in the XO family (see Figure 13.2). SO is mainly found in eukaryotes and is located in the mitochondrial intermembrane space where it catalyzes the physiologically vital oxidation of sulfite to sulfate (Equation 13.7), which is the terminal reaction in the oxidative degradation of the sulfur-containing amino acids cysteine and methionine.

Another member of this enzyme family, nitrate reductase, is actually catalyzing the electronically reverse type of reaction (Equation 13.8), accomplishing the reduction of nitrate to nitrite with an oxidation of the active site molybdenum from Mo(IV) to Mo(VI) in the reductive half-reaction.

$$SO_3^{2-} + H_2O \rightleftarrows SO_4^{2-} + 2H^+ + 2e^- \tag{13.7}$$

$$NO_3^- + 2e^- + 2H^+ \rightleftarrows NO_2^- + H_2O \tag{13.8}$$

13.4.2.1 Crystal Structures and Active Site Composition of SO-Family Enzymes

From a crystal structure of SO from chicken liver [68] it was derived that the molybdenum atom is fivefold coordinated by one oxo group at a distance of 1.75 Å, two sulfur atoms at 2.4 Å (from the MPT dithiolene moiety), one sulfur atom from a cysteine at 2.5 Å, and one water or hydroxo ligand at 2.2 Å in a pseudo-square-pyramidal coordination geometry. By EXAFS (extended X-ray absorption fine structure) experiments with the oxidized form of the enzyme [44, 69, 70], the coordination sphere around the molybdenum was shown to contain two oxygen atoms at a distance of 1.68 Å (which suggests that these are both oxo ligands) and three sulfur atoms at 2.41 Å. In the reduced form, however, three sulfur ligands (2.38 Å) and only one oxo ligand (1.69 Å) were found to be coordinated to molybdenum. This implies that the crystal structure of SO from chicken liver was determined in its reduced form even though the enzyme had been isolated in its oxidized form. Such a reduction may occur in enzymes when exposed to high energy radiation such as X-rays. For assimilatory nitrate reductase, EXAFS results revealed that in the oxidized state the molybdenum center is coordinated to two terminal oxygen atoms at 1.71 Å and two to three sulfur atoms at 2.44 Å, whereas a single terminal oxygen at 1.67 Å and a set of sulfurs at 2.37 Å were found in its fully reduced form.

Carbon monoxide dehydrogenase is another rather unusual member of this family. It has puzzled experimental and theoretical researchers for some time and it still does. Overall, its structure is similar to that of other enzymes of this family, although it carries an extraordinary S–Cu moiety inserted into the molybdenum cysteine interaction, which is normally a direct Mo–S bond in SO [71]. This copper was not identified as such in the first X-ray structure of this enzyme, because it seemed just not feasible that such a structural motif was possible. A selenium atom was refined instead because selenium in the active site of an oxidoreductase is not that unusual. This Mo–S–Cu motive makes this enzyme's active site very difficult to model and structural–functional investigations in this regard are mostly of a theoretical nature [72].

13.4.2.2 Mechanistic Implications

In the reductive half-cycle of the reaction catalyzed by SO, molybdenum is reduced from the Mo(VI) to the Mo(IV) state, coupled to the oxidation of sulfite to sulfate. According to the overall mechanism (Scheme 13.3), originally (and quite early) proposed by Hille [73–75], the initial binding of sulfite at the active site is most likely a nucleophilic attack of the lone pair of sulfite to the electrophilic equatorial Mo=O moiety, generating a transient two-electron reduced oxo-Mo(IV) sulfate species (B). Water then displaces sulfate to complete the OAT reaction, generating species C. Subsequently, the reduction of 2 equiv of

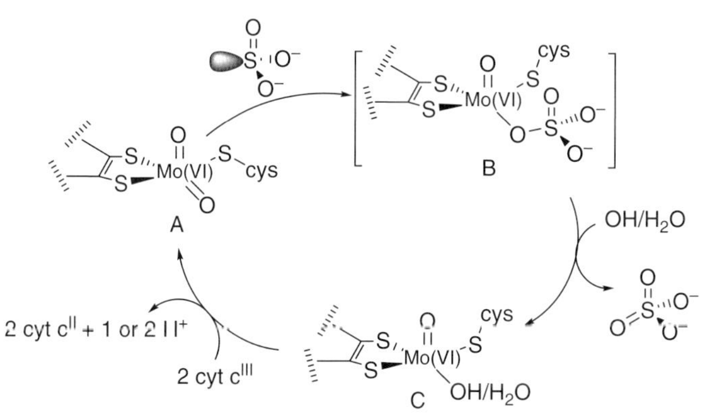

Scheme 13.3 Proposed reaction mechanism for the conversion of sulfite to sulfate by sulfite oxidase. The reaction mechanism shows the formation of the Mo(IV)-product bound intermediate followed by displacement of product by solvent and transfer of electrons to the physiological electron acceptor, cytochrome c, through the b5-type heme.

cytochrome c and loss of two protons follows in the process, which returns the active site to its original state (A) wherein the oxygen atom incorporated into the substrate is ultimately derived from water.

13.4.2.3 Model Complexes for the Sulfite Oxidase Family

The synthetic requirement for accurately modeling an enzyme of the SO family regarding its structure is to combine a Mo(VI)O_2 core with three thiolate donors, two of which are part of a dithiolene ligand and the remaining an analog of cysteine. This is complicated by the fact that monodithiolene coordination is quite uncommon in dithiolene complex chemistry (with the exception of organometallic compounds) [42, 56]. In addition, one should remember that the three thiolate donors in SO are lying in the equatorial plane. The five atoms around the metal constitute a square-pyramidal geometry with one oxo ligand at the apical position.

A large number of synthetic analogs for the SO family of enzymes are reported in the literature. Many of them have been structurally characterized and screened for their catalytic oxo-transfer properties. A majority, in particular, of the earlier model complexes for the SO family of enzymes, failed to mimic the coordination environment with three thiolate and two terminal oxo groups in their oxidized form, which is quite a challenging composition. Making matters worse, the reaction of the reduced and oxidized forms (Mo(VI)O_2/Mo(IV)O) of the OAT reactive molybdenum core with each other (see Equation 13.9) by comproportionation commonly generates a dimeric $Mo(V)_2O_3$ moiety, which is catalytically dead. The formation of this inactive species obstructs the catalytic cycle for good, unless it is in equilibrium with the monomeric reduced and oxidized species, in which case the catalytic activity is still dramatically decreased.

$$\mathrm{Mo(IV)O + Mo(VI)O_2 \rightleftarrows Mo(V)_2O_3} \quad (13.9)$$

In an attempt to prevent this formation of the $Mo(V)_2O_3$ core by dimerization, the groups of (predominantly) Holm and Cervilla [76–80] designed several model complexes such as $Mo(VI)O_2(L\text{-}NS_2)$, $[NH_4]_2[Mo(VI)O_2(O_2CC(S)Ph_2)_2]\cdot 2H_2O$, and $Mo(VI)O_2({}^t BuL\text{-}NS)_2$ (Figure 13.9a–c) using sterically bulky ligands. The aim was to minimize or prohibit a direct contact between the two metal centers appearing at both ends of the catalytic turnover. The Mo=O bonds in all complexes are sterically shielded by bulky Ph/Ph-p-^{t}Bu groups in the direction of a potential Mo–O–Mo bond in order to prevent dimerization. All these complexes efficiently catalyzed the OAT from dimethylsulfoxide (DMSO) onto the well-known proxy substrate PPh_3 forming dimethyl sulfide (DMS) and PPh_3O. This reaction, which does not take place without a catalyst at ambient temperature, is the classical test reaction for OAT catalytic ability of complexes developed by Berg and Holm [77]. In contrast to the natural processes, both the oxidation and the reduction of the molybdenum center proceed via OAT. However, it has to be noted that phosphane, based on the significant oxo-philicity of the phosphorus atom, is comparably easy to oxidize. For substrates such as sulfite, additional mediators would be required. Consequently, most model studies regarding reactivity or catalytic potential use nonnatural substrates for the oxidation reactions.

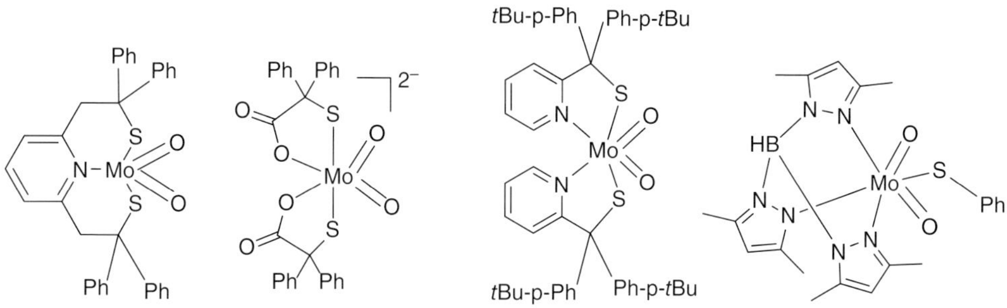

$Mo(VI)O_2(L\text{-}NS_2)$ (a) $[Mo(VI)O_2(O_2CC(S)Ph_2)_2]^{2-}$ (b) $Mo(VI)O_2(tBuL\text{-}NS)_2$ (c) $LMo(VI)O_2(SPh)$ (d)

Figure 13.9 (a–d) Functional model complexes for the SO family of molybdenum enzymes with sterically bulky ligands minimizing or preventing unwanted dimerization of reduced and fully oxidized species.

A structurally much simpler model complex $[MoO_2(mnt)_2]^{2-}$ ($mnt^{2-} = 1,2$-dicyanoethylenedithiolate) synthesized by Sarkar and coworkers [81, 82] also showed OAT catalytic activity by oxidizing hydrogensulfite to hydrogensulfate, in an almost nature-like transformation. Interestingly, this oxidation reaction exhibits "enzyme-like" kinetics, that is, a Michaelis–Menten type of kinetic behavior was observed as shown in Scheme 13.4. On the basis of this observation, it had been contemplated if hydrogensulfite was not the actual substrate for SOs. Experimental and theoretical evidence, however, strongly points to sulfite as the substrate, just as initially proposed [83, 84].

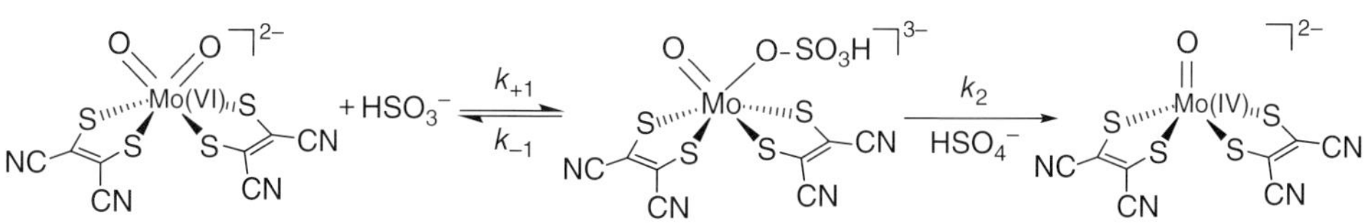

Scheme 13.4 "Enzyme-like" kinetic behavior of model complex $[MoO_2(mnt)_2]^{2-}$ in the oxidation of hydrogensulfite to hydrogensulfate.

It was further established that molecules structurally related to sulfite such as sulfate, dihydrogenphosphate, and dihydrogenphosphite coordinate the molybdenum center competitively and, by this, prevent the oxo transfer to sulfite. This is a rare example of competitive binding inhibition of a model complex similar to processes taking place at an enzyme's active site [85].

Not surprisingly, the same complex ($[MoO_2(mnt)_2]^{2-}$) also exhibited oxo-transfer activity toward phosphanes [86]. By changing the substituents on phosphorus, it was shown that the steric bulkiness of the substituents does not significantly influence the rate of OAT. It is rather the basicity of the substrate that determines the rate ($PEt_3 > PEt_2Ph > PEtPh_2 > PPh_3$). This is consistent with a binding of the incoming substrate with the oxo ligand rather than the molybdenum center.

Aside from the transformation of the natural substrate sulfite, the second challenging reactivity of the active sites is the regeneration of the enzymes' cofactors through two consecutive PCETs, which was first achieved with a model complex in 1992 [87]. The transformation of a Mo(IV)O core to the Mo(VI)O_2 core by two distinguished one-electron transfer steps, accurately mimicking the natural system, was observed with the model complex LMo(VI)O_2(SPh) (where L = hydrotris(3,5-dimethyl-1-pyrazolyl)borate; see Figure 13.9d) [87]. The sterically bulky facial tridentate pyrazolylborate ligand can inhibit the dimerization of reduced and fully oxidized species and restricts all chemistry of the complex to the three remaining facial sites. In dimethyl formamide (DMF) or MeCN, oxygen atom abstraction by PPh_3 from LMo(VI)O_2(SPh) leads to the formation of LMo(IV)O(SPh) or weakly solvated LMo(IV)O(SPh)(solvent). These can be trapped as mononuclear species in pyridine as LMo(IV)O(SPh)(py) or after oxidation, as mononuclear LMo(V)OCl(SPh) in CH_2Cl_2. In dry toluene, in contrast to its behavior in dichloromethane and pyridine, LMo(IV)O(SPh) forms the unwanted and catalytically dead dinuclear [LMo(V)O(SPh)]$_2$O species (see Equation 13.10).

$$\text{LMo(IV)O(SPh)} + \text{LMo(VI)O}_2\text{(SPh)} \rightleftarrows [\text{LMo(V)O(SPh)}]_2\text{O} \qquad (13.10)$$

The LMo(VI)O_2(SPh) complex is a catalyst for the oxidation of PPh_3 to PPh_3O and, consequently, addition of DMSO to the reduced complex LMo(IV)O(SPh) regenerates the catalytically active site LMo(VI)O_2(SPh) as does the presence of water and oxygen in toluene solution. The complex reaction behavior and the interchange of species are summarized in Scheme 13.5.

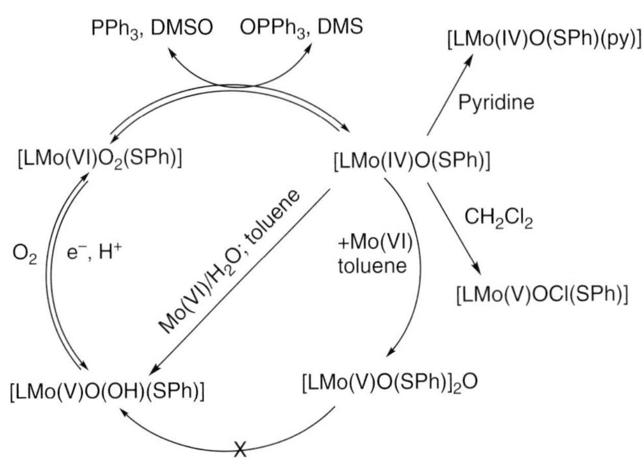

Scheme 13.5 Scheme summarizing the observed reactions for the [LMo(VI)O_2(SPh)] complex.

Another strategy to avoid dimerization through comproportionation is the immobilization of the molybdenum centers on a polymer, which is able to separate efficiently the individual reaction sites [88]. These are consequently also more challenging to characterize and it is not clear beyond doubt what actually goes on at the metal centers.

The above-discussed model complexes show to some extent the accurately mimicking catalytic activity for the SO enzymes but are structurally more or less distinct from the natural active sites. None bears exactly one dithiolene ligand if a dithiolene ligand is present at all. The coordination chemistry of Mo(VI) typically demands a sixfold coordination for the sake of stability. Consequently, for a $Mo(VI)O_2$ core, having two positions occupied by two terminal oxo ligands, the remaining four positions prefer being occupied by two dithiolene ligands, which is the most common coordination motive found throughout the literature about respective dithiolene complexes. For the oxidation state Mo(IV), it is similarly the combination of two dithiolene ligands with one oxo ligand. These general chemical problems were successfully tackled by Holm and coworkers [89] by two very different synthetic strategies. The first method leads to the formation of monodithiolene Mo(V) complexes analogous to the high pH form of SO by starting off with the bis-dithiolene complex, which is shown in Scheme 13.6. The resulting complex $[MoOCl(SC_6H_2\text{-}2,4,6\text{-}Pr^i{}_3)(bdt)]^{1-}$ shows still the closest structural similarity to the active site of SO in its high pH form with one oxo, one thiol, one dithiolene, and the chloro ligand, which is in place of the natural hydroxide ligand. The second synthetic method (Scheme 13.7) results in an excellent model for the oxidized form of SO with both oxo ligands being present, a thiolate in place of cysteinate and benzenedithiolene mimicking MPT. Disappointingly, this complex was unable to show any OAT reactivity, neither with the biological substrate (SO_3^{2-}/HSO_3^-) nor with the well-known proxy substrate PPh_3.

For the unusual enzyme CO dehydrogenase, only few structural models are reported currently, all of which possess a moiety in which a molybdenum or tungsten atom is bridged via two sulfide ligands to a copper atom [90, 91]. The coordination sphere of Mo/W is completed by an oxo ligand and benzenedithiolene, whereas copper is further coordinated by a bulky thiophenolate for instance. Catalytic reactivity has not been reported for those two models; neither was there any catalytic transformation of CO to CO_2 by molybdenum or tungsten compounds.

In view of the models discussed, it appears that at the moment the SO family model chemistry needs to choose between functional and structural mimicry of the natural enzymes. Combining both remains one of the big challenges in this field.

13.4.3 The DMSO Reductase (DMSOR) Family

The DMSOR family of molybdenum (and tungsten) enzymes is a considerably larger and more diverse group of enzymes compared to the two previously discussed ones but restricted to the prokaryotic regime. The enzymes of this

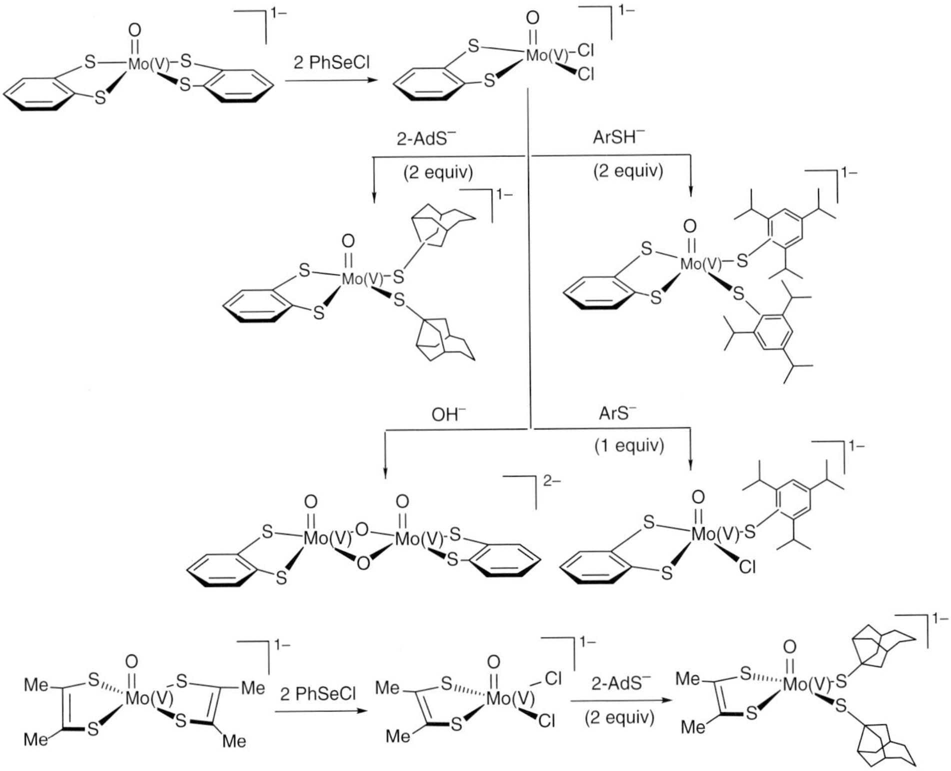

Scheme 13.6 Synthesis of monodithiolene Mo(V) complexes from bis-dithiolene complexes by Holm and coworkers [89]. (Adapted from Ref. [92], (© 2011), with permission from Elsevier [92].)

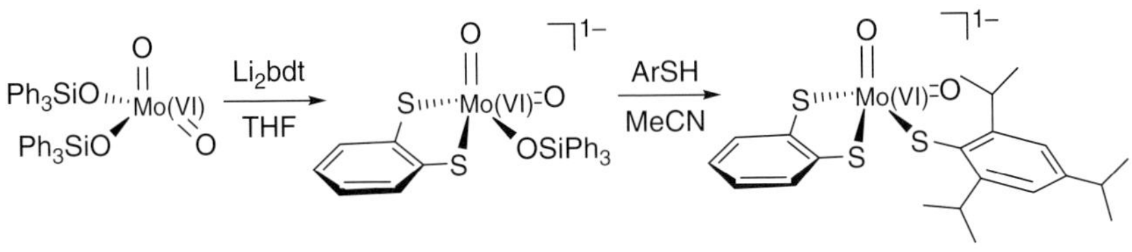

Scheme 13.7 Synthesis of the structurally most accurate model for the oxidized active site of sulfite oxidase. (Adapted from Ref. [92], (© 2011), with permission from Elsevier [92].)

family bear a $L_2Mo(VI)(O)R$ core in the oxidized state and a $L_2Mo(IV)(R)$ core in the reduced state (with L = MPT and R = a ligand most often contributed by the polypeptide). The family name giving enzymes catalyze the reduction of DMSO to DMS, as shown in Equation 13.11.

$$\text{DMSOR-Mo(IV)} + (CH_3)_2S = O + 2H^+ \rightleftarrows \text{DMSOR-Mo(VI)}(CH_3)_2S + H_2O \tag{13.11}$$

Considering its diversity, the DMSOR family has been further divided into three subfamilies [32, 93, 94]:

- *Type I*: dissimilatory nitrate reductases, FDH
- *Type II*: respiratory nitrate reductases, ethylbenzene dehydrogenase (EBDH)
- *Type III*: DMSOR and trimethylamine *N*-oxide (TMAO) reductase.

13.4.3.1 Crystal Structures and Active Site Composition of the DMSOR Family Enzymes

Currently, the crystal structures of several enzymes of the DMSOR family are known. At the active site, the metal atom is coordinated by two molybdopterin guanine dinucleotide moieties (MGD, the guanine being bound by a phosphate ester bond to MPT), and different additional ligands (oxo or sulfido plus an amino acid residue) as shown in Figure 13.2. The amino acid coordinated to the molybdenum atom varies throughout the several enzymes of the family: for instance, cysteine-thiolate coordination in periplasmic dissimilatory nitrate reductase and acetylene hydratase, selenocysteine-selenolate coordination in FDH, and serine-alcoholate coordination in DMSOR [95]. The exceptional enzyme arsenite oxidase is not covalently bound to the peptide with one oxo and two MPT ligands only. The known FDHs are capable of using either molybdenum or tungsten at their active sites. Aside from these two unusual enzymes, all others bear cofactors with very similar composition.

The first crystal structures reported for members of this family were of enzymes from *Rhodobacter sphaeroides* [96] and from *Rhodobacter capsulatus* [97]. Even at present, the structures of the oxidized and reduced forms of these DMSORs are not understood in all detail, as the crystal structure determination results conflict with results obtained from EXAFS studies. According to the quite old crystal structure of *R. sphaeroides*, one of the MGDs exhibit coordination modes to the molybdenum in the oxidized and in the reduced state, which are distinct. In the oxidized state, three of the dithiolene sulfurs are at a shorter distance and the fourth one at a longer but still bonding distance. In the reduced structure, one MPT remains tightly bound, while the other apparently has one weakly coordinated and one non-coordinated sulfur. In the oxidized form, there is also an oxo group and an oxygen from serine giving a six-coordinate molybdenum-mono-oxo active site, whereas in the reduced form the serine remains coordinated and the oxo group has been removed giving a molybdenum des-oxo center (see Figure 13.10a).

In contrast to these observations, only one of the MPTs is coordinated in the structure of DMSOR from *R. capsulatus* as determined by Schneider *et al.* [97], whereas the other MGD is present but not coordinated (Figure 13.10b). The coordination sphere of the oxidized molybdenum further includes two oxo ligands, and the oxygen from a serine residue. In other X-ray structures [98, 99] of the *R. capsulatus* enzyme, both MPTs were found to be coordinated, even upon dithionite or DMS reduction (Figure 13.10c). Here, the molybdenum is also coordinated by two oxo ligands, and the oxygen from a serine residue in the oxidized form, generating an unusual sevenfold coordination sphere around Mo(VI).

In the crystal structure of oxidized trimethylamine *N*-oxide reductase (TMAOR) from *Shewanella massilia*, molybdenum was found to be ligated by four sulfur atoms from the two MGDs, two oxo groups, and the oxygen of Ser149

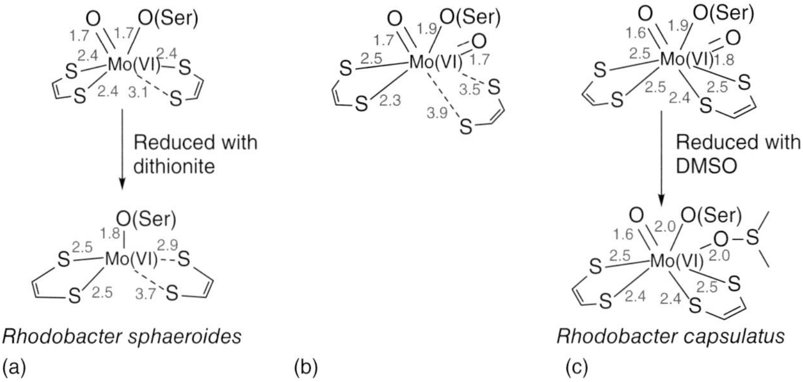

Figure 13.10 (a–c) Active site structures and bond lengths in angstrom for DMSO reductase enzymes. (Adapted from Ref. [30], (© 2000), with permission from Elsevier [30].)

(serine) which would similarly constitute an unfavorable sevenfold coordination [100]. A re-examination of these crystal structure as interpreted in their original publications, by Mo–K edge EXAFS experiments of the molybdenum site of recombinant *Escherichia coli* TMAOR offers a different picture [101], revealing the presence of four thiolate ligands at a rather close distance of 2.43 Å, one terminal oxygen at 1.71 Å, and one more oxygen ligand at 1.83 Å. This is similar to the crystal structure of the oxidized form of DMSOR from *R. sphaeroides*, only with a longer and much more likely length for the serine oxygen to molybdenum bond.

13.4.3.2 **Mechanistic Implications**

In the protein crystal structure of DMSOR, a large funnel-like cavity (25 Å deep) leads from the protein surface to the center of the protein, which bears the cofactor and most likely forms the entrance for substrates into the active site. The reduced form of the enzyme in comparison provides more space for the oxygen atom of a DMSO molecule to bind with the Mo(IV) center, which significantly weakens the S=O bond. A subsequent shift of the sulfur atoms in the MPT toward molybdenum triggers the liberation of the reaction product DMS by a combination of electronic and steric factors. Molybdenum is now left in its Mo(VI) oxidation state by establishing the Mo=O bond and the oxidative half-cycle is completed. In the reductive half-reaction, the subsequent transfer of two electrons from a cytochrome and two protons liberates a molecule of water, removing this oxygen atom from the active site molybdenum. This brings the molybdenum center back to its active des-oxo Mo(IV) state. The catalytic cycle of the DMSOR, as proposed in accordance with most of the experimental evidence [102–105], is illustrated in Scheme 13.8. It involves the breaking of a molybdenum–sulfur bond of one MPT ligand as indicated by the observation of at least one unusually long Mo–S distance in the crystal structures.

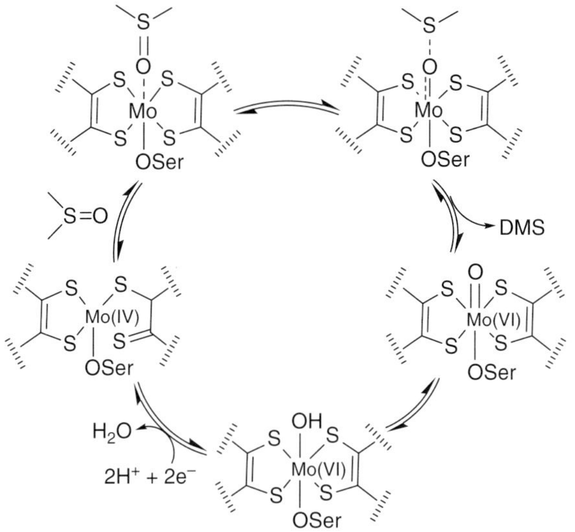

Scheme 13.8 Proposed catalytic cycle for the DMSO reductase. The generated water is initially bound to molybdenum and then replaced by the incoming substrate DMSO.

13.4.3.3 Model Complexes for the DMSOR Family

Bearing in mind the active site structures of the large majority of the DMSOR family enzymes, it is clear that mono-oxo, rather than di-oxo, sixfold coordinated Mo(VI) compounds are required when modeling enzymes of this family but such complexes are rarely reported. The most frequently reported form in the early synthetic model chemistry was the *cis*-di-oxo moiety $Mo(VI)O_2$ [106]. At the same time, of these early mono-oxo Mo(VI) compounds, a large number was even sevenfold coordinated [107–109]. In 1988, Mondal *et al.* reported complexes of the type [MoO(cat)(Sap)] (with cat, catecholate (Cat^{2-}), naphthalene-2,3-diolate ($Naphcat^{2-}$), or 3,5-di-*tert*-butylcatecholate ($DTBcat^{2-}$), and Sap, the Schiff base dianion *N*-salicylidene-2-aminophenolate). These are the first examples of mononuclear Mo(VI)O centers with a sixfold coordination (Figure 13.11) [110]. Following this, several mono-oxo Mo(VI) coordination complexes were reported and some of them were successfully screened for understanding the redox potentials and oxo-transfer properties [111–115]. Missing in these complexes, however, is an MPT-like dithiolene coordination. In fact, even sulfur ligands are rarely present.

Before the publication of the crystal structure of *R. sphaeroides* DMSOR, Wedd and coworkers [116] synthesized $[Mo(V)O(SPh)_4]^-$, which can be considered

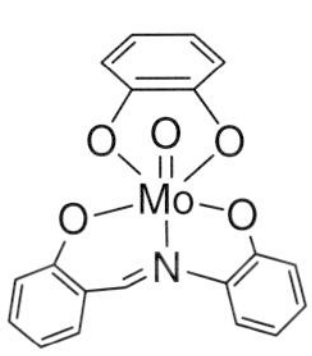

Figure 13.11 Chemical structure of MoO(cat)(Sap).

as one of the earliest and simplest developed model complexes in this series. After that, several complexes having such $[Mo(V)OS_4]^-$ coordination have been synthesized and extensively studied, both theoretically and experimentally with respect to geometry, electronic structure, redox properties, and bonding situation [117–120].

The first oxo-bis-(dithiolene) complexes such as $[Mo(IV)O(bdt)_2]^{2-}$ and $[Mo(V)O(bdt)_2]^-$ were synthesized and characterized by X-ray crystallography by the Garner group [22]. The former reduced species was shown to reduce (via OAT) trimethylamine-*N*-oxide (TMNO), a substrate that is also transformed by enzymes of the DMSOR family [100]. The tungsten analog of this complex was similarly reactive [121]. Because it was found that this reaction proceeds quite smoothly and in a rather controlled manner, TMNO is often used to generate the oxidized M(VI) species from the M(IV) species in molybdenum and tungsten model chemistry [41, 122].

After that, several bis-dithiolene Mo(IV)O complexes were developed using symmetrical and unsymmetrical dithiolenes.

In order to achieve an even better resemblance to the MPT moiety of DMSOR enzymes, the first molybdenum complex $[MoO(fdt)_2]^{2-}$ (Figure 13.12a) was developed with a ligand system that not only includes the dithiolene moiety but also the pyrane feature of the enzymatic MPT [123]. Later in the same year, Sugimoto *et al.* synthesized similar types of complexes such as $[MoO(L(X))_2]^{2-}$ (where X = CH_2, S, O) (Figure 13.12b) using a pyrane dithiolene molybdenum moiety. The crystal structures of these complexes show that the oxo ligand together with the four sulfur atoms from the dithiolene function constitute a square pyramidal geometry with a comparably weak Mo=O bond character [25]. A molybdenum complex

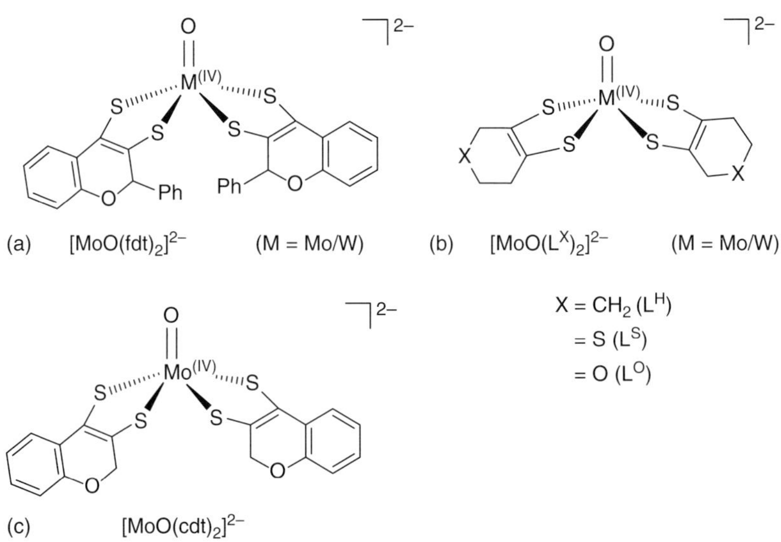

Figure 13.12 (a–c) Molybdenum and tungsten complexes with pyrane dithiolene coordination.

related to the first pyran-dithiolene complex, of which no crystal structure was available, was eventually structurally characterized. Interestingly, it crystallized as cis-isomer (Figure 13.12c) [13], which is unusual because, typically, the crystallization of the more symmetric isomer is favored. This actually points to a substantial problem associated with the use of nonsymmetrical dithiolene ligands: the formation of a mixture of geometric isomers and explains why only few crystal structures of such complexes are available.

As mentioned, bis-(dithiolene)mono-oxo Mo(IV) complexes can be converted to the corresponding dioxo Mo(VI) complexes through controlled oxidation by Me_3NO to receive model compounds for both the reduced and the oxidized oxidation state of the enzymes.

Structurally very close models for $Mo(IV)O(dt)_2/Mo(VI)O_2(dt)_2$ (dt, dithiolene) couples have been prepared by silylation of $[MO(bdt)_2]^{2-}$ and $[MO_2(bdt)_2]^{2-}$ (M = Mo, W), respectively, by Donahue *et al.* [24] (Scheme 13.9).

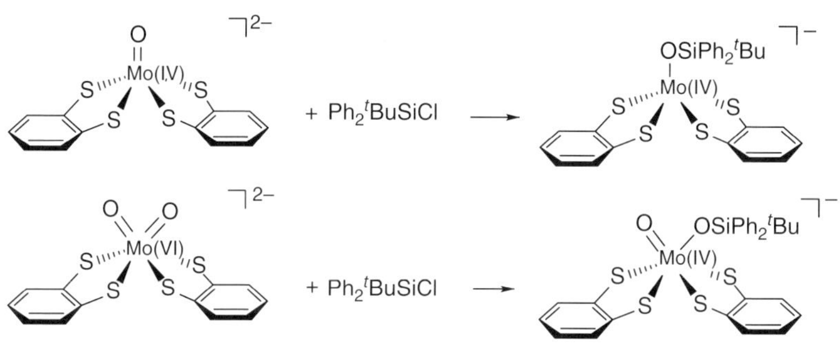

Scheme 13.9 Synthesis of $[MO(OSiPh_2{}^tBu)(bdt)_2]^-$ and *des*-oxo $[M(OSiPh_2{}^tBu)(bdt)_2]^-$ (bdt, benzenedithiolate).

When replacing the bdt ligand by its 3,6-dichloro relative (i.e., two chlorine substituents both ortho to the thiol functions), the resulting compounds are reactive in a stoichiometric, but not catalytic reaction, which converts arsenite to arsenate in aqueous solution, mixed with acetonitrile though [124]. Particularly remarkable in this case is that the entire process of the natural enzyme arsenite oxidase from the DMSOR family [125, 126] was artificially copied including the PCET reactions.

Similar model systems with the $Me_2C_2S_2{}^{2-}$ ligand instead of the aromatic dithiolenes were shown to reduce selenate ($SeO_4{}^{2-}$) to selenite ($SeO_3{}^{2-}$) [127] in a reaction analogous to the enzymatic transformation of selenate by selenate reductase [128], another member of the DMSOR family.

For the catalytic reduction of DMSO to DMS, many and structurally diverse examples can be found throughout the literature [76–78, 122, 129–133]. This is, to some extent, based on the fact that the standard test reaction for OAT catalytic activity is the reduction of DMSO and oxidation of PR_3 by the complexes [77], which can be very conveniently followed by ^{31}P NMR spectroscopy and is employed quite frequently. It is also a reaction that is not too difficult to achieve and does not require a specific set of ligands. Consequently, the respective active complexes do not necessarily carry dithiolene ligands. Substrates slightly more difficult to reduce than DMSO are TMAO (to trimethylamine), which is preferred as equimolar oxidizing agent because of the controlled way the reaction proceeds, and nitrate (to nitrite). In these two cases, mostly dithiolene complexes (i.e., molybdenum but also tungsten centers with sulfur coordination) have been shown to be capable of transforming these substrates in enzyme-like conversions [12, 122, 134–136]. These processes are typically less efficient than the reduction of DMSO, most likely due to potential inhibition of the active species by coordination of substrates or products and/or a less favorable energetic reaction profile.

All these reactions of the DMSOR family enzymes and respective model compounds involve the transfer of a doubly bonded terminal oxygen to or from the substrate.

13.5 Selected Tungsten-Containing Enzymes and Relevant Model Chemistry

In contrast to molybdenum enzymes, tungsten enzymes are mostly found in ancient organisms and those that live in nonambient habitats [28]. From the very closely related chemical, structural, and reactivity properties, it is, however, obvious that they have evolved from each other, with the distribution of the respective enzymes among the organisms pointing to an evolution from tungsten to the molybdenum enzymes [28, 137]. Owing to this very close relation and in order to avoid repetition, in the following only the most prominent or peculiar representatives of tungsten enzymes are discussed in detail.

13.5.1 The Aldehyde Ferredoxin Oxidoreductase (AOR) Family

These tungsten enzymes catalyze the OAT net reaction according to Equation 13.12 with proposed involvement of W(VI) as the catalytic center for the oxidation of aliphatic aldehydes. The reduced W(IV) center is oxidized back to the W(VI) species by ferredoxin centers (iron sulfur clusters), hence the name of the enzyme. Just like for the molybdenum proteins, which analogously catalyze the oxidation of aldehydes to carboxylic acids (see Section 13.4.1, Equation 13.4), water is the source of the oxygen and the regeneration of the active site proceeds via PCET.

$$\begin{aligned}\mathrm{RCHO} + \mathrm{H_2O} + 2\ \text{oxidized ferredoxin} \rightarrow \mathrm{RCOOH} + 2\mathrm{H^+}\\ + 2\ \text{reduced ferredoxin}\end{aligned} \tag{13.12}$$

The vast majority of tungsten enzymes known at present actually belong to the AOR family. In analogy to the molybdenum enzymes of the DMSOR family, two MPT ligands were found to bind to the central tungsten through dithiolene-sulfur coordination. The tungsten atom and the two pairs of dithiolene sulfurs are arranged in a distorted square-pyramidal structure. In the literature, there is a lot of ambiguity regarding the additional coordination sites, but despite what has been debated for some time, it is likely that the oxidized enzyme has a $(mpt)_2W(VI)O(OH)$ core and the reduced enzyme has a $(mpt)_2W(IV)(OH)$ core [31]. An earlier EXAFS study by George *et al.* [138] of *Pyrococcus furiosus* AOR already indicated accordingly the presence of an oxo group at 1.7 Å coordinated to the tungsten atom and an additional O (or possibly N; they could not rule nitrogen out at the time the experiments took place) atom present at 2.1 Å.

13.5.1.1 **Model Chemistry for the AOR Family**

Tungsten-mediated oxo transfer and related synthetic chemistry from a bioinorganic point of view has progressed far less than the respective molybdenum chemistry [40, 139]. As in the case of molybdenum model chemistry, the $W(IV)O(dithiolene)_2$ and $W(VI)O_2(dithiolene)_2$ complexes have been investigated comprehensively based on their comparably facile preparation and intended or unintended occurrence in inorganic chemistry laboratories. Significant targeted bioinorganic dithiolene chemistry with tungsten started only in 1992 with the synthesis of a set of complexes, $[W(V)O(bdt)_2]^{1-}$, $[W(IV)O(bdt)_2]^{2-}$, and $[W(VI)O_2(bdt)_2]^{2-}$ [121] by researchers, having experience with related molybdenum chemistry. After another 4 years followed the naphthalene-2,3-dithiolate complexes, $[W(IV)O(ndt)_2]^{2-}$ and $[W(VI)O_2(ndt)_2]^{2-}$ by the same group [140]. $(PPh_4)_2[W(IV)O(bdt)_2]$ and $(NEt_4)_2[W(IV)O(bdt)_2]$ were synthesized by borohydride reduction of $(PPh_4)[W(V)O(bdt)_2]$ or $(NEt4)[W(V)O(bdt)_2]$, having been obtained by a simple ligand exchange reaction between $[WO(SPh)_4]^-$ and 2 equiv of bdt-H_2. These complexes readily undergo oxidation by triethylamine-*N*-oxide to give the corresponding dioxotungsten(VI) complex, $[W(VI)O_2(bdt)_2]^{2-}$ in a reaction that is itself a typical OAT.

$[Et_4N]_2[W(IV)O(mnt)_2]$ was prepared by oxo transfer between $[Et_4N]_2[W(VI)O_2(mnt)_2]$ and PPh_3 [141], again a reaction similar to the enzymatic OATs. This property was frequently exploited in particular in tungsten but also in molybdenum model chemistry for catalytic purposes and for the preparation of reduced or oxidized species from their redox-counterparts.

Extraordinarily, the mnt complex $[Et_4N]_2[W(IV)O(mnt)_2]$ was shown to efficiently reduce CO_2/HCO_3^- (at pH 7.5) to yield $HCOO^-$ and $[Et_4N]_2[W(VI)O_2(mnt)_2]$ according to Equation 13.13, mimicking tungsten-formate dehydrogenase (W-FDH) activity [141] with unprecedented accuracy. This is a reaction of quite some significance, as it transforms formally the greenhouse gas CO_2 to a molecule (formate) that can be used in organic synthesis. Optimizing the complex composition and hence catalytic activity or developing similarly active models for the molybdenum and tungsten cofactors are certainly worthwhile tasks presently on their way in the laboratories of researchers working in this field.

$$[W(IV)O(mnt)_2]^{2-} + HCO_3^- \rightarrow [W(VI)O_2(mnt)_2]^{2-} + HCOO^- \quad (13.13)$$

The synthetic accomplishments listed, which set the foundation for the bioinorganic tungsten chemistry, were followed by the synthesis of a series of complexes, $[W(IV)O(S_2C_2R_2)_2]^{2-}$ (R = phenyl, pyridin-2-yl, pyridin-3-yl, pyridin-4-yl, or quinoxalin-2-yl) by Garner and coworkers [142]. In these models, the un-symmetric nature of MPT and the presence of N-hetrocyclic functions were successfully addressed.

More recently, other important tungsten compounds with promising ligand systems involving pyrane dithiolene species have been synthesized. $[W(IV)O(fdt)_2]^{2-}$ was the first in this series, but no crystal structure could be obtained because of the very poor crystallization behavior (Figure 13.12a in previous chapter) [123]. Later, Sugimoto and Sugimoto [143] synthesized $[WO(pdt)_2]^{2-}$ and $[WO_2(pdt)_2]^{2-}$ as the first structurally characterized tungsten complexes with pyrane dithiolene ligands (Figure 13.12b).

A series of various bis(benzenedithiolene)tungsten(IV,VI) complexes were designed and prepared by Holm and coworkers in a synthetic approach to structurally more accurate models for the active sites of tungstoenzymes [2]. Using easily accessible $[W(IV)O(bdt)_2]^{2-}$ as precursor, silylation with Me_3SiCl, ${}^tBuMe_2SiCl$, and ${}^tBuPh_2SiCl$ transforms the oxo ligand to a silyloxide ligand, affording $[W(IV)(bdt)_2(OSiMe_3)]^{1-}$, $[W(IV)(bdt)_2(OSi^tBuMe_2)]^{1-}$, and $[W(IV)(bdt)_2(OSi^tBuPh_2)]^{1-}$ respectively. Oxidation of the desoxo complex, $[W(IV)(bdt)_2(OSi^tBuPh_2)]^{1-}$ with Me_3NO gives the corresponding mono-oxo tungsten(VI) complex $[W(VI)O(bdt)_2(OSi^tBuPh_2)]^{1-}$. Direct silylation of the oxidized W(VI) species $[W(VI)O_2(bdt)_2]^{2-}$ with ${}^tBuPh_2SiCl$ affords the same complex. Further silylation of $[W(VI)O(bdt)_2(OSi^tBuPh_2)]^{1-}$ with Me_3SiCl gives $[W(VI)O(bdt)_2Cl]^{1-}$ from which the (unfortunately unstable) species $[W(VI)O(bdt)_2L]^{1-}$ (L = ${}^tBuO-$, PhS−) were generated in solution. Complexes $[W(IV)(bdt)_2(OSiMe_3)]^{1-}$, $[W(IV)(bdt)_2(OSi^tBuMe_2)]^{1-}$, and $[W(IV)(bdt)_2(OSi^tBuPh_2)]^{1-}$, based on the similarity, can be regarded as tungsten analogs of the molybdenum model complexes for the reduced DMSOR family enzymes. However, the elusive $[W(VI)O(OH)(dt)_2]$ and $[W(IV)(OH)(dt)_2]$ complexes would be still better mimics of the active sites of tungsten AOR family enzymes considering the current knowledge of the coordination environment of these enzymes, which was not available at the time of the preparation of the complexes. Further development of this chemistry by the Holm group [122, 144] led to similar models with less unnatural (Si containing) mono-anionic ligands ($[W(VI)O(L)(dt)_2]^-/[W(IV)(L)(dt)_2]^-$; LH = PhOH, PhSH, PhSeH, *i*PrOH). The coordination environment of the above-mentioned complexes $[W(VI)O(OR)(dt)_2]^-/[W(IV)(OR)(dt)_2]^-$ can be regarded as closely analogous to that of the tungsten center in the AOR family of enzymes. More recently, the synthesis of $[W(IV)(OMe)(S_2C_2Me_2)]^-$ was reported [127]. Here, the substituent on the alcoholate ligand is only a small methyl group, which exhibits the closest resemblance to the hydroxide ligand in the enzymes' active sites, having been achieved today. Consequently, the immediate coordination environment of the tungsten center in this complex is the most similar for this group of tungsten enzymes known so far.

Even though many of the above-mentioned complexes do catalyze the OAT from DMSO to phosphanes as in the OAT test reaction, only one

($[W(VI)O(S_2)(mnt)_2]^{2-}$) has been reported to show activity regarding the oxidation of aldehydes to carboxylic acids as in this enzyme family [145]. This complex has a very unusual composition with the disulfide coordinated to tungsten. It is reduced by formaldehyde to $[W(VI)O(mnt)_2]^{2-}$. When using crotonaldehyde instead in MeCN containing H_2O (5% v/v) or in CH_2Cl_2, the complex was found to be similarly reduced but also to yield crotonic acid in a reaction similar to that of the active AOR enzyme of *P. furiosus*.

The scarcity of functional models may be because the majority of complexes has not been tested or because these complexes are typically inactive in this respect (negative results are rarely reported). This is a topic researchers in this field may wish to look into with more effort, as this type of reactivity could be certainly useful for organic synthetic purposes or, for instance, for treating exhaust gases containing aldehydes.

13.5.2
The Formate Dehydrogenase (FDH) Family of Enzymes

The two prominent members of the FDH family are FDH and *N*-formylmethanofuran dehydrogenase. The former catalyzes the conversion of formate to CO_2 (Equation 13.14) in a reaction that is reverse to the one observed with a tungsten model complex (see Equation 13.13), and the latter catalyzes the conversion of *N*-formylmethanofuran to methanofuran and CO_2 (i.e., removal of a formyl substituent).

$$HCOO^- \rightleftarrows CO_2 + H^+ + 2e^- \tag{13.14}$$

The reactions of both enzymes are potentially reversible depending only on the relative concentrations of substrates [146]. This is an exception among the molybdenum and tungsten enzymes, which are usually strictly oxidases or reductases only.

The crystal structure of the FDH from *D. gigas* was solved in 2002 [147]. In this enzyme, four sulfur atoms from two MGDs coordinate to tungsten in addition to the selenium atom of a selenocysteine and one hydroxyl or sulfide ligand as the resolution of the data (1.8 Å) was not sufficient to unambiguously distinguish between oxygen and sulfur. Interestingly, in a recent reevaluation of the crystallographic data of the molybdenum-containing *E. coli*, FDH originally recorded by Boyington *et al.* [148] Raaijmakers and Romão [149] found that the apical ligand was better refined as a sulfur atom (=S or –SH, not H_2O, or –OH) at the molybdenum site. This would actually be analogous to the W-FDH from *D. gigas*. Yet the available resolution is not high enough to reach a conclusion based on structural data to ultimately decide between oxygen and sulfur for this apical position. There is, however, strong chemical evidence for the presence of a sulfur ligand. In an inactivation experiment with the FDH from *Methanobacterium formicicum* using cyanide, it was observed that the incubation of the oxidized form of the enzyme with cyanide resulted in the release of equimolar amounts of thiocyanate and the subsequent deactivation of the enzyme [150]. This observation strongly points to the presence of a sulfur ligand (most likely a SH-group) at the sixth coordination site in W-FDH [4]. This would mean that the coordination environment around tungsten consists of only sulfur and selenium atoms, which is a quite unusual coordination motive.

13.5.2.1 Model Chemistry for the W-FDH Family

In order to model the FDH family of enzymes, a sulfur ligand is required. Complicating matters, it is not known with certainty if it actually is a sulfide or rather a thiolate that is bound to tungsten at the active site. Not so long ago, Groysman and Holm [151] were successful in synthesizing $(Et_4N)[W(VI)S(SeAd)(S_2C_2Me_2)_2]$ as the first FDH family analog with the coordination of a sulfide ligand to tungsten. With this complex being the only example for a closely related model of the immediate coordination sphere, it is, however, obvious that the tungsten model chemistry as a whole is still not advanced enough to address all vital structural and reactivity issues. Regarding functional models for the two types of reactions associated with the enzymes of the W-FDH family, the best representative is still the model complex for the AOR family, which was more or less accidentally found to catalyze the reverse reaction of the FDH. Overall, the bioinorganic model chemistry even for the two well-known families of tungsten enzymes (AOR and FDH) remains a challenge for bioinorganic chemists working in this field.

13.5.3 Acetylene Hydratase

The enzyme acetylene hydratase was not known at the time when Hille first classified the molybdenum and tungsten enzymes and therefore originally only two tungsten families were proposed. Acetylene hydratase is different from these two families of tungsten enzymes and the MPT-dependent enzymes in general as it does not catalyze a redox reaction by which the substrate as a whole is oxidized or reduced. Instead, it catalyzes the conversion of acetylene to acetaldehyde, which is in fact the addition of water to acetylene (Equation 13.15). Interestingly, tungsten is not even believed to change its oxidation state during the catalytic turnover. However, it has to be pointed out that in fact a redox process takes place, as one carbon atom of acetylene is oxidized and the other reduced. The crystal structure of acetylene hydratase from *Pelobacter acetylenicus* [152] revealed a distorted octahedral geometry of the tungsten site coordinated by four sulfur atoms from the dithiolene groups of two MGDs, the fifth site being occupied by a sulfur atom of a cysteine residue and the sixth by a water molecule.

$$HC \equiv CH + H_2O \leftrightarrow CH_3CHO \tag{13.15}$$

Its reactivity overall is rather peculiar and the discovery of this enzyme is comparably recent. The respective model chemistry is therefore in its very infant stage. Interestingly, the same model complex $[W(IV)O(mnt)_2]^{2-}$, which was able to transform CO_2 to formate, is also able to catalyze the reaction of acetylene to acetaldehyde in an aqueous acetonitrile solution [135]. Its oxidized di-oxo analog was shown to be inactive but became active after reduction with dithionite. A major obstacle for small compounds catalyzing this reaction is the potential formation of acetylene adducts and it has been proposed that dithiolene coordination is necessary for a respective catalyst to maintain its activity [153].

Interestingly, it is again a rather small and simple model with the dithiolene ligand, which is the most facile to prepare and handle (mnt), that was successfully employed as a functional model.

The question remains, though, if we would not be able to generate even more efficient and/or stable catalytic systems, when trying to model more of the actual active sites of the enzymes. We have certainly not yet learnt and understood all of nature's tricks applied to this fascinating group of enzymes. Despite the efforts of almost half a centenary of many research groups, there is still a lot to learn about molybdenum- and tungsten-dependent oxidoreductases and a lot to gain by understanding them and mimicking them in more detail.

References

1. Enemark, J.H., Cooney, J.J.A., Wang, J.-J., and Holm, R.H. (2003) *Chem. Rev.*, **104**, 1175–1200.
2. Lorber, C., Donahue, J.P., Goddard, C.A., Nordlander, E., and Holm, R.H. (1998) *J. Am. Chem. Soc.*, **120**, 8102–8112.
3. Stiefel, I. (1997) *J. Chem. Soc. Dalton Trans.*, 3915–3924.
4. Bevers, L.E., Hagedoorn, P.L., and Hagen, W.R. (2009) *Coord. Chem. Rev.*, **253**, 269–290.
5. Stiefel, E.I., Eisenberg, R., Rosenberg, R.C., and Gray, H.B. (1966) *J. Am. Chem. Soc.*, **88**, 2956–2966.
6. Eisenberg, R., Stiefel, E.I., Rosenberg, R.C., and Gray, H.B. (1966) *J. Am. Chem. Soc.*, **88**, 2874–2876.
7. Alvarez, S., Vicente, R., and Hoffmann, R. (1985) *J. Am. Chem. Soc.*, **107**, 6253–6277.
8. Kapre, R.R., Bothe, E., Weyhermuller, T., DeBeer George, S., and Wieghardt, K. (2007) *Inorg. Chem.*, **46**, 5642–5650.
9. Ray, K., Debeer George, S., Solomon, E.I., Wieghardt, K., and Neese, F. (2007) *Chemistry*, **13**, 2783–2797.
10. Milsmann, C., Patra, G.K., Bill, E., Weyhermuller, T., DeBeer George, S., and Wieghardt, K. (2009) *Inorg. Chem.*, **48**, 7430–7445.
11. Sproules, S., Banerjee, P., Weyhermüller, T., Yan, Y., Donahue, J.P., and Wieghardt, K. (2011) *Inorg. Chem.*, **50**, 7106–7122.
12. Enemark, J.H., Cooney, J.J.A., Wang, J.-J., and Holm, R.H. (2003) *Chem. Rev.*, **104**, 1175–1200.
13. Samuel, P.P., Horn, S., Döring, A., Havelius, K.G.V., Reschke, S., Leimkühler, S., Haumann, M., and Schulzke, C. (2011) *Eur. J. Inorg. Chem.*, **2011**, 4387–4399.
14. Joshi, H.K., Inscore, F.E., Schirlin, J.T., Dhawan, I.K., Carducci, M.D., Bill, T.G., and Enemark, J.H. (2002) *Inorg. Chim. Acta*, **337**, 275–286.
15. Joshi, H.K., Cooney, J.J.A., Inscore, F.E., Gruhn, N.E., Lichtenberger, D.L., and Enemark, J.H. (2003) *Proc. Natl. Acad. Sci. U.S.A.*, **100**, 3719–3724.
16. Joshi, H.K. and Enemark, J.H. (2004) *J. Am. Chem. Soc.*, **126**, 11784–11785.
17. Ryde, U., Schulzke, C., and Starke, K. (2009) *J. Biol. Inorg. Chem.*, **14**, 1053–1064.
18. Baba, K., Okamura, T.A., Suzuki, C., Yamamoto, H., Yamamoto, T., Ohama, M., and Ueyama, N. (2006) *Inorg. Chem.*, **45**, 894–901.
19. Sugimoto, H., Tarumizu, M., Tanaka, K., Miyake, H., and Tsukube, H. (2005) *Dalton Trans.*, 3558–3565.

20. Oku, H., Ueyama, N., Kondo, M., and Nakamura, A. (1994) *Inorg. Chem.*, **33**, 209–216.
21. Ueyama, N., Oku, H., Kondo, M., Okamura, T., Yoshinaga, N., and Nakamura, A. (1996) *Inorg. Chem.*, **35**, 643–650.
22. Boyde, S., Ellis, S.R., Garner, C.D., and Clegg, W. (1986) *J. Chem. Soc., Chem. Commun.*, 1541–1543.
23. Davies, E.S., Beddoes, R.L., Collison, D., Dinsmore, A., Docrat, A., Joule, J.A., Wilson, C.R., and Garner, C.D. (1997) *J. Chem. Soc., Dalton. Trans.*, 3985–3995.
24. Donahue, J.P., Goldsmith, C.R., Nadiminti, U., and Holm, R.H. (1998) *J. Am. Chem. Soc.*, **120**, 12869–12881.
25. Sugimoto, H., Harihara, M., Shiro, M., Sugimoto, K., Tanaka, K., Miyake, H., and Tsukube, H. (2005) *Inorg. Chem.*, **44**, 6386–6392.
26. Burgess, B. K. and Lowe, D. J. (1996) **Chem. Rev.**, **96**, 2983-3011.
27. Hille, R. (1996) *Chem. Rev.*, **96**, 2757–2816.
28. Johnson, M.K., Rees, D.C., and Adams, M.W. (1996) *Chem. Rev.*, **96**, 2817–2840.
29. Schulzke, C. (2011) *Eur. J. Inorg. Chem.*, **2011**, 1189–1199.
30. Thapper, A., Lorber, C., Fryxelius, J., Behrens, A., and Nordlander, E. (2000) *J. Inorg. Biochem.*, **79**, 67–74.
31. Hille, R. (2002) *Trends Biochem. Sci.*, **27**, 360–367.
32. Romão, M.J. (2009) *Dalton Trans.*, 4053–4068.
33. Hine, F.J., Taylor, A.J., and Garner, C.D. (2010) *Coord. Chem. Rev.*, **254**, 1570–1579.
34. Collison, D., Garner, C.D., and Joule, J.A. (1996) *Chem. Soc. Rev.*, **25**, 25–32.
35. Brody, M.S. and Hille, R. (1995) *Biochim. Biophys. Acta, Protein Struct. Mol. Enzymol.*, **1253**, 133–135.
36. Schultz, B.E., Hille, R., and Holm, R.H. (1995) *J. Am. Chem. Soc.*, **117**, 827–828.
37. McMaster, J., Tunney, J.M., and Garner, C.D. (2004) *Dithiolene Chemistry*, John Wiley & Sons, Inc, Hoboken, NJ, pp. 539–583.
38. McCleverty, J.A. (2007) *Progress in Inorganic Chemistry*, John Wiley & Sons, Inc., pp. 49–221.
39. Holm, R.H. (1987) *Chem. Rev.*, **87**, 1401–1449.
40. Holm, R.H. (1990) *Coord. Chem. Rev.*, **100**, 183–221.
41. Sugimoto, H., Tatemoto, S., Suyama, K., Miyake, H., Mtei, R.P., Itoh, S., and Kirk, M.L. (2010) *Inorg. Chem.*, **49**, 5368–5370.
42. Majumdar, A. and Sarkar, S. (2011) *Coord. Chem. Rev.*, **255**, 1039–1054.
43. Brondino, C.D., Romão, M.J., Moura, I., and Moura, J.J.G. (2006) *Curr. Opin. Chem. Biol.*, **10**, 109–114.
44. Cramer, S.P., Wahl, R., and Rajagopalan, K.V. (1981) *J. Am. Chem. Soc.*, **103**, 7721–7727.
45. Bordas, J., Bray, R.C., Garner, C.D., Gutteridge, S., and Hasnain, S.S. (1980) *Biochem. J.*, **191**, 499–508.
46. Cramer, S.P. and Hille, R. (1985) *J. Am. Chem. Soc.*, **107**, 8164–8169.
47. Turner, N.A., Bray, R.C., and Diakun, G.P. (1989) *Biochem. J.*, **260**, 563–571.
48. Romão, M.J., Archer, M., Moura, I., Moura, J.J.G., LeGall, J., Engh, R., Schneider, M., Hof, P., and Huber, R. (1995) *Science*, **270**, 1170–1176.
49. Huber, R., Hof, P., Duarte, R.O., Moura, J.J., Moura, I., Liu, M.Y., LeGall, J., Hille, R., Archer, M., and Romão, M.J. (1996) *Proc. Natl. Acad. Sci. U.S.A.*, **93**, 8846–8851.
50. Enroth, C., Eger, B.T., Okamoto, K., Nishino, T., Nishino, T., and Pai, E.F. (2000) *Proc. Natl. Acad. Sci. U.S.A.*, **97**, 10723–10728.
51. Hille, R. and Sprecher, H. (1987) *J. Biol. Chem.*, **262**, 10914–10917.
52. Okamoto, K., Eger, B.T., Nishino, T., Pai, E.F., and Nishino, T. (2008) *Nucleosides Nucleotides Nucleic Acids*, **27**, 888–893.
53. Rebelo, J.R., Dias, J.D., Huber, R.H., Moura, J.M., and Romão, M.R. (2001) *J. Biol. Inorg. Chem.*, **6**, 791–800.
54. Okamoto, K., Matsumoto, K., Hille, R., Eger, B.T., Pai, E.F., and Nishino, T. (2004) *Proc. Natl. Acad. Sci. U.S.A.*, **101**, 7931–7936.

55. João Romão, M., Knäblein, J., Huber, R., and Moura, J.J.G. (1997) *Prog. Biophys. Mol. Biol.*, **68**, 121–144.
56. Schulzke, C. and Samuel, P.P. (2011) *Molybdenum and Tungsten Cofactor Model Chemistry*, Nova Science Publishers Inc., New York.
57. Sugimoto, H. and Tsukube, H. (2008) *Chem. Soc. Rev.*, **37**, 2609–2619.
58. Müller, A., Diemann, E., Jostes, R., and Bögge, H. (1981) *Angew. Chem. Int. Ed. English*, **20**, 934–955.
59. Bristow, S., Collison, D., Garner, C.D., and Clegg, W. (1983) *J. Chem. Soc. Dalton Trans.*, 2495–2499.
60. Wieghardt, K., Hahn, M., Weiss, J., and Swiridoff, W. (1982) *Z. Anorg. Allg. Chem.*, **492**, 164–174.
61. Faller, J.W. and Ma, Y. (1989) *Organomet.*, **8**, 609–612.
62. 62.Wilson, G.L., Greenwood, R.J., Pilbrow, J.R., Spence, J.T., and Wedd, A.G. (1991) *J. Am. Chem. Soc.*, **113**, 6803–6812.
63. Thapper, A., Donahue, J.P., Musgrave, K.B., Willer, M.W., Nordlander, E., Hedman, B., Hodgson, K.O., and Holm, R.H. (1999) *Inorg. Chem.*, **38**, 4104–4114.
64. Doonan, C.J., Nielsen, D.J., Smith, P.D., White, J.M., George, G.N., and Young, C.G. (2005) *J. Am. Chem. Soc.*, **128**, 305–316.
65. Laughlin, L.J., Eagle, A.A., George, G.N., Tiekink, E.R., and Young, C.G. (2007) *Inorg. Chem.*, **46**, 939–948.
66. Zama, K., Fukuoka, A., Sasaki, Y., Inagaki, S., Fukushima, Y., and Ichikawa, M. (2000) *Catal. Lett.*, **66**, 251–253.
67. Nomiya, K., Yagishita, K., Nemoto, Y., and Kamataki, T.-A. (1997) *J. Mol. Catal. A: Chem.*, **126**, 43–53.
68. Kisker, C., Schindelin, H., Pacheco, A., Wehbi, W.A., Garrett, R.M., Rajagopalan, K.V., Enemark, J.H., and Rees, D.C. (1997) *Cell*, **91**, 973–983.
69. George, G.N., Kipke, C.A., Prince, R.C., Sunde, R.A., Enemark, J.H., and Cramer, S.P. (1989) *Biochemistry*, **28**, 5075–5080.
70. George, G.N., Garrett, R.M., Prince, R.C., and Rajagopalan, K.V. (1996) *J. Am. Chem. Soc.*, **118**, 8588–8592.
71. Dobbek, H., Gremer, L., Kiefersauer, R., Huber, R., and Meyer, O. (2002) *Proc. Natl. Acad. Sci. U.S.A.*, **99**, 15971–15976.
72. Siegbahn, P.E.M. and Shestakov, A.F. (2005) *J. Comput. Chem.*, **26**, 888–898.
73. Hille, R. (1999) *Essays Biochem.*, **34**, 125–137.
74. Hille, R. (1994) *Biochim. Biophys. Acta, Bioenerg.*, **1184**, 143–169.
75. Brody, M.S. and Hille, R. (1999) *Biochemistry*, **38**, 6668–6677.
76. Berg, J.M. and Holm, R.H. (1985) *J. Am. Chem. Soc.*, **107**, 917–925.
77. Berg, J.M. and Holm, R.H. (1985) *J. Am. Chem. Soc.*, **107**, 925–932.
78. Palanca, P., Picher, T., Sanz, V., Gomez-Romero, P., Llopis, E., Domenech, A., and Cervilla, A. (1990) *J. Chem. Soc., Chem. Commun.*, 531–533.
79. Gheller, S.F., Schultz, B.E., Scott, M.J., and Holm, R.H. (1992) *J. Am. Chem. Soc.*, **114**, 6934–6935.
80. Schultz, B.E., Gheller, S.F., Muetterties, M.C., Scott, M.J., and Holm, R.H. (1993) *J. Am. Chem. Soc.*, **115**, 2714–2722.
81. Das, S.K., Chaudhury, P.K., Biswas, D., and Sarkar, S. (1994) *J. Am. Chem. Soc.*, **116**, 9061–9070.
82. Chaudhury, P.K., Das, S.K., and Sarkar, S. (1996) *Biochem. J.*, **319**, 953–959.
83. Thapper, A., Deeth, R.J., and Nordlander, E. (1999) *Inorg. Chem.*, **38**, 1015–1018.
84. Hernandez-Marin, E. and Ziegler, T. (2009) *Inorg. Chem.*, **48**, 1323–1333.
85. Pacheco, A., Hazzard, J.T., Tollin, G., and Enemark, J.H. (1999) *J. Biol. Inorg. Chem.*, **4**, 390–401.
86. Lorber, C., Rosaria Plutino, M., Elding, L.I., and Nordlander, E. (1997) *J. Chem. Soc., Dalton Trans.*, 3997–4004.
87. Xiao, Z.G., Young, C.G., Enemark, J.H., and Wedd, A.G. (1992) *J. Am. Chem. Soc.*, **114**, 9194–9195.
88. Heinze, K. and Fischer, A. (2007) *Eur. J. Inorg. Chem.*, **2007**, 1020–1026.
89. Lim, B.S., Willer, M.W., Miao, M., and Holm, R.H. (2001) *J. Am. Chem. Soc.*, **123**, 8343–8349.
90. Takuma, M., Ohki, Y., and Tatsumi, K. (2005) *Inorg. Chem.*, **44**, 6034–6043.

91. Groysman, S., Majumdar, A., Zheng, S.-L., and Holm, R.H. (2009) *Inorg. Chem.*, **49**, 1082–1089.
92. Holm, R.H., Solomon, E.I., Majumdar, A., and Tenderholt, A. (2011) *Coord. Chem. Rev.*, **255**, 993–1015.
93. McDevitt, C.A., Hugenholtz, P., Hanson, G.R., and McEwan, A.G. (2002) *Mol. Microbiol.*, **44**, 1575–1587.
94. McEwan, A.G., Ridge, J.P., McDevitt, C.A., and Hugenholtz, P. (2002) *Geomicrobiol. J.*, **19**, 3–21.
95. Moura, J.G., Brondino, C., Trincão, J., and Romão, M. (2004) *J. Biol. Inorg. Chem.*, **9**, 791–799.
96. Schindelin, H., Kisker, C., Hilton, J., Rajagopalan, K.V., and Rees, D.C. (1996) *Science*, **272**, 1615–1621.
97. Schneider, F., Lowe, J., Huber, R., Schindelin, H., Kisker, C., and Knablein, J. (1996) *J. Mol. Biol.*, **263**, 53–69.
98. McAlpine, A.S., McEwan, A.G., Shaw, A.L., and Bailey, S. (1997) *J. Biol. Inorg. Chem.*, **2**, 690–701.
99. McAlpine, A.S., McEwan, A.G., and Bailey, S. (1998) *J. Mol. Biol.*, **275**, 613–623.
100. Czjzek, M., Dos Santos, J.-P., Pommier, J., Giordano, G., Méjean, V., and Haser, R. (1998) *J. Mol. Biol.*, **284**, 435–447.
101. Zhang, L., Nelson, K.J., Rajagopalan, K.V., and George, G.N. (2008) *Inorg. Chem.*, **47**, 1074–1078.
102. Rajagopalan, K.V. (1993) *Molybdenum Enzymes, Cofactors, and Model Systems*, vol. **535**, American Chemical Society, pp. 38–49.
103. George, G.N., Hilton, J., and Rajagopalan, K.V. (1996) *J. Am. Chem. Soc.*, **118**, 1113–1117.
104. Garton, S.D., Hilton, J., Oku, H., Crouse, B.R., Rajagopalan, K.V., and Johnson, M.K. (1997) *J. Am. Chem. Soc.*, **119**, 12906–12916.
105. George, G.N., Nelson, K.J., Harris, H.H., Doonan, C.J., and Rajagopalan, K.V. (2007) *Inorg. Chem.*, **46**, 3097–3104.
106. Bruce, A., Corbin, J.L., Dahlstrom, P.L., Hyde, J.R., Minelli, M., Stiefel, E.I., Spence, J.T., and Zubieta, J. (1982) *Inorg. Chem.*, **21**, 917–926.
107. Liebeskind, L.S., Sharpless, K.B., Wilson, R.D., and Ibers, J.A. (1978) *J. Am. Chem. Soc.*, **100**, 7061–7063.
108. Marabella, C.P., Enemark, J.H., Newton, W.E., and McDonald, J.W. (1982) *Inorg. Chem.*, **21**, 623–627.
109. Bristow, S., Enemark, J.H., Garner, C.D., Minelli, M., Morris, G.A., and Ortega, R.B. (1985) *Inorg. Chem.*, **24**, 4070–4077.
110. Mondal, J.U., Schultz, F.A., Brennan, T.D., and Scheidt, W.R. (1988) *Inorg. Chem.*, **27**, 3950–3956.
111. Sinclair, L., Mondal, J.U., Uhrhammer, D., and Schultz, F.A. (1998) *Inorg. Chim. Acta*, **278**, 1–5.
112. Mondal, J.U., Zamora, J.G., Kinon, M.D., and Schultz, F.A. (2000) *Inorg. Chim. Acta*, **309**, 147–150.
113. Mondal, J.U., Zamora, J.G., Siew, S.-C., Garcia, G.T., George, E.R., Kinon, M.D., and Schultz, F.A. (2001) *Inorg. Chim. Acta*, **321**, 83–88.
114. Głowiak, T., Jerzykiewicz, L., Sobczak, J.M., and Ziółkowski, J.J. (2003) *Inorg. Chim. Acta*, **356**, 387–392.
115. Mondal, J.U., Almaraz, E., and Bhat, N.G. (2004) *Inorg. Chem. Commun.*, **7**, 1195–1197.
116. Boyd, I., Dance, I., Murray, K., and Wedd, A. (1978) *Aust. J. Chem.*, **31**, 279–284.
117. McNaughton, R.L., Tipton, A.A., Rubie, N.D., Conry, R.R., and Kirk, M.L. (2000) *Inorg. Chem.*, **39**, 5697–5706.
118. McMaster, J., Carducci, M.D., Yang, Y.-S., Solomon, E.I., and Enemark, J.H. (2001) *Inorg. Chem.*, **40**, 687–702.
119. Bradbury, J.R., Wedd, A.G., and Bond, A.M. (1979) *J. Chem. Soc., Chem. Commun.*, 1022–1025.
120. Wang, X.-B., Inscore, F.E., Yang, X., Cooney, J.J.A., Enemark, J.H., and Wang, L.-S. (2002) *J. Am. Chem. Soc.*, **124**, 10182–10191.
121. Ueyama, N., Oku, H., and Nakamura, A. (1992) *J. Am. Chem. Soc.*, **114**, 7310–7311.
122. Lim, B.S. and Holm, R.H. (2001) *J. Am. Chem. Soc.*, **123**, 1920–1930.
123. Schulzke, C. (2005) *Dalton Trans.*, 713–720.

124. Sugimoto, H., Tarumizu, M., Miyake, H., and Tsukube, H. (2006) *Eur. J. Inorg. Chem.*, **2006**, 4494–4497.
125. Ellis, P.J., Conrads, T., Hille, R., and Kuhn, P. (2001) *Structure*, **9**, 125–132.
126. Conrads, T., Hemann, C., George, G.N., Pickering, I.J., Prince, R.C., and Hille, R. (2002) *J. Am. Chem. Soc.*, **124**, 11276–11277.
127. Wang, J.-J., Tessier, C., and Holm, R.H. (2006) *Inorg. Chem.*, **45**, 2979–2988.
128. Maher, M.J., Santini, J., Pickering, I.J., Prince, R.C., Macy, J.M., and George, G.N. (2003) *Inorg. Chem.*, **43**, 402–404.
129. Alphonse, F.-A., Karim, R., Cano-Soumillac, C., Hebray, M., Collison, D., Garner, C.D., and Joule, J.A. (2005) *Tetrahedron*, **61**, 11010–11019.
130. Andrade, C.E.A., Ma, X.L., Meyer-Klaucke, W., and Schulzke, C. (2010) *Polyhedron*, **29**, 664–668.
131. Ma, X., Schulzke, C., Schmidt, H.-G., and Noltemeyer, M. (2007) *Dalton Trans.*, 1773–1780.
132. Ma, X.L., Schulzke, C., Yang, Z., Ringe, A., and Magull, J. (2007) *Polyhedron*, **26**, 5497–5505.
133. Sung, K.-M. and Holm, R.H. (2001) *J. Am. Chem. Soc.*, **123**, 1931–1943.
134. Majumdar, A., Pal, K., and Sarkar, S. (2006) *J. Am. Chem. Soc.*, **128**, 4196–4197.
135. Jiang, J. and Holm, R.H. (2005) *Inorg. Chem.*, **44**, 1068–1072.
136. Majumdar, A., Pal, K., and Sarkar, S. (2008) *Inorg. Chem.*, **47**, 3393–3401.
137. Döring, A. and Schulzke, C. (2010) *Dalton Trans.*, **39**, 5623–5629.
138. George, G.N., Prince, R.C., Mukund, S., and Adams, M.W.W. (1992) *J. Am. Chem. Soc.*, **114**, 3521–3523.
139. Tucci, G.C., Donahue, J.P., and Holm, R.H. (1998) *Inorg. Chem.*, **37**, 1602–1608.
140. Oku, H., Ueyama, N., and Nakamura, A. (1996) *Bull. Chem. Soc. Jpn.*, **69**, 3139–3150.
141. Sarkar, S. and Das, S.K. (1992) *Proc. Indian Acad. Sci., Chem. Sci.*, **104**, 533–534.
142. Davies, E.S., Aston, G.M., Beddoes, R.L., Collison, D., Dinsmore, A., Docrat, A., Joule, J.A., Wilson, C.R., and Garner, C.D. (1998) *J. Chem. Soc., Dalton Trans.*, 3647–3656.
143. Sugimoto, H. and Sugimoto, K. (2008) *Inorg. Chem. Commun.*, **11**, 77–80.
144. Lim, B.S., Donahue, J.P., and Holm, R.H. (1999) *Inorg. Chem.*, **39**, 263–273.
145. Das, S.K., Biswas, D., Maiti, R., and Sarkar, S. (1996) *J. Am. Chem. Soc.*, **118**, 1387–1397.
146. Reda, T., Plugge, C.M., Abram, N.J., and Hirst, J. (2008) *Proc. Natl. Acad. Sci. U.S.A.*, **105**, 10654–10658.
147. Raaijmakers, H., Macieira, S., Dias, J.M., Teixeira, S., Bursakov, S., Huber, R., Moura, J.J.G., Moura, I., and Romão, M.J. (2002) *Structure*, **10**, 1261–1272.
148. Boyington, J.C., Gladyshev, V.N., Khangulov, S.V., Stadtman, T.C., and Sun, P.D. (1997) *Science*, **275**, 1305–1308.
149. Raaijmakers, H.A. and Romão, M. (2006) *J. Biol. Inorg. Chem.*, **11**, 849–854.
150. Barber, M.J., May, H.D., and Ferry, J.G. (1986) *Biochemistry*, **25**, 8150–8155.
151. Groysman, S. and Holm, R.H. (2007) *Inorg. Chem.*, **46**, 4090–4102.
152. Seiffert, G.B., Ullmann, G.M., Messerschmidt, A., Schink, B., Kroneck, P.M.H., and Einsle, O. (2007) *Proc. Natl. Acad. Sci. U.S.A.*, **104**, 3073–3077.
153. Yadav, J., Das, S.K., and Sarkar, S. (1997) *J. Am. Chem. Soc.*, **119**, 4315–4316.

Part V
Applicative Perspectives

Bioinspired Catalysis: Metal-Sulfur Complexes, First Edition.
Edited by Wolfgang Weigand and Philippe Schollhammer.

14
Electrode Materials and Artificial Photosynthetic Systems

Phong D. Tran, Marc Fontecave, and Vincent Artero

14.1
Introduction

In this concluding chapter, we show how the homogeneous molecular catalysts developed through the bio-inspired approach can be exploited in a technological context. This is made possible thanks to surface science and nanotechnologies that provide means to strongly attach these molecules onto electroconductive support so that they become part of the electrode. Besides corresponding to a shift from electro-assisted catalysis to supported electrocatalysis, such a transition is required for technological applications, both to minimize the amount of catalytic material engaged in devices and to optimize electronic connections between electrode support, catalytic material, and electrolytes or membranes. As an alternative to this molecular engineering approach, the direct design of metal–sulfur materials as bio-inspired solid-state catalysts for the hydrogen evolution reaction (HER) is an exploding field that has produced remarkable breakthrough in a very short time since 2005. Both types of noble-metal free electrode materials constitute alternatives to the use of noble metals such as platinum in electrolyzers. These technological devices are able to produce a fuel (hydrogen) from water and electricity, which can be produced from renewable sources such as sunlight and wind. The grand challenge for this century is however to directly produce a solar fuel such as hydrogen in a photoelectrochemical (PEC) device. Such devices already exist but they are either too expensive or too fragile for industrial applications. We show how the bio-inspired approach based on the current understanding of photosystems (PSs) and hydrogenases can provide part of the solution to the energetic transition toward a solar-fuel-based economy.

14.2
Electrode Materials for Hydrogen Evolution

In this section, we first discuss how a homogeneous molecular catalyst can be immobilized on electrode surface to engineer catalytic materials. We then discuss

Bioinspired Catalysis: Metal-Sulfur Complexes, First Edition.
Edited by Wolfgang Weigand and Philippe Schollhammer.

recent advances in the development of solid-state metal-sulfides HER catalysts as well as electrode materials engineered from these materials.

14.2.1
Electrode Materials Based on Bio-Inspired Molecular Catalyst

There have been some very interesting reports dealing with the application of hydrogenases to replace Pt, since the demonstration by Armstrong that hydrogenases adsorbed on carbon electrodes [1] are comparable in activity to Pt nanoparticles [2]. The main disadvantages of hydrogenases as electrocatalysts reside in (i) their large size, which results in very low density of catalytic centers loaded on surfaces, translating into low current densities; (ii) the low accessibility of the active sites, which are buried inside the protein, making it difficult to achieve a good electronic coupling to the electrode in some cases; (iii) the cost for their production that in addition can hardly be scaled up; and (iv) their great sensitivity with regard to traces of oxygen, which makes them difficult to use in practical situations. With the same objective of designing cheap and robust electrodes for H_2 production, a number of groups have extended their studies of H_2-evolving metal catalysts (Chapters 4 and 7) to their immobilization on surfaces and characterization of these novel materials. In general, these complexes display much lower catalytic activities than hydrogenases themselves but they provide a solution to several limitations of the enzyme. They are much less difficult to prepare and thus are less expensive materials; they show, under certain conditions, larger stabilities, and in particular they can be handled in air, which is critical to their exploitation in practical technological devices; their smallest size allows more catalytic material to be bound to the electrode surface and better electron coupling between the electrode and the active metal site.

The construction of such electrode materials require that appropriate methods are developed to achieve an efficient and stable grafting of the molecular catalyst to the electrode surface. The different strategies reported in the literature are discussed here and concern standard carbon surfaces, carbon nanotubes (CNTs), and semiconductor materials.

14.2.1.1 Entrapment of Catalyst within a Polymeric Material

Originally, Spiro [3] reported that stable catalytic materials can be simply obtained by incorporating positively charged derivatives of the cobalt porphyrin [Co(TPP)] (Figure 14.1) into a Nafion film laid onto a glassy carbon electrode, however with a low electroactivity (26 turnover number (TON) achieved after 90 min at −0.95 V vs saturated calomel electrode (SCE) in 0.1 mol l^{-1} aqueous trifluoroacetic acid (TFA) solution) because of the poor electron-transfer characteristics of Nafion films. When coated on a bare pyrolytic graphite electrode, neutral [Co(TPP)] (Figure 14.1) incorporated in a Nafion film can reduce protons but only with a large overpotential (−0.7 V vs Ag/AgCl; pH = 1) and a quite low turnover frequency (TOF) value (70 h^{-1}) [4]. Better catalytic activity ($2 \cdot 10^5$ h^{-1} TOF) was observed with an applied potential of −0.90 V versus Ag/AgCl and

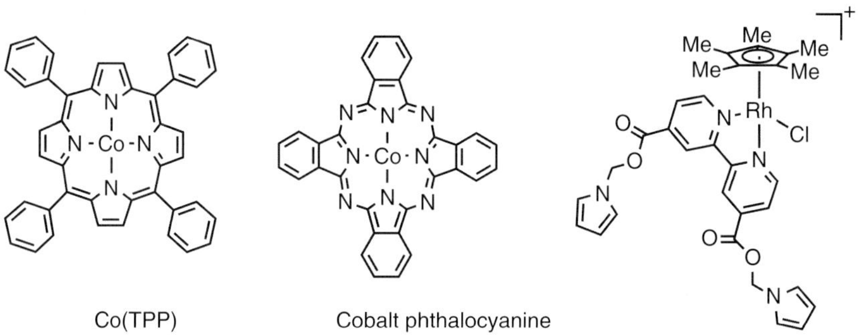

Figure 14.1 Selected cobalt [3–5] and rhodium-based [6] H_2-evolving catalysts that have been immobilized onto electrode surfaces by incorporation into a polymeric support layer.

at pH = 1 for a cobalt phthalocyanine (Figure 14.1) incorporated in a poly(4-vinylpyridine-*co*-styrene) film deposited on a graphite electrode [5]. In both cases, the catalytic proton reduction reaction was limited by the electron transfer within the matrix.

Electrodes can also be coated with conducting polymers obtained via an electropolymerization process. Electropolymerization of [Cp*Rh(L)Cl](BF_4) (L = bis-[(4,4′-bispyrrol-1-ylmethyl)methoxycarbonyl]-2,2′-bipyridyl) leads to a stable film capable of proton electroreduction at −0.31 V versus normal hydrogen electrode (NHE) at pH below 4 (Figure 14.1). Quantitative current efficiency corresponding to 353 turnovers was observed during a 14 h electrolysis experiment at pH 1 using a carbon-felt electrode coated with the electropolymerized rhodium complex [6]. More recently, an electropolymeric material incorporating a hydrogenase model compound was provided by Pickett and coworkers [7]. To achieve that goal, a dinuclear thiolate-bridged Fe complex containing CO ligands, thus combining many structural features of the active site of [FeFe]-hydrogenases, was synthesized using a dithiolate-bridging ligand carrying a nucleophilic group. This complex was reacted with a glassy carbon electrode coated with a functionalized polypyrrole film, containing electrophilic groups (Figure 14.2). This material has been shown to electrocatalyze proton reduction at potentials slightly more positive than those for the complex-free polymer, however still with a too large overpotential for practical applications. A similar strategy has been developed by Pool and DuBois [8] with a mononuclear Ni complex with diphosphine ligands carrying electropolymerizable thiophene groups (Figure 14.2). While the molecular complex displays interesting electrocatalytic properties when assayed in solution (limited overpotential of 280 mV and TOF of 56 s^{-1}), the electropolymerized film obtained by electrooxidation of the complex is not electrocatalytically active for still unknown reasons.

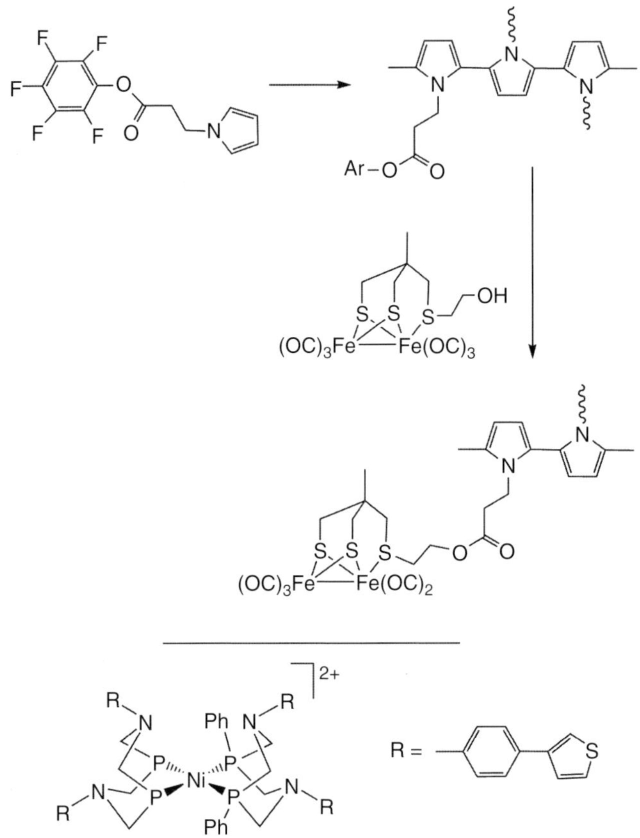

Figure 14.2 Bio-inspired complexes immobilized onto electrode surfaces by covalent incorporation into an electrodeposited polypyrrole (top) [7] or polythiophene (bottom) film [8].

14.2.1.2 Covalent Attachment of Catalyst to Electrode Material

In the case of hydrogenases, covalent grafting is the method of choice for the design of stable electrocatalytic materials. Spiro initially reported covalent grafting of a series of cobalt porphyrins on glassy carbon electrodes via an amide link with surface carboxylic acid groups. These electrodes displayed high activity for H_2 production associated with low overpotential in neutral aqueous solution but were found to be quite fragile. Film instability or disrupting processes at either film-electrode or film-electrolyte interfaces were postulated as possible reasons of this fragility [3].

Similar methodologies have been used by Gloaguen and colleagues [9] to graft models of [FeFe]-hydrogenase active sites on highly ordered pyrolytic graphite electrode surfaces. The electrode was first polished in the presence of ammonia, resulting in the functionalization with amino groups [10]. Then a diiron complex with a dithiolate bridge derivatized with an activated ester group was anchored, but again no electrocatalytic activity was observed using the grafted electrode.

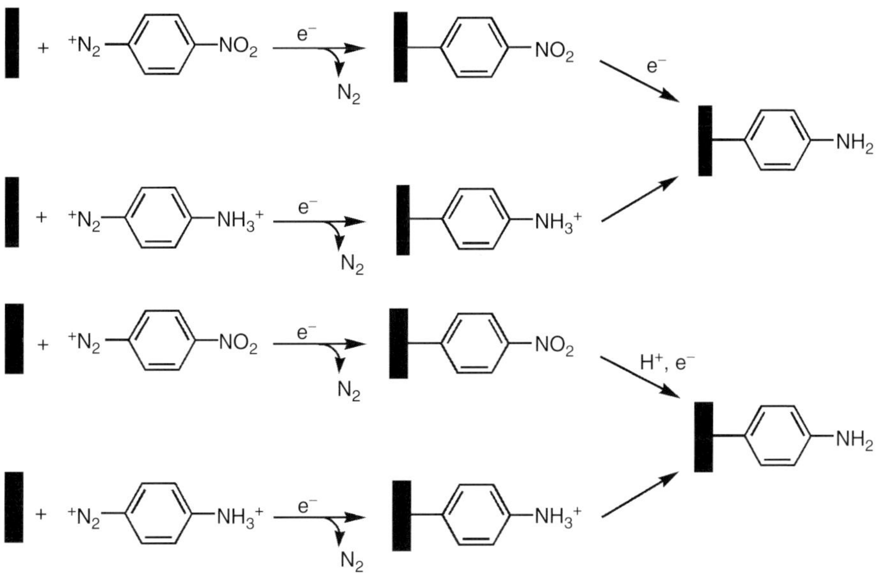

Figure 14.3 Functionalization of conductive surfaces by an amino-polyphenylene layer via diazonium salt reduction [11].

The electroreduction of diazonium salts (Figure 14.3) proved to be an efficient alternative method to form covalent linkages between a model compound and a carbon electrode [11]. This was applied by Darensbourg's group (Figure 14.4a) [12], Gloaguen and colleagues [9] (Figure 14.4b), and ourselves [13] (Figure 14.4c) to dinuclear Fe complexes modeling [FeFe]-hydrogenase active sites: dithiolate-bridged diiron complexes with activated succinimide ester groups have been grafted on glassy carbon electrodes modified with either aminophenyl or *p*-(2-aminoethyl)phenyl groups. Unfortunately, these materials displayed very weak electrocatalytic responses because of a great instability, due to the loss of CO ligands in the reduced state of the catalyst as shown by X-ray photoelectron spectroscopy (XPS) and attenuated total reflection infrared (ATR-IR) studies [13].

A major breakthrough was reported in our laboratory in 2009 regarding novel Ni-based CNT-coated electrodes with remarkable electrocatalytic activities both for H_2 evolution and uptake [14]. The molecular Ni complex used in this study was efficiently covalently attached to multiwall carbon nanotubes (MWCNTs) using the diazonium strategy thanks to a fruitful collaboration with Bruno Jousselme and Serge Palacin in CEA Saclay (France). Clearly, diazonium salt reduction appears as an increasingly popular method to functionalize carbon-based materials, and in particular CNTs, as it allows decorating the surface with a variety of chemical functions such as carboxylate derivatives, amine, azide, or alkyne [11, 15, 16].

Among hydrogenase bio-inspired catalysts, DuBois' nickel bisdiphosphine complexes are probably the ones with the most outstanding catalytic properties.

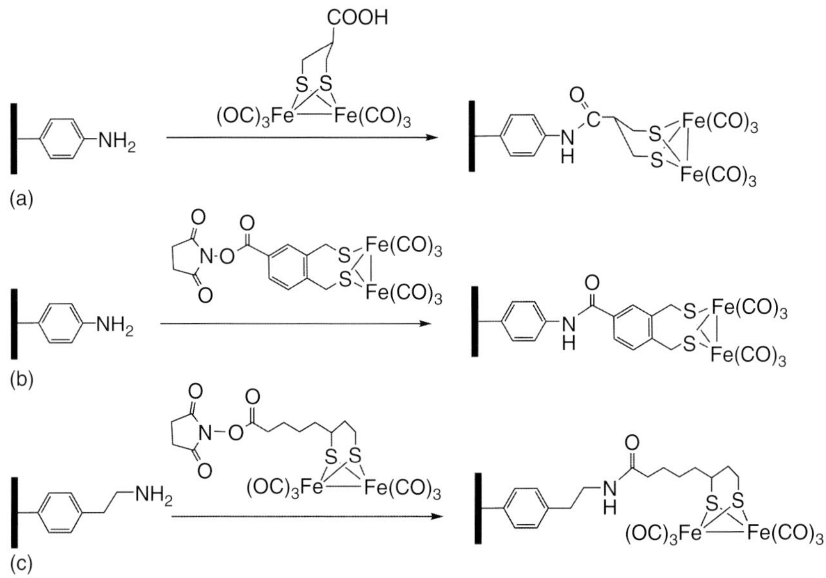

Figure 14.4 (a–c) Covalent immobilization of diiron mimics via an amide linkage (see text for references).

In addition, their synthetic versatility allowed us to synthesize an analog **1**$(BF_4)_2$ (Figure 14.5), in which activated ester functionalities were introduced. An amidation reaction then afforded covalent attachment of the catalyst (Figure 14.6) with CNTs previously decorated with amine groups at their surface through electrochemical reduction of the 4-(2-aminoethyl)phenyldiazonium salt (Figure 14.5) [14].

Deposition of these electroactive Ni-functionalized CNTs onto a gas diffusion layer, developed for proton exchange membrane (PEM) applications and made of carbon fibers, and coated with Nafion generated a cheap, stable, air-resistant electrocatalytic material (Figure 14.7) with unique performances especially under the strongly acidic conditions required in the expanding PEM technology. Hydrogen indeed evolves from aqueous sulfuric acid solution at the thermodynamic equilibrium (no overpotential requirement) and the material proved remarkably stable: up to 100 000 turnovers could be measured during a 10 h experiment without loss of activity. Interestingly, this Pt-free catalyst is also very efficient for hydrogen oxidation under the same conditions with current densities similar to those observed for hydrogenase-based materials (1–2 mA cm^{-2}) under mass-transport limitation. The maximum current density that can be measured in the absence of such a limitation (i.e., using a rotating disk electrode on which the catalytic material has been deposited) reaches 40 mA cm^{-2}. Bidirectional catalysis was completely unexpected from solution studies. This shows how modification of the distant environment of a metal complex, for instance, its third coordination sphere, may also allow to tune its catalytic properties. Years after, DuBois' group could design a catalyst showing similar bidirectional catalytic activity but in the

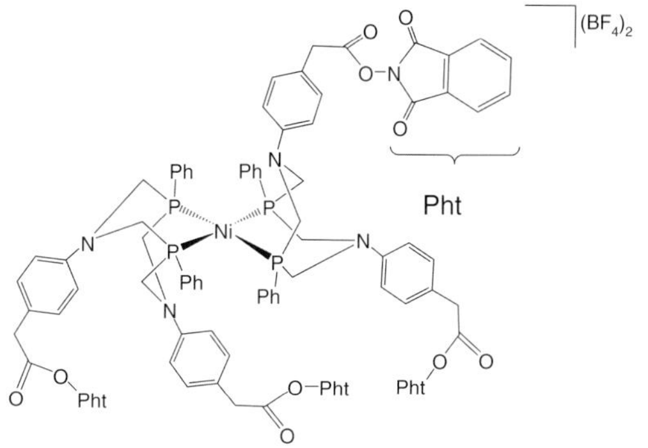

$\mathbf{1}(BF_4)_2$

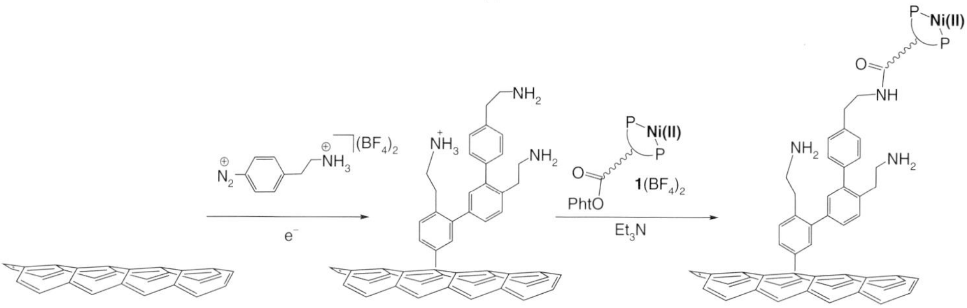

Figure 14.5 Electrografting of amine groups on MWCNTs and post-functionalization with complex **1** $(BF_4)_2$ through amide links [14]. This is a simplified representation of the structure of the material for the sake of clarity: the number of phenylene residues is indeed arbitrary and attachment of the nickel complex to two or more surface amine groups not excluded.

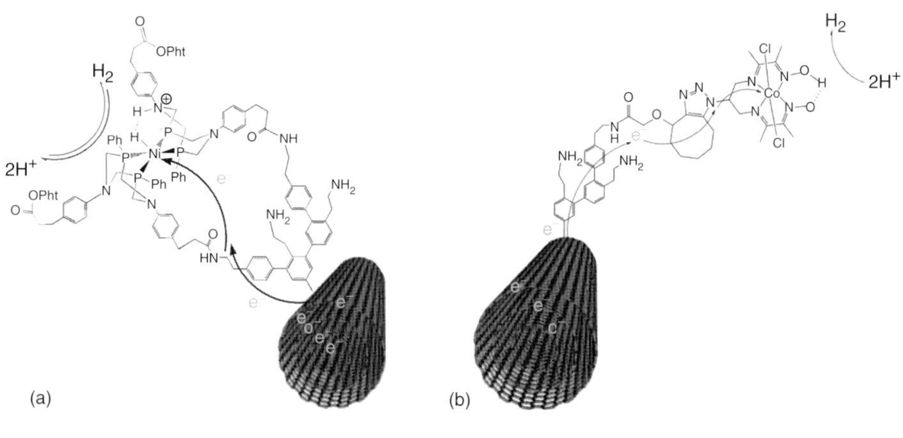

Figure 14.6 Schematic representation of the structure and reactivity of the bio-inspired H_2-evolving nickel (a) [14] and cobalt (b) [17] catalyst grafted on a carbon nanotube. Electrons are exchanged between the carbon nanotube and the metallic centers where H^+ is reduced to H_2 or H_2 oxidized to H^+.

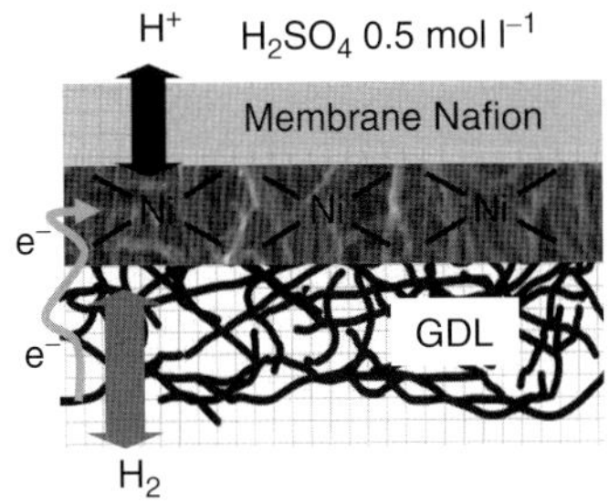

Figure 14.7 Structure of the membrane-electrode assembly used for the electrocatalytic characterization of the Ni-functionalized carbon nanotubes under conditions relevant for PEM applications.

homogeneous bulk phase. This was possible through a subtle modification of the substituents at the P_2N_2 ligand [18]. To the best of our knowledge, the Ni-functionalized CNTs were the first molecular-engineered and noble-metal free electrode material capable of achieving hydrogen evolution/oxidation without overpotential requirement (Figure 14.6).

The same methodology was used to stabilize a diimine–dioxime cobalt catalyst at the surface of an electrode made of CNTs (Figure 14.6). Again, this yields a very active electrocatalytic cathode material, mediating H_2 generation with a TON of 55 000 in 7 h from pure aqueous solutions (acetate buffer pH 4.5) at relatively low overpotentials (onset potential for H_2 evolution is −350 mV vs reversible hydrogen electrode (RHE)) [17]. Scanning electron microscopy was used to exclude the formation of any particle at the surface of CNTs and XPS confirmed the preservation of the electronic structure of the grafted molecular complex after extensive catalytic turnover. This material is thus remarkably stable, allowing extensive cycling without degradation, which contrasts with the limited stability of the same molecular catalyst under bulk electrolysis conditions. This clearly indicates that grafting provides a largely increased protection of these cobalt catalysts against modification and allows preserving their molecular structure and the related activity.

14.2.1.3 Noncovalent Attachment of Catalyst to Electrode Material Via π–π Stacking Interaction

Among the available noncovalent CNT functionalization approaches, the π–π stacking of ligands on the nanotube surface is found to be a viable approach. Pyrene derivatives are well known for their ability to adsorb onto graphite via a π–π stacking interaction. We successfully used this strategy for the immobilization of DuBois' nickel bisdiphosphine H_2-evolving catalysts on CNTs. Pyrene derivatives of nickel bisdiphosphine complexes (Figure 14.8) can be easily immobilized onto CNTs, deposited on gas diffusion layers. This one-step process is practically a much easier method than the covalent grafting as described, which necessitates three steps: (i) deposition of CNTs onto the electrode support; (ii) modification of the nanotubes with the electrografting process; and (iii) post-functionalization reaction. In addition using the π-stacking method,

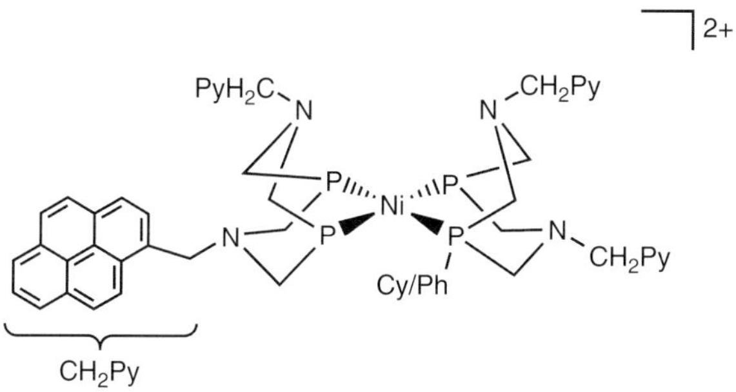

Figure 14.8 Schematic structure of nickel(II) bisdiphosphine complex containing four pyrene π-stacking anchor sites.

surface catalyst loading can be easily tailored by varying the amount of CNTs deposited on the electrode. For example, a catalyst surface concentration as high as $8 \cdot 10^{-9}$ mol cm^{-2} was obtained with a CNT deposit of 0.2 mg cm^{-2}. For comparison, the covalent method relying on diazonium electroreduction followed by an amidation post-functionalization step affords a limited surface coverage of $1.5 \cdot 10^{-9}$ mol cm^{-2} [19]. Finally, when stacked onto CNTs, the nickel bisdiphosphine complexes show catalytic performances comparable with those displayed by the complexes covalently attached to CNTs. We thus believe that this methodology will be useful to elaborate more efficient electrocatalytic materials.

In these two Ni-functionalized CNTs materials, the Ni-molecular catalyst is located at the crossroads of the three interpenetrated networks allowing percolation of protons (the Nafion membrane), hydrogen (the pores in the gas diffusion layer), and electrons (the carbon fibers of the gas diffusion layer relayed by the conducting CNTs). In a way and even if it is not as well defined as in the protein, the catalyst environment in this membrane-electrode assembly reproduces that found in the active sites of hydrogenases buried into the polypeptidic framework but connected to the surface of the protein via a gas diffusion channel, a network of hydrogen-bonded amino acids for proton transport and the array of electron-transferring iron–sulfur clusters.

At this point we note that the bio-inspired approach, coupled with nanochemical tools, allowed for the preparation of the first noble-metal free catalytic materials for H^+/H_2 evolution working in both directions and without overpotential requirement. However, the current densities displayed by these materials are in the 1–10 mA cm^{-2} range, depending on temperature, while Pt-based active layers currently developed for electrolyzers and fuel-cells exceed 1 A cm^{-2}. There is thus still room for improvement. Both the optimization of the TOF of the molecular catalysts and the increase in the catalyst loading within the tridimensional architecture of the electrode will allow improvement of the performances of such bio-inspired materials.

14.2.2
Electrode Materials Based on Bio-Inspired All-Inorganic Catalysts

14.2.2.1 Metal-Sulfide Catalysts

All-inorganic materials, which do not contain any fragile organic ligands, are also developed as potentially more robust alternatives to metal-organic molecular catalysts. Promising non-noble metal-oxides-based electrocatalysts have been reported for water oxidation [20–25]. For hydrogen evolution catalysis, great interest has recently been paid to metal sulfides, particularly those of molybdenum (MoS_2) and tungsten (WS_2). Indeed, [Mo–S] and [W–S] clusters are found in the catalytic active sites of CO dehydrogenases and nitrogenases, which could generate hydrogen under certain conditions (Chapters 7, 8, and 12) [26].

NiS_x, CoS_x, and FeS_x were known for long as H_2-evolving electrocatalysts in alkaline solution [27, 28]. Owing to the pronounced ionic nature of the metal–sulfur bond in these materials, they are not stable under acidic and release H_2S. Mo–S and W–S bonds are more covalent, which makes MoS_2 and WS_2 chemically robust even in strong acidic solution. These two sulfides have a layered morphology consisting of S–M–S (M = Mo or W) hexagonal packed layers, which are held together by van der Waals interactions with a characteristic layer-to-layer distance of about 0.6–0.7 nm. These materials are well known as heterogeneous catalysts for hydrodesulfurization [29]. Recently, they were also identified as attractive electrocatalysts for hydrogen evolution, with better catalytic activities displayed by MoS_2 as compared to WS_2 [30]. Chorkendorff and colleagues [30, 31] showed that the Mo edges of the MoS_2 layers are the catalytically active sites for hydrogen evolution. The electrocatalytic current density related to hydrogen evolution is thus proportional to the density of edge sites. Hence, a solution to improve catalytic activities of MoS_2 materials relies on increasing the edge surface area, which can be achieved by depositing MoS_2 in an amorphous form. Electrochemically deposited amorphous MoS_2 as reported by Hu [32] is the most active MoS_2 electrocatalysts described so far. Another exciting approach was reported by Dai [33] by employing graphene sheets as a platform to selectively grow MoS_2 nanoclusters (few layers), which expose high edge surface area. The MoS_2/graphene composite was found to be more active than a pure MoS_2 equivalent, displaying significantly higher H_2 evolution current densities at a given potential. Besides the high surface area of edge sites in such a composite, the high electronic conductivity within graphene sheets, which allows for a fast electron transfer to the MoS_2 active sites, might also play an important role in enhancing electrocatalytic activities. A similar positive effect was reported when we immobilized nickel bisdiphosphine or cobalt diimine–dioxime catalysts onto CNT surfaces [14, 17, 19].

The intrinsic electrocatalytic activities of metal-sulfide materials can also be significantly enhanced by introducing a promoter, such as Ni and Co ions, within their structure [30, 34]. Such a modification is accompanied by morphology changes [35]. However, the real structure of these promoted materials is still not clear, because the materials are generally amorphous. From density functional theory (DFT) calculations, it was proposed that promoters activate the catalytically inactive S-edges of the metal sulfide [30]. Further efforts are required to gain more insights into the structure and catalytic mechanisms at work at the surface of these promoted materials. However, the promoting effect also raised the question whether introducing a promoter does create novel bimetallic catalytic centers [promoter-S_x-M] that might resemble those found in enzymes.

The concept of metal-sulfide electrocatalysts for hydrogen evolution can indeed be extended to heterobimetallic ternary sulfides [36]. By a simple precipitation process between $[Cu^{I}(MeCN)](BF_4)$ and $[M^{VI}S_4]^{2-}$ (M = Mo, W) precursors in dimethylformamide (DMF)/acetonitrile solvent, crystalline Cu_2MS_4 ternary sulfides can be obtained. Their structure is shown in Figure 14.9 [36, 37]. Interestingly, the $\{Cu_4S_4M\}$ building block of these materials possesses several structural similarities to the active sites of molybdenum CO-dehydrogenase [26] and of hydrogenases [38–40], suggesting that they could be considered as solid-state bio-inspired equivalents of the active sites of these enzymes. Besides, molybdenum CO-dehydrogenase is known to display hydrogenase activity under some conditions [41]. Thanks to such a structural similarity, Cu_2MoS_4 is a good catalyst for H_2 evolution over a wide range of pH in aqueous solution (0–12) [36]. In a pH 7 phosphate buffer solution, Cu_2MoS_4 catalyzes hydrogen evolution with a relatively low overpotential requirement (about 135 mV), apparent exchange current density of 0.04 mA cm^{-2}, and quantitative current-to-hydrogen efficiency. The tungsten analog, Cu_2WS_4, is also active for H_2 evolution but requires about 200 mV additional overpotential to achieve a similar catalytic current density. Nevertheless, both Cu_2MoS_4 and Cu_2WS_4 catalysts are robust and remain intact under extended catalytic turnover. Hence, these solid-state catalysts represent an attractive alternative to platinum and bio-inspired molecular catalysts [36].

14.2.2.2 Electrode Engineering from Metal-Sulfide Catalysts

The simplest way to prepare electrode materials from metal-sulfide solid-state catalysts consists in drop-casting a suspension of these catalysts (in water or volatile organic solvent) on electrode surface [42]. A similar strategy was applied to load clusters containing a $[Mo_3S_4]$ core onto a Si electrode surface from a solution in

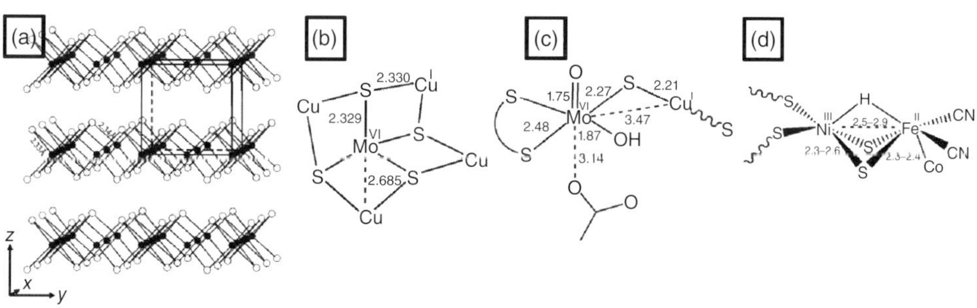

Figure 14.9 Crystalline structure of Cu_2MoS_4 (a); $[Cu_4S_4Mo]$ building block (b); catalytic centers of molybdenum CO-dehydrogenase (c); and [NiFe]-hydrogenase (d) [36].

CH_2Cl_2/CH_3OH [43]. By doing so, electrodes with low morphological homogeneity are obtained. Adopting spray-casting techniques was reported to be valuable to improve the homogeneity of the electrode [42]. Another issue is the low mechanical stability of the electrode and detachment of the catalyst from the electrode substrate is expected to occur upon generation of hydrogen bubbles. The mechanical properties of electrodes could be enhanced by employing a polymeric binder such as Nafion, which is also a proton-conducting matrix [33, 36]. However, mass transport (electron and proton transport) could be an issue for these electrodes.

As earlier discussed in the section dealing with immobilization of molecular catalysts on electrode surface, covalent grafting is the most appropriate approach for engineering a robust electrode. Indeed, the first example of covalently grafting of $[Mo_3S_4]$-containing clusters on an Au(111) surface via formation of Au–S chemical bond was reported by Chorkendorf's group [44]. This monolayer self-assembling methodology is exclusively applied for Au electrode and might be only viable for the specific $[Mo_3S_4]$ incomplete cubane-like clusters that contain the S^{2-} ligands in its structure.

Metal-sulfide catalysts can also be directly synthesized onto electrode substrates. MoS_2/Toray carbon paper electrodes can be obtained through the impregnation of Toray carbon paper with a solution of ammonium heptamolybdate precursor followed by drying and further annealing under a H_2S/H_2 atmosphere [30]. More attractive approaches rely on the direct electrodeposition of the MoS_2 catalyst onto electrode substrates from aqueous solutions of $[MoS_4]^{2-}$ precursor. Controlled potential anodic (Equation 14.1) or cathodic (Equation 14.2) deposition resulted in an amorphous MoS_2 film [45]. The film obtained by an anodic deposition usually contained some S_8 impurities. Therefore, it was also referred to as *MoS_3 film*. More recently, Hu and colleagues [32] reported that a dynamic potential deposition also resulted in an amorphous MoS_3 film. Interestingly, the authors reported that a MoS_3 film obtained either by anodic deposition or by dynamic potential deposition was catalytically more active for hydrogen evolution than a MoS_2 film obtained by cathodic deposition, although in both cases, MoS_2 is the real catalytically active species.

$$[MoS_4]^{2-} \rightarrow MoS_2 + (1/8)S_8 + 2e^- \tag{14.1}$$

$$[MoS_4]^{2-} + 2e^- + 2H_2O \rightarrow MoS_2 + 2SH^- + 2OH^- \tag{14.2}$$

Electrodeposition was also found to be valuable to incorporate Co, Ni, and Fe ions in MoS_2 and improve their activity for electrocatalytic hydrogen evolution [34].

The possibility of electrodepositing metal-sulfide catalysts paves a new and attractive way to combine these catalysts with appropriate semiconductor via a so-called photoassisted electrodeposition process and engineer photoelectrode materials or photocatalyst for hydrogen generation [46]. We discuss this methodology in more detail in the following section.

14.3 Photoelectrode Materials for Hydrogen Evolution

With efficient catalytic materials in hand, the next step toward the construction of a PEC cell is the combination with a light-harvesting agent that converts the photon energy into an electrochemical potential, finally used by the catalyst to generate the fuel. Here also, two solutions can be used, whether the light-harvesting element is a solid-state semiconductor material or a molecular photosensitizer.

14.3.1 All-Inorganic Photocatalysts Composed of Solid-State Semiconductor and Solid Inorganic Catalyst

Since the pioneering work on light-driven water splitting reported by Fujishima and Honda [47], several families of semiconductors (silicon, metal oxide, metal sulfide, etc.) have been developed and/or evaluated in the context of photocatalytic water oxidation or hydrogen generation. Current challenges in developing semiconductor materials for water-splitting application concern (i) the search for novel components which could exploit a larger part of the visible spectrum, (ii) the enhancement of PEC stability of semiconductors especially those with a narrow band gap, (iii) the nanostructuration of semiconductor materials to enhance their charge separation properties as well as the transport of charges (electrons and holes) to the surface where they can drive chemical reactions, and (iv) their optimal combination with appropriate electrocatalysts accelerating the chemical reaction. In this section, we discuss the last point dealing with engineering of photocatalyst for hydrogen generation by interfacing an electrocatalyst with a solid-state light absorber.

As earlier stated, platinum is so far and by far the best electrocatalyst for H_2 evolution in terms of catalytic activity and robustness. It has therefore been interfaced with many semiconductor materials such as Si [48–50], Cu_2O [51, 52], or Cu(In)(Ga)S(Se) [53, 54]. However, recent demonstrations have been made regarding the use of more abundant and therefore much cheaper H_2-evolving catalysts.

The Lewis group [48] in Caltech evidenced that a Si microarray electrode decorated with a Ni or a NiMo alloy as HER electrocatalyst displayed comparable photocatalytic activities to those achieved for a similar Si electrode but decorated with platinum. However, Ni and its alloy are not chemically stable in nonalkaline solution, while Si suffers from oxidation under alkaline to neutral conditions. Therefore, more robust electrocatalysts are required. Molybdenum-sulfide-based catalysts, for example, incomplete cubane-like clusters with the $[Mo_3S_4]$ core or MoS_2 were found to be such good candidates. Decorating Si nanopillar electrode with $[Mo_3S_4]$-containing clusters (likely used as a precursor of a solid-state catalyst with a MoS_2 composition) resulted in a huge anodic shift (500 mV) of the onset potential for H_2 evolution under irradiation, which is still the most important shift ever recorded so far for a Si electrode [43]. At 0 V versus NHE, in a pH 0 electrolyte solution and under visible light illumination, the Si/[Mo_3S_4] photoelectrode shows an impressive photocurrent density of 9.0 mA cm^{-2} which corresponds to a solar-to-hydrogen conversion efficiency of 10%. MoS_2 and WS_2 are other H_2-evolving catalysts having been used for interfacing with semiconductor materials for the construction of photocatalytic materials for hydrogen generation [46, 55–58]. Li's group [55] reported that a CdS/MoS_2 catalyst displayed higher photocatalytic activities compared with a CdS/Pt equivalent. Recently, we found that a Si/MoS_2 electrode is comparable to a Si/Pt electrode for light-driven hydrogen evolution (Figure 14.10a) [46]. Interestingly, MoS_2 can be loaded onto an Si surface via a photoassisted electrodeposition process. In such a process, photogenerated electrons on the Si conduction band migrate to the surface where they reduce $[MoS_4]^{2-}$ producing MoS_2. As a result, only a moderate external bias is required and MoS_2 is deposited under mild conditions, that is, from an aqueous solution, under visible light irradiation, at room temperature, and under atmospheric pressure. This process is therefore much more attractive than the conventional thermal decomposition method, which involves the decomposition of $[MoS_4]^{2-}$ at high temperature (400–600 °C) under toxic atmosphere of H_2S/H_2 mixture [30, 55].

It is worth mentioning that spontaneous Si oxidation, forming a nonconductive SiO_2 layer, leads to degradation of this photoelectrode. To avoid the formation of such a native oxide layer, the dissolved oxygen concentration should be lower than 15 ppb, which is not feasible in a PEC device [43]. In practice, Si oxidation can be prevented through the deposition of a conducting protective layer such as indium tin oxide (ITO) or Ti [59, 60]. Using this approach, Chorkendorf's group reported on the design of multiple layer photocathode Si/Ti/MoS_2 (Figure 14.10b). In this design, the thin layer of Ti acts as an effective protective layer, while the electrodeposited MoS_2 porous layer acts as catalyst for hydrogen evolution.

In this configuration, the Si/Ti/MoS_2 electrode is photocatalytically less active than a Si/Ti/Pt electrode equivalent. The former electrode required about 140 mV more negative potential to generate a photocurrent similar to that achieved employing the latter platinum-decorated electrode [59]. This result is in line with

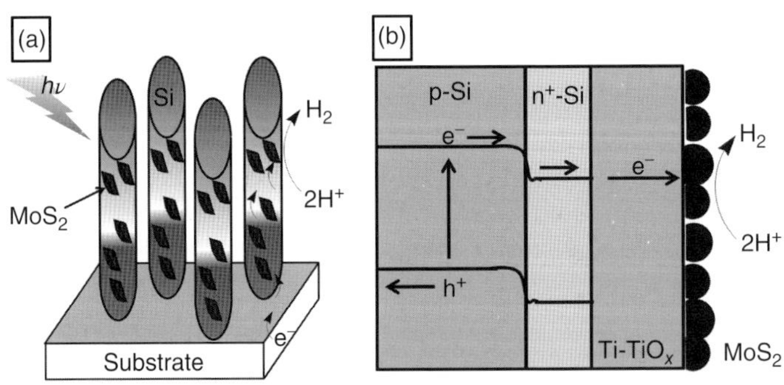

Figure 14.10 (a,b) Photocathode engineering for hydrogen generation composed of Si absorber and molybdenum-sulfide-based catalyst [43, 46, 59].

the fact that the best MoS_2-based material is not as good as an electrocatalyst when compared to platinum [30–32, 42]. However, when MoS_2 is directly interfaced with semiconductors, the composite semiconductor/MoS_2 display similar [43, 46] or even higher [55] photocatalytic activities than the composite semiconductor/Pt electrode. This is due to a better interfacial interaction between MoS_2 and the semiconductor material, which allows for a more efficient electron injection from the conduction band of the semiconductor to MoS_2 [55]. Another possible explanation could reside in a distinct bending of the conduction and valence bands of the semiconductor at the interface with MoS_2.

In these architectures, the robustness of the inorganic catalyst toward decomposition or hydrolysis warrants a large chemical stability under operation. However, the catalysts were often simply deposited onto the semiconductor surface without any robust linker. Therefore, the detachment of the catalyst into solution during catalytic hydrogen generation often occurs. For example, the detachment of Pt nanoparticles from Cu_2O photocathode resulted in a decrease of the catalytic activity, and frequent redecoration of the electrode with Pt is required [51, 61]. Similarly, irreversible leaching of the $[Mo_3S_4]$-based catalyst, which was physically absorbed on Si surface, to the solution was observed [43].

14.3.2
Solid-State Semiconductor and Molecular Catalyst

Until very recently, the only photoelectrodes for hydrogen photoproduction based on functional molecular H_2-evolving catalysts were based on semiconductors for light harvesting and charge separation: coating of a p-type silicon material with polystyrene functionalized with 1-methyl-[1.1]ferrocenophane (Figure 14.11a) provided a H_2-evolving photocathode: under illumination, the voltage photogenerated within the silicon layer allows proton reduction to occur with an underpotential of 300 mV, that is, the proton reduction occurs at a potential 300 mV more positive than the standard apparent potential of the H^+/H_2 couple [62].

A similar approach has been recently reported by Nann, Pickett, and coworkers aiming at building a H_2-evolving photoelectrode (Figure 14.11b). Nanocrystals of InP, a semiconductive material absorbing a significant part of the visible spectrum, have been deposited and connected, through interactions with 1,4-benzenedithiolate, on a gold electrode using a layer-by-layer procedure. $[Fe_2(\mu\text{-}S)_2(CO)_6]$ has then been adsorbed as a H_2-evolving catalyst on this material and photocurrents ($\sim$250 nA cm^{-2}), associated with H_2 evolution from neutral aqueous solution, could be measured under illumination with 395 nm light emitting diode (LED), while the electrode was poised at −500 mV versus Ag/AgCl, thus 250 mV more positive than the standard potential of the H^+/H_2 couple [63].

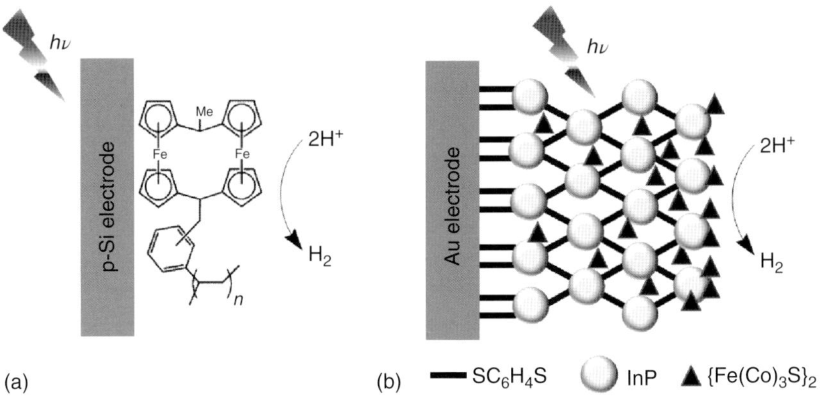

Figure 14.11 Structure of the H_2-evolving photocathode based on a p-type Si electrode covered with a [1.1] ferrocenophane derivative [62] (a) or a diiron mimic of the active site of FeFe hydrogenase (b) [63].

14.3.3 All-Molecular-Based Electrode Materials

In photosynthetic organisms, the whole photocatalytic system is molecular. The biochemical knowledge now at our disposal regarding the manipulation of the natural systems, PS II [64–66] and hydrogenases [67] allows copying of the natural working scheme and construction of a H_2-evolving photocathode by combining a PS and a hydrogenase enzyme in an artificial way on the surface of a conductive substrate.

A photocatalytic bio-construct (Figure 14.12) was designed by Heberle and colleagues using the membrane-bound hydrogenase from *Ralstonia eutropha* H16 fused with an extrinsic subunit (PsaE) of a histidine-tagged form of PS I [68]. This system has been immobilized at the surface of a nickel-functionalized gold electrode through a nickel His–Tag interaction. When this electrode is poised at −0.09 V versus NHE, so as to reduce the soluble electron carrier *N*-methylphenazonium methyl sulfate (PMS), and under visible illumination (λ_{max} = 700 nm), photocurrents are observed (85 nA cm^{-2}), 30% of which could be assigned to hydrogen evolution [69].

Molecular photosensitizers and bio-inspired catalysts can also be used instead of enzymes [70]. Early achievements in this field have been made with multicomponent systems, where photosensitizers and catalysts are free in solution. More recently, a series of supramolecular photocatalytic H_2-evolving systems have been reported [71] as a first step toward the construction of fully molecular photocathode materials. In that context, bio-inspiration can also lead to the design of smart linkers that may favor proton-coupled electron transfers within the system [72–74]. Other supramolecular systems that have been reported on the basis of mononuclear cobalt catalysts, mainly cobaloximes, are shown in Figure 14.13 [71, 75].

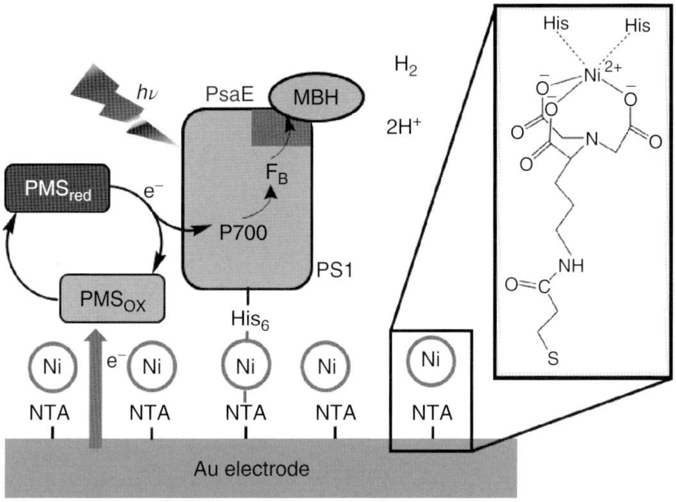

Figure 14.12 Structure of the biochemical H_2-evolving photoelectrode designed by Heberle and coworkers [69].

Despite the efforts of a number of groups, no photocathode could be obtained so far through the simple grafting of such supramolecular photocatalyst onto transparent conducting oxide materials. Recently, however, the group of Sun showed how it is possible to take benefit from the knowledge accumulated in the context of dye-sensitized solar cells (DSSCs) [81] to design a first proof-of-concept for a fully molecular H_2-evolving photocathode [82]. The assembly shown in Figure 14.14 is based on a dye-sensitized photocathode originated from research on tandem DSSCs [83, 84]. This photocathode relies on efficient hole injection from an excited push–pull organic chromophores into a p-type semiconducting transparent oxide, NiO. This electrode has been coated with a Nafion film embedding a cobaloxime catalyst. Upon irradiation, the resulting photoelectrode generates a photocurrent (5–15 $\mu A\,cm^{-2}$), associated with H_2 evolution, that rapidly vanished because of the leaching of the cobaloxime catalyst out of the solution. While both the photocurrent value and the stability are low, this is the first demonstration that the bio-inspired molecular approach works for the production of hydrogen from sunlight and water in a PEC device. Similar approaches have been also developed to construct O_2-evolving photoanodes by coating a dye-sensitized TiO_2 photoanode with molecular catalysts for water oxidation embedded into Nafion [85, 86].

14.4 Artificial Photosynthetic Systems

At this point, the final question resides in the possibility of coupling both H_2- and O_2-evolving photocatalytic systems to form a device. The seminal work on that

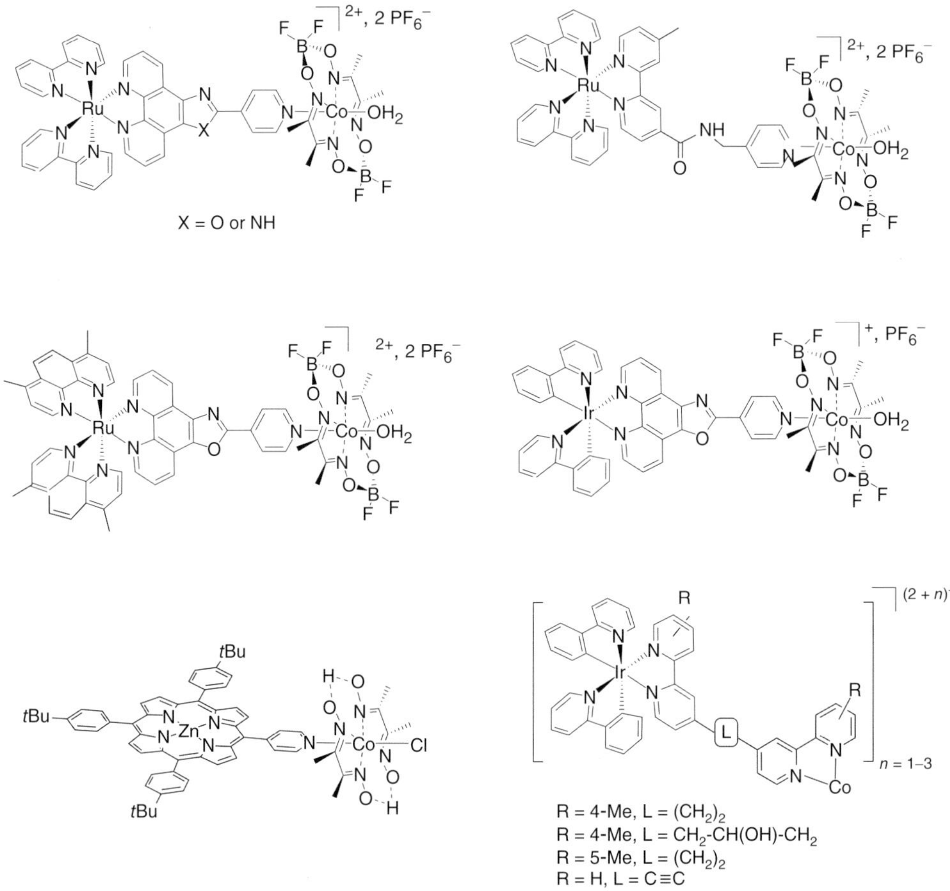

Figure 14.13 Structure of selected supramolecular photocatalysts for light-driven hydrogen production [76–80].

topic was reported by Honda and Fujishima in 1972 [47], with a water-splitting cell made from a TiO_2 photoanode and a Pt cathode. These results are however questioned as H_2 generation has not been directly proved in that case and the cathode reaction is likely a reduction of O_2 generated at the TiO_2 photoanode [87]. Nevertheless, this concept has been further developed by Domen, who reported active microheterogeneous systems for overall water splitting. An example of such a system contains two semiconducting materials such as ZrO_2/TaON and WO_3. Their low band-gaps allow them to absorb visible light. The particles of ZrO_2/TaON and WO_3 are decorated with smaller particles of platinum as catalysts for H_2 evolution, and O_2 evolution, respectively. Both Pt/ZrO_2/TaON and Pt/WO_3 systems are suspended in water in the presence of the IO_3^-/I^- couple that act as a redox mediator between them [88]. This system achieves overall water splitting with a quantum yield of 6.3% under irradiation by 420.5 nm monochromatic light. The major issue

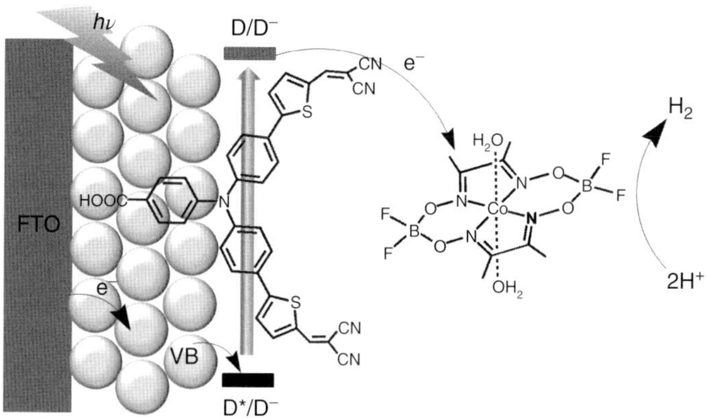

Figure 14.14 Structure of the fully organic photocathode designed by Sun and colleagues [82].

associated with such a process is that O_2 and H_2 are produced in the same compartment and in stoichiometric amounts so that efficiency may be associated to explosion risks.

These safety issues disappeared when both gases were produced in distinct compartments, as in a PEC cell. Several examples of self-operating PEC cells (i.e., without any external energy source except sunlight) are known in the literature. A first series could be described as a combination of a fully inorganic photovoltaic component (generating a photovoltage at solid-state junctions between inorganic semiconductors) coated with catalysts for H_2 and O_2 evolution at both sides (Figure 14.15a) [60, 89, 90]. These devices are wireless forms of the brute force approach consisting in coupling a photovoltaic cell with a water electrolyzer. The most recent example [60] uses noble-metal free catalysts for both water oxidation and hydrogen evolution, that is, the anodically deposited cobalt oxide (O_2-CoCat) reported by Nocera in 2008 [20, 91] and an electrodeposited NiMoZn alloy. A second series still contains a photovoltaic component but in combination with a photoelectrode, either for H_2 photoproduction ($GaInP_2$/Pt in the monolithic photovoltaic–PEC device reported by Turner) [92] or O_2 evolution (α-Fe_2O_3 or WO_3 in the devices reported by Miller and Rocheleau) [93–96] (Figure 14.15b,c). These first examples display quite good efficiencies for energetic conversion (12.4% for the cell reported by Turner) but are very expensive because of the fabrication of the photovoltaic assembly. The so-called tandem cell developed by Grätzel at EPFL (Switzerland) belongs to the second series but includes a cheap DSSC as the photovoltaic component feeding a WO_3/Pt or a Fe_2O_3/Pt PEC cell (Figure 14.15d). It develops 3.5 mA cm^{-2} with an overall yield of ~5% [81]. The cell described by Dismukes and Spiccia also belongs to this class. Its initial design took advantage of a synthetic $Mn_4O_4L_6$ cubane-like complex, bearing diarylphosphinate L ligands (Figure 14.16), which experiences conversion of its bridging oxides into O_2 upon photooxidation in the gas phase [97]. This photocatalyst was entrapped in Nafion polymer and deposited onto a TiO_2

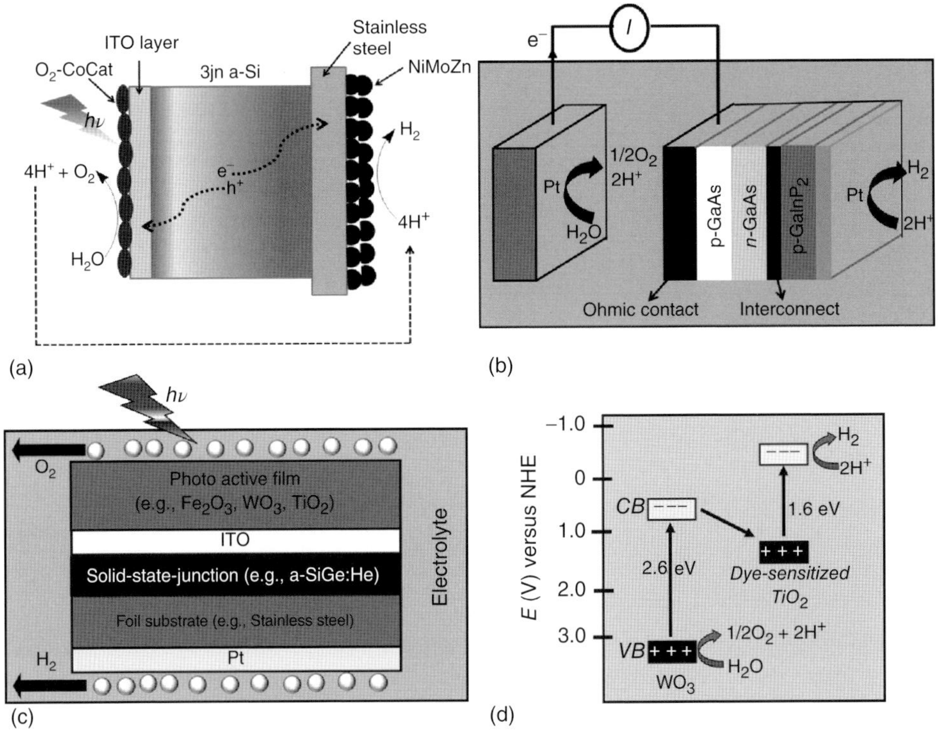

Figure 14.15 Selected devices for photocatalytic water splitting: (a) Miller [89] and Nocera's [60] artificial leaf, (b) Turner's monolithic photoelectrochemical–photovoltaic design [92], (c) Hawaï hybrid planar multijunction photoelectrode [93–96], and (d) principle of Grätzel's tandem cell [81].

electrode sensitized with a Ru-diimine dye. This photoanode, which uses two photons to sustain water oxidation, was coupled to a platinum electrode for the construction of a photoelectrocatalytic cell achieving overall water splitting without external bias (Figure 14.16) [85, 98]. While a reinvestigation of this system by Spiccia [99] finally demonstrates that the catalytic species is not the molecular cubane complex but a manganese oxide phase which evolves O_2 upon light absorption, this is an important result owing to the simple construction of this device as compared to Grätzel's tandem cell.

14.5 Toward Photoelectrode Materials for CO_2 Reduction

One of the most exciting and challenging reactions, also related to artificial photosynthesis, is the photoreduction of CO_2 into carbon-based fuels. As for water splitting, elaboration and optimization of a photoelectrode for CO_2 reduction are requested. A number of homogeneous systems, based on a photosensitizer,

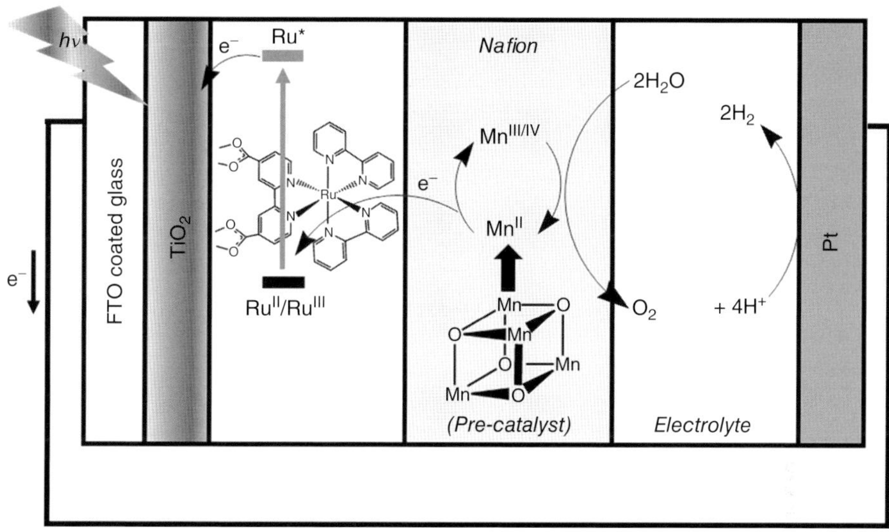

Figure 14.16 Schematic representation of the photoelectrochemical device for overall water splitting reported by Spiccia and colleagues (see text for details about the true active species in this device) [85].

in most cases ruthenium tris-bipyridine $[Ru(bipy)_3]^{2+}$ complex, because of its stability and its interesting photophysical and redox properties, and a molecular catalyst, such as $[Re(bipy)(CO)_3\{P(OEt)_3\}]^+$ [100] or $[Ru(bipy)_2(CO)_2]^{2+}$ [101] complexes, have been studied. The two moieties can be covalently associated into a multifunctional system such as the one shown in Figure 14.17, which is among the most active ones (several hundreds of catalytic cycles and quantum yields in the range of 20%) [100].

Besides this homogeneous approach, CO_2 photoreduction has also been carried out in heterogeneous systems, mainly using a variety of semiconductors associated with metallic catalysts such as Pd, Pt, or Cu [102].

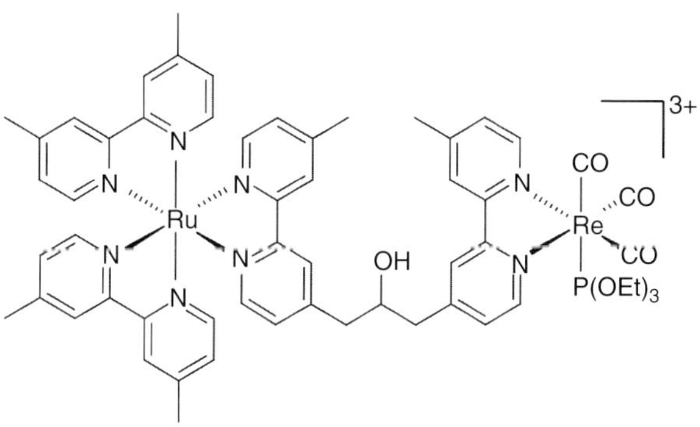

Figure 14.17 Supramolecular photocatalyst for light-driven CO_2 reduction, designed by Ishitani' group [100].

In the following, we focus on the very few examples in which a semiconductor has been associated with a molecular catalyst, either an enzyme or a synthetic complex. The first example of such a device was reported by Parkinson in 1984. The photoelectrode was made from a combination of a p-type semiconductor, namely Zn-doped indium phosphide, and a formate dehydrogenase, which catalyzes the reversible CO_2/HCO_2H interconversion, adsorbed at the surface of solid material. Upon illumination of the photocathode with visible light in the presence of methyl viologen as a redox mediator in phosphate aqueous buffer, CO_2 is efficiently converted into formate (the photovoltaic conversion efficiency of the cell is ~6–7%) [103]. Surprisingly, such a system has not been further investigated, most probably because this class of enzyme is highly unstable especially with regard to oxygen.

Examples of semiconductors coupled to molecular catalysts for CO_2 reduction are rare. The first one uses p-type hydrogenated Si as the photocathode and Re(bipy-Bu^t)$(CO)_3$Cl as the catalyst. Photoassisted electrochemical reduction of CO_2 to CO on this photocathode was achieved at a potential 600 mV lower than that required with a Pt electrode in the dark, with a quantitative electrical-to-chemical energy conversion efficiency and an overall light-to-chemical energy conversion efficiency of 10%. The system is stable, lasting for hours without any sign of degradation [104, 105]. The p-type InP solid-state semiconductor has also been used to build a hybrid photocatalytic system for CO_2 reduction. A ruthenium complex embedded in a polypyrrole film coating the surface of InP was used as the catalyst for the conversion of CO_2 into formate. Remarkably, this photocathode could be efficiently combined with TiO_2, used as a photoanode for water oxidation, so that an artificial photosynthetic system able to carry out light-assisted reduction of CO_2 by H_2O to formate, with a selectivity for formate of 70% and a solar-to-chemical energy conversion efficiency of 0.04%, could be achieved [106].

14.6 Conclusion and Perspective

Within a few years, remarkable progress has been obtained for the construction of novel electrodes and photoelectrodes for H_2 evolution and water oxidation, the two sides of water splitting. The stability and turnover frequencies of molecular and solid-state catalysts have been significantly improved and it has been demonstrated that molecular catalysts can be integrated into stable electrode materials. New methods for nanostructuring electrode materials and for grafting molecular compounds at their surface have appeared. Catalysts have been successfully interfaced with semiconducting materials or molecular photosensitizers and a diversity of synthetic supramolecular assemblies combining a catalyst unit and a photosensitizer have been reported. Finally, while the pieces of the puzzle are elaborated in different laboratories all over the world, awareness of the importance of international collaboration has increased and it is likely that, through collaborative international effort, an efficient and thus an economically viable photoelectrocatalytic system, absolutely free of noble metals, for overall water splitting will appear in the very next future.

References

1. Pershad, H.R., Duff, J.L.C., Heering, H.A., Duin, E.C., Albracht, S.P.J., and Armstrong, F.A. (1999) *Biochemistry*, **38**, 8992–8999.
2. Jones, A.K., Sillery, E., Albracht, S.P.J., and Armstrong, F.A. (2002) *Chem. Commun.*, 866–867.
3. Kellett, R.M. and Spiro, T.G. (1985) *Inorg. Chem.*, **24**, 2378–2382.
4. Abe, T., Taguchi, F., Imaya, H., Zhao, F., Zhang, J., and Kaneko, M. (1998) *Polym. Adv. Technol.*, **9**, 559–562.
5. Zhao, F., Zhang, J., Abe, T., Wohrle, D., and Kaneko, M. (1999) *J. Mol. Catal. A*, **145**, 245–256.
6. Cosnier, S., Deronzier, A., and Vlachopoulos, N. (1989) *J. Chem. Soc., Chem. Commun.*, 1259–1261.
7. Ibrahim, S.K., Liu, X.M., Tard, C., and Pickett, C.J. (2007) *Chem. Commun.*, 1535–1537.
8. Pool, D.H. and DuBois, D.L. (2009) *J. Organomet. Chem.*, **694**, 2858–2865.
9. Vijaikanth, V., Capon, J.F., Gloaguen, F., Schollhammer, P., and Talarmin, J. (2005) *Electrochem. Commun.*, 7, 427–430.
10. Anne, A., Blanc, B., Moiroux, J., and Saveant, J.M. (1998) *Langmuir*, **14**, 2368–2371.
11. Pinson, J. and Podvorica, F. (2005) *Chem. Soc. Rev.*, **34**, 429–439.
12. Thomas, C.M., Rudiger, O., Liu, T., Carson, C.E., Hall, M.B., and Darensbourg, M.Y. (2007) *Organometallics*, **26**, 3976–3984.
13. Le Goff, A., Artero, V., Metayé, R., Moggia, F., Jousselme, B., Razavet, M., Tran, P.D., Palacin, S., and Fontecave, M. (2010) *Int. J. Hydrogen Energy*, **35**, 10719–10724.
14. Le Goff, A., Artero, V., Jousselme, B., Tran, P.D., Guillet, N., Metaye, R., Fihri, A., Palacin, S., and Fontecave, M. (2009) *Science*, **326**, 1384–1387.
15. Evrard, D., Lambert, F., Policar, C., Balland, V., and Limoges, B. (2008) *Chem. Eur. J.*, **14**, 9286–9291.
16. Belanger, D. and Pinson, J. (2011) *Chem. Soc. Rev.*, **40**, 3995–4048.
17. Andreiadis, E.S., Jacques, P.-A., Tran, P.D., Leyris, A., Chavarot-Kerlidou, M., Jousselme, B., Matheron, M., Pécaut, J., Palacin, S., Fontecave, M., and Artero, V. (2012) *Nat. Chem.* doi: 10.1038/NCHEM.1481
18. Smith, S.E., Yang, J.Y., DuBois, D.L., and Bullock, R.M. (2012) *Angew. Chem. Int. Ed.*, **51**, 3152–3155.
19. Tran, P.D., Le Goff, A., Heidkamp, J., Jousselme, B., Guillet, N., Palacin, S., Dau, H., Fontecave, M., and Artero, V. (2011) *Angew. Chem. Int. Ed.*, **50**, 1371–1374.
20. Kanan, M.W. and Nocera, D.G. (2008) *Science*, **321**, 1072–1075.
21. Jiao, F. and Frei, H. (2009) *Angew. Chem. Int. Ed.*, **48**, 1841–1844.
22. Jiao, F. and Frei, H. (2010) *Chem. Commun.*, **46**, 2920–2922.
23. Najafpour, M.M., Ehrenberg, T., Wiechen, M., and Kurz, P. (2010) *Angew. Chem. Int. Ed.*, **49**, 2233–2237.
24. Wiechen, M., Zaharieva, I., Dau, H., and Kurz, P. (2012) *Chem. Sci.*, **3**, 2330–2339.
25. Zaharieva, I., Chernev, P., Risch, M., Klingan, K., Kohlhoff, M., Fischer, A., and Dau, H. (2012) *Energy Environ. Sci.*, **5**, 7081–7089.
26. Dobbek, H., Gremer, L., Kiefersauer, R., Huber, R., and Meyer, O. (2002) *Proc. Natl. Acad. Sci. U.S.A.*, **99**, 15971–15976.
27. Banica, F.G., Spataru, N., and Spataru, T. (1997) *Electroanalysis*, **9**, 1341–1347.
28. Sabela, R. and Paseka, I. (1990) *J. Appl. Electrochem.*, **20**, 500–505.
29. Topsoe, H., Lausen, B.S., and Massoth, F.E. (1996) in *Hydrotreating Catalysis-Science and Technology*, vol. 11 (eds J.R. Anderson and M. Boudart), Springer-Verlag, New York.
30. Bonde, J., Moses, P.G., Jaramillo, T.F., Norskov, J.K., and Chorkendorff, I. (2009) *Faraday Discuss.*, **140**, 219–231.

31. Jaramillo, T.F., Jørgensen, K.P., Bonde, J., Nielsen, J.H., Horch, S., and Chorkendorff, I. (2007) *Science*, **317**, 100–102.
32. Merki, D., Fierro, S., Vrubel, H., and Hu, X. (2011) *Chem. Sci.*, **2**, 1262–1267.
33. Li, Y., Wang, H., Xie, L., Liang, Y., Hong, G., and Dai, H. (2011) *J. Am. Chem. Soc.*, **133**, 7296–7299.
34. Merki, D., Vrubel, H., Rovelli, L., Fierro, S., and Hu, X. (2012) *Chem. Sci.*, **3**, 2515–2525.
35. Brorson, M., Carlsson, A., and Topsøe, H. (2007) *Catal. Today*, **123**, 31–36.
36. Tran, P.D., Nguyen, M., Pramana, S.S., Bhattacharjee, A., Chiam, S.Y., Fize, J., Field, M.J., Artero, V., Wong, L.H., Loo, J., and Barber, J. (2012) *Energy Environ. Sci.*, **5**, 8912–8916.
37. Pruss, E.A., Snyder, B.S., and Stacy, A.M. (1993) *Angew. Chem., Int. Ed. Engl.*, **32**, 256–257.
38. Volbeda, A., Charon, M.-H., Piras, C., Hatchikian, E.C., Frey, M., and Fontecilla-Camps, J.C. (1995) *Nature*, **373**, 580–587.
39. Peters, J.W., Lanzilotta, W.N., Lemon, B.J., and Seefeldt, L.C. (1998) *Science*, **282**, 1853–1858.
40. Nicolet, Y., Piras, C., Legrand, P., Hatchikian, C.E., and Fontecilla-Camps, J.C. (1999) *Structure*, **7**, 13–23.
41. Santiago, B. and Meyer, O. (1996) *FEMS Microbiol. Lett.*, **136**, 157–162.
42. Vrubel, H., Merki, D., and Hu, X. (2012) *Energy Environ. Sci.*, **5**, 6136–6144.
43. Hou, Y., Abrams, B.L., Vesborg, P.C.K., Björketun, M.E., Herbst, K., Bech, L., Setti, A.M., Damsgaard, C.D., Pedersen, T., Hansen, O., Rossmeisl, J., Dahl, S., Nørskov, J.K., and Chorkendorff, I. (2011) *Nat. Mater.*, **10**, 434–438.
44. Kristensen, J., Zhang, J.D., Chorkendorff, I., Ulstrup, J., and Ooi, B.L. (2006) *Dalton Trans.*, 3985–3990.
45. Ponomarev, E.A., Neumann-Spallart, M., Hodes, G., and Lévy-Clément, C. (1996) *Thin Solid Films*, **280**, 86–89.
46. Tran, P.D., Pramana, S.S., Kale, V.S., Nguyen, M., Chiam, S.Y., Batabyal, S.K., Wong, L.H., Barber, J., and Loo, J. (2012) *Chem. Eur. J.*, **44**, 13994–13999.
47. Fujishima, A. and Honda, K. (1972) *Nature*, **238**, 37–38.
48. McKone, J.R., Warren, E.L., Bierman, M.J., Boettcher, S.W., Brunschwig, B.S., Lewis, N.S., and Gray, H.B. (2011) *Energy Environ. Sci.*, **4**, 3573–3583.
49. Boettcher, S.W., Warren, E.L., Putnam, M.C., Santori, E.A., Turner-Evans, D., Kelzenberg, M.D., Walter, M.G., McKone, J.R., Brunschwig, B.S., Atwater, H.A., and Lewis, N.S. (2011) *J. Am. Chem. Soc.*, **133**, 1216–1219.
50. Oh, I., Kye, J., and Hwang, S. (2011) *Nano Lett.*, **12**, 298–302.
51. Paracchino, A., Laporte, V., Sivula, K., Gratzel, M., and Thimsen, E. (2011) *Nat. Mater.*, **10**, 456–461.
52. Tran, P.D., Batabyal, S.K., Pramana, S.S., Barber, J., Wong, L.H., and Loo, S.C.J. (2012) *Nanoscale*, **4**, 3875–3878.
53. Tabata, M., Maeda, K., Ishihara, T., Minegishi, T., Takata, T., and Domen, K. (2010) *J. Phys. Chem. C*, **114**, 11215–11220.
54. Yokoyama, D., Minegishi, T., Maeda, K., Katayama, M., Kubota, J., Yamada, A., Konagai, M., and Domen, K. (2010) *Electrochem. Commun.*, **12**, 851–853.
55. Zong, X., Yan, H., Wu, G., Ma, G., Wen, F., Wang, L., and Li, C. (2008) *J. Am. Chem. Soc.*, **130**, 7176–7177.
56. Zong, X., Wu, G., Yan, H., Ma, G., Shi, J., Wen, F., Wang, L., and Li, C. (2010) *J. Phys. Chem. C*, **114**, 1963–1968.
57. Zong, X., Han, J., Ma, G., Yan, H., Wu, G., and Li, C. (2011) *J. Phys. Chem. C*, **115**, 12202–12208.
58. Tang, M.L., Grauer, D.C., Lassalle-Kaiser, B., Yachandra, V.K., Amirav, L., Long, J.R., Yano, J., and Alivisatos, A.P. (2011) *Angew. Chem. Int. Ed.*, **50**, 10203–10207.
59. Seger, B., Laursen, A.B., Vesborg, P.C.K., Pedersen, T., Hansen, O., Dahl, S., and Chorkendorff, I. (2012) *Angew. Chem. Int. Ed.*, **51**, 9128–9131.
60. Reece, S.Y., Hamel, J.A., Sung, K., Jarvi, T.D., Esswein, A.J., Pijpers, J.J.H., and Nocera, D.G. (2011) *Science*, **334**, 645–648.
61. Paracchino, A., Mathews, N., Hisatomi, T., Stefik, M., Tilley, D., and Gratzel, M. (2012) *Energy Environ. Sci.*, **5** (9), 8673–8681.

62. Muellerwesterhoff, U.T. and Nazzal, A. (1984) *J. Am. Chem. Soc.*, **106**, 5381–5382.
63. Nann, T., Ibrahim, S.K., Woi, P.-M., Xu, S., Ziegler, J., and Pickett, C.J. (2010) *Angew. Chem. Int. Ed.*, **49**, 1574–1577.
64. Ferreira, K.N., Iverson, T.M., Maghlaoui, K., Barber, J., and Iwata, S. (2004) *Science*, **303**, 1831–1838.
65. Kruse, O., Rupprecht, J., Mussgnug, J.H., Dismukes, G.C., and Hankamer, B. (2005) *Photochem. Photobiol. Sci.*, **4**, 957–970.
66. Umena, Y., Kawakami, K., Shen, J.R., and Kamiya, N. (2011) *Nature*, **473**, 55–65.
67. Fontecilla-Camps, J.C., Amara, P., Cavazza, C., Nicolet, Y., and Volbeda, A. (2009) *Nature*, **460**, 814–822.
68. Schwarze, A., Kopczak, M.J., Rogner, M., and Lenz, O. (2010) *Appl. Environ. Microbiol.*, **76**, 2641–2651.
69. Krassen, H., Schwarze, A., Friedrich, B., Ataka, K., Lenz, O., and Heberle, J. (2009) *ACS Nano*, **3**, 4055–4061.
70. Esswein, M.J. and Nocera, D.G. (2007) *Chem. Rev.*, **107**, 4022–4047.
71. Andreiadis, E.S., Chavarot-Kerlidou, M., Fontecave, M., and Artero, V. (2011) *Photochem. Photobiol.*, **87**, 946–964.
72. Lachaud, T., Quaranta, A., Pellegrin, Y., Dorlet, P., Charlot, M.F., Un, S., Leibl, W., and Aukauloo, A. (2005) *Angew. Chem. Int. Ed.*, **44**, 1536–1540.
73. Moore, G.F., Hambourger, M., Gervaldo, M., Poluektov, O.G., Rajh, T., Gust, D., Moore, T.A., and Moore, A.L. (2008) *J. Am. Chem. Soc.*, **130**, 10466–10467.
74. Pan, J.X., Xu, Y.H., Benko, G., Feyziyev, Y., Styring, S., Sun, L.C., Akermark, B., Polivka, T., and Sundstrom, V. (2004) *J. Phys. Chem. B*, **108**, 12904–12910.
75. Artero, V., Chavarot-Kerlidou, M., and Fontecave, M. (2011) *Angew. Chem. Int. Ed.*, **50**, 7238–7266.
76. Fihri, A., Artero, V., Razavet, M., Baffert, C., Leibl, W., and Fontecave, M. (2008) *Angew. Chem. Int. Ed.*, **47**, 564–567.
77. Fihri, A., Artero, V., Pereira, A., and Fontecave, M. (2008) *Dalton Trans.*, 5567–5569.
78. Li, C., Wang, M., Pan, J.X., Zhang, P., Zhang, R., and Sun, L.C. (2009) *J. Organomet. Chem.*, **694**, 2814–2819.
79. Jasimuddin, S., Yamada, T., Fukuju, K., Otsuki, J., and Sakai, K. (2010) *Dalton Trans.*, **46**, 8466–8468.
80. Zhang, P., Wang, M., Li, C., Li, X., Dong, J., and Sun, L. (2010) *Chem. Commun.*, **46**, 8806–8809.
81. Grätzel, M. (2001) *Nature*, **414**, 338–344.
82. Li, L., Duan, L.L., Wen, F.Y., Li, C., Wang, M., Hagfeld, A., and Sun, L.C. (2012) *Chem. Commun.*, **48**, 988–990.
83. Li, L., Gibson, E.A., Qin, P., Boschloo, G., Gorlov, M., Hagfeldt, A., and Sun, L.C. (2010) *Adv. Mater.*, **22**, 1759–1762.
84. Qin, P., Zhu, H.J., Edvinsson, T., Boschloo, G., Hagfeldt, A., and Sun, L.C. (2008) *J. Am. Chem. Soc.*, **130**, 17629.
85. Brimblecombe, R., Koo, A., Dismukes, G.C., Swiegers, G.F., and Spiccia, L. (2010) *J. Am. Chem. Soc.*, **132**, 2892–2894.
86. Chen, Z., Concepcion, J.J., Luo, H., Hull, J.F., Paul, A., and Meyer, T.J. (2010) *J. Am. Chem. Soc.*, **132**, 17670–17673.
87. Bard, A.J. (2010) *J. Am. Chem. Soc.*, **132**, 7559–7567.
88. Maeda, K., Higashi, M., Lu, D.L., Abe, R., and Domen, K. (2010) *J. Am. Chem. Soc.*, **132**, 5858–5868.
89. Rocheleau, R.E., Miller, E.L., and Misra, A. (1998) *Energy Fuel*, **12**, 3–10.
90. Yamada, Y., Matsuki, N., Ohmori, T., Mametsuka, H., Kondo, M., Matsuda, A., and Suzuki, E. (2003) *Int. J. Hydrogen Energy*, **28**, 1167–1169.
91. Kanan, M.W., Surendranath, Y., and Nocera, D.G. (2009) *Chem. Soc. Rev.*, **38**, 109–114.
92. Khaselev, O. and Turner, J.A. (1998) *Science*, **280**, 425–427.
93. Miller, E.L., Marsen, B., Paluselli, D., and Rocheleau, R. (2005) *Electrochem. Solid-State Lett.*, **8**, A247–A249.
94. Miller, E.L., Paluselli, D., Marsen, B., and Rocheleau, R.E. (2005) *Sol. Energy Mater. Sol. Cells*, **88**, 131–144.

95. Miller, E.L., Rocheleau, R.E., and Khan, S. (2004) *Int. J. Hydrogen Energy*, **29**, 907–914.
96. Miller, E.L., Rocheleau, R.E., and Deng, X.M. (2003) *Int. J. Hydrogen Energy*, **28**, 615–623.
97. Ruttinger, W. and Dismukes, G.C. (1997) *Chem. Rev.*, **97**, 1–24.
98. Service, R.F. (2009) *Science*, **325**, 1201.
99. Hocking, R.K., Brimblecombe, R., Chang, L.-Y., Singh, A., Cheah, M.H., Glover, C., Casey, W.H., and Spiccia, L. (2011) *Nat. Chem.*, **3**, 461–466.
100. Sato, S., Koike, K., Inoue, H., and Ishitani, O. (2007) *Photochem. Photobiol. Sci.*, **6**, 454–461.
101. Tamaki, Y., Morimoto, T., Koike, K., and Ishitani, O. (2012) *Proc. Natl. Acad. Sci. U.S.A.*, **109**, 15673–15678.
102. Yui, T., Tamaki, Y., Sekizawa, K., and Ishitani, O. (2011) *Photocatalysis*, **303**, 151–184.
103. Parkinson, B.A. and Weaver, P.F. (1984) *Nature*, **309**, 148–149.
104. Kumar, B., Smieja, J.M., and Kubiak, C.P. (2010) *J. Phys. Chem. C*, **114**, 14220–14223.
105. Kumar, B., Smieja, J.M., Sasayama, A.F., and Kubiak, C.P. (2012) *Chem. Commun.*, **48**, 272–274.
106. Sato, S., Arai, T., Morikawa, T., Uemura, K., Suzuki, T.M., Tanaka, H., and Kajino, T. (2011) *J. Am. Chem. Soc.*, **133**, 15240–15243.

Index

Bioinspired Catalysis: Metal-Sulfur Complexes, First Edition.
Edited by Wolfgang Weigand and Philippe Schollhammer.